W9-AQE-478

ITALIAN PHYSICAL SOCIETY

PROCEEDINGS

OF THE

# INTERNATIONAL SCHOOL OF PHYSICS « ENRICO FERMI »

## COURSE LVII

edited by C. WEINER

Director of the Course

VARENNA ON LAKE COMO

VILLA MONASTERO

31st JULY - 12th AUGUST 1972

# *History of Twentieth Century Physics*

1977

*ACADEMIC PRESS • NEW YORK AND LONDON*

***A SUBSIDIARY OF HARCOURT BRACE JOVANOVICH, PUBLISHERS***

SOCIETA' ITALIANA DI FISICA

# RENDICONTI
DELLA
# SCUOLA INTERNAZIONALE DI FISICA «ENRICO FERMI»

## LVII Corso

a cura di C. Weiner
Direttore del Corso

VARENNA SUL LAGO DI COMO
VILLA MONASTERO
31 LUGLIO - 12 AGOSTO 1972

# *Storia della fisica del XX secolo*

1977

*ACADEMIC PRESS • NEW YORK AND LONDON*
*A SUBSIDIARY OF HARCOURT BRACE JOVANOVICH, PUBLISHERS*

ACADEMIC PRESS INC.
111 FIFTH AVENUE
NEW YORK, N. Y. 10003

*United Kingdom Edition*
Published by
ACADEMIC PRESS INC. (LONDON) LTD.
24/28 OVAL ROAD, LONDON N. W. 1

*Library of Congress Catalog Card Number: 74-408*

PRINTED IN ITALY

# INDICE

# Introduction.

C. WEINER

*School of Humanities and Social Science*
*Massachusetts Institute of Technology - Cambridge, Mass.* 02139

Several dimensions of the history of twentieth-century physics are explored in this volume by historians as well as by physicists who played significant roles in the developments of the last five decades. Together they take into account the philosophical and historical roots of the concepts and techniques of physics, the styles of individuals and institutions, and the influences of social and political environments. The authors were among the lecturers at the two-week school on history of physics at the Villa Monastero at Varenna on Lake Como in the Summer of 1972. The occasion was specially significant because it was the first time that the Italian Physical Society's renowned International School of Physics « Enrico Fermi » had devoted a session to history of physics rather than to a field on the forefront of contemporary physics research. It was also significant because the faculty included scholars concerned with the history, philosophy and politics of physics as well as physicists who were major participants in the events being studied. They learned from one another and benefited from the rich discussions.

In this volume, much as in the original lectures, the first-person accounts are mingled with the other presentations. Personal recollections by P. A. M. DIRAC, H. B. G. CASIMIR, Victor F. WEISSKOPF, Edoardo AMALDI and Lew KOWARSKI are interspersed with studies by Martin J. KLEIN, John L. HEILBRON, Joan BROMBERG, Paolo ROSSI, Yehuda ELKANA, Gerald HOLTON, Martin J. SHERWIN and Walter GOLDSTEIN.

Another important feature of the Varenna summer school was the intense and spirited discussion that engaged a large proportion of the faculty and students in informal evening sessions. The school took place during the Vietnam War, and the lectures on the social and political history of physics gave rise to concern about the role of physics in contemporary history. These discussions culminated in a statement, drafted by some of the participants, condemning the war and the use of physics to prosecute it. The statement was approved by most of the school's participants.

I want to express my gratitude to G. JONA-LASINIO, who was Secretary for the summer course on history of physics and who worked with me in my role as Director to organize and conduct it. Appreciation is also due to G. TORALDO DI FRANCIA who, as President of the Società Italiana di Fisica, played a key role in establishing the summer course and participated in all of its aspects during its two-week duration. My thanks also go to the staff members of the Società who administered the course so effectively and helped to provide a pleasant environment for our work.

I am sure that all of the participants join with me in dedicating this volume to the late Léon ROSENFELD, whose lectures and contributions to the discussions and spirit of the school were enormously important. His life and his scholarship as physicist, historian, and philosopher will continue to be deeply appreciated by all of us.

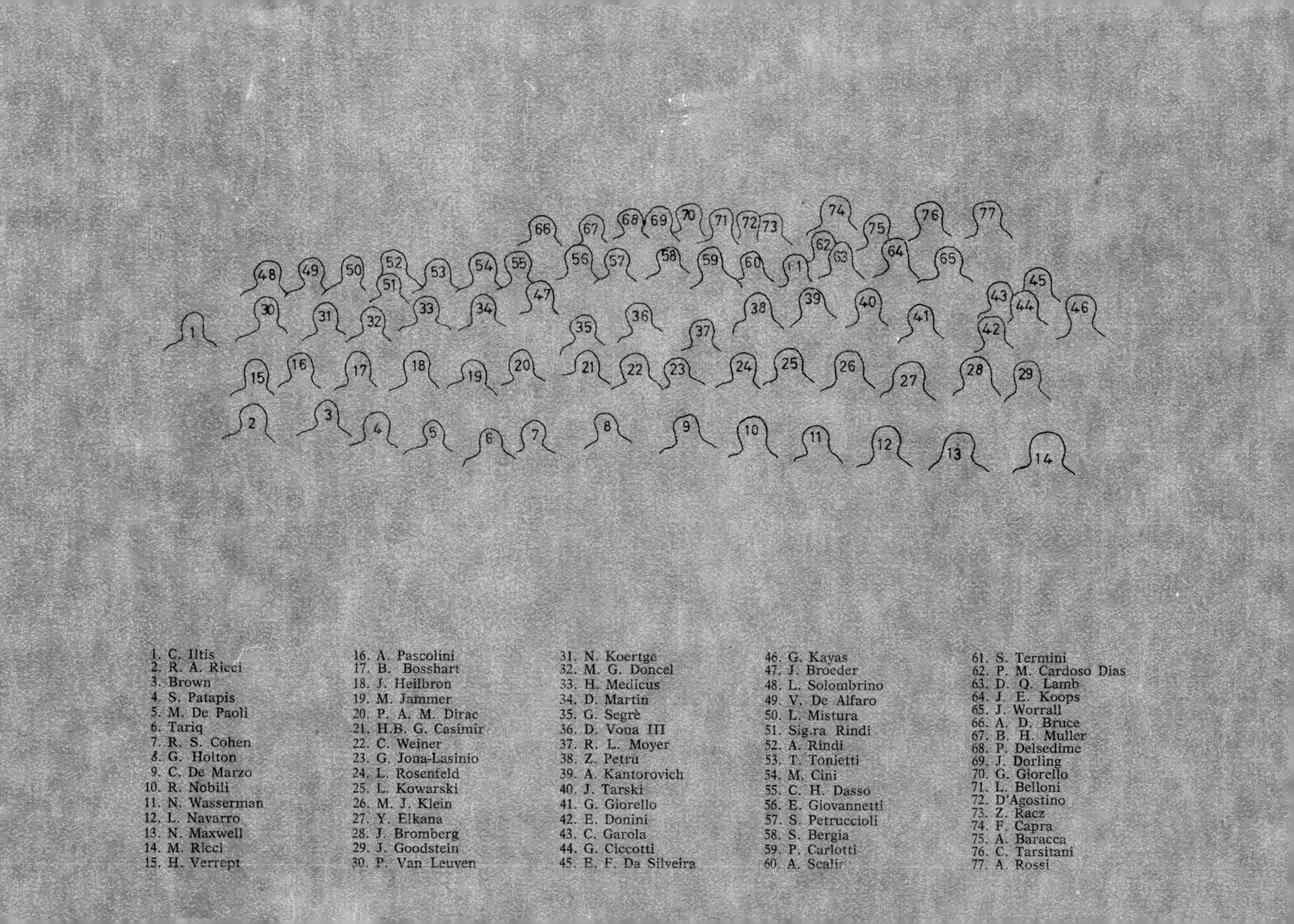

1. C. Iltis
2. R. A. Ricci
3. Brown
4. S. Patapis
5. M. De Paoli
6. Tariq
7. R. S. Cohen
8. G. Holton
9. C. De Marzo
10. R. Nobili
11. N. Wasserman
12. L. Navarro
13. N. Maxwell
14. M. Ricci
15. H. Verrept
16. A. Pascolini
17. B. Bosshart
18. J. Heilbron
19. M. Jammer
20. P. A. M. Dirac
21. H.B. G. Casimir
22. C. Weiner
23. G. Jona-Lasinio
24. L. Rosenfeld
25. L. Kowarski
26. M. J. Klein
27. Y. Elkana
28. J. Bromberg
29. J. Goodstein
30. P. Van Leuven
31. N. Koertge
32. M. G. Doncel
33. H. Medicus
34. D. Martin
35. G. Segrè
36. D. Vona III
37. R. L. Moyer
38. Z. Petru
39. A. Kantorovich
40. J. Tarski
41. G. Giorello
42. E. Donini
43. C. Garola
44. G. Ciccotti
45. E. F. Da Silveira
46. G. Kayas
47. J. Broeder
48. L. Solombrino
49. V. De Alfaro
50. L. Mistura
51. Sig.ra Rindi
52. A. Rindi
53. T. Tonietti
54. M. Cini
55. C. H. Dasso
56. E. Giovannetti
57. S. Petruccioli
58. S. Bergia
59. P. Carlotti
60. A. Scalir
61. S. Termini
62. P. M. Cardoso Dias
63. D. Q. Lamb
64. J. E. Koops
65. J. Worrall
66. A. D. Bruce
67. B. H. Muller
68. P. Delsedime
69. J. Dorling
70. G. Giorello
71. L. Belloni
72. D'Agostino
73. Z. Racz
74. F. Capra
75. A. Baracca
76. C. Tarsitani
77. A. Rossi

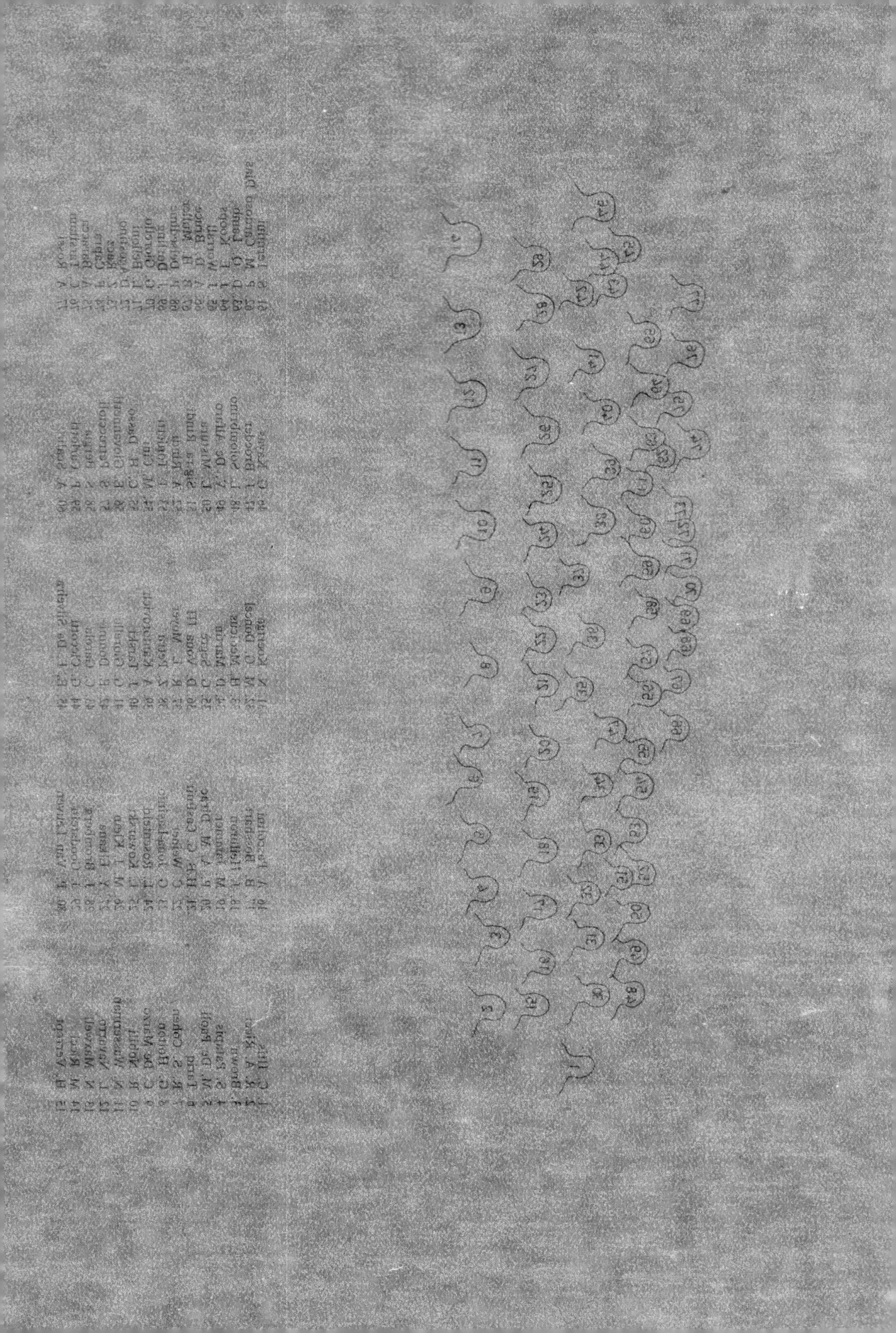

SOCIETÀ ITALIANA DI FISICA

SCUOLA INTERNAZIONALE DI FISICA « E. FERMI »

LVII CORSO - VARENNA SUL LAGO DI COMO - VILLA MONASTERO - 31 Luglio - 12 Agosto 1972

# The Beginnings of the Quantum Theory.

M. J. KLEIN

*Department of History of Science and Medicine, Yale University - New Haven, Conn.*

## Introductory note.

The version of my Varenna lectures given here includes a more restricted subject matter than I presented originally, but I have dealt with the material in more detail than was possible in the actual lectures. In preparing this text for publication I have drawn freely on several of my previously published papers: « Max Planck and the Beginnings of the Quantum Theory », *Archive for History of Exact Sciences*, **1**, 459-479 (1962); « Thermodynamics and Quanta in Planck's Work », *Physics Today*, **19**, no. 11, 23-32 (1966); « Einstein's First Paper on Quanta », *The Natural Philosopher*, **2**, 59-86 (1963); « Einstein, Specific Heats, and the Early Quantum Theory », *Science*, **148**, 173-180 (1965). The reader is referred to these papers for more detailed bibliographic information. I thank the copyright owners for permission to quote from my work. I should also like to acknowledge the support for my work provided by the National Science Foundation.

## 1. – Planck and the quantization of energy.

In his later years PLANCK often expressed his deep conviction that « the search for the absolute » was « the loftiest goal of all scientific activity » [1]. The context of his remarks clearly indicated that he saw the two laws of thermodynamics as a prototype of that « loftiest goal. » For PLANCK had formed himself as a physicist by his self-study of the writings of CLAUSIUS, that lucid but rather argumentative man who first distinguished and formulated the two laws of thermodynamics, and it was thermodynamics as seen by CLAUSIUS that set the pattern of Planck's scientific career. He devoted the first fifteen years or so of that career to clarifying, expounding and applying the second law of thermodynamics and especially the concept of irreversibility. Planck's solid and successful work in this field did not bring him all the satisfaction he might properly have expected. One reason was that he learned, too late, that some of his results had been anticipated a few years earlier in the memoirs of GIBBS. More disturbing was the rise of a powerful school of thought, the « energeticists », led by OSTWALD and HELM, which rejected the clear distinctions made by CLAUSIUS and offered a new master theory that would have replaced the elegant mathematical structure of thermodynamics by a confused and inconsistent tangle. PLANCK later described his failure to persuade the energeticists of the errors of their ways as « one of the most painful experiences of my entire scientific life ».

As a disciple of CLAUSIUS, PLANCK looked upon the second law of thermodynamics as having absolute validity: processes in which the total entropy decreased were to be strictly excluded from the natural world. He did not care to follow CLAUSIUS in pursuing « the nature of the motion which we call heat », or in searching for a mechanical explanation of the second law of thermodynamics. And he most certainly did not follow BOLTZMANN in his reformulation of the second law of thermodynamics as a statistical law. Boltzmann's statistical mechanics made the increase of entropy into a highly probable rather than an absolutely certain feature of natural processes, and this was not in keeping with Planck's own commitments. The statistical interpretation of entropy is conspicuously absent from the papers PLANCK wrote in the early 1890's under such titles as « General Remarks on Modern Developments in the Theory of Heat » and « The Essence of the Second Law of Thermodynamics » [2].

One should not think, however, that PLANCK was content to keep thermodynamics a completely independent subject, separate from the rest of physics. He preferred the rigorous arguments of pure thermodynamics to the difficult but approximate treatment of molecular models in kinetic theory, but he also felt strongly the need to relate the irreversibility described by the second law to the other fundamental laws governing the basic conservative processes. He rejected Boltzmann's approach because it rested on statistical assumptions, and PLANCK wanted to avoid these. He hoped that the principle of increasing entropy could be preserved intact as a rigorous theorem in some more comprehensive and more fundamental theory.

In March 1895 PLANCK presented a paper to the Academy of Sciences at Berlin that seemed to represent a basic shift in his interests [3]. He had just put aside his usual thermodynamic concerns to discuss the problem of the resonant scattering of plane electromagnetic waves by an oscillating dipole of dimensions small compared to the wavelength. A careful reader would have noticed, however, that at the end of the paper PLANCK admitted that this study was only undertaken as a preliminary to tackling the problem of black-body radiation. The scattering process offered a way of understanding how the equilibrium state of the radiation in an enclosure at fixed temperature could be maintained. The thermodynamics of radiation was the underlying problem, and Planck's attention may have been drawn to it by Wien's paper of 1894 which presented the displacement law.

The following February PLANCK had further results to report to the Academy [4]. He had extended his studies to the radiation damping of his charged oscillators, and he was impressed by the difference between radiation damping and damping by means of the ordinary resistance of the oscillator. Radiation damping was a completely conservative mechanism that did not require one to invoke the transformation of energy into heat, or to supply another characteristic constant of the oscillator in order to describe its damping. PLANCK thought this could have far-reaching implications for the fundamental question of irreversibility and the second law. As he put it, « The study of conservative damping seems to me to be of great importance, since it opens up the prospect of a possible general explanation of irreversible processes by means of conservative forces—a problem that confronts research in theoretical physics more urgently every day. »

One year later, in February 1897, he communicated the first of what would become a series of five papers, extending over a period of more than two years, on irreversible phenomena in radiation [5]. The extended introduction itself indicated that PLANCK was planning a major work. He began by asserting that no one had yet successfully explained how a system governed by conservative interactions could proceed irreversibly to a final state of thermodynamic equilibrium. He explicitly discounted Boltzmann's $H$-theorem as an unsuccessful attempt in this direction, citing the criticisms recently raised by ZERMELO, Planck's own student, against Boltzmann's analysis. PLANCK then announced his own program for deriving the second law of thermodynamics for a system consisting of radiation and charged oscillators in an enclosure with reflecting walls. He would introduce no damping other than radiation damping, but would take the basic mechanism for irreversibility to be the alteration of the form of an electromagnetic wave by the scattering process—its apparently irreversible conversion from incident plane to outgoing spherical wave. The ultimate goal of this program would be the explanation of reversibility for conservative systems and, as a valuable by-product, the determination of the spectral distribution of black-body radiation.

PLANCK had high hopes. His goal was precisely right for a disciple of CLAUSIUS. It would have been a splendid conclusion to his work in thermodynamics, and it would have put an end, once and for all, to claims that the second law was merely a matter of probability. How was PLANCK to know that he was headed in a very different direction, that he had started on what he would later call « the long and multiply twisted path » to the quantum theory?

There was, unfortunately, a fundamental flaw in Planck's proposal and it was promptly pointed out by BOLTZMANN [6]. The equations of electrodynamics could not produce a monotonic approach to equilibrium any more than could the equations of mechanics; both needed to be supplemented by appropriate statistical assumptions. Nothing in the equations of electrodynamics would, for example, forbid the inverse of Planck's scattering process. (It is reasonable to suppose that BOLTZMANN was, at the least, not deterred from pointing out this error by Planck's negative comments on his own work. Planck's support of ZERMELO did not help matters either, since BOLTZMANN had found Zermelo's criticism particularly irksome.)

PLANCK finally granted that a statistical assumption was necessary and introduced what he called the hypothesis of « natural radiation », the appropriate analogue of Boltzmann's hypothesis of « molecular chaos », the hypothesis underlying the $H$-theorem. With the help of this hypothesis PLANCK was able to complete his program, in a sense, and he reported his work in the last paper of the series in June, 1899 [7].

PLANCK showed that the spectral distribution $\varrho(\nu, T)$ of the radiation was simply related to the average energy $\overline{u}_\nu(T)$ of a charged harmonic oscillator of frequency $\nu$ at temperature $T$. Here $\varrho(\nu, T)\,d\nu$ is the energy of the radiation per unit volume in the frequency interval $\nu$ to $\nu + d\nu$. (He had continued to treat harmonic oscillators, not because they were thought to be a realistic model for matter, but rather because Kirchhoff's theorem ensured that the equilibrium radiation distribution was independent of the system with which the radiation interacted, and oscillators were simplest to treat [8].) By equating the emission and absorption rates of the oscillator in equilibrium with natural radiation PLANCK derived the result

$$\varrho(\nu, T) = (8\pi\nu^2/c^3)\,\overline{u}_\nu(T)\,, \tag{1}$$

where $c$ is the velocity of light.

It is evident from eq. (1) that PLANCK needed only to determine $\overline{u}_\nu(T)$, the average energy of a harmonic oscillator at temperature $T$, in order to have the explicit form for the distribution law. Remarkably enough, although classical statistical mechanics provided a « well known » and simple answer for $\overline{u}_\nu(T)$ from the equipartition theorem, PLANCK made no use of it, nor did he then or for some years indicate that he realized its existence. Instead PLANCK took what he later referred to as a « thermodynamic » approach, looking for a suitable relationship between the energy and the entropy of the oscillator, rather than one between the energy and the temperature. He introduced this relationship by means of a definition

$$S = -\frac{u}{\beta\nu}\ln\frac{u}{ae\nu}\,, \tag{2}$$

where $S$ is the entropy of the oscillator, $u$ is its energy (previously called $\overline{u}_\nu(T)$), $\beta$ and $a$ are constants and $e$ is the base of the natural logarithms. In the original paper PLANCK did not motivate this definition when he introduced it, but from his discussion later in that paper and also from his later comments on this work, it seems most likely that he was guided by the form of Wien's distribution law

$$\varrho(\nu, T) = \alpha\nu^3 \exp[-\beta\nu/T]. \tag{3}$$

WIEN had proposed this form for the distribution law in 1896, but his derivation was less than convincing. Lord RAYLEIGH commented: « Viewed from the theoretical side, the result appears to me to be little more than a conjecture ». The important thing about Wien's distribution in the late 1890's was not Wien's derivation, but rather the fact that it gave an adequate account of all the experimental results on the energy distribution in black-body radiation which were then available. It seemed reasonable to suppose that a fundamental theory of radiation, such as PLANCK proposed to develop, would have to conclude with an adequate grounding for the Wien distribution law, if the theory were to be in accord with experiment. In any event the energy $u$ of the oscillator can be found from Wien's distribution law with the help of eq. (1). Knowing $u$ as a function of temperature one finds the entropy by an integration, since $T^{-1} = \partial S/\partial u$. The entropy so obtained has the form of Planck's « definition », eq. (2), with $a$ equal to $\alpha c^3/8\pi$.

With the entropy of an oscillator defined by eq. (2), PLANCK could then determine the entropy of the radiation in equilibrium with it and go on to prove that the total entropy was a monotonically increasing function of time, just the property required of the entropy by the second law. That PLANCK could also demonstrate that the equilibrium distribution was the Wien law of eq. (3) is hardly remarkable, since that result followed inevitably from his choice of the entropy expression given by eq. (2), as already indicated. PLANCK wrote that he was impressed by the simplicity of the relationship expressed in eq. (2) and particularly by the fact that $\partial^2 S/\partial u^2$, which entered directly into his calculation, was simply proportional to minus $u^{-1}$. He was, nevertheless, aware that his choice of a particular expression for entropy as a function of energy determined the resulting distribution law, and he gave arguments which seemed to make that choice uniquely determined by the requirements of consistency with the displacement law and the second law of thermodynamics. His conclusion was expressed in the following sentence: « I believe that it must therefore be concluded that the definition given for the entropy of radiation, and also the Wien distribution law for the energy which goes with it, is a necessary consequence of applying the principle of entropy increase to the electromagnetic theory of radiation, and that the limits of validity of this law, should there be any, therefore coincide with those of the second law of thermodynamics. Further experimental test of this law naturally acquires all the greater fundamental interest for this reason ».

These arguments and remarks were made in a paper presented to the Academy on May 18, 1899, and were repeated verbatim in Planck's article in the *Annalen der Physik* which summarized the series of five Academy papers [9]. The *Annalen* paper was received by the editors on November 7, 1899, and it appeared in print early in 1900. By the time PLANCK corrected the proofs of this paper, the «further experimental tests» which he had called for were in progress. The results of these tests which had already been published caused him to add a note in proof remarking that experimental deviations from the Wien distribution had been observed. As PLANCK had already indicated, deviations from Wien's law created a serious problem indeed from the standpoint of his theory, and PLANCK proceeded to reconsider his arguments in some detail. In a paper received by the editors of the *Annalen* on March 22, 1900, he reported his new considerations [10]. The experimental situation was still unclear, but Planck's review of the assumptions and reasoning in his former work led him to propose new arguments, this time attempting to derive the form he had previously assumed for the energy-entropy relationship. PLANCK found, however, that, in order to satisfy the second law of thermodynamics, *i.e.* in order to have the entropy increase monotonically in time, it would suffice to have $\partial^2 S/\partial u^2$ be any negative function of the energy $u$; the specific form $-(g(\nu)/u)$, which was equivalent to Wien's law, was not a thermodynamic necessity. Nevertheless, PLANCK concluded that Wien's distribution law could still be deduced, if he made a very plausible assumption on the functional dependence of the time derivative of the entropy. Once again he ended his arguments with the Wien distribution, even if they did not have the full weight of thermodynamic reasoning to support them.

By October 1900, however, the experimental picture had changed considerably. The very careful work of RUBENS and KURLBAUM with long waves over a wide range of temperatures had shown beyond any doubt that Wien's distribution law was inadequate. These new measurements also indicated clearly that, for very long wavelengths, the distribution function $\varrho(\nu, T)$ approached a very different form, becoming proportional to the absolute temperature $T$. PLANCK had been informed of these results by RUBENS and KURLBAUM several days before they were reported to the German Physical Society on October 19, 1900, so that he had the opportunity to reflect on the results and to prepare an extended «remark» for the discussion after KURLBAUM delivered the paper. This «discussion remark» was devoted to «An Improvement of the Wien Distribution», the improvement being a new distribution law, now universally known as the Planck distribution law [11]. The arguments for this new distribution law and the discussion of its immediate results properly belong to the next stage of our discussion.

I have already mentioned Wien's displacement law, published in 1894, but its content has not yet been discussed [12]. On the basis of thermodynamics and electrodynamics WIEN had rigorously proved that the distribution function $\varrho(\nu, T)$ is essentially a function of only one variable, the ratio $\nu/T$. More precisely, Wien's displacement law requires that $\varrho(\nu, T)$ have the form

$$\varrho(\nu, T) = \nu^3 f(\nu/T) , \tag{4}$$

where $f$ is a function of the single variable. This severe restriction on the distribution function proved to be a very powerful aid to Planck's later arguments.

Planck's problem in October 1900 was to determine a distribution law which was consistent both with the positive results of his own work and with the new experimental findings of RUBENS and KURLBAUM. Since the quantity $\partial^2 S/\partial u^2$ had figured prominently in his earlier analysis of how the entropy increased in time, it was natural for PLANCK to center his attention on the form of this function. We have already seen that the negative reciprocal of $\partial^2 S/\partial u^2$ is simply proportional to $u$ when the Wien distribution is valid. The next simplest possibility is to take $\partial^2 S/\partial u^2$, or rather its negative reciprocal, proportional to $u^2$. It is easy to see that, when this is done, $u$, and therefore $\varrho(\nu, T)$, will be proportional to $T$, just as RUBENS and KURLBAUM had found to be the case in the long-wavelength limit. The proper limiting forms for low and high frequencies could then be preserved by taking $-(\partial^2 S/\partial u^2)^{-1}$ proportional to $u(\gamma + u)$, where $\gamma$ is a (frequency dependent) constant. On these grounds, of simplicity and proper behavior in the limit, PLANCK proposed the distribution law

$$\varrho(\nu, T) = \frac{A\nu^3}{\exp[B\nu/T] - 1} , \tag{5}$$

where $A$ and $B$ are constants. This is the law which follows from the assumption just mentioned for $\partial^2 S/\partial u^2$, where the frequency dependence is fixed by the displacement law together with eq. (1).

The adequacy of Planck's proposed distribution law was confirmed immediately. As PLANCK described it later, « The very next morning I received a visit from my colleague Rubens. He came to tell me that after the conclusion of the meeting, he had that very night checked my formula against the results of his measurements and found a satisfactory concordance at every point.... Later measurements, too, confirmed my radiation formula again and again—the finer the methods of measurement used, the more accurate the formula was found to be » [13].

Now one of the key points in determining Planck's choice of the distribution formula was that it agreed with the experimental results in being proportional to $T$ in the limit of small $\nu$. Planck's formula was not the first in the literature to show this property, and he referred in a footnote to an empirical formula proposed by LUMMER and JAHNKE which had the same property. He did not, however, refer to another paper in which a distribution law proportional to $T$ for long wavelengths was not only proposed, but was also derived from fundamental principles, Rayleigh's « Remarks upon the Law of Complete Radiation ». This two-page note had appeared in the June, 1900 issue of the *Philosophical Magazine* [14]. In it Rayleigh had shown that, if the equipartition theorem of statistical mechanics could be applied to « the modes of aetherial vibration », then the distribution law for black-body radiation is uniquely determined to have a form radically different from that of the Wien distribution. Rayleigh was well aware of the conditional nature of this conclusion saying, « The question is one to be settled by experiment; but in the meantime I venture to suggest a modification of [the Wien distribution], which appears to me more probable *a priori*. Speculation upon this subject is hampered by the difficulties which attend the Boltzmann-Maxwell doctrine of the partition of energy. According to this doctrine, every mode of vibration should be alike favoured; and although, for some reason not yet explained, the doctrine fails in general, it seems possible that it may apply to the graver modes ».

Rayleigh's method for arriving at the radiation distribution law was essentially different from Planck's. His argument concerned itself directly with the radiation and did not need to refer to a material system with which it was in equilibrium. The number of standing waves or allowed modes of electromagnetic vibration of the enclosure, whose frequencies lie in the interval from $\nu$ to $\nu + \mathrm{d}\nu$, is proportional to $\nu^2 \mathrm{d}\nu$, by reasoning which was practically second nature to the author of *The Theory of Sound*. According to « the Boltzmann-Maxwell doctrine », *i.e.* the equipartition theorem, the average energy of every one of these modes, regardless of its frequency, would be proportional to $T$ at thermal equilibrium, with a universal proportionality constant. It follows at once that the distribution function $\varrho(\nu, T)$ must have the form

$$\varrho(\nu, T) \propto \nu^2 T \,. \tag{6}$$

As Rayleigh remarked, this is in accord with the displacement law.

Although Rayleigh did not trouble to point it out explicitly in this note, it must have been quite obvious to him that a distribution law of this form could not possibly hold for all frequencies, since it would lead to an infinite concentration of energy at the high frequencies; the integral of $\varrho(\nu, T)$ over frequency would diverge. Rayleigh was in any case well aware of the many problems associated with the equipartition theorem, and had written on that subject earlier in the year. For these reasons, presumably, he added a remark immediately after deriving eq. (6): « If we introduce the exponential factor, the complete expression will be

$$\varrho(\nu, T) \propto \nu^2 T \exp[-\beta\nu/T] \,. \tag{7}$$

Whether [this equation] represents the facts of observation as well as [the Wien distribution] I am not in a position to say. It is to be hoped that the question may soon receive an answer at the hands of the distinguished experimenters who have been occupied with this subject ».

This last equation of Rayleigh's, clearly intended as only a guess at how the consequence of the equipartition theorem could be made convergent, was apparently the only thing in his paper that attracted the notice of those to whom it was addressed. This is evident from the paper which RUBENS and KURLBAUM presented to the Prussian Academy on October 25, 1900, less than a week after Planck's new distribution had come to their attention [15]. RUBENS and KURLBAUM made a systematic comparison of their results with five different formulae which had been proposed for the distribution law: those of WIEN (eq. (3)), PLANCK (eq. (5)), RAYLEIGH (eq. (7)) and two others due to THIESEN and to LUMMER and JAHNKE. They concluded that only Planck's formula and that proposed by LUMMER and JAHNKE were in agreement with their results, and they gave their preference to Planck's formula on grounds of simplicity. (The Lummer-Jahnke formula contained $\nu/T$ to the power 1.3 in the exponent, where the number 1.3 was chosen solely for fitting the results.)

The point to be stressed is that RUBENS and KURLBAUM discussed Rayleigh's work in the same tone of voice, so to speak, that they used in dealing with strictly *ad hoc* formulae which had no theoretical foundations. They failed to grasp the fundamental importance of the fact that their results did show that $\varrho$ was proportional to $T$ for low frequencies, in complete agreement with Rayleigh's conclusion that equipartition should apply « to the graver modes ». When Rayleigh's paper was reprinted in his *Scientific Papers* two years later, he took the opportunity to remark on the proportionality of $\varrho$ to $T$ for low frequencies. « This is what I intended to emphasize. Very shortly afterwards the anticipation above expressed was confirmed by the important researches of RUBENS and KURLBAUM who operated with exceptionally long waves. »

RUBENS and KURLBAUM were not the only ones who missed this central point in Rayleigh's paper, which he had probably not underlined sufficiently. I have already mentioned that PLANCK made no reference to RAYLEIGH in his October, 1900, communication. Nor did he refer to RAYLEIGH in his papers introducing the quantum concept which appeared a few months later. Although it might appear from this lack of reference that PLANCK did not know of Rayleigh's work, I find this a rather unlikely hypothesis. PLANCK had been devoting virtually all of his efforts to the radiation problem for over three years by this time, and he is not likely to have missed something on this subject written by a leading thinker and published in a major journal. Furthermore, we know that RUBENS and KURLBAUM had seen Rayleigh's paper and must have referred to it in preparing their own report during the week after PLANCK proposed his new equation to them. We also know that PLANCK kept in close touch with the experimenters' attempts to fit their results with his and other distribution laws. Finally, in his paper of December 14, 1900, PLANCK referred explicitly to the Rubens and Kurlbaum paper in which Rayleigh's work was quoted. It seems hard to believe, therefore, that PLANCK was not aware of

Rayleigh's article. That PLANCK did not react to Rayleigh's work, even if he did know of it, is not so surprising as it may seem at first. There was nothing in Planck's background that would have led him to expect universal validity for the equipartition theorem, a statistical result, even when it was applied to mechanical systems. RAYLEIGH had applied it directly to the aether modes, however, and nothing in his short note indicated that he took its failure very seriously himself.

« But even if the absolutely precise validity of the radiation formula is taken for granted, so long as it had merely the standing of a law disclosed by a lucky intuition, it could not be expected to possess more than a formal significance. For this reason, on the very day when I formulated this law, I began to devote myself to the task of investing it with a true physical meaning. This quest automatically led me to study the interrelation of entropy and probability—in other words, to pursue the line of thought inaugurated by BOLTZMANN. » « After a few weeks of the most strenuous work of my life, the darkness lifted and an unexpected vista began to appear. » [16]

These are the words PLANCK used in his Nobel Lecture to describe his efforts from October 19 to December 14, 1900, on which date he presented his results to the German Physical Society in a paper entitled « On the Theory of the Energy Distribution Law in the Normal Spectrum » [17]. During those two months PLANCK had adopted a method he had never previously used—Boltzmann's combinatorial, nonkinetic, statistical interpretation of the entropy as developed in his great memoir of 1877 [18]. In addition PLANCK had taken a step that was eventually to change the most basic features of physical theory. These two aspects of what he had done are almost inextricably woven together in both the paper already referred to and in the more complete paper published soon afterwards in the *Annalen der Physik* [19]. We must, nevertheless, try to separate them here, if we are to appreciate the innovations which PLANCK made.

Planck's earlier work had shown that only one more key step was necessary for the theory of the radiation spectrum: a sound theoretical determination of the relationship between the energy $u$ and the entropy $S$ of a harmonic oscillator of frequency $\nu$. Once this was known, the average energy of the oscillator could be found, and the distribution function $\varrho(\nu, T)$ would then be fixed with the help of eq. (1), which relates $u$ and $\varrho(\nu, T)$. (It might be well to re-emphasize the point that harmonic oscillators entered the picture only because PLANCK chose them as the simplest material system which could be in equilibrium with the electromagnetic radiation, availing himself of the freedom given by Kirchhoff's theorem.) Planck's previous attempts at fixing the relationship between $S$ and $u$ on general thermodynamic grounds supplemented by plausibility arguments had failed, as we have seen. He was quite sure that he knew what this relationship had to be (from the work discussed in the last Section); if his conjectured distribution law, eq. (5), were correct, as it seemed to be, then it implied the equation

$$S = \frac{A'}{B}\left[\left(1+\frac{u}{A'\nu}\right)\ln\left(1+\frac{u}{A'\nu}\right)-\frac{u}{A'\nu}\ln\frac{u}{A'\nu}\right], \tag{8}$$

where $A$ and $B$ are the constants of eq. (5), and $A' = Ac^3/8\pi$. In order to establish eq. (8), new methods were necessary, and, as already mentioned, PLANCK found these in Boltzmann's work.

According to BOLTZMANN, the entropy of a system in a given state is proportional to the logarithm of the probability of this state. This probability in turn is to be found as the number of complexions, the number of distinct microscopic arrangements, compatible with the given state. Planck's task then, once this general approach was accepted, was to find a method for determining $W$, the number of complexions of his set of oscillators. It is sufficient to consider $N$ oscillators, all of frequency $\nu$, whose total energy $U_N$ is then $N$ times the average energy $u$ of one oscillator. Since the entropy is an additive function, $S_N$ will also be $N$ times $S$, the entropy of one oscillator

$$U_N = Nu, \tag{9}$$

$$S_N = NS. \tag{10}$$

The entropy of the $N$-oscillator system $S_N$ is now set equal to a proportionality constant $k$ times the logarithm of $W$

$$S_N = k \ln W, \tag{11}$$

where the additive constant which might appear is set equal to zero. PLANCK then made the same assumption that BOLTZMANN had made, that any one complexion of the system is as likely to occur as any other, so that $W$ can be obtained by counting the number of complexions. *But*, in order to carry out this counting procedure, it is essential that the energy to be shared among the $N$ oscillators must not be considered as a continuously varying, infinitely divisible quantity. It must instead be treated as consisting of an integral number of finite equal parts, if meaningful, finite values for $W$ are to be obtained. PLANCK refers to these as elements of energy $\varepsilon$ and writes

$$U_N = P\varepsilon, \tag{12}$$

where $P$ is a (large) integer representing the total number of elements of energy.

With this assumption made, it is evident that there is a finite number of complexions equal to the number of ways in which the $P$ elements of energy can be divided up among the $N$ oscillators. This number $W$ is given by combinatorial analysis as

$$W = \frac{(N+P-1)!}{P!\,(N-1)!} \simeq \frac{(N+P)^{N+P}}{P^P N^N}\,, \tag{13}$$

where the second, approximate form comes from dropping the 1 compared to the large numbers $N$ and $P$ and using Stirling's approximation for the factorials. From eqs. (11) and (13) the entropy is given by the equation

$$S = k\{(N+P)\ln(N+P) - P\ln P - N\ln N\}\,. \tag{14}$$

When $P/N$ is replaced by $u/\varepsilon$, and $S_N/N$ by $S$, according to eqs. (9), (10) and (12), the entropy $S$ of one oscillator in terms of its average energy $u$ has the form

$$S = k\left\{\left(1+\frac{u}{\varepsilon}\right)\ln\left(1+\frac{u}{\varepsilon}\right) - \frac{u}{\varepsilon}\ln\frac{u}{\varepsilon}\right\}. \tag{15}$$

At this stage in the argument, the size of the energy elements $\varepsilon$ is completely arbitrary. In fact, however, $S$ must depend on the frequency of the oscillators as well as on $u$ in a way prescribed by Wien's displacement law, and since $k$ is a universal constant, this frequency dependence must come into $\varepsilon$. The displacement law actually requires that the entropy have the form

$$S = g(u/\nu)\,, \tag{16}$$

so that the energy element $\varepsilon$ must be proportional to the frequency of the oscillator

$$\varepsilon = h\nu\,, \tag{17}$$

where $h$ is the second universal natural constant in the theory. The expression for $S$ as a function of $u$ is now fully determined in terms of the constants $h$ and $k$:

$$S = k\left\{\left(1+\frac{u}{h\nu}\right)\ln\left(1+\frac{u}{h\nu}\right) - \frac{u}{h\nu}\ln\frac{u}{h\nu}\right\}. \tag{18}$$

This result has exactly the same form as eq. (8), and the distribution law that goes with it must therefore be

$$\varrho(\nu, T) = \frac{8\pi\nu^2}{c^3}\,\frac{h\nu}{\exp[h\nu/kT]-1}\,. \tag{19}$$

This was Planck's derivation of the distribution law obtained, as he claimed, by «pursuing the line of thought inaugurated by BOLTZMANN». I shall examine that claim in the Appendix. But what did PLANCK really think about this derivation, and what did he think, in particular, about those mysterious elements of energy that he had introduced? The only answer we have was written long after the event.

Thirty years later, in a letter to WOOD, PLANCK described what he had done as « an act of desperation », undertaken against his naturally peaceful and unadventurous disposition. « But, » he went on, « I had already been struggling with the problem of the equilibrium of matter and radiation for six years (since 1894) without success; I knew that the problem is of fundamental significance for physics; I knew the formula that reproduces the energy distribution in the normal spectrum; a theoretical interpretation *had* to be found at any cost, no matter how high. ». He described himself as ready to sacrifice any of his previous convictions except the two laws of thermodynamics. When he found that the hypothesis of energy quanta would save the day he considered it « a purely formal assumption, and I did not give it much thought except for this: that I had to obtain a positive result, under any circumstances and at whatever cost » [20].

Planck's remark has the ring of truth; he was not a man to make such a personal and emotional statement lightly. But, we can still ask, what made PLANCK take his result seriously? The energy elements may have been only « a purely formal assumption » leading to the distribution law PLANCK knew to be correct, and yet he referred to « other relationships » deduced from his theory, « which seem to me to be of considerable importance for other fields of physics and chemistry. » PLANCK was referring to the far-reaching importance of the two constants in the distribution law $k$ and $h$. He had his own reasons for taking them seriously indeed, but we must go back to his earlier work to understand those reasons.

The final section of Planck's fifth paper on irreversible radiation phenomena, reprinted verbatim in the *Annalen* paper early in 1900, dealt with the significance of the radiation constants. These were the constants $a$ and $\beta$ in his expression for the entropy of an oscillator, eq. (2), or equivalently the constants $\alpha$ and $\beta$ in the Wien distribution law, eq. (3), which PLANCK thought he had just shown to be a necessary consequence of the second law of thermodynamics. PLANCK evaluated these two universal constants from the then available data on black-body radiation, finding for $\beta$ the value $0.4818 \cdot 10^{-10}$ s °K and for $a$ the value $6.885 \cdot 10^{27}$ erg s. He then proceeded to discuss these results in a remarkable passage.

PLANCK observed that these two constants together with the velocity of light $c$ and the gravitational constant $G$ could be used to define new units of mass, length, time and temperature and that these units properly deserved the title of « natural units ». All systems of units previously employed owed their origins to the accidents of human life on this earth, wrote PLANCK. The usual units of length and time derived from the size of the Earth and the period of its orbit, those of mass and temperature from the special properties of water, the Earth's most characteristic feature. Even the standardization of length using some spectral line would be quite as arbitrary, as anthropomorphic, since the particular line, say the sodium $D$-line, would be chosen to suit the convenience of the physicist. The new units that he was proposing would be truly « independent of particular bodies or substances, would necessarily retain their significance for all times and for all cultures, including extra-terrestrial and nonhuman ones », and therefore deserved the name of « natural units ». That they were of awkward sizes ($10^{-33}$ cm, $10^{-42}$ s, etc.) was obviously of no importance. « These quantities preserve their natural significance so long as the laws of gravitation, and the propagation of light in vacuum, and the two laws of thermodynamics retain their validity. » [21]

I have referred earlier to Planck's conviction that the search for the absolute was the physicist's proper goal. The universal constants as well as the most general physical laws belonged to that category of the absolute for him. As he put it in an essay written in his ninetieth year, « The endeavor to discover [the absolute constants] and to trace all physical and chemical processes back to them is the very thing that may be called the ultimate goal of scientific research and study » [22]. He had obviously felt the same way half a century earlier.

Although PLANCK was already prepared to assign an absolute significance to the radiation constants in 1899, he could not obtain any specific physical results from them yet. In his new theory there were again two constants $h$ and $k$, closely related to the former $a$ and $\beta$. (The constant $h$ differed from $a$ because the Wien and Planck laws did not have the same analytical form and also because PLANCK worked from newer data in December 1900. A similar remark holds for the comparison of $k$ and $a/\beta$ [23].) But now PLANCK could discuss in detail the physical significance of at least one of his constants. He pointed out that, since $k$ is a universal proportionality factor connecting entropy and $\ln W$, it follows from Boltzmann's work on the entropy of a gas that

$$k = R/N_0 , \tag{20}$$

where $R$ is the gas constant which appears in the macroscopic equation of state of an ideal gas, and $N_0$ is Avogadro's number, the number of molecules in a gram-mole. Since the value of $R$ was well established, Planck's computation of $k$ meant that he had also determined Avogadro's number. Planck's value for it was $6.175 \cdot 10^{23}$ molecules per mole. Other quantities follow directly, such as Loschmidt's constant, the number of molecules per cubic centimeter of gas at standard conditions, and the mean kinetic energy of a molecule. Less obvious, especially in 1900, was the fact that the elementary unit of electric charge $e$ was now also determined as essentially the ratio of the macroscopic Faraday constant to Avogadro's number. Planck's value for $e$ was $4.69 \cdot 10^{-10}$ e.s.u.

PLANCK thoroughly appreciated the importance of these determinations of the basic natural constants which his theory had made possible. As he said, « All these relationships can lay claim to absolute, not approximate validity, so long as the theory is really correct.... Their test by more direct methods will be a problem (for further research) as important as it is necessary ». These words appear at the conclusion of the paper we have been discussing, and, when PLANCK rewrote this work a few weeks later for the *Annalen der Physik*, he separated these considerations on the natural constants from the principal argument concerning the radiation distribution in order to give them the emphasis they well deserved. In his later writings, PLANCK carefully pointed out on several occasions that, although $k$ was understandably referred to as Boltzmann's constant, BOLTZMANN had never attached any great significance to it nor had he ever made any estimate of its numerical value.

Planck's values of $N_0$ and $e$ were by far the best estimates of these basic quantities which had yet appeared in the literature. There had, in fact, been no direct determination of $N_0$; the only estimates of $N_0$ available were very indirect ones based on oversimplified models from the kinetic theory of gases. It was not until 1908 that PERRIN began his series of experiments which were to give an essentially direct determination of $N_0$, but one which was less reliable than Planck's [24]. As for his value of $e$, $4.69 \cdot 10^{-10}$ e.s.u., the early attempts at its measurement gave results from 1.3 to 6.5 in the same units.

Unfortunately, Planck's contemporaries did not properly appreciate these results; the handbooks went on printing crude determinations of Avogadro's number, ignoring Planck's value. The first experimentalist to quote Planck's value of $e$ seems to have been RUTHERFORD, in 1908, probably because he and GEIGER had obtained essentially the same value, $4.65 \cdot 10^{-10}$ e.s.u., from the charge on the alpha-particle and were glad to have a confirmation of a result 50 % higher than Thomson's current best determination.

PLANCK himself laid heavy emphasis on these concrete results of his theory, both in his papers and in his *Lectures on the Theory of Heat Radiation* published in 1906, where he first interpreted $h$ as the quantum of action. I am convinced that, with Planck's particular sensitivity to the importance of the natural constants, it was these results that assured him that quanta were more than an *ad hoc* hypothesis, useful only for arriving at the radiation law. Of course $h$, the second constant in his equation, the essentially new constant in the theory, was as yet unexplored. He remarked in his *Lectures* at several points that $h$ must have some direct electrodynamic meaning, that this meaning must be found before the theory of radiation could be considered fully satisfactory, but that a lot more research would be needed before this meaning was revealed.

The kind of electrodynamic meaning that PLANCK had in mind for $h$ was suggested in a letter he wrote to EHRENFEST in July, 1905 [25]. EHRENFEST was engaged in an analysis of Planck's assumptions and had written to PLANCK asking several questions about them. In his answer PLANCK pointed out that the existence of a discrete unit of electric charge imposed certain limitations on the electromagnetic field. He went on to write: « Now it seems to me not completely impossible that there is a bridge from this assumption (of the existence of an elementary quantum of electric charge $e$) to the existence of an elementary quantum of energy $h$, especially since $h$ has the same dimensions and also the same order of magnitude as $e^2/c$. But I am not in a position to express any definite conjecture about this ». PLANCK never published this remark, so far as I can tell.

It is fair to say that PLANCK was fully committed to the quantum itself, but not necessarily to a quantum theory.

## APPENDIX

Let us now examine the question: how did PLANCK depart from Boltzmann's methods, which he claimed to be following? I think the question is best approached by comparing Planck's analysis with that given in Boltzmann's memoir of 1877, the memoir in which the statistical interpretation of entropy is set forth at length, separated from the difficulties of the kinetic treatment of the approach to equilibrium. The very first problem which BOLTZMANN treats in this paper, as a simple introduction to his concepts and methods, bears a remarkable resemblence to Planck's own problem.

BOLTZMANN considers a simple model of a gas consisting of $N$ molecules, in which the energy of each individual molecule can take on only certain discrete values which form an arithmetic progression $0, \varepsilon, 2\varepsilon, \ldots, M\varepsilon$. His comments on this model are revealing. « This fiction does not, to be sure, correspond to any realizable mechanical problem, but it is indeed a problem which is much easier to handle mathematically and which goes over directly into the problem to be solved, if one lets the appropriate quantities become infinite. If this method of treating the problem seems at first sight to be very abstract, it nevertheless is generally the quickest way of getting to one's goal in such problems, and if one considers that everything infinite in nature never has meaning except as a limiting process, one cannot understand the infinite manifold of possible energies for each molecule in any way other than as the limiting case which arises when each molecule can take on more and more possible velocities. »

BOLTZMANN then turns his attention to the possible states of this gas model when it is assigned a total energy of $P\varepsilon$, where $P$ is a large integer. Any such state is characterized by the set of integers $w_0, w_1, \ldots, w_M$ which give the number of molecules having energy $0, \varepsilon, \ldots, M\varepsilon$. The $w$'s are subject to the two constraints

$$\sum_{r=1}^{M} w_r = N \tag{21}$$

and

$$\sum_{r=1}^{M} r w_r = P \,, \tag{22}$$

which express the fixed number of particles and the fixed total energy. Each of the states, characterized by a set of numbers $\{w_r\}$, can be achieved in many ways, which BOLTZMANN refers to as complexions, depending on *which* of the molecules are found with energies $0, \varepsilon, \ldots$. The number of complexïons for a given distribution $\{w_r\}$ is readily recognized to be given by the expression

$$W_B = \frac{N!}{w_0!\, w_1! \ldots} \,. \tag{23}$$

BOLTZMANN makes the basic assumption that any particular complexion (in which, for example, molecule number one has energy $17\varepsilon$, molecule number two has energy $3\varepsilon$, etc.) is as likely to occur as any other particular complexion. (He makes this assumption plausible by a comparison to the game of lotto, or bingo.) It follows, then, that the probability of occurrence of a state characterized by the set $\{w_r\}$ is equal to $W_B$ for this state divided by $\sum W_B$, where the sum is over all sets $\{w_r\}$ compatible with eqs. (21) and (22).

Now it is essential in Boltzmann's procedure that he asks for that state, *i.e.* that set of numbers $\{w_r\}$, for which $W_B$ is a maximum, since it is that most probable state which he will identify with the state of thermodynamic equilibrium. It is not necessary to repeat here the calculation in which $W_B$ is maximized and the $\{w_r\}$ in the equilibrium state are shown to obey the « Boltzmann distribution », *i.e.* $w_r$ in the equilibrium state is proportional to $\exp[-\beta' r\varepsilon]$, where the constant $\beta'$ is shown to be proportional to $T^{-1}$. It is important, however, to recall that at an appropriate stage in the calculation BOLTZMANN takes the limit in which $\varepsilon$ goes to zero and $M$ goes to infinity in such a way that the molecules can really take on *all* values for their energy. For BOLTZMANN the $\varepsilon$ is an artifice which makes the calculation possible (and makes the continuum intelligible!).

We can now compare Boltzmann's procedure with Planck's way of handling his problem. (The formal similarity in the problems has been deliberately stressed by using a common notation.) The first evident difference between the two procedures is in the meaning attached to the quantity $W$. In Boltzmann's discussion, the quantity $W_B$ in eq. (23) is the number of complexions, detailed assignments of the energies of the individual molecules, compatible with a given distribution. It is proportional to the probability of one state $\{w_r\}$ compared to another. PLANCK, on the other hand, never introduces quantities analogous to the $\{w_r\}$. For him the $W$ of eq. (13) is a probability *by definition*: he has no « model » for understanding this probability in any sense analogous to Boltzmann's. Thus when PLANCK sets $k \ln W$ equal to the entropy, he says, « In my opinion, this stipulation basically amounts to a definition of the probability $W$; for we have absolutely no point of departure, in the assumptions which underlie the electromagnetic theory of radiation, for talking about such a probability with a definite meaning ».

PLANCK, then, does not find the number of complexions which belong to the most probable state at all. Instead he takes for his $W$ a quantity which is the total number of complexions for *all* sets $\{w_r\}$ which satisfy the constraints of eqs. (21) and (22). In other words, Planck's $W$ is equal to the quantity BOLTZMANN called $\sum W_B$, and the equivalent of eq. (13) appears in Boltzmann's memoir when he calculates $\sum W_B$, the normalization factor for his probabilities.

One might well ask why PLANCK deviated from Boltzmann's procedure at this particular stage. Had he carried on with Boltzmann's method, he would have arrived at exactly the same result for the average energy of an oscillator, namely $\varepsilon/\{\exp[\varepsilon/kT]-1\}$. ROSENFELD [26] suggests that PLANCK actually started with eq. (8) for the entropy of an oscillator required by his conjectured distribution law, and went from that to the corresponding form for $S_N$, the entropy of $N$ oscillators. If $S_N$ were to be given by an expression of the form $k\ln W$, the form of $W$ was then determined to be something like $(N+P)^{N+P}/N^N P^P$. This last result could then be recognized as a legitimate approximation to $(N+P-1)!/(N-1)!\,P!$, a standard formula of the theory of combinations, which, as we have just seen, actually appeared in the Boltzmann paper to which PLANCK was referring. This rather plausible conjecture is well confirmed by what PLANCK himself says in his *Naturwissenschaften* article written in 1943 [27].

There is one other aspect of Planck's combinatorial procedure which deserves a comment here. Some years after Planck's work, when EINSTEIN had driven the theory a long step further by showing that radiation itself could behave as if it consisted of energy quanta, physicists tried to reinterpret Planck's reasoning along this same line. This attempt to consider Planck's energy elements as, in some sense, particles of energy seemed a plausible one, but it was quite inconsistent with the combinatorial treatment which had to be given these « particles » if one were to obtain the Planck distribution law. This was pointed out by EHRENFEST in 1911 and again, in more detail, in 1914. (The latter paper is especially noteworthy as it contains the simple and graphic derivation of the basic combinatorial formula, eq. (13), which is now universally given.) What EHRENFEST showed, in effect, was that « particles » which have to be counted according to eq. (13) are not independent particles in any ordinary sense [28]. (They are, in fact, particles which obey the Bose-Einstein statistics, but that concept could not be clarified until many years later.)

This first, combinatorial, deviation from BOLTZMANN is less striking than the one we shall now consider. As we have already seen, BOLTZMANN too used « energy elements » $\varepsilon$ in order to carry out his combinatorial procedure, but BOLTZMANN was always ready to take the limit $\varepsilon\to 0$, once the discreteness was no longer necessary to the analysis. It is obviously of the very essence of Planck's work that $\varepsilon$ could not be allowed to vanish, if the proper distribution law were to be reached. PLANCK apparently did not even consider the possibility of taking this limit. This is undoubtedly related to Planck's apparent unawareness of the equipartition theorem and all it implied, which we have already seen.

This aspect of Planck's work seems to have been recognized first in 1905, and it came out very clearly in the course of an exchange between JEANS and Lord RAYLEIGH in the columns of *Nature* [29]. In the May 18, 1905 issue, RAYLEIGH repeated his calculation of five years before, but this time he took care to include all of the proportionality constants which he had not bothered with earlier. The distribution law he obtained was

$$\varrho(\nu, T) = (8\pi\nu^2/c^3)(kT)\,, \tag{24}$$

and, as he pointed out, this is exactly the same as the form that Planck's law, eq. (19), takes in the limit of low frequencies (*i.e.* when $h\nu/kT \ll 1$), a limit which PLANCK had not explicitly considered. It followed that the quantity $k$, and therefore Avogadro's number $N_0$, could in principle be determined from the experimental value of $\varrho(\nu, T)$ at low frequencies by using eq. (24) without any reference to the Planck distribution itself. RAYLEIGH went on to say « A critical comparison of the two processes [*i.e.* his own and Planck's] would be of interest, but not having succeeded in following Planck's reasoning, I am unable to undertake it. As applying to all wavelengths, his formula would have the greater value if satisfactorily established. On the other hand, the reasoning which leads to [eq. (24)] is very simple, and this formula appears to me to be a necessary consequence of the law of equipartition as laid down by BOLTZMANN and MAXWELL. My difficulty is to understand how another process, also based upon Boltzmann's ideas, can lead to a different result ».

Actually RAYLEIGH had made an error of a factor of eight in his calculation, which was soon pointed out by JEANS. (Equation (24) is, however, correct, and it is the limiting form of Planck's law.) RAYLEIGH readily admitted his error, but returned to the same point: « But while the precise agreement of results in the case of very long waves is satisfactory so far as it goes, it does not satisfy the wish expressed in my former letter for a comparison of processes. In the application to waves that are not long, there must be some limitation on the principle of equipartition ».

Rayleigh's repeated request for a critical discussion was finally met, at least in part, by JEANS in *Nature* for July 27, 1905. JEANS undertook a severe criticism of Planck's arguments on two principal points. These points are just the two we have been discussing, where PLANCK broke with BOLTZMANN. Thus JEANS attacked Planck's use of $W$ as a « probability », pointing out that no population was given from which probabilities could be calculated, and that one could not introduce such a population with an *a priori* probability law consistent with Planck's arguments. Jeans's second point was that PLANCK had no right to refrain from taking the limit in which $\varepsilon$ is zero. If this were done, Planck's expression for the average energy of an oscillator would reduce to $kT$ in accord with the equipartition theorem. JEANS recognized that PLANCK had fixed $\varepsilon$ as equal to $h\nu$ by the use of Wien's displacement law, but he argued that nothing in the displacement law determined the value of $h$, « whereas statistical mechanics gives us the further information that the true value of $h$ is $h = 0$ ». If by « true » one means in agreement with the equipartition theorem, then JEANS was correct. JEANS erred only in supposing that « the methods of both are in effect the methods of statistical mechanics and of the theorem of equipartition of energy ». That was Jeans' method, but it was certainly not Planck's.

## 2. – Einstein and the early quantum theory.

In June 1905 the *Annalen der Physik* published an article by EINSTEIN under the weighty title « On a Heuristic Viewpoint Concerning the Production and Transformation of Light » [30]. This is the paper commonly referred to by physicists as Einstein's paper on the photoelectric effect, but that is hardly an adequate description of it. EINSTEIN himself characterized it in a letter to a friend as « very revolutionary », and that phrase, together with the paper's title, correctly suggest that much more was at stake than just the photoelectric effect. In this paper EINSTEIN set himself against the strong tide of nineteenth-century physics and dared to challenge the highly successful wave theory of light, which was one of its most characteristic features. He argued instead that light can, and for many purposes must, be considered as composed of a collection of independent particles (quanta) of energy that behave like the particles of a gas. This hypothesis of light quanta, the « heuristic viewpoint » of the title, meant a revival and modernization of the corpuscular theory of light, which had been buried under the weight of all the evidence accumulated for the wave theory during almost a century. The power of the hypothesis was shown immediately by the ease with which EINSTEIN could account for a series of phenomena, including the photoelectric effect, that had not yielded to the electromagnetic wave theory of light. But granting the success of the hypothesis, what led him to take this extreme position, so far outside the pattern of current ideas?

EINSTEIN devoted the major part of his paper to answering just this question, that is to presenting the arguments which had suggested his new « heuristic viewpoint ». These arguments, at once fundamentally simple and incredibly daring, demonstrate the essential features of Einstein's whole approach to physics. They had their roots in his earlier profound studies of thermodynamics and statistical mechanics, and they grew naturally into his fruitful investigations of the quantum theory during the years which followed. The insight into the structure of radiation that was developed in these arguments gave EINSTEIN the confidence to maintain his hypothesis of light quanta against the overwhelming support for the wave theory. The power of Einstein's reasoning did not, however, compel conviction in others. Very few were willing or able to follow him in accepting the startling idea of light quanta on the strength of deductions that were based on the statistical interpretation of the second law of thermodynamics. Even in 1913, in a letter which proposed EINSTEIN for membership in the Prussian Academy and for a research professorship and which extolled his work and his genius, PLANCK could still include the remark: « That he may sometimes have missed the target in his speculations, as for example, in his hypothesis of light quanta, cannot really be held against him » [31]. And MILLIKAN, describing his experimental confirmation of Einstein's equation for the photoelectric effect in 1916, could say of the same hypothesis « I shall not attempt to present the basis for such as assumption, for, as a matter of fact, it had almost none at the time » [32]. EINSTEIN had not been mistaken when he called this work « very revolutionary! ».

EINSTEIN was fundamentally concerned with a disturbing dualism in the foundations of physics. The old idea that mechanics was the basic science—that understanding a phenomenon meant explaining it in mechanical terms, that mechanics could serve as the basis for a fully developed world view—had proved inadequate. Electromagnetic phenomena had never really been reduced to an acceptable mechanical form, nor did EINSTEIN accept the then current efforts to reverse the procedure and make electromagnetism the basic science, the foundation of an electromagnetic world view. One was left with two kinds of fundamental theories, very different in character. EINSTEIN began his paper by pointing to this fact.

« There is a profound formal difference, » he wrote, « between the theoretical ideas that physicists have formed concerning gases, and other ponderable bodies, and Maxwell's theory of electromagnetic processes in so-called empty space. » EINSTEIN was referring to the contrast between the essentially discrete atomic theory of matter, in which a finite number of mechanical quantities completely specified the state of a system, and the essentially continuous electromagnetic-field theory, in which a set of continuous functions was needed to specify the state of the field. This dualism between particle and field, between mechanics and electromagnetism, was the starting point for his considerations.

What EINSTEIN suggested was that physicists investigate the consequences of assuming that the energy of light is distributed discontinuously in space, that is that the energy of light consists of a finite number of energy quanta, localized at various points of space, and that these quanta can be produced or absorbed only as units. Such an assumption seemed to be excluded at once by the thorough experimental confirmation of the electromagnetic-wave theory of light, but, EINSTEIN pointed out, this evidence was not quite conclusive: all optical observations yielded only time averages and did not fix the instantaneous values of the quantities in question. It was still conceivable, at least to EINSTEIN, that the wave theory might fail in its attempts to explain phenomena involving the emission of light or its transformation from one frequency to another. EINSTEIN suggested that for such phenomena as black-body radiation and the photoelectric effect one might do well to consider replacing the wave theory of light by the hypothesis of light quanta which he was advancing. This modest proposal of Einstein's, which « might prove useful to some investigators in their researches », needed justification, and he proceeded at once to offer it.

As the first step in his argument EINSTEIN pointed out the kind of serious problem that arose from the incompatibility of the two disparate fundamental theories, mechanics and electromagnetism, when they had to be brought to bear together. The « profound formal difference » could actually lead to a disaster. This disaster, to which EHRENFEST later gave the dramatic name « the ultraviolet catastrophe », has become so commonplace for the authors of textbooks on modern physics that we must remind ourselves of how very far it was from being common knowledge in 1905. Nobody who had written on the problem of black-body radiation before that date, except Lord RAYLEIGH, had even attacked the problem from the quarter in which this difficulty appeared. We have already seen that RAYLEIGH did not emphasize the significance of his result, and that it created no stir among the small group actively concerned with the problem of radiation. The « ultraviolet catastrophe », recognized and presented as a failure of classical physics, actually made its first appearance in the paper of Einstein's that we are now discussing.

The situation that EINSTEIN described was very simple. He considered a volume, enclosed by reflecting walls, that contained a gas and, in addition, a number of harmonically bound electrons. These electrons, acting as charged harmonic oscillators, would emit and absorb electromagnetic radiation and, when the system was in thermodynamic equilibrium, this radiation would be identical with the black-body radiation. Since these linear oscillators could also exchange energy with the freely moving molecules of the gas, the laws of kinetic theory, and the equipartition theorem in particular, required that the average energy $u$ of such an oscillator have the value

$$u = (R/N_0)\,T\,, \tag{25}$$

where $T$ is the absolute temperature of the gas, $R$ is the universal gas constant and $N_0$ is Avogadro's number. By requiring that the oscillators be in thermodynamic equilibrium with the radiation field, one could also relate $u$ to the spectral density of the radiation. This had already been done by PLANCK in 1899 as we have seen, with the result given previously in eq. (1):

$$\varrho(\nu,\, T) = (8\pi\nu^2/c^3)\,u\,. \tag{1}$$

By equating these two expressions for $u$, one a result of statistical mechanics, the other of electrodynamics, one was led inevitably to the result

$$\varrho(\nu,\, T) = (8\pi\nu^2/c^3)(R/N_0)\,T\,. \tag{26}$$

This result, as EINSTEIN pointed out, was not only in conflict with experiment, but it also meant that the theory did not lead to a definite distribution of energy between matter and radiation in the enclosure; for, if one tried to calculate the total energy per unit volume of the radiation by integrating $\varrho(\nu, T)$ over all frequencies, the result obtained from eq. (26) was clearly infinite. Such a failure was not an absolute one, as EINSTEIN went on to show, since the unacceptable $\varrho(\nu, T)$ of eq. (26) was simply related to another expression for $\varrho(\nu, T)$ that did account for all the experimental results. This successful $\varrho(\nu, T)$ was Planck's distribution law, which had the form

$$\varrho(\nu,\, T) = \frac{\alpha\nu^3}{\exp[\beta\nu/T] - 1}\,, \tag{27}$$

where $\alpha$ and $\beta$ are constants. In the high-temperature, long-wavelength limit, when $\nu/T$ was sufficiently small, Planck's law went over to the form

$$\varrho(\nu, T) = (\alpha/\beta)\nu^2 T , \tag{28}$$

whose dependence on frequency and temperature coincided with that of eq. (26). Not only did the functional form of the « catastrophic » $\varrho(\nu, T)$ come out of this procedure, but the constant coefficient also seemed to check. If one took the values for $\alpha$ and $\beta$ that PLANCK had calculated from the experimental data on black-body radiation and used the known values of the gas constant and the velocity of light, then one could calculate Avogadro's number from the equation

$$N_0 = (\beta/\alpha)(8\pi R/c^3) = 6.17 \cdot 10^{23} . \tag{29}$$

The resulting value of $N_0$ was identical with that obtained by PLANCK by quite a different line of reasoning; and it agreed with the relatively crude determinations of Avogadro's number that had been made by other methods prior to this time. EINSTEIN concluded that, although arguments based on the electromagnetic theory of light are sound for long wavelengths and high radiation densities, such arguments fail completely for short wavelengths and low radiation densities.

If the electromagnetic theory of light could not be trusted to give sound results, how was one to proceed? Einstein's answer was, in effect, « Boldly! ». Since the theory did not produce an explanation of experimental findings, why not turn the procedure around and see what could be learned about the structure of radiation from the well-established experimental facts, « without assuming any picture » of the basic processes, as EINSTEIN put it. One must not misinterpret: EINSTEIN was certainly not proposing to apply a naive empiricism; nothing could have been further removed from his way of approaching physics. What he did propose was the use of the experimentally established law of the radiation spectrum in combination with that most sweeping generalization, the second law of thermodynamics in its statistical form [33].

The statistical interpretation of the second law of thermodynamics had been Boltzmann's greatest contribution to science, but it was also the subject of Einstein's first major publications. EINSTEIN had written three papers during the years 1902 to 1904 in which he extended and developed Boltzmann's ideas, reworking the foundations of the subject with his own characteristic originality [34]. These papers of Einstein's did not attract much attention when they appeared, but I call attention to them here to emphasize that Einstein's thinking in 1905 was solidly established in the statistical thermodynamics which he had made his own.

The key concept in thermodynamics is the entropy, and EINSTEIN opened his new attack on the radiation problem by relating the entropy of the radiation to the spectral distribution function $\varrho(\nu, T)$. Since the radiation in one frequency interval could be considered as independent of that in any other interval, the entropy $S$ of the radiation contained in a volume $v$ could be expressed in the form

$$S = v \int_0^\infty \varphi(\varrho, \nu)\, d\nu \,. \tag{30}$$

Black-body radiation is radiation in thermodynamic equilibrium, which means that the entropy must be a maximum for a given energy. From this condition and the definition of the absolute temperature $T$ as the reciprocal of the derivative of entropy with respect to energy at constant volume, it was easy to show that $\partial\varphi/\partial\varrho$ had to satisfy the equation

$$\frac{\partial \varphi}{\partial \varrho} = \frac{1}{T} \,. \tag{31}$$

This general result made no use of any particular form for the spectral distribution function $\varrho(\nu, T)$. In order to go further, and to obtain $\varphi$ as a function of $\varrho$ and $\nu$, and then the entropy itself, EINSTEIN had to introduce an explicit form for $\varrho$. For this purpose he chose not Planck's distribution function, eq. (27), but the older Wien distribution function

$$\varrho(\nu, T) = \alpha\nu^3 \exp\left[-\beta\nu/T\right] . \tag{3}$$

Wien's law had been thoroughly confirmed by experiment in the region of large values of $\nu/T$, where it is the limiting form of Planck's distribution. EINSTEIN based his calculations on this Wien distribution, perhaps because of its greater simplicity, recognizing that any conclusions drawn from it would necessarily be limited in their validity to those situations (large values of $\nu/T$) where Wien's law did apply.

Once Wien's law was assumed, it was a straightforward matter to obtain an explicit form for $\partial\varphi/\partial\varrho$ by solving eq. (3) for the reciprocal temperature

$$\frac{\partial \varphi}{\partial \varrho} = \frac{1}{T} = -\frac{1}{\beta\nu} \ln\left(\varrho/\alpha\nu^3\right) , \tag{32}$$

and then, by integrating, to find the function $\varphi$ in the form

$$\varphi(\varrho, \nu) = -\left(\varrho/\beta\nu\right)\{\ln\left(\varrho/\alpha\nu^3\right) - 1\} \,. \tag{33}$$

This equation for $\varphi$ led immediately to the entropy; for radiation with frequencies in the interval from $\nu$ to $\nu + d\nu$, whose energy $E$ could be expressed as $v\varrho\, d\nu$, the expression for $S$ was the following:

$$S = v\varphi \, \mathrm{d}\nu = -(E/\beta\nu)\{\ln (E/\alpha v\nu^3 \, \mathrm{d}\nu) - 1\} . \tag{34}$$

The physical significance of the entropy always emerges most clearly when one calculates the entropy change associated with some process carried out by the system. In this case EINSTEIN considered the entropy change that occurred when the volume was changed from $v_0$ to $v$, keeping the energy of the (monochromatic) radiation fixed at the value $E$. This entropy change was readily calculated from eq. (34) to be of the form

$$S - S_0 = (E/\beta\nu) \ln (v/v_0) . \tag{35}$$

The change in the entropy of monochromatic radiation, expressed in eq. (35), had exactly the same dependence on the volume as did the entropy change of an ideal gas or a dilute solution in an isothermal process, a striking result which demanded further analysis.

In order to draw the far-reaching conclusion suggested by eq. (35), EINSTEIN had to show that this logarithmic dependence of the entropy on the volume had roots which went much deeper than any special assumption about the mechanics of gases or dilute solutions. His analysis had to rest directly on the statistical interpretation of the second law of thermodynamics. The cornerstone of this statistical interpretation is Boltzmann's principle: the logarithmic relationship between entropy and probability. According to BOLTZMANN the entropy difference $S - S_0$ between two states of a thermodynamic system is proportional to the logarithm of the relative probability $W$ of the occurrence of these two states:

$$S - S_0 = (R/N_0) \ln W \tag{36}$$

(the universal proportionality constant $R/N_0$ had been fixed by the analysis of an ideal gas). EINSTEIN proceeded to apply this principle to a collection of $n$ particles moving freely in a volume $v_0$, a system with a definite entropy $S_0$. EINSTEIN assumed only that the motion of the particles showed no preference for one subvolume of $v_0$ compared to another and that the particles moved independently of one another. No restriction was imposed on the laws of motion or on the nature of any other matter which might also be present in $v_0$. It was then in order to ask: « What is the probability $W$ that all $n$ of the particles... accidentally find themselves in the subvolume $v$ at a randomly chosen instant of time? ». From the assumptions stated the answer was evidently given by the equation

$$W = (v/v_0)^n . \tag{37}$$

Applying Boltzmann's principle, EINSTEIN could then write down the entropy difference between this fluctuation state and the original equilibrium state in the form

$$S - S_0 = n(R/N_0)\ln(v/v_0)\,. \tag{38}$$

This argument established the basis of the entropy equation (38) for a system of particles: only the independence of their motions and the homogeneity of these motions with respect to the original volume were necessary. Einstein's next step was to reverse the argument and apply it to the radiation. *Since* the entropy difference between the corresponding states of the radiation was given by eq. (35), and *since* eqs. (35) and (38) are structurally identical, the probability of finding all the radiant energy $E$ (of frequency between $\nu$ and $\nu + d\nu$) in the subvolume $v$ *must* be given by the equation

$$W = (v/v_0)^{n'}\,, \tag{39}$$

where the exponent $n'$ is just $E$ divided by $(R/N_0)\beta\nu$. EINSTEIN drew what for him was the inescapable conclusion: « Monochromatic radiation of low density (within the region of validity of the Wien distribution law) behaves with respect to thermal phenomena as if it were composed of independent energy quanta of magnitude $(R/N_0)\beta\nu$ ».

But how seriously was one to take this conclusion? Did it really amount to anything more than an analogy, with the « as if » the essential phrase in its statement? Here is Einstein's answer: « Now if monochromatic radiation (of sufficiently low density) behaves like a discontinuous medium with respect to the dependence of its entropy on volume, a discontinuous medium consisting of energy quanta of magnitude $(R/N_0)\beta\nu$, this suggests investigating whether the laws of production and transformation of light are also of the kind they would be if light consisted of energy quanta of such a nature ». The conclusion, in other words, *was* to be taken seriously, and EINSTEIN immediately exploited this « suggestion » as to the nature of radiation, tenous as it might (and did) seem to others, pressing it in directions that might yield experimentally verifiable consequences.

The history of physics took one of its most ironic turns in 1887 when, in the course of the very experiments that brilliantly confirmed the correctness of Maxwell's electromagnetic theory of light, HERTZ discovered the photoelectric effect [35]. For the peculiar properties of the photoelectric effect proved to be impossible to understand on the basis of Maxwell's theory. Most remarkable among these properties was the fact, brought out by Lenard's experiments in 1902 [36], that the energies of the electrons emitted from a metal surface under irradiation by ultraviolet light were independent of the intensity of the incident light. Since the intensity of any wave phenomenon is a measure of the energy transported by the wave, how was one to understand the existence of a maximum energy for the photoelectrons that was independent of the incident intensity?

Einstein's proposal that light be considered as composed of independent energy quanta gave a direct answer to this question. The process of photoelectric emission could then be viewed as a combination of independent events, the simplest of which is the absorption of such a quantum of energy by an electron in the metal surface, and its conversion into kinetic energy of the electron which is thereby set free. The maximum energy of such a photoelectron would then be determined by the energy of one light quantum, and on Einstein's hypothesis this energy is $(R/N_0)\beta\nu$; in other words it would be the frequency of the incident light rather than its intensity that fixes the energy of the photoelectrons. Even in this simplest case, however, the kinetic energy of the free electron would be less than the energy of the quantum absorbed, since a certain amount of work $P$ is required to remove the electron from the metal in which it is normally bound. The resulting equation for the maximum kinetic energy of the photoelectrons would therefore have the form

$$(\text{K.E.})_{\max} = (R/N_0)\beta\nu - P\,. \tag{40}$$

If the energy of a quantum were shared among several electrons, or if the electrons receiving energy from the incident light were in the interior of the metal, then these electrons would emerge with energies less than the maximum given by eq. (40). An increase in the intensity of the incident light, interpreted in this way, simply meant more quanta of the same energy striking the metal, and gave rise to more photoelectrons with the same distribution of energies, in agreement with Lenard's observations.

As EINSTEIN pointed out in his paper, this theory of the photoelectric effect had definite experimental consequences that had not yet been studied. The maximum kinetic energy of the photoelectrons is obtained experimentally by measuring the stopping potential $V$, that electrostatic potential which will just prevent any photoelectrons from reaching the collecting electrode, and so will cut off the photoelectric current. Since $V$ times the electronic charge $e$ must be the maximum energy of the photoelectrons, the basic equation (40) can be rewritten in the form

$$V = (R/N_0)(\beta/e)\,\nu - \Phi\,, \tag{41}$$

where $\Phi$ is just $P/e$. EINSTEIN remarked, somewhat laconically, on this equation: « If the formula derived is correct, then $V$ must be a straight-line function of the frequency of the incident light, when plotted in Cartesian co-ordinates, whose slope is independent of the nature of the substance investigated ». The implication was even stronger—not only should the slope of the predicted straight line be a universal constant, but its value would be the ratio of the basic radiation constant $(R/N_0)\beta$ to the electronic charge $e$. This basic radiation constant $(R/N_0)\beta$, which determines the magnitude of the energy quanta, will have been recognized by the reader as Planck's constant $h$, though I have refrained from calling it that for reasons to be discussed below.

The prediction that EINSTEIN made in eq. (41) was a bold one, almost as bold as the theory that led to it. Nothing at all was known about the frequency dependence of the stopping potential in 1905, and EINSTEIN was predicting both its form and the precise value of the essential constant in the equation. It actually took almost a decade of difficult experimentation before all features of Einstein's equation could be fully tested. At the end of that period MILLIKAN was able to summarize his extensive experiments with the sentence: « Einstein's photoelectric equation has been subjected to very searching tests and it appears in every case to predict exactly the observed results » [37].

Although I do not propose to follow the story of the experimental work on the photoelectric effect here, it will be worth our while to look at Millikan's attitude toward the subject of his beautiful experiments. MILLIKAN made no secret of this attitude. In a paper published in 1949 and addressed to *Albert Einstein on His Seventieth Birthday*, MILLIKAN wrote, referring to the photoelectric equation, « I spent ten years of my life testing that 1905 equation of Einstein's, and, contrary to all my expectations, I was compelled in 1915 to assert its unambiguous experimental verification in spite of its unreasonableness since it seemed to violate everything that we knew about the interference of light » [38]. He had been just as forthright in 1916: « We are confronted, however, by the astonishing situation that these facts were correctly and exactly predicted nine years ago by a form of quantum theory which has now been pretty generally abandoned ». It was in this paper too that MILLIKAN referred to Einstein's « bold, not to say reckless, hypothesis of an electromagnetic light corpuscle of energy $h\nu$, » which « flies in the face of the thoroughly established facts of interference ». MILLIKAN was more outspoken than most of his colleagues, but his opinion of the light quantum hypothesis was very widely shared. MILLIKAN also suggested that Einstein's « reckless hypothesis » was apparently made solely to account for the fact that the energy of photoelectrons is independent of the intensity of the light while it does depend on its frequency. This certainly ignores the arguments that EINSTEIN himself advanced for his hypothesis, arguments that grew out of the basic features of Einstein's approach to physics.

In this connection it must not be forgotten that EINSTEIN applied this hypothesis to more than just the photoelectric effect, even in his first paper on quanta. He showed how the quantum hypothesis accounted in a simple way for Stokes's rule for photoluminescence, for example. This rule stated that the frequency of the fluorescent light is always less than or equal to that of the light which excites the luminescence, which becomes a direct consequence of the law of conservation of energy once one grants that the energy of a quantum is proportional to its frequency. In a similar fashion EINSTEIN also analysed the inverse photoelectric effect, in which radiation is produced by electron bombardment, and the process of photoionization of gases.

There is one final point about Einstein's analysis of these phenomena that shows how closely the light quantum hypothesis was tied to its origins in the arguments discussed above. EINSTEIN had used the Wien distribution law for black-body radiation in his calculations, recognizing its limitation to high frequencies and low densities of radiation. The light quantum hypothesis had emerged from these calculations, and EINSTEIN explicitly pointed out that all conclusions drawn from it were subject to the same limitations as the Wien distribution itself. These qualifying statements of Einstein's would be sufficient by themselves to show that the light quantum hypothesis was in no sense an *ad hoc* hypothesis invented just to explain the photoelectric effect and kindred phenomena.

It is commonly believed that EINSTEIN developed his hypothesis of light quanta as an extension of Planck's theory of black-body radiation. That belief is not supported by a careful reading of the work of both physicists. Einstein's argument for light quanta shows no trace of the reasoning PLANCK had used five years earlier, as discussed in the previous Section. EINSTEIN did refer to PLANCK in his paper, but both references are to be found in the earlier sections of Einstein's paper where he was discussing the inadequacy of classical electromagnetic theory for these problems. One of the references appealed to Planck's work of 1899 for a result of electromagnetic theory, eq. (1). In his second reference to PLANCK, EINSTEIN did quote Planck's distribution law, but only as an equation that adequately described all the experimental information on the radiation spectrum. Not a word was said about Planck's assumption that the oscillators interacting with the radiation could take on only those discrete energies that were integral multiples of $h\nu$. EINSTEIN used neither Planck's distribution law nor his discrete, quantized oscillator energies in his own arguments. It is certainly significant that EINSTEIN always wrote the magnitude of his light quanta as $(R/N_0)\beta\nu$ and did not use Planck's form $h\nu$. This is not merely a matter of notation, since PLANCK had laid emphasis on the importance of $h$ as a basic natural constant, and Einstein's preference for the form $(R/N_0)\beta$ suggests that he had not accepted Planck's views.

But one need not appeal to such subtleties as Einstein's choice of notation in order to show that EINSTEIN was not building on Planck's work. One has only to read the paper that EINSTEIN wrote the following year [39]. Here are the first two paragraphs of that paper in which EINSTEIN summarized his own view of the 1905 paper and its relationship to Planck's work.

« In an article that appeared last year I have shown that Maxwell's theory of electricity in combination with the electron theory leads to results that are in contradiction with the experiments on black-body radiation. I was led, by a route set forth in that article, to the view that light of frequency $\nu$ can only be absorbed and emitted in quanta of energy $(R/N_0)\beta\nu$, where $R$ denotes the absolute gas constant per mole, $N_0$ is the number of real molecules in a mole, and $\beta$ is the exponential coefficient of the Wien (or Planck) radiation formula. This relation was developed for a region corresponding to the region of validity of Wien's radiation formula.

« At that time it seemed to me as though Planck's theory of radiation formed a contrast to my work in a certain respect. New considerations, which are given in the first section of this paper, demonstrated to me, however, that the theoretical foundation on which Planck's radiation theory rests differs from the foundation that would result from Maxwell's theory and the electron theory, and indeed differs exactly in that Planck's theory implicitly makes use of the hypothesis of light quanta just mentioned. »

The new considerations to which EINSTEIN referred started from the question: how did PLANCK arrive at a distribution law different from that required by the classical theory? (Just this same question was raised by Lord RAYLEIGH [29] in 1905 and by EHRENFEST [40] in 1906, each of the three questioners apparently unaware of the others, and each coming at the question in a somewhat different way. It is even fair to say that PLANCK raised the same question himself for the first time at this period in his lectures on the theory of heat radiation.) Einstein's new arguments set Planck's theory in the framework of statistical mechanics.

Einstein's calculation showed that Planck's entropy formula could only be obtained if the energy of an oscillator of frequency $\nu$ were restricted to the values $n(R/N_0)\beta\nu$, where $n$ is an integer. As EINSTEIN pointed out, however, Planck's theory involved a second assumption, in addition to the discreteness of the energy. PLANCK also needed to assume that the connection between the spectral density of the radiation and the average energy of an oscillator, expressed in eq. (1), must continue to hold, even though the basis for its derivation had been removed when the oscillator's energy was quantized. This second assumption was not a trivial one, as it had to apply even when the average energy of the oscillator was small compared to the quantum of energy.

EINSTEIN summarized his conclusion this way: « In my opinion the preceding considerations do not by any means refute Planck's theory of radiation; they seem to me rather to demonstrate that, in his radiation theory, PLANCK introduced a new hypothetical principle into physics—the hypothesis of light quanta ».

Einstein's attitude toward the concept of quanta differed sharply from Planck's. PLANCK had quantized the energy of a charged oscillator interacting with electromagnetic radiation in order to justify, to derive, the radiation formula he had proposed, a formula he already knew to be in agreement with experiment. The energy quanta, whose magnitude had to be proportional to the frequency of the oscillator for thermodynamic reasons, gave PLANCK a first suggestion of what the constant $h$ in his radiation formula might signify.

PLANCK saw the quantization of the oscillator's energy as the way to achieve the radiation law: he would have avoided even that radical step if he could have, but he certainly did not want to abandon any more of the established structure of theory than be absolutely had to.

EINSTEIN, on the contrary, had argued *from* the empirically confirmed radiation law *to* the existence of energy quanta. Where PLANCK saw quantization as a sufficient condition for obtaining the radiation law, EINSTEIN claimed that the radiation law demanded the existence of quanta as a necessary consequence. And to EINSTEIN quanta represented a basic aspect of the structure of radiation, rather than just a particular property of oscillators of a certain type. Once EINSTEIN had recognized the significance of the « Rayleigh-Jeans catastrophe », he never stopped probing and pondering the implications of Planck's radiation law, searching for a clue that might suggest the ideas that could replace classical theory [41]. The paper entitled « Planck's Theory of Radiation and the Theory of Specific Heat », which EINSTEIN sent to the *Annalen der Physik* in November 1906, reported an entirely new set of connections that he had found in the course of his probing [42].

He had been reworking Planck's derivation of the expression for the average energy of one of the oscillators that absorb and emit electromagnetic radiation. PLANCK, a novice in statistical mechanics in 1900, had not made clear just how his work was related to Boltzmann's methods. It was far from obvious to Planck's readers what he had really done to arrive at the radiation law. EINSTEIN began his paper with a new derivation of the equation for the average energy, going back to the fundamentals of statistical mechanics as he had independently redeveloped them a few years earlier. He showed again that a consequent treatment by the classical methods gave the equipartition result for the average energy $u$ of an oscillator:

$$u = (R/N_0)\,T\,. \tag{25}$$

This result had its roots in the basic classical assumption that equal regions of phase space should be given equal weights in the averaging process. In Einstein's own way of interpreting the probabilities used in statistical mechanics this meant that the system spent equal fractions of any long time interval in regions of equal phase volume. To avoid the equipartition result and to arrive at Planck's expression for the average energy of an oscillator one had to drop this assumption and replace it with another: only those regions of phase space in which the energy took on the discrete values $0$, $\varepsilon$, $2\varepsilon$, ..., $n\varepsilon$, ... were to have nonzero weights, and these integral multiples of the unit energy $\varepsilon$ of the oscillator were to be weighted equally. Under this new assumption the average energy could readily be calculated and had the value

$$u = \frac{\varepsilon}{\exp[N_0\,\varepsilon/RT]-1}\,. \tag{42}$$

The quantum of energy $\varepsilon$ is set equal to $(R/N_0)\beta\nu$, where $\beta$ is the constant $h/k$ in Planck's notation, and $\nu$ is the frequency of the oscillator, in order to satisfy the displacement law. The average energy can then be written in the form

$$u = \frac{(R/N_0)\beta\nu}{\exp[\beta\nu/T]-1}\,. \tag{43}$$

This result, in combination with eq. (1), leads directly to Planck's radiation law.

EINSTEIN proceeded to comment on this argument and its implications. It indicated, above all, the point at which the kinetic theory of heat had to be modified in order that it be in accord with the radiation law. This, in turn, raised a major point of principle.

« While up to now molecular motions have been supposed to be subject to the same laws that hold for the motions of the bodies we perceive directly (except that we also add the postulate of complete reversibility), we must now assume that, for ions which can vibrate at a definite frequency and which make possible the exchange of energy between radiation and matter, the manifold of possible states must be narrower than it is for the bodies in our direct experience. We must in fact assume that the mechanism of energy transfer is such that the energy can assume only the values 0, $(R/N_0)\beta\nu$, $2(R/N_0)\beta\nu$, ..., $n(R/N_0)\beta\nu$, .... »

This was by no means all, for EINSTEIN went on to write:

« I now believe that we should not be satisfied with this result. For the following question forces itself upon us: If the elementary oscillators that are used in the theory of the energy exchange between radiation and matter cannot be interpreted in the sense of the present kinetic molecular theory, must we not also modify the theory for the other oscillators that are used in the molecular theory of heat? There is no doubt about the answer, in my opinion. If Planck's theory of radiation strikes to the heart of the matter, then we must also expect to find contradictions between the present kinetic molecular theory and experiment in other areas of the theory of heat, contradictions that can be resolved by the route just traced. In my opinion this is actually the case, as I try to show in what follows ».

These remarks show how inadequately this paper of Einstein's is described by those who refer to it as simply an application of the quantum theory to solids. It would be more to the point to say that the paper was written to show that there was, or would have to be, a quantum *theory*, and that the range of phenomena which could be clarified by such a theory included the properties of matter as well as those of radiation. EINSTEIN was showing in a new way how deeply the foundations of classical physics had been undermined.

The contradictions to which EINSTEIN referred in the passage quoted above concerned the violations of the equipartition theorem that were exhibited in the specific heats of solids. The early calorimetric measurements of DULONG and PETIT had shown that the heat capacities of the elements in the solid state had a common value, if these heat capacities were always taken for a gram-atomic weight. This Dulong-Petit rule provided a rough method for estimating atomic weights and generally served as one of the few early indications that a kinetic molecular theory of solids might also be possible. The Dulong-Petit rule found a simple explanation if the thermal motions of the atoms in the solid were taken to be simple harmonic oscillations about positions of equilibrium. Each atom would have three independent vibrations of this type, and, since the average energy of such a simple harmonic oscillation is just $(R/N_0)\,T$ from the equipartition theorem, the total energy of one mole of the solid would be $3N_0(R/N_0)\,T$. The heat capacity per mole is the temperature derivative of this expression $3R$, or about 6 calories per degree.

So far there is no contradiction, of course. But this explanation of the Dulong-Petit rule proved too much, since the rule is only a rule and a number of elements have heat capacities much smaller than the Dulong-Petit value. These exceptions occur particularly among the lightest elements such as beryllium, boron and carbon. It was also well known before 1900 that these same elements had heat capacities that varied rapidly with temperature and that approached the Dulong-Petit value at temperatures well above room temperature [43]. The difficulty was to find « some escape from the destructive simplicity of the general conclusion », as RAYLEIGH expressed it in a very similar connection [44].

The situation was, however, even more disturbing, as EINSTEIN pointed out after describing the facts I have just summarized. By 1906 there were good reasons to believe that atoms had an internal structure and that they contained, in some way, electrons. EINSTEIN referred in particular to Drude's work on dispersion which indicated that, while the infra-red absorption frequencies of solids could be assigned to ionic vibrations, ultraviolet absorption frequencies seemed to be associated with electronic vibrations. But, if this were the case, then once again the equipartition theorem would demand too much, since it would require a full contribution of $R/N_0$ from each electronic vibration, and the heat capacity would have to be far greater than the Dulong-Petit value.

EINSTEIN had displayed the contradictions; he now proceeded to resolve them with one stroke. For, if his view of the universality of the quantum hypothesis was correct (« if Planck's theory strikes to the heart of the matter »), then the average energy of any oscillator is not given by the equipartition value $(R/N_0)\,T$, but rather by the expression of eq. (43). In this case, however, the energy and specific heat depend on the frequencies of the atomic vibrations in the solid. EINSTEIN made the simplest possible assumption here, recognizing explicitly that he was probably over-simplifying: he took all atomic vibrations to be independent and of the same frequency $\nu$. The energy $U$ of 1 mole of the solid would then be given by the equation

$$U = \frac{3R\beta\nu}{\exp[\beta\nu/T] - 1}; \tag{44}$$

the specific heat follows at once by differentiating $U$ with respect to the temperature. If the specific heat is plotted as a function of temperature, or rather of $T/\beta\nu$, one obtains a curve that rises smoothly and monotonically from zero at the origin and approaches the equipartition value $3R$ asymptotically when $T/\beta\nu$ becomes large. Roughly speaking, the heat capacity is negligibly small when $T/\beta\nu$ is less than 0.1, and has about the equipartition value when $T/\beta\nu$ is appreciably greater than one. Since light atoms would be expected to vibrate at higher frequencies than heavier ones, other things being equal, this result already gave a qualitative insight into why the light elements had anomalously low heat capacities at room temperature.

The implications of Einstein's specific-heat equation went much further than these qualitative remarks. EINSTEIN took it for granted that the vibrations which contribute to the heat capacity included those whose frequencies could be measured by a study of the optical absorption of the solid, at least in certain cases. From the known value of the constant $\beta$, he readily estimated that, unless the optical absorption occurred at wavelengths greater than several microns, the corresponding vibration would make no contribution to the heat capacity at room temperature. Only when the wavelength absorbed was greater than about 50 microns (well into the infra-red) would the full equipartition value of the specific heat be observed at room temperature. The data available to ENSTEIN were consistent with these results, and his estimates of infra-red absorption frequencies from specific heat values were remarkably good, considering how over-simplified a model of the solid he had used.

Even more striking than this unexpected new relationship between optical and thermal properties was the general theorem implied by Einstein's equations: the specific heat of all solids must become vanishingly small at sufficiently low temperatures. The exceptions to the Dulong-Petit rule were not to be considered as exceptional at all; they were just substances that exhibited the universal decrease of specific heat with decreasing temperature at relatively high temperatures, because of their light atoms and correspondingly high vibrational frequencies. Diamond, for example, had a specific heat that did not approach the Dulong-Petit value until it was heated to temperatures over 1000 °C, and its specific heat fell off to almost a tenth of that value when it was cooled to only 50 °C. A test of the theory for other materials, particularly for the large class that did obey the Dulong-Petit rule, would, however, require experiments at low temperature. Just such experiments were even then being planned and would soon be carried out at Berlin, but not for the purpose of testing Einstein's ideas.

The zero of the absolute temperature scale introduced by Kelvin is the only temperature with an absolute significance, but the absolute zero seemed to have no particular interest for physicists prior to 1905. In December of that year Nernst proposed a new theorem, which eventually took on the enviable status of a new law of thermodynamics, that established an essential relation between the thermal behavior of matter at temperatures near absolute zero and problems of pressing and even practical interest to chemists [45]. I do not intend to review Nernst's reasoning here, but I must point out that Nernst was concerned with chemical equilibria in gases at high temperatures. Thermodynamics left one without a method for calculating the essential constant in the condition for equilibrium, and Nernst found that this gap could be filled if he postulated that entropy differences between all states of a system disappear at absolute zero. Evidence available to Nernst made this look plausible, but much work had to be done before « the new heat theorem » would rest on secure foundations.

Nernst discussed this theorem in his Silliman lectures at Yale in the fall of 1906 and remarked on its implications for calorimetry:

« For the specific heats of liquids or solids at the absolute zero, our hypothesis requires that every atom shall have a definite value for the atomic heat, independent of the form, crystallized or liquid (*i.e.* amorphous), and of whether it is in chemical combination with other atoms. Numerous measurements by different experimenters have shown, in full agreement with each other, that the atomic heats in the solid state decrease greatly at low temperatures, but at the present time it is impossible to calculate the limiting value toward which they tend. For want of a better assumption I believe we can set for the present the value of the atomic heats at absolute zero for all elements equal to 1.5. Of course it is somewhat unsatisfactory to calculate with such a doubtful value; but, on the one hand, we are obliged for the sake of the following calculations to make some assumption, and, on the other hand, it makes little difference for the following purposes what value the atomic heat has between the limits 0 and 2. » [46]

This uncertainty in the behavior of the specific heats at low temperatures had to be removed in order to test the theorem and then to use it freely. As Nernst pointed out in a paper read to the Prussian Academy a month later, it would be enough to follow the specific heats down to the boiling point of hydrogen, or in many cases only to the boiling point of oxygen, in order to observe the limiting behavior [47].

These measurements presented a major experimental problem. Earlier workers had been content to measure average values of the specific heat over wide temperature intervals, and Nernst had to develop new methods in order to determine the specific heat at definite temperatures where the magnitudes are small. It was not until February 1910 that Nernst began to report his results [48]. He and his co-workers had studied a wide variety of elements and compounds from room temperature down to liquid-air temperatures: all had shown a marked decrease in specific heat as the temperature was lowered. NERNST remarked that « one gets the impression that the specific heats are converging to zero as required by Einstein's theory ». This seems to be Nernst's first reference to Einstein's work: he reported qualitative agreement with Einstein's equation, and announced that his co-workers LINDEMANN and MAGNUS were in the process of examining the degree to which there was also quantitative agreement.

NERNST had more to say about this quantitative agreement a month or so later when he lectured on his work to the French Physical Society [49]. This time he quoted Einstein's equation, described its connection with « that old enigma », the Dulong-Petit rule, and reported that the data so far obtained agreed very well with Einstein's specific-heat formula. The support that Einstein's result gave to the new heat theorem did not escape Nernst's attention, either. But he admitted freely that Einstein's theory gave to specific-heat measurements an intrinsic interest that he himself had not been aware of when he planned his experimental program. What is most striking in Nernst's remarks about Einstein's work is the glaring omission of any reference to the quantum theory. In April 1910 NERNST was obviously convinced of the importance of Einstein's result, but he was not ready yet to accept or at least to comment on the theory that had led to this result.

This reluctance did not persist much longer. NERNST apparently turned his attention almost immediately to the twin problems of extending his measurements to liquid-hydrogen temperatures and acquiring a full grasp of the quantum theory behind the specific-heat formula.

One key aspect of the situation was taken up by LINDEMANN, the young English physicist who was Nernst's student and collaborator during this period. The single parameter in Einstein's equation for the specific heat of a solid was the vibrational frequency; once this was fixed, the value of the specific heat was determined for all temperatures. EINSTEIN had already argued that this vibrational frequency must be identical with the optical absorption frequency as determined by the method of « residual rays ». He had also pointed out that not all thermal vibrations are optically active, since the vibrating particle could be a neutral atom rather than a charged ion. EINSTEIN had not, however, given a general way of relating the vibrational frequency, which determined the thermal behavior, to other measurable properties of the solid. In June 1910 LINDEMANN submitted a paper to the *Physikalische Zeitschrift* which offered a method of filling this gap [50]. His reasoning was based on a very simple and plausible physical assumption. At the melting point of the solid its structure is disrupted, and so LINDEMANN assumed that the amplitude of atomic vibrations at the melting temperature must be some definite fraction of the interatomic distance in the crystal. This assumption allowed him to express the vibrational frequency simply in terms of the melting temperature, the molecular weight and the density. He was led to results that agreed well with optical absorption frequencies where they had been measured, and also with the frequencies deduced from the specific-heat data by means of Einstein's formula. This work was done with Nernst's « constant helpful interest »; Lindemann's results surely served in turn to fortify Nernst's growing belief in Einstein's work.

The new data that NERNST obtained on specific heats down to liquid-hydrogen temperatures had the same effect. All the materials measured behaved in accordance with theoretical expectations; even lead, whose specific heat had not fallen more than 10 percent in going from room temperature, showed almost a 50-percent drop in the additional 60 degrees of cooling down to liquid-hydrogen temperature. Nernst's experimental curves had the shape and structure required by Einstein's theory, and they departed from the theory only at very low values of the specific heat, where the theoretical predictions were definitely below the measured values.

NERNST described these results in several papers written early in 1911 [51]. He was now thoroughly convinced not only that Einstein's result was essentially correct, but also that its verification was a strong argument for the quantum theory that lay behind it. He wrote,

« I believe that nobody who has acquired, by long years of practice, a reasonably reliable sense for the experimental test of a theory (never by any means a simple matter) will be able to contemplate these results without becoming convinced of the mighty logical power of the quantum theory, which immediately clarifies all the essential features. »

His lecture « On Modern Problems in Thermodynamics », delivered to the Prussian Academy of Sciences on 26 January 1911, gave NERNST the opportunity to discuss the matter at greater length and also to become even more eloquent on the subject of the quantum theory [52]. He now described Planck's introduction of the hypothesis of energy quanta as an innovation in the same class as those due to NEWTON and to DALTON. The quantum theory, he said, was, to be sure, still only a rule for calculation—« a very odd rule, one might even say a grotesque one »—but it had so proven its fruitfulness in Planck's work on radiation and Einstein's on molecular mechanics that it was the duty of science to take it seriously and investigate it from as many sides as possible.

Nernst's enthusiastic support for the quantum theory was soon translated into practical terms. It was NERNST who persuaded SOLVAY, the wealthy Belgian industrial chemist, to convene a select group of physicists to discuss the problems of radiation and quanta. This first Solvay Conference took place in the fall of 1911 and it marked the coming of age for the problems associated with the concept of quanta [53].

## REFERENCES


[1] M. PLANCK: *Scientific Autobiography and Other Papers* (New York, N. Y., 1949), p. 35.

[2] Planck's papers are collected in a three-volume edition under the title *Physikalische Abhandlungen und Vorträge* (Braunschweig, 1958). This edition will be referred to simply as *Papers*. See M. PLANCK: *Papers* I, pp. 372, 437.

[3] M. PLANCK: *Papers* I, p. 445.

[4] M. PLANCK: *Papers* I, p. 466.

[5] M. PLANCK: *Papers* I, p. 493.

[6] L. BOLTZMANN: *Wissenschaftliche Abhandlungen* (Leipzig, 1909; reprinted New York, N. Y., 1968). This three-volume collection will be referred to as L. BOLTZMANN: *Papers*. See L. BOLTZMANN: *Papers* III, pp. 615, 618, 622.

[7] M. PLANCK: *Papers* I, p. 560.

[8] For a thorough history of the radiation problem extending through Planck's work of 1901 see H. KANGRO: *Vorgeschichte des Planckschen Strahlungsgesetzes* (Wiesbaden, 1970). Kangro's book contains detailed bibliography and an especially careful treatment of the experimental side of the development. Also see M. JAMMER: *The Conceptual Development of Quantum Mechanics* (New York, N. Y., 1966) for a concise history of the developments through 1927.

[9] M. PLANCK: *Papers* I, p. 614.

[10] M. PLANCK: *Papers* I, p. 668.

[11] M. PLANCK: *Papers* I, p. 687.

[12] W. WIEN: *Ann. der Phys.*, **52**, 132 (1894).

[13] M. PLANCK: *Scientific Autobiography*, p. 40.

[14] Lord RAYLEIGH: *Phil. Mag.*, **49**, 539 (1900).

[15] H. RUBENS and F. KURLBAUM: *S.-B. Preuss. Akad. Wiss.* (1900), p. 929.

[16] M. PLANCK: *A Survey of Physical Theory* (New York, N. Y., 1960), p. 102; *Papers* III, p. 121.

[17] M. PLANCK: *Papers* I, p. 698. This paper and the one cited in note [11] have been reprinted with English translations in an edition edited by H. KANGRO with the title *Planck's Original Papers in Quantum Physics* (London and New York, N. Y., 1972).

[18] L. BOLTZMANN: *Papers* II, p. 164. See also my article on BOLTZMANN in *Acta Phys. Austriaca, Suppl.*, **10**, 53 (1973).

[19] M. PLANCK: *Papers* I, p. 717.

[20] M. PLANCK to R. W. WOOD: 7 October 1931. This letter is in the collections of the Center for History of Physics, American Institute of Physics, New York City. The full text of the letter is reprinted in A. HERMANN: *Frühgeschichte der Quantentheorie* (1899-1913) (Mosbach in Baden, 1969), pp. 31-32. The first chapter of Hermann's book deals with Planck's introduction of quanta, and the whole book contains much valuable material on the period it treats. The English translation, *The Genesis of Quantum Theory* (1899-1913) (Cambridge, Mass., 1971), omits some of the extensive bibliography of the German original.

[21] M. PLANCK: *op. cit.*, note [9].

[22] M. PLANCK: *Scientific Autobiography*, p. 78.

[23] H. KANGRO: *op. cit.*, pp. 144-148.

[24] For a general discussion see M. J. NYE: *Molecular Reality* (London and New York, N. Y., 1972).

[25] M. PLANCK to P. EHRENFEST: 6 July 1905. This letter is in the Ehrenfest collection of the National Museum for the History of Science at Leyden.
[26] L. ROSENFELD: *Osiris*, **2**, 149 (1936). Also see his article in *Max-Planck-Festschrift, 1958* (Berlin, 1958), p. 203. The *Osiris* article is an excellent survey of the early history of the quantum theory.
[27] M. PLANCK: *Naturwissenschaften*, **31**, 153 (1943); *Papers* III, p. 255.
[28] See M. J. KLEIN: *Paul Ehrenfest* (Amsterdam, 1970), p. 255-257.
[29] See *Nature*, **72**, 54, 243, 293 (1905).
[30] A. EINSTEIN: *Ann. der Phys.*, **17**, 132 (1905). An English translation by A. B. ARONS and M. B. PEPPARD can be found in *Amer. Journ. Phys.*, **33**, 367 (1965).
[31] Quoted in C. SEELIG: *Albert Einstein, A Documentary Biography* (London, 1956), p. 143-145.
[32] R. A. MILLIKAN: *The Electron* (Chicago, Ill., 1917), p. 238.
[33] For further discussion of Einstein's early work along these lines, see M. J. KLEIN: *Science*, **157**, 509 (1967).
[34] A. EINSTEIN: *Ann. der Phys.*, **9**, 417 (1902); **11**, 170 (1903); **14**, 354 (1904).
[35] H. HERTZ: *Ann. der Phys.*, **31**, 983 (1887).
[36] P. LENARD: *Ann. der Phys.*, **8**, 149 (1902).
[37] R. A. MILLIKAN: *Phys. Rev.*, **7**, 355 (1916).
[38] R. A. MILLIKAN: *Rev. Mod. Phys.*, **21**, 343 (1949).
[39] A. EINSTEIN: *Ann. der Phys.*, **20**, 199 (1906).
[40] P. EHRENFEST: *Phys. Zeits.*, **7**, 528 (1906).
[41] See M. J. KLEIN: *The Natural Philosopher*, **3**, 1 (1964); R. MCCORMMACH: *Historical Studies in the Physical Sciences*, **2**, 41 (1970).
[42] A. EINSTEIN: *Ann. der Phys.*, **22**, 180 (1907).
[43] See, for example, W. NERNST: *Theoretische Chemie*, 3rd Ed. (Stuttgart, 1900), p. 175.
[44] Lord RAYLEIGH: *Phil. Mag.*, **49**, 98 (1900).
[45] W. NERNST: *Gött. Nachr.* (1906), p. 1. See also Nernst's book, *Die theoretischen und experimentellen Grundlagen des neuen Wärmesatzes* (Halle, 1918).
[46] W. NERNST: *Thermodynamics and Chemistry* (New Haven, Conn., 1913), pp. 63-64.
[47] W. NERNST: *S.-B. Preuss. Akad. Wiss.* (1906), p. 933.
[48] W. NERNST: *S.-B. Preuss. Akad. Wiss.* (1910), p. 262.
[49] W. NERNST: *Journ. Phys. Théor. Appl.*, **9**, 721 (1910).
[50] F. A. LINDEMANN: *Phys. Zeits.*, **11**, 609 (1910).
[51] W. NERNST: *S.-B. Preuss. Akad. Wiss.* (1911), p. 306; *Zeits. Elektrochem.*, **17**, 265 (1911).
[52] W. NERNST: *S.-B. Preuss. Akad. Wiss.* (1911), p. 65.
[53] The proceedings of the first Solvay Conference were published under the editorship of P. LANGEVIN and M. DE BROGLIE: *La théorie du rayonnement et les quanta* (Paris, 1912). See also M. DE BROGLIE: *Les premiers congrès de physique Solvay* (Paris, 1951).

# Lectures on the History of Atomic Physics 1900-1922.

J. L. HEILBRON

*Department of History, University of California - Berkeley, Cal.*

## I. - ATOMIC PHYSICS IN 1900.

The theory of atomic structure came into being with the discovery of the electron in 1897. To be sure, it had then long been suspected that the indivisible atom of the chemists was not competent to explain the origin of spectral lines or the basis of the periodic classification of the elements, which had strongly reinforced the old idea that different atoms consist of different amounts of the same basic substance or « protyle ». But no substantial progress was made in the design of atomic models before the commonest part of the atom, the electron, had been chipped off, measured and recognized for what it was.

Many physicists contributed to the experimental work immediately preceding the identification of the electron and a few, including particularly WIECHERT, deserve honorable mention as quasi-discoverers [1]. But it was only THOMSON, Maxwell's successor once removed as Cavendish Professor of Physics in Cambridge, who had the boldness and imagination to recognize in the cathode rays the radiator of spectral lines, the long-sought protyle of the chemical elements, and, into the bargain, the carrier of the elementary unit of electrical charge. His uncommon breadth of vision no doubt owed something to his unique training: he had studied chemistry and obtained an engineering degree before 1876, when he went to Cambridge to read mathematics; experimental physics came into his repertoire only in 1884, when, to everyone's great surprise and to the annoyance of many, he was elected Rayleigh's successor in the Cavendish chair [2].

Thomson's insight into the nature of the cathode-ray particle, alias the electron, found quick confirmation from two independent lines of research. First, Lorentz' successful theory of the normal Zeeman effect showed that particles with one characteristic in common with the electron—namely the charge-to-mass ratio $e/m$—appeared to be present in intact atoms and, as Thomson supposed, to be responsible for the emission of spectral lines [3]. The importance of the Zeeman effect for grounding and extending the theory of atomic strcuture can scarcely be overrated [4].

The second chief corroboration of Thomson's views came from the study of the rays from radioactive substances. In 1898 RUTHERFORD showed that the rays from uranium consisted of a short-range component, which he called $\alpha$, and a more penetrating fraction, $\beta$ [5]. The following year several physicists succeeded in deflecting the $\beta$'s with a magnetic field and thereby determined their characteristic $e/m$ [6]. When the value came out approximately equal to that for the Lorentz-Zeeman ion, for the cathode-ray corpuscle and for the particles liberated from metals by ultraviolet light, few wished to dispute Thomson's claim that the electron was a fundamental and universal constituent of atoms [7].

Knowing the size of the bricks does not tell much about the building. How many bricks, or rather electrons, are there in an atom? What determines their number? How are they arranged? How do they interact with one another and with the positive charge supposed necessary to keep the atom neutral? What properties do they possess singly and in the aggregate that explain the periodicity of the elements, the nature of chemical combination, the phenomena of radioactivity and the laws of spectral series? I shall call the effort to answer these questions « Thomson's program », for he not only set the questions, but he and his students at first were almost the only physicists working at their solution. There were several reasons why his program, despite its grandeur, initially had little general appeal.

## 1. – The status of model making.

Scientists at the turn of the century recognized an indelible difference between the practices of English and of continental physicists. « A Briton » —I quote from one of them, FITZGERALD— « A Briton wants emotion in his science, something to raise enthusiasm, something with human interest » [8]. He has in mind mechanical models, *i.e.* detailed representations of physical phenomena, especially light and electromagnetism, in terms of the motions and interactions of hypothetical particles or media. Three aspects of these representations deserve notice.

First, the range of acceptable ingredients was quite large. One did not need to restrict oneself to orthodox mechanism, admitting only push-pull or contact forces; for had MAXWELL himself, in his kinetic theory of gases, not hesitated to suppose a direct action at a distance, depending on the inverse fifth power of the separation of the interacting molecules? [9]. The single effective constraint was that only those classes of forces with which physicists had become familiar since the time of NEWTON should be admitted; and it went without saying that the resultant description had to be continuous in space and time.

Second, the representations were not meant or taken literally. To quote FITZGERALD again [10]:

To suppose that (electromagnetic) aether is at all like the model I am about to describe (which was made from tennis balls and rubber bands) would be almost as bad a mistake as to suppose a sphere at all like $x^2 + y^2 + z^2 = r^2$ and to think that it must, in consequence, be made of paper and ink.

In the same vein THOMSON, in his *Recent Researches in Electricity*, urges his readers to shop about for a model of electrodynamic interactions with which they feel comfortable: « The question as to which particular method (of illustration) the student should adopt is for many purposes of secondary importance, provided that he does adopt one » [11]. The same physicist might on different occasions use different and even conflicting pictures of the same phenomena, a practice much approved by Lord KELVIN who, of course, repeatedly insisted that his models were not « true of nature » [12], but « a disjointed series of tableaux which appeal to the imagination » [13].

The third point about the representations is, so to speak, their institutionalization. As one gathers from the remark of THOMSON quoted earlier, the principal English pedagogues of physics considered a theory incomplete without an accompanying model or analogy, ideally elaborated to the last detail. Such pictures, they believed, fixed ideas, trained the imagination, and suggested further applications of theory. According to LODGE, who thought that English models of the aether were the greatest glory of 19th-century civilization, these advantages impel us to make use of whatever satisfactory models we can devise; otherwise, he said [14]:

We must (either) become first-rate mathematicians, able to live wholly among symbols, dispensing with pictorial images and much adventitious aid; or we must remain in hazy ignorance of the stages which have been reached, and of present knowledge.

This pedagogical bias was built into the training of British physicists, and particularly of their elite corps, the Cambridge wranglers, who sat in 18 of the 37 chairs for physics in the United Kingdom [15]. Wranglers were those who survived with first-class honors a stiff week-long examination in mathematics held in an unheated room in January. When THOMSON took this examination in 1880, it covered all branches of higher mathematics, including applications to dynamics and electromagnetism; later some specialization became possible, but these refinements need not concern us. What matters is that the examination required, and the course of study inculcated, great facility at designing and analysing just the types of models prized by MAXWELL and KELVIN, who were high wranglers themselves [16].

The dominance of the wranglers over late Victorian physics is nicely illustrated by Schuster's candidacy for a chair of mathematical physics established at the University of Manchester in 1881. SCHUSTER had submitted testimonials to his mathematical competence from continental masters like KIRCHHOFF, but none from Cambridge men. His campaign manager recognized that, with such backing, SCHUSTER would certainly lose; and he instructed him to obtain endorsements from friendly high wranglers, for their opinion, at Manchester, would count for more than that of a thousand Kirchhoffs. Fortunately SCHUSTER knew a few wranglers from St. John's College, who were delighted to certify him, especially because his chief competitor for the Manchester chair was THOMSON, a Trinity man, the natural enemy of all Johnians [17].

The wranglers began to lose their hegemony after the turn of the century, when experimental physicists trained at the Canvendish and at the Universities of Manchester and London began to acquire chairs; but even as late as 1920 Cambridge mathematicians held one-quarter of the important relevant professorships. Plainly one would expect that, in the design of atomic models, the English would take the lead.

In fact they provided almost all the manpower. The leading continental theoreticians—the Kirchhoffs, the Helmholtz', the Poincarés—did not hold English model making in high esteem. KIRCHHOFF, according to HERTZ, found it « painful... to see atoms and their vibrations willfully stuck in the middle of a theoretical discussion » [18]. HELMHOLTZ, who much admired KELVIN, freely confessed his want of taste and talent for doing physics in the English manner. As for POINCARÉ, in his experience all Frenchmen were oppressed by « a feeling of discomfort, even of distress », at their first encounter with the writings of MAXWELL [19].

I do not wish to give the impression that continental physicists depreciated mechanical reductionism. Quite the contrary. On this point we have the testimony of that same KIRCHHOFF who is pained by the constructions of the English [20]:

It has been postulated that all physical phenomena ... have their origin in motion, that all physics therefore is to be reduced to mechanics. Should this succeed ... the highest conceivable goal will be reached, in view of the simplicity of the representation; the reduction in question is therefore to be sought in the fullest measure.

Similar sentiments may be culled in any desired quantity from the prefaces of French and German physics books of the latter 19th century. And yet continental usage differed altogether from the English. The continentals, who rightly considered reductionism a distant and perhaps an impossible goal, did not pursue piecemeal analogies or provisional illustrative models; they might accept the chemical atom, the gas-theoretical molecule and the electromagnetic aether, but, in general, they did not care to form a circumstantial picture of their modes of operation. They preferred general formulations of the thermodynamic type and abstract or even axiomatic presentations [21].

In sum: one reason that Thomson's program developed slowly was that, in 1900, almost all continental physicists—and that means about 60 percent of all physicists then in existence—had no taste for it. « One wonder(s) where (atomic) physics would have been »—I am quoting words an American professor, a great admirer of the English style, wrote in 1908—« One wonder(s) where (atomic) physics would have been if it had not been for British common sense » [22].

We should not leave this topic without knocking on the head the common misconception—doubtless derived from over-crediting the few but highly vocal agitators centered on OSTWALD—that continental scientists at the end of the last century had all but abandoned the atomic theory of matter [23]. OSTWALD advocated a system called « energetics », which drew its inspiration partly from positivism and partly from the great success of thermodynamics in delivering up the principal relations of physical chemistry without resorting to molecular models. OSTWALD and his followers held that most of the ills of fin-de-siècle physics, and particularly certain difficulties in the kinetic theory of gases, all could be avoided by the simple expedient of discarding the model, and indeed all models, and reducing physics to an account of the transformations of energy [24].

In 1895 OSTWALD and his principal disciple, the mathematician HELM, gave invited talks to the annual meeting of the Gesellschaft deutscher Naturforscher und Ärzte. The invitations seem to have been issued at the recommendation of BOLTZMANN, who not only did not fear the energeticists, but expected that, with a good airing, they would expire without a shot. And so it was. HELM demanded that not only mechanical models, but mechanics itself must go: the laws of motion, and even the behaviour of mass points, were to be derived solely from energy considerations. For this he was so roundly cudgelled by BOLTZMANN and others that he demanded and received a public apology [25]. Then OSTWALD arose to deliver a speech « On Overcoming Scientific Materialism », by which he meant the doctrine that all « physical phenomena should be reduced to matter in motion, to the mechanics of atoms » [26].

After emphasizing failures of the kinetic theory of gases, and especially the problem of reversibility, he said:

> We must therefore renounce the hope of representing the physical world by referring natural phenomena to a mechanics of atoms. « But »—I hear you say—« but what will we have left to give us a picture of reality if we abandon atoms »? To this I reply: « Thou shalt not take unto thee any graven image, or any likeness of anything ». Our task is not to see the world through a dark and distorted mirror, but directly, so far as the nature of our minds permits. The task of science is to discern relations among *realities, i.e.* demonstrable and measurable quantities ... it is not a search for forces we cannot measure, acting between atoms we cannot observe.

This call was no more successful than Helm's. As OSTWALD recalls in his autobiography, he was surrounded by enemies at this meeting; never before, he says, but often thereafter, was he faced with such united opposition. When he disclosed the title of his proposed address to the President of the Society, he was relegated to the end of the program, and another chemist was enlisted to speak first, in favor of atoms. Even Ostwald's closest chemical colleagues refused to follow him: « they said that I might be right, but that they could not work out their problems without the help of the atomic picture » [27].

The small following the antiatomists did acquire soon dissipated and OSTWALD himself, with his wonted frankness, later announced that the new discoveries in physics, and particularly the experiments on Brownian motion of PERRIN, had convinced him of the reality of atoms [28].

## 2. – Anticipated difficulties in the design of atomic models.

The second cause for delay in the development of atomic physics was the difficulty of the undertaking, which recommended itself only to the hardiest even among the British. As one might suspect, the brotherhood of wranglers had no trouble anticipating the obstacles which the atomic architect would face; and it is a great tribute to them that, even before the turn of the century, they had identified and discussed most of the pertinent problems. I would like to bring six of these problems briefly to your attention: the nature of the positive charge deemed necessary to neutralize and to retain the atomic electrons; the number $n$ of these electrons; the fixing of the size of the atom; the prevention of radiative collapse; the form of the spectral series; and the identity of the spectral radiator.

The *positive charge* was perhaps the most pressing of these problems. THOMSON initially planned to make do with electrons alone, to suppose that they managed to act together, when assembled in an atom, as if the space they occupied contained a positive charge equal to the sum of their charges. Only negative electrons were to have what he called a « real existence » [29]. But the single-fluid theory of electricity, as his own work was to show, could not be sustained, and atom builders had perforce to guess at the nature of the positive charge.

A popular early view supposed the existence of positive electrons, identical in all respects but charge to their certified negative siblings. LARMOR had used them in his work on the foundations of electrodynamics and LODGE, who was much given to symmetry, lobbied strongly for them on the grounds of reasonableness and simplicity. Experiment nonetheless declined to furnish the Larmor-Lodge particle, which provided the basis for a few models, none highly elaborated, and then dropped from sight [30]. This did not bring atom builders much further, for the rejection of the symmetric view of the missing positive charge did not imply any particular substitute.

The second problem concerned $n$, the *number of electrons in an atom.* It was formulated as a search for the dependence of $n$ on what was presumed to be a more fundamental quantity, the atomic weight $A$. The solution depended critically on the assumed character of the positive charge; if, for example, one postulated the positive electron, one concluded that $n$ approached 1000 times $A$; if one followed THOMSON and FITZGERALD, and tried to make do with a single-fluid theory of electricity, one obtained twice that number. The early atomic models all assumed huge electronic populations [31]; and it was not until THOMSON devised ways to make $n$ an object of experimental investigation and not merely of speculation, that atomic theory began a steady advance.

The presumption in favor of great populations, based upon the relative masses of atoms and electrons, was strengthened by consideration of the rich spectra given off by even the lightest atoms. Since one assumed in 1900 that spectral lines originated in the oscillations of negative electrons each with three degrees of freedom, it followed for many physicists that, at a minimum, the $n$ of an atom had to equal one-third the number of its spectral lines. As several thousand lines were then known in the iron spectrum, the value of $n$ inferred on optical grounds agreed well with that obtained from the most plausible guess about the nature of the positive charge [32].

The third difficulty, the very *definition of the atom*, arose from the circumstance that (as is now well known) one cannot obtain a length from the constants characterizing electrons and their interactions. Or perhaps one should say that no such length emerges from the classical electrodynamics; relativity theory produces one, the so-called radius of the electron, $e^2/mc$, but this does not help fix the separations of electrons bound into atoms. In 1900 LARMOR drew attention to this difficulty by observing that « given any existing steady system of electrons, the same system, altered to any other scale of linear magnitude, is possible if there are none but electric actions ». He concluded, quite rightly, that he and his contemporaries were « hardly on the threshold of the structure of the atom » [33].

In fact the existence of atoms with structures as well defined as the sharpness of spectral lines required was a great puzzle for the old statistical mechanics, a point (as we shall see) that seems to have bothered BOHR very early in his career. His resolution of the problem required a concept, the quantum of action, not yet introduced into physics when LARMOR issued his warning.

The fourth difficulty, the eventual *collapse* of any system of moving electrons owing to their radiation, varied in seriousness according to the details of the atomic model and the radiation mechanism. All agreed in referring the electronic motions responsible for spectral emission to oscillations about equilibrium; but was that equilibrium static or dynamic? In the latter case, the more attractive because it could be realized under the action of inverse-square forces alone, the problem of collapse would seem to be acute, for the electrons must be accelerated in order to describe closed orbits, and while accelerated they are obliged to radiate.

The first discussion of this problem we also owe to LARMOR, who showed that, if the equilibrium motions of the electrons are properly harmonized, one need not disturb oneself over radiation loss from an unexcited atom. Assuming that the algebraic sum of the accelerations of all the electrons vanishes, LARMOR found that the total radiation of the system is to that produced by a single particle with the same average absolute acceleration as the square of the ratio of the atomic radius $a$ to the wavelength $\lambda$ of the radiated light [34]. Since this ratio was known to be about $10^{-3}$, it appeared that an appropriate adjustement of the accelerations would yield a radiation one-millionth that of a single particle.

The argument does not apply in the case of one electron, and closer analysis shows that the radiation from the equilibrium state in general depends very sensitively on the number of electrons in the atom. In fact, if the positive charge exerts a direct-distance force on the electrons, the radiation is very much smaller than that given by LARMOR, for it diminishes not as $(a/\lambda)^2$ but as $(a/\lambda)^{2n}$, where $n$ is the number of electrons whose total acceleration is adjusted to zero [35]. For large $n$ the radiation loss from such a system can be negligible: once again $n$ emerges as the critical quantity for the problem of atomic structure.

*The form of the spectral formulae.* Although for the moment one could consider the atom saved from radiation collapse, one might still doubt that any arrangement of electrons could execute a system of vibrations at all like that given by the spectroscopic formulae. This difficulty was urged with great vigor by Lord RAYLEIGH who considered general analytical expressions for the forces acting on vibrating systems, and deduced that, except in quite special cases, one would always obtain simple relations involving the *square* of the vibration frequencies. Now the formulae of BALMER and of KAYSER and RUNGE give simple expressions for the first power of the frequency and the constant doublet separations in the spectra of the alkalis cannot easily be expressed in terms of the squares of the frequencies. RAYLEIGH ended by doubting whether « the analogy of radiating bodies is to be sought at all in ordinary mechanical systems vibrating about equilibrium » [35*a*].

But one could always suppose that for some special system the root might be easily extracted; and RAYLEIGH himself pointed out that FITZGERALD had once discussed just such a system, a linear sequence of magnets executing small vibrations about the line passing through their centers [36]. In such a case

$$\nu = \text{const} \times \cos \frac{s\pi}{2M} \simeq \text{const} \times \left(1 - \frac{\pi^2 s^2}{8M^2}\right),$$

where $s$ is the integer specifying the mode of vibration. One consequently can obtain a close analogy to Balmer's formula if one supposes that hydrogen lines arise from different magnetic systems ($M$ variable) vibrating in the same mode ($s$ constant).

The upshot of Rayleigh's analysis therefore was a general doubt that mechanical systems of a certain kind could be found and the specification of an exceptional case with the desired property. One did not know quite what conclusion to draw, and, like THOMSON, generally resolved the matter by making no attempt to derive the spectral formulae from a model.

This course of action was further recommended by our last difficulty, the *nature of the spectral radiator.* A great mass of spectroscopic evidence—like the different responses of lines in the same spectrum to a magnetic field or to changes in the conditions in the discharge tube—suggested that several, perhaps many, different vibrating systems must be associated in the production of a single spectrum [37]. THOMSON obtained a striking confirmation of this view from Lorentz' theory of dispersion by gases, which, in the case of long waves, predicts a simple dependence of the index of refraction $\mu$ on the natural frequencies $\nu_i$ of the electrons responsible for the scattering:

$$\mu^2 - 1 = \sum \frac{c^2 e^2}{4\pi m} \frac{p_i}{\nu_i^2},$$

where $p_i$ is the number of effective electrons in unit volume. THOMSON computed what should have been far too low an estimate of $\mu$ by considering only one resonance frequency $p$ and by setting $p = N$, the number of molecules in unit volume [38]. The rock-bottom estimate turned out to be 20 to 100 times too large! It appeared, as he observed, that «in a luminous gas the spectral lines are not given out by every molecule, but only by a comparatively small number of systems» which might well be different for each line. THOMSON inclined to identify these systems with «aggregates of greater complexity than the molecule»; others, especially LOCKYER and the dean of spectroscopists, KAYSER, favored subatomic complexes [39]. In either case one knew nothing about the nature of the vibrator, or, consequently, how spectral measurements should be utilized in the theory of atomic structure.

## 3. – The physics profession in 1900.

The third reason retarding growth of atomic theory, and one that might be thought sufficient in itself, is that, in its early days, atomic theory did not exist. By this paradox I mean to call attention to the fact that at the turn of the century there were no institutes dedicated to its cultivation, no congresses to discuss it or any of its departments and—most interesting of all—no rubric for it in any of the major subject classifications or abstracting journals. Thomson's atom, for example, is noticed under «physical chemistry» in the *Fortschritte der Physik*, and under «electrical theory» in the *Beiblätter*; Rutherford's appears under «Becquerel rays» and under «radioactivity» [40]. English reviews had a more appropriate but vaguer rubric, the « constitution of matter », which became current in Europe only after 1910 [41].

When atomic physics crystallized about 1910 it of course drew on lines of experimental work—particularly on electrical discharges through gases, spectral analysis and radioactivity—that were recognized subjects at the turn of the century. By far the largest part of this work took place in the academic institutions of Europe and the United States. Some colleagues and I at Berkeley are studying the personnel, funding and productivity of academic physics at the turn of the century. Our results make possible an estimate of the amount of men and treasure devoted to what might be called the proto-atomic physics of 1900.

There were some 700 physicists in the higher schools of the leading physics-producing nations in 1900, distributed as in Table I.

TABLE I. – *Academic physicists c.* 1900.

| | Faculty and assistants | Per $10^6$ of population |
|---|---|---|
| Austria/Hungary | 64 | 1.5 |
| Britain | 114 | 2.7 |
| France | 105 | 2.8 |
| Germany | 145 | 2.9 |
| Universities | 109 | |
| Techn. Hochsch. | 36 | |
| Italy | 63 | 1.8 |
| United States | 215 | 2.9 |

Here faculty includes all ranks from professor to Privatdozent. It appears that in the entire world in 1900 there were no more than 1000 academic physicists. To fix ideas, the staff of CERN in 1966 was 2490, of whom 717 were classed as « technical personnel » and 413 as scientists.

We have also been able to estimate the moneys expended on these physicists for salary, laboratory and new plant. We express sums in German marks which, in 1900, had about the same purchasing power as dollars did in 1966 (see Table II).

One sees that, in the leading countries, the total annual investment in academic physics was about 8.5 million marks. Again, for comparison, CERN's budget for 1966, exclusive of capital expenditures, was about $ 25 000 000. Allowing liberally for inflation, that represents about two and one-half times the world-wide expenditure on academic physics in 1900. Note further that, in 1900, all the leading European countries spent about the same proportion of their national incomes on academic physics, namely 0.004 percent, which is just under half the percentage of their incomes they now give towards maintaining a single facility, CERN.

The capital investment in laboratory buildings and fixtures is more difficult to assess; Table II gives under «new plant» our estimate of the average annual capital outlay in the years around 1900. A large, up-to-date teaching and research laboratory like the one opened at Manchester in 1900 cost about

TABLE II. – *Annual investment in Academic Physics c.* 1900 (in 1000's of marks).

| | Salaries and fees | Laboratory expenditures | New plant | Total per Academic Physicist | Total/$10^5$ marks of Nat. Income | Cont. to CERN 1966/ Nat. Income ($\cdot 10^{-5}$) |
|---|---|---|---|---|---|---|
| Austria/Hungary | 305 | 155 | 100 | 8.8 | 8.0 | 8.9 ([a]) |
| Britain | 815 | 225 | 610 | 14.5 | 4.6 | 9.4 |
| France | 635 | 290 | 240 | 11.1 | 4.5 | 9.0 |
| Germany | 765 | 385 | 340 | 10.3 | 4.4 | 9.0 ([b]) |
| Universities | 615 | 300 | 240 | 10.6 | — | — |
| Techn. Hochsch. | 150 | 85 | 100 | 9.3 | — | — |
| Italy | 260 | 180 | 80 | 8.3 | 3.4 | 8.1 |
| United States | 1430 | 660 | 900 | 14.0 | 4.8 | — |

(*a*) Austria only. (*b*) Federal Republic of Germany.

750 000 marks for plant and equipment; the new science laboratories of the Sorbonne, finished in 1898, cost roughly 1.8 million marks; while the most elaborate of all facilities, the building and testing and research instruments of the Physikalisch-Technische Reichsanstalt had consumed 7.2 million marks by 1900, of which about one-third went to physics *per se*. I have not been able to obtain a very good estimate for the capital investment at CERN; but to 1964 the synchro-cyclotron and proton synchrotron had cost over $ 40 000 000, which is probably not far from half the world-wide investment in plant and facilities for physics in 1900.

My concern here is to estimate the fraction of personnel and of *research* money devoted to proto-atomic physics in 1900. The results can only be approximate: one must guess at the proportion of laboratory expenses going into research, and one must decide, sometimes arbitrarily, the proto-atomic qualifications of particular fields. Taking the latter problem first, radioactivity, a good part of the electron theory including magneto-optics and a portion of spectral analysis plainly qualify, while crystallography, most X-ray research, singing flames and the kinetic theory of gases do not. Without rehearsing the fine points of our definitions, I give you the results: about 20 percent of academic physicists in 1900 worked on proto-atomic fields; and about one-fifth of the papers published in the leading journals, *Annalen der Physik*, *Comptes Rendus* and *Philosophical Magazine*, concern such fields, as do slightly less than 20 percent of the invited contributions to the International Congress of Physics held in Paris in 1900 [43].

Accurate determination of research expenditures is not possible because most laboratory or institute budgets of 1900 do not distinguish between funds provided for instruction and those devoted to research. Specialized instruments were often acquired and justified as necessities for instruction, although they might be applied solely to research, and demonstration or teaching equipment frequently wound up as temporary research apparatus. One thinks of LANGEVIN forced to interrupt his experiments to free a table for beginning students, or THOMSON obliged to disassemble his apparatus as bits of it were wanted for laboratory exercises [44]. The machine shop and mechanics attached to the larger laboratories likewise present a tough knot to the accountant, for they made specialized research equipment as well as instructional apparatus. These formidable difficulties have not deterred us. The estimates of the fraction of laboratory expenditures for academic physics which went to research in the four biggest physics-producing nations in 1900 are shown in Table III.

TABLE III. – *Resources devoted to research.*

| | Percentage of laboratory expenditures | Amount (1000's of marks) |
|---|---|---|
| Britain | 50 | 113 |
| France | 38 | 110 |
| Germany | 57 | 220 |
| United States | 40 | 260 |
| Total | | 703 |

We find that about 30 percent of this sum, or 230000 marks, went to proto-atomic physics; and since about 20 percent of the physicists in the Big Four, or 120 men, consumed it, it appears that each proto-atomic academic physicist of 1900 had roughly 2000 marks/year for his work.

To put this sum in perspective, a good vacuum pump then cost 1000 marks, a 5-inch Rowland grating 800 to 1000, and a liquid-air machine some 5000. It was said that a standard piece of research cost at least 1500 marks [45]. The proto-atomic physicist, who was somewhat better funded (because concentrated in the larger centers) than his colleagues in other work, had just about enough to keep his research going. There were, however, many local and even national imbalances: France fell below the average of the Big Four; Italy enjoyed large institutes well staffed with mechanics but barren of instruments [46]; Britain and the United States profited and suffered from the vagaries of private philanthropy; and even in Germany, a RÖNTGEN, after winning the Nobel Prize, could complain that his research budget did not enable him to buy the X-ray tubes he needed for his experiments [47].

We are left with the impression of a small, lively enterprise, sometimes held back for want of relatively inexpensive facilities, equipment or supplies. Consideration of the material limitations of the time makes one admire all the more the achievements of the first generation of atom builders.

## II. - The Age of Thomson.

The atomic models of the age of THOMSON—by which I mean the years from the discovery of the electron to the acceptance of the Bohr theory—do not earn their authors high marks for invention. On the contrary, what excites our admiration is the steady progress, culminating in the nuclear theory of RUTHERFORD and the doctrine of atomic number, that THOMSON and his school made with methods and models of the homeliest character.

THOMSON proposed his well-known dynamical representation of the atom, irreverently (and, as we shall see, inappropriately) called a plum-pudding, in 1903. Two patently unsatisfactory alternatives had by then been discussed and discarded. One alternative [48], a static atom composed of equal quantities of negative and hypothetical positive electrons might seem attractive, both for its symmetry and for its freedom, in its equilibrium position, from the radiation drain. But, as we know from the theorem of the Reverend EARNSHAW [49], no static equilibrium is possible for a system of point charges interacting according to the law of squares: it will either disperse or coagulate. Inventors of static atoms had therefore to endow electrons with forces besides Coulomb's. When the Bohr atom had made him desperate even THOMSON, as we shall see, stooped to such a step; but in 1904, when Coulombian electrons had begun to bring order to much of physics and chemistry, the assumption of supernumerary forces seemed cowardly and retrogressive. One may recall that Lord Kelvin's appeal to Boscovichian forces to explain the properties of radium was greeted, as it deserved to be, as a relic from a dead—or perhaps I should say a classical—age [50].

The second alternative superseded by Thomson's pudding was the perennial Saturnian model, in which electrons, arranged in concentric rings, circulate about a positively charged central body or nucleus. Of the several early forms of this model the best known and most detailed was that of NAGAOKA [51], Professor of Physics at the University of Tokyo. NAGAOKA does not entirely escape the generalization that the early contributors to the theory of atomic structure were Cambridge men. He had learned physics at Tokyo under KNOTT, the disciple of an old senior wrangler, TAIT of Edinburgh; and he had visited the Cavendish during an extended study trip to Europe in the middle nineties [52].

NAGAOKA tried to associate certain spectral regularities with the frequencies of the small vibrations the electrons execute about their equilibrium orbits under slight external disturbances. But no more than in the static case can such a dynamical system exist under the action of electrodynamic forces alone: in any Saturnian atom with more than one electron there always exists at least one unstable mode of disturbed oscillation about the equilibrium orbits. The amplitudes of the unstable modes grow until the model flies apart [52*a*].

NAGAOKA missed this unpleasant consequence largely because he blundered in adapting Maxwell's prize-winning investigation of the stability of Saturn's rings to the electromagnetic case. The forces between Maxwell's gravitating particles are attractive; those between Nagaoka's electrons repulsive; the difference in sign accounts for the radical instability of the Saturnian atom. These lugubrious facts were brought to Nagaoka's attention by another of Thomson's students, a former wrangler appropriately called SCHOTT [53]; and after a few effete attempts at repair the nuclear model was dropped until resurrected by RUTHERFORD for an application NAGAOKA had not considered.

Note that the instability in question is *mechanical*: it is not derived from that radiation drain that plays so important a part in the standard history —or perhaps I should say mythology—of physics. The radiation drain, as noticed earlier, can be reduced to acceptable limits in a well-populated atom. The mechanical instability, however, will not voluntarily remove itself; when BOHR adopted Rutherford's model he eliminated the instability by fiat, by exempting from destruction orbits which satisfied a certain quantum condition. This legislation, which amounted to introducing unspecified and perhaps unspecifiable nonclassical forces, gravely violated the presuppositions of Thomson's school.

## 1. – Thomson's goals and style.

Thomson's greatest service to atomic theory was to initiate a promising research program. Already in his first paper announcing the discovery of the electron he had pointed to Mayer's magnets (Fig. 1), which arrange themselves in concentric circles under an external magnetic field, as a pledge that the periodic properties of the elements could be traced to atoms constructed of identical parts [54]. He did not doubt that these parts were electrons, but he did not, at first, perceive how to arrange them in a mathematically tractable manner. The clue to the plum-pudding came from Kelvin's bizarre representation of a radium atom as a diffuse sphere of positive electricity within which a few electrons sat in static equilibrium [55]. The diffuse sphere acts upon electrons within it by a direct distance force, as one immediately sees by comparing the gravitational case; and, if the number of contained electrons is not too large, a ring of them is stable against small disturbances about its equilibrium position. In Kelvin's application the electrons, struck by a hypothetical penetrating cosmic radiation, absorbed energy, vibrated with larger and larger amplitudes, and shot out of the atom to appear in the world as β-rays; and the diffuse sphere itself, interacting with other atoms *via* an appropriate Boscovichian force, sailed through matter as an α-particle.

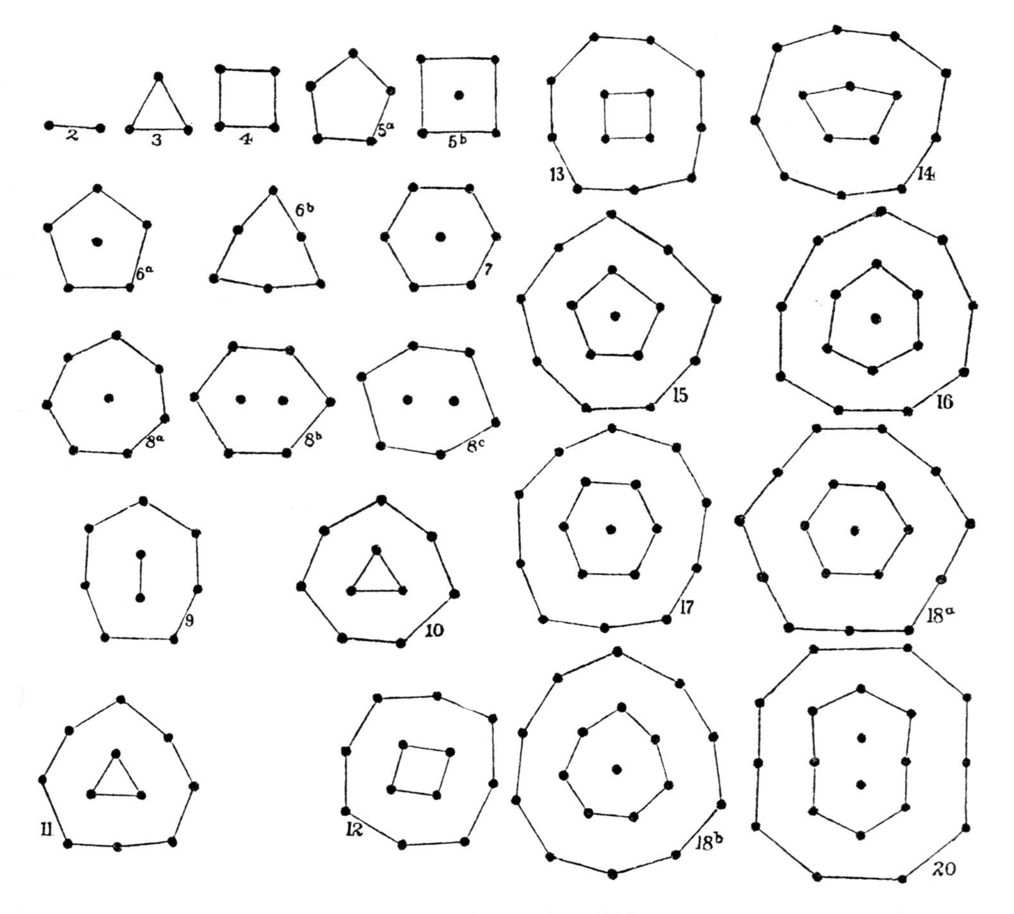


Fig. 1. – Mayer's magnets, from *Am. Journ. Sci.*, **116**, 248-249 (1878). The diagrams were made by pressing a piece of paper against the inked tips of the floating magnets.

THOMSON adopted the diffuse sphere as a convenient mathematical artifice and made the equilibrium of the electrons *dynamical*: his scheme is a Saturnian atom whose central charge has expanded to engulf the circulating electrons, and which consequently does not resemble a proper quiescent plum-pudding. A preliminary paper in the style of LARMOR showed that one need not worry about the radiation loss from a ring of several electrons; indeed THOMSON made a virtue of necessity and associated radioactivity with ancient atoms whose internal motions had decayed by radiation to the point of instability and explosion [56]. Since at this time (1904) THOMSON still thought that the atom contained a great many electrons—on the order of 1000 times the atomic weight—he did not lack particles to populate his rings and disarm the radiation drain.

The heart of Thomson's analysis [57] is the calculation of the frequencies of the perturbed oscillations of the electrons in a single-ring atom as a function of their number, say $p$. Not that he cared to make spectroscopic associations in the manner of NAGAOKA; indeed, he never tried to compute the frequencies of spectral series from any model. THOMSON hoped rather to learn from the frequencies how large $p$ might be before mechanical instability set in. It turned out to be six. To accomodate more electrons in a single ring the rate of change of the restoring force with distance had to be greater than that afforded by the diffuse charge alone. Now nothing could be simpler than increasing this rate: one need merely put one or more electrons ($q$ in all, say) at the atom's center. THOMSON calculated the values of $q$ necessary for, or consistent with, the stability of an outer ring of $p$ electrons. The $q$ electrons themselves must be distributed in rings; and for each value of the total population $n = p + q$ the distribution is unique.

This distribution, which produces an electronic parallel to Mayer's magnets, is far more suggestive. Thomson shows that, if $p = 20$, $q$ must lie between 59 and 67, inclusive. If $q$ is close to the minimum, the atom could increase its stability by losing one of its 20 outer electrons; such an atom would act electropositively. Similarly, if $q$ is near a maximum, the atom might tend to gain an electron, or act electronegatively. The models characterized by $p = 20$ consequently offer a striking analogy to the elements of the second or third periods of Mendeleev's table [57*a*].

It was this elucidation of the periodic table, expanded and translated into German, that brought to continental physicists an inkling that something might come from the theory of atomic structure. In 1909 BORN thought Thomson's model sufficiently promising to take it as the subject of his Habilitationsvortrag; and in 1911 Sommerfeld's physics colloquium studied it with the help of floating magnets [58].

Three points about Thomson's analogy deserve attention First, he has introduced the fundamental idea that atoms of successive elements in the periodic table differ from one another by the addition of a single electron. Second, he has, from a modern point of view, interchanged the roles of core and valence electrons. The atoms of each period are characterized by the same number of external electrons, and differ only in the populations of their inner rings. Chemical and optical properties consequently derive primarily from the deeper-lying electrons; the members of a chemical family have only internal structures in common. Likewise all the electrons in the atom, and not just the deepest, are implicated in radioactivity: and it is therefore difficult to find room in

Thomson's scheme for structures with identical chemical and different radioactive properties. The existence of isotopes, as BOHR later emphasized, could not be explained plausibly on the basis of the diffuse-sphere atom.

The third point about Thomson's analogy is that, despite the mathematical labor which secured it, it was essentially qualitative. Here we reach a perplexing and perennial characteristic of Thomson's physics. At the very beginning of his career, in 1882, he had won the prestigious Adams Prize at Cambridge for a lengthy essay on Kelvin's vortex atoms. These structures, which may be likened to smoke rings in air, were supposed to consist of and move about in a universal frictionless fluid, subject solely to the laws of hydrodynamics. To describe encounters between such atoms and the ways in which their vortex filaments might be linked in permanent or transitory ways required the most severe and rigorous calculations; to apply the quantitative results to the simplest physical or chemical phenomenon proved all but impossible, and THOMSON had to content himself with the same sort of qualitative and suggestive connections which were to recommend his electronic atom. And, as in the latter case, it was the number of primitive elements—*i.e.* vortex rings—which could remain interconnected in stable equilibrium that played the leading part in his thought [59]. But nowhere, either in the vortex or the electronic atom, did THOMSON ever identify *particular structures* with particular atoms. «Things needed not to be very exact for THOMSON», BOHR used to say, «and if it resembled a little, it was so» [60].

## 2. – The order of $n$.

Like other members of the Cambridge school, THOMSON esteemed a model or analogy not for its exactness or appropriateness but for its capacity to suggest further researches. In his hands his model atom admirably met that test: with the help of his students, particularly BARKLA, and with an eye to his model, he designed new ways for probing the structure of atoms and obtained a result—approximate of course—of the greatest importance in the early history of atomic theory. The result was, so to speak, a decimation of the atom, a 1000-fold reduction in its estimated population.

THOMSON counted the atomic electrons by computing the contributions of his model atoms to the scattering of various kinds of radiations [61]. The outermost electrons, he supposed, would disperse light of optical frequencies; hence measurements of dispersion, when compared to his formulae, ought to give a minimum value for $n$. Reference to old experiments on hydrogen made $n \simeq A$. To improve the estimate THOMSON looked to the scattering of X and $\beta$ rays, which (he expected) would penetrate the atom and agitate deep-lying electrons. In the case of X-rays he supposed the scatterers to be free; in the case of $\beta$-particles he supposed them immovable; consequently the X-rays counted loosely or moderately bound electrons, and the β's the most tightly held atomic constituents. From experiments by BARKLA and by RUTHERFORD, THOMSON deduced that the number of X-ray scatterers equalled about $2A$ and the number of β scatterers about $0.2A$. Evidently, he said, the number of electrons in an atom is of the order of its atomic weight.

Thomson's calculations, which need not occupy us here, generally do not rest upon the details of his special atomic model; the diffuse sphere only plays a part in the relatively unimportant case of optical dispersion. But I do not doubt that the model guided his thinking, especially in regard to the scattering of β-rays. Although other physicists, and particularly LENARD, had earlier recognized the value of β scattering as an atomic probe [62], THOMSON was the first to derive a quantitative theory of the phenomenon and the first to use it to obtain fruitful information about the atom. The theory depended upon the simplification, authorized by the model, that the positive charge does not contribute sensibly to the observed scattering. On the contrary, according to THOMSON, the basic mechanism of scattering is a close collision between the β-particle and a bound electron; the observed deflection arises from a very large number of such encounters, each deviating the particle's path by an amount which, in general, is itself inappreciable. This, the so-called multiple-scattering theory, flowed quite naturally from Thomson's atomic model and, as we shall see, perished with it.

The new estimate of the order of magnitude of $n$ had several important consequences. First, it revealed that most of the weight of the atom lay in its positively charged component; the diffuse sphere, which to THOMSON had been but a mathematical artifice, took on substantial existence [63]. He accordingly changed his line of experimental research to study the so-called canal rays, in the hope of detecting the positive counterpart to the electron. As one knows, these researches led ultimately to the mass spectrograph of ASTON [64].

Second, the reduction in $n$ reopened the radiation problem. Not enough electrons were now available to assign a single degree of mechanical freedom to each spectral line. Moreover, reducing the electronic population raised the possibility that radiation collapse might come sooner than the radioactivists desired. Thomson's response to these considerations was characteristic. He observed, first, that a single electron might easily, and in a perfectly classical manner, emit as many spectral lines as you please. For consider, he said, the results of Sir George DARWIN, the Astronomer Royal, on the periodic paths open to satellites subject to the simultaneous pulls of a sun and a planet [65]. It appeared that, if the satellite entered certain regions of space whose boundaries depended upon the relative masses of the sun and planet, its orbit would cease to be periodic or multiply periodic; it would then either fall upon one of the gravitating bodies or wander into space. Assume now that not one but several planets perturb the motion: the prohibited regions will become quite elaborate, and perhaps arrange themselves into a series of concentric spheroidal shells separated by concentric regions permitting periodic motions. THOMSON proposed a simple application of this fanciful system to the problem of spectral emission: the sun and the planets represent atomic constituents; the satellite is an electron in the process of being bound. When the satellite finds itself in a region of periodic motion it emits a spectral line of appropriate frequency, thereby loses energy, crosses into an aperiodic region, falls closer to the atom's center, emerges into a second periodic region, emits a second and different spectral line, and so on until it has radiated away all its binding energy. As will appear, a somewhat similar picture guided Bohr's first thoughts about spectral emission.

The faster radiation drain implied by the drastically reduced value of $n$ did not bother THOMSON. Whereas SCHOTT, for example, was worried enough to allow his atoms to recoup their radiation losses from the self-energies of their electrons, which he supposed gradually to expand in size [66], THOMSON continued to assume that enough electrons existed to restrain the radiation drain in the way he had always envisaged. In particular, he did not believe that hydrogen possessed but one electron and helium only two. The discovery of the correct *order* of $n$ brought THOMSON close to, but still some distance from the central principle of atomic theory, the doctrine of atomic number.

The chief line of research that led to the recognition of atomic number flowed naturally from Thomson's technique of deducing atomic characteristics from the scattering of rays from radioactive substances. He himself returned to the subject in 1910 when new data concerning β-rays had undermined some of his earlier assumptions [67]. The new treatment of multiple scattering tried to follow the careers of individual β-particles and introduced important considerations from the theory of probability. It also corresponded much more closely to realizable experimental conditions; when put to the test at the Cavendish Laboratory by CROWTHER, it yielded the important result $n \simeq 3A$. CROWTHER thereupon repeated the old experiments on X-rays, and discovered that they gave a somewhat higher constant of proportionality than BARKLA had found [68]. It therefore appeared, in the Summer of 1910, that Thomson's new theory had been brilliantly confirmed. An interesting application of the theory concerned the α-particle, which, as everyone then knew, is a doubly ionized helium atom [69]. Since helium, according to Crowther's results, should contain about a dozen electrons, the α-particle, according to Thomson's group, must be a structure of atomic dimensions with something on the order of 10 constituent particles.

## 3. – Rutherford's atom and the advancement of Thomson's program.

RUTHERFORD disagreed. We all know the story of the Geiger-Marsden experiment in which, to everyone's great surprise, about one in every 8000

$\alpha$-particles striking a thin gold foil emerged on the side of incidence [70]. It appeared to RUTHERFORD that to turn the path of a highly energetic $\alpha$-particle by 90° or more an atom required a more intense electric field than Thomson's flimsy models could provide. A simple way of increasing the field was to return to the Saturnian picture; for, if one concentrates the $n$ positive charges of a Thomson sphere into a point, one gets a single scatterer with not $n$ but $n^2$ times the kick of a single electron. RUTHERFORD eventually adopted just such a picture, and defended it with the beautiful and simple theory of $\alpha$ scattering with which we are all familiar. But the nuclear model is not so much the outcome of the calculation as its main ingredient; for Rutherford's approach depended upon an idiosynchratic conception of $\alpha$-particles which required him to suppose the nuclear atom for the case of helium *before* he had begun to analyse the scattering data.

The $\alpha$-particle has had a curious history. It was initially identified by RUTHERFORD as the more easily absorbed of two constituents of the radiation from uranium, the other being the penetrating $\beta$-rays [71]. When, in 1899, the $\beta$'s were first deflected by a magnetic field [72], another distinction entered the definition, and the $\alpha$'s became the «nondeviable» rays. For several years, until 1902 to be exact, RUTHERFORD and most other physicists considered the $\alpha$'s a secondary radiation, analogous or identical to X-rays, stimulated by the passage of the $\beta$'s. In that year, however, RUTHERFORD suceeded in bending the paths of $\alpha$'s in both electric and magnetic fields, and in measuring their ratio of charge to mass [73]; it appeared that in fact the $\alpha$ was a mighty individual of atomic dimensions, a complex system of electrons. In the transition from ray to particle the $\alpha$ retained an important property: it was thought to splash through matter undeviated by the atoms in its path. To use a favorite analogy of Bragg's, the collision of an $\alpha$ with an atom resembled the interpenetration of two meteor swarms; the encounter would mean nothing for either swarm, although it might well be a serious matter for an occasional pair of unlucky electrons [74].

The experiments of GEIGER and MARSDEN forced still another change in one's picture of the nature of $\alpha$-particles. It appeared that those Olympian particles could be readily knocked off course: the meteor as a whole was more sensitive to the fates of its parts than BRAGG had allowed. It was just here that Rutherford's peculiar concept of the $\alpha$-particle developed. Having begun his career by distinguishing rigorously between $\alpha$'s and $\beta$'s and then prospered by finding analogies between them, he now took the last step in their assimilation and began to regard the $\alpha$ as a point charge with respect to the process of scattering. One suspects that the assimilation grew gradually and unconsciously in his mind as he designed one experiment after another in which the thickness of the glass penetrable by $\alpha$-particles played a critical role. The best known of these experiments, and the most relevant for our purposes, was the famous demonstration of the relationship between $\alpha$-particles and helium, performed in 1908. A source of $\alpha$'s contained in a tube transparent to them was sealed into a larger evacuated vessel containing no trace of helium; as the source decayed its $\alpha$'s crossed into the vessel, captured electrons there, and gave rise to a helium spectrum where none had been before [75]. RUTHERFORD tacitly assumed that particles of atomic dimensions could not cross glass.

Since the $\alpha$-particle is a point, the helium atom may be represented most simply as a two-electron ring surrounding a doubly-charged nucleus. Evidently $n_{He} = \frac{1}{2}A_{He}$. Observe not only that the ratio is smaller than Crowther's, but, what is much more important, that it is supposed to be *exact*. Thomson's ratios, as you will recall, never pretended to exactness. But RUTHERFORD expected more; and one of the great attractions of his model, emphasized by all its chief early advocates, was just that the ratio which one necessarily assumed for helium reappeared at the opposite end of the periodic table, among the heavy metals, when one analysed the data of GEIGER and MARSDEN in accordance with the scattering theory of RUTHERFORD [76].

That theory differed from Thomson's not only in introducing a new major scattering center, namely the nucleus. It also differed by ascribing deviations acquired in traversing thin metal foils to a *single* encounter with a *single* nucleus: a charged particle whose path is shifted by an appreciable angle by such a target almost certainly received its total deflection at a single stroke. RUTHERFORD was probably brought to this conclusion—which, like the assumption of the point $\alpha$-particle, greatly simplifies the calculations—by some observations of BRAGG, who provided counsel and sympathy during the course of Rutherford's computations [77]. Only one of those calculations need detain us, namely the demonstration that on the single-scattering theory Thomson's atom is incompetent to reflect $\alpha$-particles in the amount found by GEIGER and MARSDEN.

Let me remind you of the straightforward classical formula

$$\text{tg}(\phi/2) = e_1 e_2/m' v^2 p \,, \tag{2.1}$$

relating the impact parameter $p$ and the angle $\phi$ through which the direction of motion of a particle of charge $e_2$, reduced mass $m'$ and velocity $v$ is turned in colliding with a fixed charge $e_1$. The same formula also holds for scattering by the diffuse sphere if $p$ exceeds the radius $a$.

Now to the demonstration. It is easy to show the incompetence of the positive sphere, whose maximum effect occurs when $p = a$ and $e_1 = ne$. For an $\alpha$-particle to be turned 90° in such an encounter

$$n = e_1/e = Mav^2/eE = 4.75 \cdot 10^5 \,,$$

using the numbers RUTHERFORD employed in his scattering paper of 1911 ($E/M$ is the charge-to-mass ratio of the $\alpha$-particle). Clearly $n = 475\,000$ was out of the question in 1910, although it was the order of magnitude expected before Thomson's important work of 1906.

Nor could a single electron, even one fixed in the sphere, scatter $\alpha$-particles as required by the experiments of GEIGER and MARSDEN. This failure may not be so obvious as that of the sphere, because, since both the $\alpha$-particle and the electron are points, the collision can be as close, and $\phi$ as large, as needed. One must resort to probabilities. The likelihood that $\phi$ exceeds a given value $\phi_0$ is evidently the same as the likelihood that $p$ be less than $p = p(\phi_0)$; hence if $Q$ particles fall upon the target, the number deviated by at least $\phi_0$ is

$$Q(\phi_0) = Q \cdot \pi p_0^2 \cdot nNt\,, \tag{2.2}$$

where $N$ is the number of atoms in unit volume and $t$ is the thickness of the scatterer. If we use (2.1),

$$Q(\phi_0) = nQ(e^2E^2/M^2v^4)\,ctn^2(\phi/2)\cdot Nt\,, \tag{2.3}$$

making $n \sim 40\,000$ for the Geiger-Marsden experiment, where $t = 8\cdot 10^{-5}$ cm and $Q(\pi/2)/Q = 1/8000$. Although this number was again far too large, it was not as hopeless as the earlier: a perfectly acceptable value would emerge if, instead if $n$, $n^2$ appeared in eq. (2.3); and this will occur if the scatterer, here assumed to have unit charge, is replaced by a point consisting of $n$ charges, *i.e.* a nucleus. In that case $n = 100$ for gold ($A = 197$), in close conformity with the Rutherfordian relation $n = A/2$.

Rutherford's famous paper resurrecting the nucleus appeared in May 1911. It made no splash. The few people who noticed it outside Rutherford's group tended to regard it as an alternative to the multiple-scattering theory or, rather, as a theory as adequate for $\alpha$'s as Thomson's was for $\beta$'s. Some, who remembered that the model had once before been discussed and discarded, did not bother to look further into the matter [78]. A major cause for indifference was that RUTHERFORD had not represented on his model those periodic properties of the elements which THOMSON had explained so plausibly, and that he did not extend his results in directions which now appear obvious. In particular, he did not explicitly draw the conclusion that $n_H = 1$, and nowhere hinted at the principles of isotopy and atomic number.

## 4. – Atomic number and isotopy.

These omissions were supplied during the latter half of 1912 and the first few months of 1913 primarily by Rutherford's associates, among whom we may mention BOHR, FAJANS, HEVESY, RUSSELL and his former collaborator SODDY. But the earliest statement of the doctrine of atomic number came from an entirely different and unexpected quarter, from a lawyer and amateur physicist of Amsterdam named VAN DEN BROEK [79]. It might at first appear that the intervention of this unknown Dutchman, who returned to obscurity after his inspired utterance, is an unnecessary complication, or even a blemish, in the history of physics. BOHR undoubtedly arrived independently and simultaneously at a similar view: had he had priority of publication, we would never have heard of VAN DEN BROEK and the capstone to the early development of atomic physics would have been set, as was only fitting, by a student of THOMSON and RUTHERFORD. One suspects that, despite his scrupulous observation of van den Broek's priority, BOHR was annoyed by the unnecessary contribution from Amsterdam.

The historian, however, is—or should be—gratified by this seeming irrelevance. For the Dutch lawyer, with the unhesitating logic of his trade, deduced the doctrine of atomic number by combining a well-known feature of the periodic table with what, according to the foregoing analysis, was to contemporaries the most important result of Rutherford's atomic theory. That result, the unexpectedly precise culmination of one part of Thomson's program, was the approximation $n = A/2$. The relevant feature of the periodic table, to which MENDELEEV, WERNER and RYDBERG, among others, had called attention [80], was that the average change in atomic weight $\Delta A$ between consecutive elements approximates two; indeed, as VAN DEN BROEK showed, it could be made quite exactly two by postulating the existence of a good many as yet undetected elements. From the two approximations, $n = \frac{1}{2}A$ and $\Delta A = 2$, VAN DEN BROEK deduced $\Delta n = 1$, a relation he proposed to consider rigorously true. We cannot, perhaps, commend his reasoning, but we must certainly applaud his conclusion: each element, beginning at one with hydrogen, is characterized by an atom containing a number of electrons exactly equal to the serial rank of the element in the periodic table.

VAN DEN BROEK did not, however, reach to the idea of isotopy, to which the scattering theories of THOMSON and RUTHERFORD offered no guide. Here another Rutherfordian line of research played the major part. According to the disintegration theory first put forward by RUTHERFORD and SODDY in 1902 [81], radioelements derive from one another by the release of $\alpha$ or $\beta$ particles. In the case of $\alpha$ emission, the most orthodox doctrine would require the daughter to differ chemically from the mother; for the $\alpha$ carries off a sensible portion of the atomic weight, and chemical theories made chemical properties a unique function of $A$. Chemical evidence did indeed confirm the discontinuity between mothers and daughters; but sometimes more remote descendents came to light with an uncanny chemical, though no radioactive, resemblance to their ancestors. By 1910 several cases of what were called « chemically inseparable elements » were known [82]. One worried whether the inseparability meant identity and, more generally, how the two or three dozen radioelements then known were to be accommodated in the periodic table.

Thomson's atomic model offered no help; as I observed earlier, it not only could not suggest, but probably could not digest, the idea of isotopy. Rutherford's model, on the other hand, broadly hinted at it: it was only necessary to incorporate a commonplace of atomic physics, namely that different sorts of phenomena had their seats in different regions of the atom. THOMSON, *e.g.*, assigned the production of characteristic X-rays to electrons close to the atomic center, where Mme CURIE, among others, placed the seat of radioactivity [83]. It took no great intelligence to see in the dichotomy of nucleus and satellite electrons a perfect representation of the distinction, which everyone recognized, between radioactive and optical/chemical phenomena. Moreover, the dichotomy easily accomodated those oxymorons, the chemically inseparable elements. Ever since Thomson's revaluation of the order of $n$ the positive portion of the atom was known to carry virtually all the atomic weight. One had only to discard the chemists' favorite preoccupation and admit that $A$ had nothing to do with chemical properties, which in fact were fixed by the number of electrons, by the atomic number $n$. Among the several approximations to the doctrine of isotopy which appeared in 1913 only those which assumed the nuclear atom came close to the definitive doctrine [84].

One might expect that THOMSON would have welcomed the splendid advancement of his program represented by the principles of isotopy and atomic number. He did not. His reservations did not derive from pique that his model atom was menaced by that of his most successful student. The mythology of physics has made too much of trivial and easily pictured differences in geometry. THOMSON held back because, from his point of view, the advance came with several unpalatable innovations.

## III. - THE ONE-QUANTUM ATOM.

PLANCK, as we know, discovered the quantum in 1900. But, with few exceptions, physicists did not discover the importance of his discovery until about 1910, when, as KLEIN has observed [85], the success of Einstein's theory of specific heat showed that Planck's innovation could be made fruitful outside the subject of radiant heat. As physicists began to seek new applications of the quantum they naturally turned to atomic theory; for atoms were understood to be the emitters of spectral lines, and $h$ evidently had something to do with the interaction of radiation and matter.

### 1. – The quantum and Thomson's atom.

There were several highly speculative attempts, dating from about 1910, to work the quantum into Thomson's atom. As one might expect, the speculators tended to be persons marginal to the physics profession, teachers in second-rate schools or young men on the thresholds of their careers, where they could afford to sow wild oats. One might also expect that such speculators would flourish neither in England, nor in Germany or France; for German and French physicists still did not prize atomic models and the English did not esteem the quantum. It was in Vienna, where the tradition of BOLTZMANN survived, that the first suggestive quantization of the atom was made; and that by a most marginal character, a doctoral student named HAAS, who wished to make his career in the history of science [86].

HAAS based his speculations on the Thomson model, which not only lay readiest to hand, but which also had the great advantage of providing a perfect representation of a Planckian oscillator: if there is but one electron in a Thomson sphere, it can oscillate through the center with a frequency $\nu'$ independent of amplitude, where

$$\nu' = \frac{1}{2\pi}\sqrt{e^2/ma^3}\,. \tag{3.1}$$

Now in 1910 the hopeful quantizer who had found a frequency immediately multiplied it by $h$ and looked about for an energy $E$ to which the product might plausibly be equated. In Haas' case the only choice—because unique—is the energy possessed by the electron when vibrating at its maximum amplitude, the radius $a$ of the Thomson sphere:

$$h\nu' = E = e^2/a\,. \tag{3.2}$$

At this point one can operate in two quite different ways, according to whether one takes $h$ or $a$ to be the more fundamental quantity. If $a$, then one eliminates $\nu'$ via (3.1) and obtains

$$h = 2\pi e\sqrt{ma}\,, \tag{3.3}$$

explaining away $h$ in terms of the parameters characterizing the electron and the radius of a Thomson sphere. If, on the other hand, one takes $h$ as fundamental, one can eliminate $a$ via (3.1):

$$h^3\nu' = 4\pi^2 me^4\,. \tag{3.4}$$

Once again there is a parting of the ways. One can either assume the constants $e$, $m$ and $h$ known, and (as we would find most natural) use (3.3) to calculate $\nu'$:

$$\nu' = 4\pi^2 m\,e^4/h^3\,. \tag{3.5}$$

Or, assuming $\nu'$ known, one might use (3.3) to obtain a value for one of the constants, say $e$:

$$e^5 = (h^3\nu'/4\pi^2)(e/m)\,. \tag{3.6}$$

Haas' choices are most interesting. He first explains away $h$ via (3.3); he next obtains a value for $e$ from (3.6) by setting $\nu'$ equal to the short-wave limit of the Balmer formula! He has no justification, except success, for this step; but with it he obtains $e = 3.1 \cdot 10^{-10}$ e.s.u, which agreed with the best measurements made at the Cavendish Laboratory, and cast doubt on the values, some 50 percent higher, obtained by PLANCK from his formula and by RUTHERFORD from counting $\alpha$-particles [87]. Note that HAAS did not use his model to calculate a value for the Rydberg constant $R$, which —had it been his purpose—he could easily have done from (3.5). For $\nu' = R/4$, whence

$$R = 16\pi^2 me^4/h^3 ,$$

precisely eight times the expression later found by BOHR. Of course Haas' Rydberg would have been numerically equal to Bohr's, since HAAS took $e_H = 3.1 \cdot 10^{-10}$ e.s.u. and BOHR, using Rutherford's value, took $e_B = 4.7 \cdot 10^{-10}$ e.s.u. (The Rydberg depends on the fifth power of $e$, and $(e_B/e_H)^5 = 8$.) Much, as one sees, depended upon having a sound value for $e$.

Haas' model—although not his special assumptions—interested LORENTZ, who brought it up at the Solvay Congress of 1911. It stimulated an interesting exchange of views. After LORENTZ had explained Haas' goal of reducing $h$ to mechanical quantities, SOMMERFELD observed that, although one could obtain connections between $h$ and atomic parameters in several ways, for his part he preferred « a general hypothesis about $h$ to particular models of atoms ». To this LORENTZ replied [88]:

You do not deny that some relation exists between the constant $h$ and atomic dimensions (positive Thomson spheres). One can express this in two ways: one either says, with HAAS, that $h$ is determined by these dimensions; or else one attributes to atoms dimensions which depend upon the magnitude of $h$. I do not see any great difference.

Only those who inverted Haas' procedure, who took $h$ as the fundamental quantity in the determination of atomic structure, succeeded in building a fruitful atomic theory.

## 2. – Bohr's approach to the problem of atomic structure.

It is very likely that BOHR had enrolled himself among the fundamentalists by the time he finished his doctoral thesis, which he defended in Copenhagen in May, 1911 [89]. The thesis, on the electron theory of metals, did not deal with atomic theory—indeed, BOHR specifically declined to assume special structures for the metal molecules in order to be able to develop the electron theory in the widest possible generality. But it was just this approach that enabled him to infer that certain failures in the theory betrayed the existence of « properties of bodies impossible to explain if one assumes that forces which act within the individual molecules ... are mechanical » [90]. In particular, he confirmed what many were beginning to believe, that the electron theory of metals as established especially by THOMSON and LORENTZ could not account for either the radiant or the specific heat of metals; and he discovered a new blemish, the curious fact that—contrary to the accepted hypothesis of LANGEVIN—the theory cannot explain the magnetic properties of metals if the electrons are supposed to obey the usual requirements of statistical mechanics [91]. This last enigma, all the more forceful because of his own devising, convinced BOHR of the necessity for some nonmechanical law which might fix the motion of atomic electrons in the manner apparently required for an unobjectionable version of Langevin's remarkably successful hypothesis. Very likely before completing his thesis BOHR conceived this law as a restriction like Planck's quantization rule.

BOHR spent his post-doctoral year in England chiefly at the Cavendish Laboratory, where he went to work with THOMSON. «I considered first of all Cambridge as the center of physics », BOHR later said of his decision to study there, « and THOMSON was a most wonderful man ..., a genius who showed the way for everybody » [92]. THOMSON received him politely, and promised to read his thesis, of which he had brought a rough translation. «I have just talked to J. J. THOMSON », BOHR wrote his brother after his first interview, « and I explained to him as well as I could my views on radiation, magnetism, etc. You should know what it was for me to talk to such a man. He was so very kind to me; we talked about so many things; and I think he thought there was something in what I said. He has promised to read my thesis, and he invited me to have dinner with him next Sunday at Trinity College, when he will talk to me about it ... » [93].

The exchange of views BOHR desired did not take place. THOMSON, who had long since given up active cultivation of the electron theory, probably never read Bohr's thesis, and in any case did not enjoy having his ancient errors rehearsed by a tenacious foreigner whose English he could scarcely understand. But even had language and divergent interests not been barriers one doubts that the intellectual communion that BOHR sought could have developed. For one thing, the imprecise and contradictory analogies which THOMSON fancied were inadequate for BOHR, who sought coherent and consistent models from which exact quantitative predictions about experimental results might be drawn. For another, THOMSON, though friendly and receptive to questions, worked alone; he seldom solicited his students' views on scientific questions, nor did he develop his own through extended conversations with others [94]. Bohr's life-long practice, on the other hand, was to refine his ideas in lengthy discussions, which often became monologues, with informed individuals. Whether his colloquist was a full collaborator, a sounding board or an amanuensis, he required some human contact at almost every stage of his work, even in writing. He dictated his papers, at first (as with his thesis) to his mother, then to his wife, and ultimately to a series of secretary-collaborators beginning with KRAMERS.

Thomson's indifference by no means deflected BOHR from the pursuit of the electron theory. It was the chief subject of his research throughout the eight months he spent at Cambridge, and it remained so during the first three months of his stay at Manchester, where he moved in March, 1912, in order to learn something of the experimental side of radioactivity. It is important to recognize that BOHR did not go to Manchester, Rutherford's citadel, in order to help develop the consequences of the nuclear atom. He went to take a six-week course on experimental technique, a standard service of the laboratory for beginners in radioactivity, after which they usually began a small research task proposed by RUTHERFORD [95]. Not that BOHR wished to become an experimentalist: his object was to capitalize on his time in England, and to make contact with RUTHERFORD, evidently the coming power in English physics. After finishing the laboratory work for the day he would return to the electron theory of metals.

BOHR came to atomic physics in a very casual way. The research topic RUTHERFORD had assigned him was interrupted for want of radium emanation; while he waited for more to grow he studied a paper on the absorption of $\alpha$-particles just published by DARWIN, the only mathematical physicist besides himself in Rutherford's group. BOHR found that Darwin's treatment rested on an unsatisfactory assumption about the interaction between the $\alpha$-particles and atomic electrons: in particular, DARWIN had ignored the binding forces, which BOHR, following a technique used by THOMSON, proposed to take into account by treating the interaction as a resonance phenomenon depending on the ratio of $1/\nu'$, the natural period of the electrons' vibrations about equilibrium, to the time required by an $\alpha$-particle to pass the atom [96].

BOHR anticipated an easy calculation which would quickly furnish a short note for the *Philosophical Magazine*. That was in early June, 1912. By the middle of the month he had abandoned the laboratory, shelved the electron theory, and given himself up entirely to the cultivation of Rutherford's atom. It is probable that the chief cause of his conversion was the discovery—no doubt made in connection with calculations of $1/\nu'$ required for his corrections to DARWIN—that the Saturnian atom is mechanically unstable [97]. Far from discouraging BOHR, the instability suggested to him that the Rutherford atom, in contrast to Thomson's, had just the feature that his thesis had led him to *require* of any satisfactory model: the same sort of nonmechanical constraint on the motions of the bound electrons implied by Bohr's analysis of magnetism was also necessary to prevent Rutherford's model, deduced from the wide-angle scattering of $\alpha$ particles, from tearing itself to pieces. Moreover, as I observed earlier, the nuclear atom, as against the plum-pudding, offered a natural representation of the disjunction between chemical and radioactive phenomena, a point which, we may imagine, had been on Bohr's mind at least since his migration to Manchester. It appears that BOHR had consigned radioactivity to the nucleus and hit upon the idea of isotopy by the middle of June, 1912.

We are fortunate in having a record of his early views on atomic structure in the form of a Memorandum drawn up for a discussion with RUTHERFORD in June or July. ROSENFELD has published a transcription of most of this document together with an instructive commentary in his edition of Bohr's first three papers on atomic structure [98]. The document begins by rehearsing the difficulty regarding mechanical stability, which, it says, requires a nonclassical condition to fix the motions; such a condition also has the advantage of specifying the size of the nuclear atom, which—as everyone had by then learnt—cannot be defined by the force balance alone. BOHR proposes to stabilize by fiat, to exempt from the usual operation of small perturbations an electron which describes a closed orbit satisfying the relation

$$\text{kinetic energy } (T) = K\nu' .$$

The usual force balance holds for these orbits; in addition they enjoy, as it were, a nonmechanical rigidity that prevents their deformation. As for the constant $K$, BOHR plainly considered it a quantity approximately equal to Planck's constant $h$; and from a certain numerical application, according to the ingenious interpretation of ROSENFELD, it appears that he took $K$ to be about $0.6h$.

This application suggests that BOHR had no theoretical basis for choosing $K$, and hoped to deduce a value—for which indeed he would then seek a theoretical justification—by comparison with experiment. This is precisely the program he was to follow in the derivation of the Rydberg constant after using the Balmer formula to deduce the value of $K$. As will appear, an essential step in that deduction invoked the formula

$$T = \pi^2 m e^4 Z^2 / 2K^2 , \tag{3.7}$$

which emerges immediately by elimination from the Memorandum's conditions on the orbit the force balance (3.8) and the Bohr constraint (3.9):

$$Ze/a^2 = 2T/a = m(2\pi\nu')^2 a , \tag{3.8}$$

$$T = K\nu' . \tag{3.9}$$

But there is little grist for the machinery of quantum theory in the Memorandum, which adresses itself primarily to the problems of Thomsonian atom building:

the number of electrons in an atom, atomic volumes, radioactivity, the structure and binding of simple diatomic molecules, held together by a girdle of electrons each satisfying (3.8) and (3.9). As this list suggests, BOHR at first disregarded spectra; like THOMSON, he thought them too complicated to serve as a guide to atomic structure.

BOHR came to BALMER following the same sort of casual interaction that had enlisted him in the cause of the Rutherford atom. First, in the late fall or early winter of 1912 he read in Copenhagen, where he had returned to begin his academic career, the puzzling papers of NICHOLSON [99]. NICHOLSON, an old wrangler who had revived the nuclear atom independently of RUTHERFORD, had computed the frequencies of the perturbed oscillations of a single ring of electrons vibrating *perpendicular* to their plane of motion. These perpendicular oscillations, contrary to those in the plane, are stable, at least in the cases NICHOLSON examined, which employed no more than five electrons. The perturbed frequencies are multiples of the orbital frequency $\nu'$, which can be chosen arbitrarily. NICHOLSON found that for a particular choice of $\nu'$ the perturbed frequencies all agreed very exactly with as many frequencies emitted by certain nebulae; moreover, ionized forms of the same model atoms could account for further nebular lines and a good many unattributed lines in the spectrum of the solar corona. And there is more: the angular momenta of the electrons in these atoms of «nebulium» and «coronium» came out to be integral multiples of $h/2\pi$. NICHOLSON had been inspired to compute the momenta by the report of the deliberations of the Solvay Congress. The result he anticipated, the quantization of the momenta, drew from him the following inspired guess [100]:

> The quantum theory has apparently not been put forward as an explanation of «series» spectra. ... Yet in the belief of this writer, it furnishes the true explanation in certain cases, and we are led to suppose that lines of a series may not emanate from the same atom, but from atoms whose internal angular momenta have, by radiation or otherwise, run down by various discrete amounts from some standard value. For example, on this view there are various kinds of hydrogen atoms, identical in chemical properties and even in weight, but different in their internal motions.

The success of Nicholson's atom bothered BOHR. Both models assumed a nucleus, and both obeyed the quantum; yet Nicholson's radiated—and with unprecedented accuracy—while Bohr's were, so to speak, spectroscopically mute. By Christmas, 1912, Bohr had worked out a compromise: his atoms related to the ground state, when all the allowed energy had been radiated away; Nicholson's dealt with earlier stages in the binding, in each of which a different, but exactly specifiable set of spectral lines was produced [101]. Just how a Nicholson atom reached its ground state BOHR never troubled to specify. He aimed merely to establish the compatibility of the two models; for his part the spectra were still too complex to guide the atom builder.

The compromise with NICHOLSON was to leave an important legacy to the definitive form of the theory.

BOHR continued to depreciate the spectra until March, 1913, when a colleague asked what his new atom—on which he had labored during the entire winter—had to say about them. « Nothing », BOHR replied, and doubtless added something about their complexity. The colleague advised him to have a look at BALMER. « As soon as I saw Balmer's formula », BOHR later used to say, « the whole thing was immediately clear to me » [102]. Perhaps he meant that it started a train of thought like the following.

« The running term of the Balmer formula is $R/n^2$; if I defined those upper states that NICHOLSON has found by $T_n = nK\nu'_n$, I would have, for the case of hydrogen,

$$T_n = \pi^2 me^4/2n^2 K^2 \tag{3.10}$$

(a generalization of (3.7)). Now, according to PLANCK, we should equate the Balmer frequencies with a difference in energy divided by $h$. What more natural than to try

$$T_n = Rh/n^2 \text{ ? »} \tag{3.11}$$

By inserting the values of the constants as known in 1913 into (3.10) and (3.11), BOHR would at last discover the value of $K$, namely $h/2$. All he then needed for a derivation of the Balmer formula was a justification for the relation

$$T_n = nh\nu'/2 \,.$$

The first of Bohr's fundamental papers on atomic structure, which he completed about a month after discovering BALMER, opens with just such a justification [103]. Consider the capture of a single electron by a bare nucleus, and assume that, initially, the two bodies are widely separated and mutually at rest. Consequently the initial « frequency » of the electron, $\nu'_0$, is zero. Now in reaching the $n$-th state, according to BOHR, the electron produces $n$ quanta of frequency $\nu$, which carry away its binding energy $W_n$ [104]. Since, for a circular orbit bound by an inverse-square force, the binding energy equals the kinetic energy,

$$W = T = nh\nu \,.$$

And what to take for $\nu$? What more appropriate, according to BOHR, who sought the factor $\frac{1}{2}$, than the arithmetical average of the mechanical frequencies of the initial and final states, *i.e.* $\nu = (0 + \nu')/2$? With this *ad hoc* argument, in which one discerns the germ of the correspondence principle, the needed rationalization was in hand. Later in the same paper BOHR proposed other formulations of his quantum rule, including, with full acknowledgment of Nicholson's priority, the quantization of the angular momentum.

## 3. – The one-quantum atom and the tentative fulfillment of Thomson's program.

After completing his analysis of the hydrogen spectrum, BOHR returned to the general concerns of Thomsonian atom building that had engaged him at Manchester. Two remarkable papers resulted, in which we again meet the material of the Memorandum, now updated to incorporate the newly-found value of $K$. The papers make public for the first time Bohr's views on isotopy and atomic number; they offer hints about characteristic X-rays and about the spectra of polyelectronic atoms; and, above all, they speculate about the distribution of the electrons in the atoms of the elements.

BOHR took four assumptions as his guides in this Thomsonian business [105]: first, all electrons lie in the same plane through the nucleus; second, the populations of the innermost rings increase with atomic number; third, each electron, regardless of its distance from the nucleus, possesses in its ground state the angular momentum $1 \cdot h/2\pi$; and fourth, of all possible arrangements satisfying the condition on the angular momentum, that characterized by the lowest energy represents the ground state.

The first three assumptions, which BOHR faithfully followed in assigning the ring populations, combine to produce what we may call the « one-quantum pancake atom », a Saturnian model each electron of which carries precisely the same charge, mass and angular momentum. The flatness of the atom and the crowding of electrons near the nucleus BOHR took to be consequences of the ordinary mechanics, which—if allowed the greatest possible latitude of action in the stationary states—would prevent the planes of the rings from separating and, with increasing nuclear charge, would favor the coalescence of neighboring internal rings containing equal numbers of electrons. We shall see in a moment how the scheme worked in practice; here we may note that BOHR makes these consequences plausible by very delicate reasoning, by the dexterous application of mechanical principles to situations in which they have an indefinite but restricted authority. Later, when other evidence showed that the atom could not be satisfactorily represented in two dimensions and the population of the innermost ring did not exceed two, BOHR could and did easily accomodate the new data—and circumvent his earlier reasoning—by relaxing the dominion of mechanics further than the original notion of « stationary state » required.

It was not so simple to drop the third assumption, the universal single quantum of angular momentum. BOHR seems to have accepted it from the very beginning of his development of the nuclear atom; one may recall his tacit assignment of the same value $K$ to the ratio of $T$ to $\nu'$ for every nonradiating electron. The compromise he struck with NICHOLSON greatly reinforced this assumption; for BOHR then explicitly understood the ground state to be that in which the electron has the least possible angular momentum, namely $1 \cdot h/2\pi$.

Moreover, a simple consideration shows that, in the case of a ring atom, one cannot permit the valence electrons more than one or at most two quanta of angular momentum. The radius $a_n$ of an « $n$-quantum ring »— a ring made up of electrons each of which has angular momentum $nh/2\pi$—is, on Bohr's principles,

$$a_n = Z_{\text{eff}} e^2/2W_n = n^2/4\pi^2 me^2 hZ_{\text{eff}},$$

where $Z_{\text{eff}}$ is the effective nuclear charge. For an alkali atom's lone valence electron, $Z_{\text{eff}} \sim 1$, so that $a_n \sim n^2$; consequently, if one took $n$ of the valence electron in the ground state to equal (as we now know it does) the number of the appropriate period in Mendeleev's table, one would make the lithium atom four times, and the sodium atom nine times, as large as hydrogen. But, as one then knew, alkali atoms are about the same size as those of hydrogen, a fact immediately represented by the model if $n = 1$ universally. How one escaped from this box makes an interesting story [106].

As for Bohr's fourth assumption—that the ground states of the atoms enjoy the minimum potential energy consistent with other restrictions—it was but a pious hope. Applied literally it required structures, like a single-ring lithium atom, in obvious disagreement with the chemical evidence. BOHR disregarded it most of the time, and adjusted the one-quantum atom to the periodic properties of the elements as best he could. Here are his results; in the symbol $N(n_1, n_2 \ldots)$, $N$ indicates the atomic number, $n_1, n_2, \ldots$ the electronic populations of successive rings counting outward from the nucleus:

| | | | | | |
|---|---|---|---|---|---|
| 1 (1) | 5 (2, 3) | 9 (4, 4, 1) | 13 (8, 2, 3) | 17 (8, 4, 4, 1) | 21 (8, 8, 2, 3) |
| 2 (2) | 6 (2, 4) | 10 (8, 2) | 14 (8, 2, 4) | 18 (8, 8, 2) | 22 (8, 8, 2, 4) |
| 3 (2, 1) | 7 (4, 3) | 11 (8, 2, 1) | 15 (8, 4, 3) | 19 (8, 8, 2, 1) | 23 (8, 8, 4, 3) |
| 4 (2, 2) | 8 (4, 2, 2) | 12 (8, 2, 2) | 16 (8, 4, 2, 2) | 20 (8, 8, 2, 2) | 24 (8, 8, 4, 2, 2) |

Very little of this has stood, and even in its time it carried little conviction [107]. These assignments nonetheless belong in Bohr's first papers on atomic theory because they are the link between him and his principal predecessors in the business of atomic design; and because, as will appear, they or, rather, refinements of them for which BOHR was chiefly responsible, constitute the high-watermark of the old quantum theory of the atom.

The old atom builders did not care for Bohr's theory of 1913. THOMSON in particular objected to its violation of ordinary mechanical principles which (he said) fully sufficed to explain so-called quantum phenomena. In a characteristic performance before the British Association in September, 1913, which he repeated the following month to the second Solvay Congress [108], he showed how one might account for Einstein's formula for the photoeffect

$$mv^2/2 = h\nu .$$

Assume, he said, that the usual Coulomb attraction $A/r^2$ operates only in a few pie-shaped sections in the plane of the electron's motion, and that, in addition, the electron meets an inverse-cube repulsion $B/r^3$ everywhere within the atom. The introduction of this nonelectromagnetic force makes possible a position of stable equilibrium within the pie-shaped regions at a distance $r_0$ from the atom's center, where

$$A/r_0^2 = B/r_0^3 .$$

The frequency of radial oscillation about $r_0$, $\nu'$, may be obtained from

$$m\ddot{x} = -A/(r_0 + x)^2 + B/(r_0 + x)^3 ,$$

whence

$$\nu' = \frac{1}{2\pi}\sqrt{B/mr_0^4} = \frac{1}{2\pi}\frac{B}{r_0^2}\sqrt{1/mB} .$$

Assume now that a passing light wave of frequency $\nu = \nu'$ strikes the electron, and gives it enough energy to cross from the pie-shaped region into one of uncompensated repulsion. It will be pushed out into the world with energy

$$W = \int_{r_0}^{\infty}(B/r^3)\cdot \mathrm{d}r = B/2r_0^2 .$$

This will all be kinetic energy, whence

$$\frac{1}{2}mv^2 = \frac{B}{2r_0^2} = \frac{1}{2}(2\pi\nu')\sqrt{mB} = (\pi\sqrt{mB})\,\nu' = (\pi\sqrt{mB})\,\nu .$$

We need only do as HAAS and LORENTZ, and set $h = \pi\sqrt{mB}$, to obtain Einstein's formula.

The theory was widely applauded, especially by *Nature*, which observed that it would not soon be forgotten. It had many imitators among the productions of old wranglers [109]. But nothing came of it; for THOMSON, « the genius who showed the way to everyone » [110], here lost the path.

## IV. - THE *n*-QUANTUM ATOM.

Despite the ingenious defenses of THOMSON and his allies, Bohr's theory rapidly gained ground. Support for his fundamental insights—the stationary states, the quantum condition, the quantum jumps—came from several unexpected quarters, while the principles of isotopy and atomic number, which had become an integral part of the nuclear model, were decisively confirmed. The confirmation came largely from Rutherford's associates, and particularly from MOSELEY; the unexpected support came primarily from the spectroscopy of helium, from studies of ionization by collision and from the discovery of the electrical splitting of spectral lines. I do not intend to dwell on any of these works, all of which had an interesting pre-history, but I should like to mention a few details which exhibit some of the themes we have been considering.

## 1. – Confirmation of Bohr's theory.

The upshot of Moseley's investigation, which occupied him for about six months in the winter and spring of 1913/4, was a pair of formulae for the frequencies of the chief characteristic X-ray lines:

$$\nu_K = (3/4)R(Z-1)^2\,, \qquad \nu_L = (5/32)R(Z-7.4)^2\,. \tag{4.1}$$

Except for an uncautions attempt at relating these expressions to Bohr's theory [111], MOSELEY presented his results as experimental proof that the square root of an atom's characteristic X-ray frequencies is proportional to its rank in the periodic table [112]. To this several physicists, and particularly NICHOLSON, objected [113]. Formulae (4.1) he said, contained not $Z$ but $Z-c$; indeed they might not depend upon $Z$ at all, but upon other and truer atomic numbers $Y$, where $Y=Z+x$ and the problematic factor $Z-c$ is to be understood as $Y-(c+x)$. NICHOLSON supposed that $x\neq 0$ and consequently that more elements existed than MOSELEY had allowed. His view was perfectly respectable, for in 1913 there were good reasons to expect elements lighter than hydrogen, like the «coronium» and «nebulium» whose existence NICHOLSON himself had inferred from his analysis of solar and nebular spectra [114]. Since Moseley's tests did not extend below calcium, they did not decide the question.

But weighty pertinent evidence did exist, namely the theory of RUTHERFORD, which, by requiring helium to have but two electrons, implied that $x=0$. Moseley's argument supposed Rutherford's results, to which, however, he made no reference. It was perhaps this blemish that BOHR had in mind when he later said [115]:

> And Moseley's thing, that is presented in a wrong manner, you see, because we knew the hydrogen, we knew the helium, we knew the whole beginning (of the periodic table).

The omission was soon supplied, and Moseley's results became the best advertisement for the nuclear atom. Again BOHR: «The Rutherford work was (at first) not taken seriously ... . There was no mention of it in any place. The great change came from MOSELEY » [116].

The first persuasive support developed outside Manchester was the unscrambling of the spectra of hydrogen and ionized helium. Several lines belonging to series of the form

$$\nu = R\left(1/q^2 - 1/(m + \tfrac{1}{2})^2\right)$$

were known in 1913, and, although none had ever been found in a tube containing pure hydrogen, all had been ascribed to it by analogy to Balmer's formula [117]. In his first paper on atomic constitution BOHR had observed that such formulae might also be written

$$\nu^+ = 4R\left((1/2q)^2 - 1/(2m + 1)^2\right) , \tag{4.2}$$

and might therefore derive from $He^+$, which, because of the nuclear charge of $2e$, should have a Rydberg constant four times that of hydrogen [118].

The spectroscopists examined the matter and found indeed that the lines in question could be obtained from tubes of helium carefully cleansed of hydrogen; but the observed frequencies differed from those given by (4.2) by more than the possible experimental error [119]. BOHR found the discrepancy in the neglected finite mass of the nucleus; his earlier computation for hydrogen (he said) should have reduced the mass $m$ of the electron to the classical quantity $m' = m/(1 + m/M)$, $M$ being the mass of a hydrogen atom; in calculating the Rydberg constant for $He^+$, the appropriate mass is $m'' = m/(1 + m/4M)$. With these corrections BOHR expected [120]

$$\frac{R_{\mathrm{He}^+}}{R_{\mathrm{He}^+}} = \frac{4m''}{m'} = \frac{4(1 + m/M)}{1 + m/4M} = 4.00160 .$$

Measurement made the ratio 4.00163, a most striking agreement. You will note that to obtain it BOHR had to take the operation of mechanics in the stationary states with the utmost literalness.

After the triumph with helium came the experiments of FRANCK and HERTZ, an equally striking support, not least because they began without reference to BOHR and at first appeared to be strong evidence against his theory. The experiments initially aimed at measuring the ionization potential of gases, for which FRANCK and HERTZ used a deceptively straightforward apparatus [121]. Electrons produced from the plate $P$ (Fig. 2) are accelerated towards the gauze $D$ but kept from reaching $F$ by the large potential difference $V_0$; the electrometer remains charged until the variable potential $V$ between $P$ and $D$ is made great enough to enable the electrons to produce positive ions at $D$ which travel to $F$.

Just before BOHR came upon the scene FRANCK and HERTZ had altered this apparatus in order to obtain the ionization potential without having to know the absolute value of the potential $V$: they brought $D$ close to $P$ and so reduced $V_0$ that electrons could easily reach $F$ [122]. They aimed at measuring the current of these electrons $I$ as a function of $V$, expecting that $I$ would

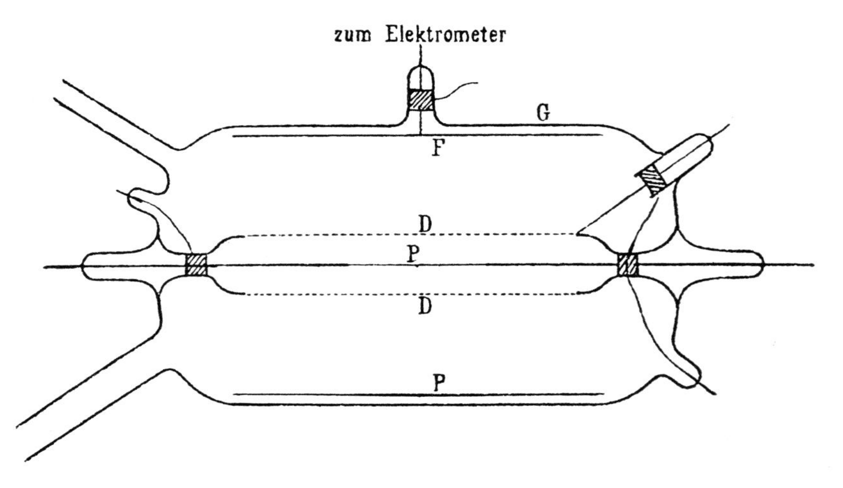


Fig. 2. – The apparatus with which FRANCK and HERTZ set out to measure « ionization » potentials. The electrons are accelerated by a potential $V$ between the platinum wire $P$ and the platinum gauze $D$; an opposing potential of $V_0 = V + 10$ volt between $D$ and the platinum foil $F$ keeps them from reaching $F$ and the electrometer, which was therefore supposed to register positive charges only. From *Verh. deut. phys. Ges.*, **15**, 2, 35 (1913).

fall rapidly when the electrons possessed a multiple of the energy just sufficient to ionize the gas, which in this case was mercury vapor. The « ionization potential » could then be obtained as the difference in voltage corresponding to successive dips in the electron current (Fig. 3). They obtained 4.9 eV.

In keeping with the spirit of the time, they looked around for a $\nu$ to relate to their « ionization potential » via Planck's formula, and found a number very close to the frequency of the most prominent line in the resonance spectrum of mercury. They looked for this ray in their apparatus; it appeared whenever, but not before the « ionization potential » was reached. Now according to Bohr's theory, which FRANCK and HERTZ had just run across, resonance must occur *before* ionization. They naturally concluded that their experiments —for which they were to win the Nobel Prize—disproved the theory [123].

BOHR turned their challenge as deftly as he had the spectroscopists': FRANCK and HERTZ had indeed measured a resonance potential but, misled by the investigation of the ionization potential by their earlier method, they had misintepreted their results. For neither apparatus, according to BOHR, measured ionization; what FRANCK and HERTZ had taken to be a stream of positive ions from the gas to the plate $F$ was in fact a current of photoelectrons produced in the plate by the resonance radiation of the gas and driven off by the high negative potential [124]! Electron impact experiments soon became a strong corroboration of the existence of stationary states [125].

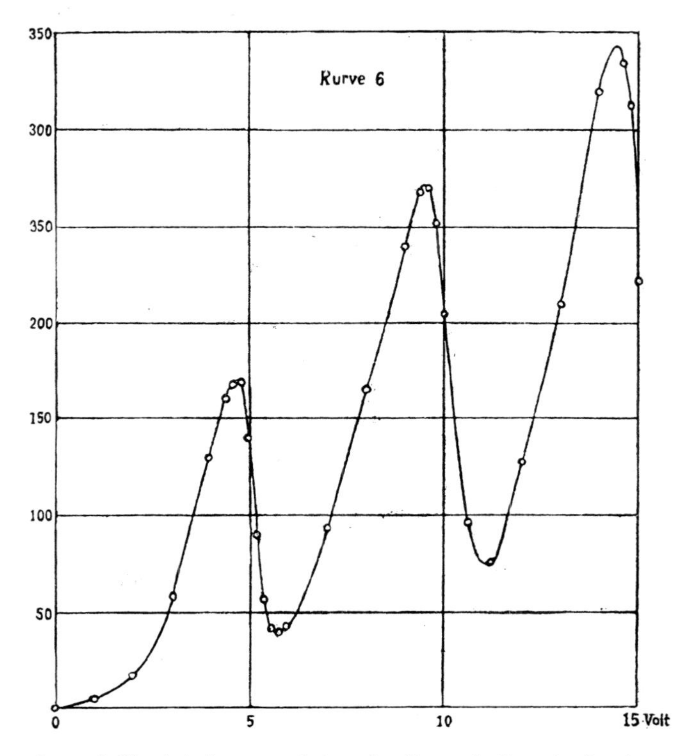


Fig. 3. – Franck and Hertz' famous determination of the «ionization» (in fact the first excitation) potential of mercury vapor. From *Verh. deut. phys. Ges.*, **16**, 462 (1914).

Finally among unexpected—and in this case unwilling—supporters of Bohr's theory came STARK, who made it a principle to reject almost all theories on which his colleagues had reached consensus [126]. One of these theories taught that a detectable splitting of spectral lines could not be obtained with electric fields of accessible strengths [127]. Since everybody else believed the resolution to be impossible, STARK did not; and, following an analogy based upon his bizarre notions of atomic structure [128], he obtained the Stark effect in canal rays of hydrogen [129] and the Nobel Prize. BOHR, who shared the consensus, immediately declared that the effect, since it could not be classical, must depend upon the quantum, and by an ingenious though arbitrary extension of his principles succeeded in computing the separation of the extreme components in the Stark splitting of the Balmer lines [130].

The success did not please STARK: having championed the quantum theory when it was unpopular [131], he had naturally withdrawn his support when it began to prosper [132].

All these developments had occurred, or were well in hand, before the outbreak of World War I. Without the war Bohr's atom would doubtless have been widely cultivated in many directions at once. As it happened, the mobilization of physicists left only two important centers for the development of atomic theory—Copenhagen and Munich—each of which produced fundamental work of a different but characteristic type. Consequently, when peacetime physics resumed, atomic physicists found themselves facing a well-defined theory, whose coherence, however, was a little artificial, an artifact of the restricted focus forced by the war. Its scope is, however, more than wide enough for us, and we must be content to notice a few developments which, taken together, undermined the one-quantum atom.

## 2. – Sommerfeld's extensions.

Here the chief sapper was SOMMERFELD, Professor of Theoretical Physics at the University of Munich. One might think him an unlikely leader of atomic physics. He had never been a friend of Thomson's program; you may recall his remark at the Solvay Congress to the effect that he preferred general hypotheses about $h$ to atomic models in the style of HAAS [133]. But SOMMERFELD had no deep methodological scruples in the matter, and his «scepticism» —to use a word he himself once chose to characterize his attitude towards pre-Bohr models [134]—derived primarily from their barrenness of quantitative results. For SOMMERFELD was, so to speak, a mathematical mercenary; he placed his very great mathematical facility in the service of any theory from which quantitative predictions about realizable experiments could be deduced. At the beginning of his career he was as much at home in pure mathematics and even number theory as in mathematics applied to technological problems, like lubrication or telegraphy [135]. As for mathematical physics, he started with an elaborate mechanical model of the electromagnetic ether [136]; when the stiff electron came along he rushed to develop its theory, which he abandoned as quickly for relativity, to which he made important contributions [137]. Similarly he accepted, without a murmur, the nonclassical character of the quantum and, somewhat later, wave mechanics [138]. Unlike THOMSON, he could jettison classical conceptions if that seemed the way to advance; and, unlike BOHR, it was progress itself—the mathematical development of a theory—that interested him, and not the exploration of fundamental physical questions.

Or perhaps one should say that Sommerfeld's concept of fundamental differed markedly from that of most physicists. While BOHR, *e.g.*, sought via the correspondence principle to extract from successful mechanical models some hints toward the construction of a theory which would accommodate both classical and quantum physics [139], SOMMERFELD employed these same models as calculational devices for obtaining the simple arithmetical relations which to him represented the « truths » of atomic physics. « Today », he wrote in 1919, in the preface to his great work, *Atombau und Spektrallinien*, « today, when we listen to the language of spectra, we hear a true atomic music of the spheres, a concordance of integral relations, an increasing order and harmony amid the greatest variety ». Or again, in 1925, in his *Lectures on Atomic Physics*: « BOHR believed that he would be able to understand the (structure of the) atom by classical calculation in the sense of the correspondence principle. That was an error. The question is much simpler and much more fundamental. It is decided not by classical mechanics but by quantum arithmetic » [140].

Bohr's lieutenant, KRAMERS, once undertook to win SOMMERFELD to the views of his master [141]. In the ensuing correspondence SOMMERFELD was pressed to define his form of the quantum theory, which required one to begin by writing an action integral for each degree of freedom of the electronic motions. BOHR had criticized this procedure on the ground that it often introduced more quantum numbers than necessary, if one required, as he did, that each quantum number correspond to one and only one independent *frequency* of the motion. Bohr's procedure demanded a clear picture of the mechanical motions, which was also needed for an application of the correspondence principle. To this SOMMERFELD in his turn objected in words which throw a flood of light on the introduction and development of his remarkably fruitful method of phase integrals [142]:

Every theory must ultimately proceed dogmatically and deductively. Compare the state of Maxwell's theory with MAXWELL, who set out the « physical » point of view (which for him meant mechanical analogies), and its state with HERTZ, who developed the theory formally and mathematically. Today everyone agrees that his procedure is the valid one. So also some day the quantum theory must ultimately be cast into a unified system of formal rules from which everything follows deductively. You can say of course that we are not nearly so far along in quantum theory as in electrodynamics, and that I certainly concede. But the action integrals $J$ are exactly those quantities that will be required in the complete version of the theory. Therefore it is very hard for me not to place them at the beginning.

He had done so first in 1915, when he replaced Bohr's single condition on the motion of an electron with the well-known pair [143]

$$\oint p_\theta \mathrm{d}\theta = kh\,, \qquad \oint p_r \mathrm{d}r = n'h\,. \tag{4.3}$$

For a strict Coulomb force, the first integral yields

$$W = (2\pi^2 me^4/h^2)\cdot(1-\varepsilon^2)/k^2\,, \tag{4.4}$$

which appears to destroy the quantization of the energy, since $\varepsilon$, the eccentricity, may so far take on any value between zero and one. But the second integral quantizes the eccentricity in such a way as to eliminate it altogether from the result:

$$1-\varepsilon^2=k^2/(n'+k)^2\,,$$

whence, from (4.4),

$$W=Rh/n^2\,,$$

where $n=n'+k$. SOMMERFELD thought this recovery of the BALMER formula to be most remarkable, and a strong confirmation of the conditions—or rather axioms—from which he began. Note that it associates $n$ orbits, $k=1, 2 \dots n$, with each value of $n$, and that an $n_k$ orbit is the less eccentric the closer the values of $n$ and $k$, $n_n$ always being a circle (Fig. 4).

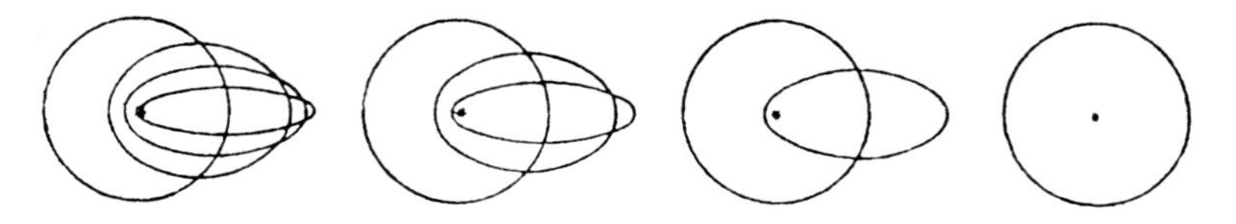

Fig. 4. – Orbits associated with the principal quantum numbers $n=1, 2, 3$ and 4. From SOMMERFELD: *Atombau und Spektrallinien* (Braunschweig, 1921²), p. 272.

Perhaps this is not to advance very far. Sommerfeld's triumph came through the application of the method to an electron bound to a point nucleus by a force deviating slightly from the inverse square. Such a situation occurs if one takes into account the relativistic variation of mass; for substituting $m/\sqrt{1-\beta^2}$ for $m$ in the standard equation of motion under a central force

$$\frac{\mathrm{d}^2u}{\mathrm{d}\theta^2}+u=\frac{Ze^2m}{p^2}\,,$$

obtained from the Kepler equation $m(\ddot{r}-p^2/r^3)=-Ze^2/r^2$ by the substitution $u=1/r$, one obtains

$$\text{(4.5)}\qquad \frac{\mathrm{d}^2u}{\mathrm{d}\theta^2}=-\left\{1-\left(\frac{Ze^2}{pc}\right)^2\right\}u+\frac{Ze^2}{p^2c^2}(-W+mc^2)\,.$$

The root $m/\sqrt{1-\beta^2}$ has been eliminated via the energy equation

$$-W=-Ze^2u+mc^2\left(\frac{1}{\sqrt{1-\beta^2}}-1\right).$$

Equation (4.5) integrates to

$$u = \text{const} \times (1 + \varepsilon \cos \gamma\theta)\,,$$

where $\gamma = 1 - (Ze/pc)^2$ is slightly less than unity. The orbit is an open ellipse with precessing perihelion (Fig. 5).

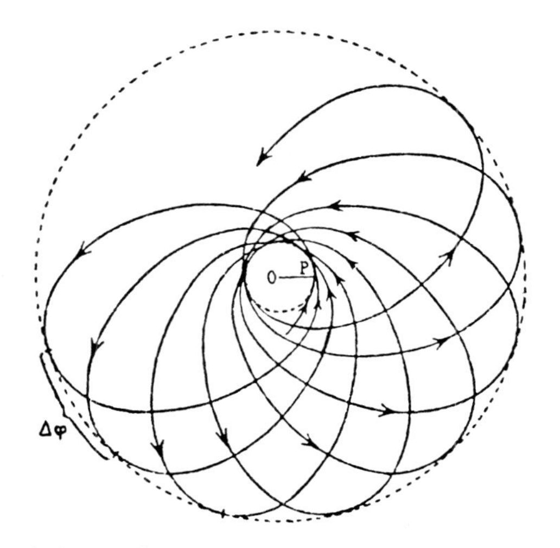


Fig. 5. – Precession of the perihelion of a Sommerfeld ellipse, from *Atombau*, ed. cit., p. 325.

Now quantization, which fixes the values of the eccentricity and of the energy, $W(n_k)$, gives

$$W(n_k) = \frac{RhZ^2}{n^2}\left[1 + \alpha^2 Z^2 \cdot \frac{1}{n}\left(\frac{1}{k} - \frac{3}{4n}\right) + O(\alpha^4)\right], \tag{4.6}$$

where $\alpha = 2\pi e^2/hc = 0.73 \cdot 10^{-2}$ = the fine-structure constant, which here made its first appearance in the literature [144]. The formula calls for $n_1 \cdot n_2$ lines corresponding to a Balmer transition $n_1 \rightarrow n_2$ since each value of $n$ defines $n$ energy levels. In fact far fewer than $n_1 n_2$ lines were found in the careful measurements of PASCHEN on the fine structure of $He^+$ [145]. SOMMERFELD destroyed the superfluous lines, and brought his formulae into brillant agreement with Paschen's results, by simple, albeit arbitrary «selection rules». The demonstration made a great impression. Despite the war SOMMERFELD was able to send his papers to Manchester and BOHR was able to reply. «I do not think (BOHR wrote him in March 1916) that I ever have enjoyed the reading of anything more than I enjoyed the study of ... your most interesting and beautiful papers» [146].

One of these papers directly challenged the one-quantum atom. According to a theory suggested by Sommerfeld's associate, KOSSEL, in the fall of 1914, an atom emits X-rays when an electron plunges into a vacancy in an inner ring created by an external agent [147]. The hardness of the ray depends upon the distance fallen; the $L$-ring is the donor and the $K$ the recipient for $K_\alpha$; $L$ receives and $M$ gives for $L_\alpha$, etc. Moreover, according to KOSSEL, certain evidence, like the existence of a soft satellite of $K_\alpha$, $K_{\alpha'}$, and several companions of $L_\alpha$, suggested the existence of a second $L$-level. How should one understand this duplicity? Well, according to SOMMERFELD, one need only assume that in the *normal* atom the $L$-level electrons are two-quantum; they may then exist either in a common circle ($n = k = 2$) or in a system of ellipses, later represented as an *Ellipsenverein* (Fig. 6), in which the particles,

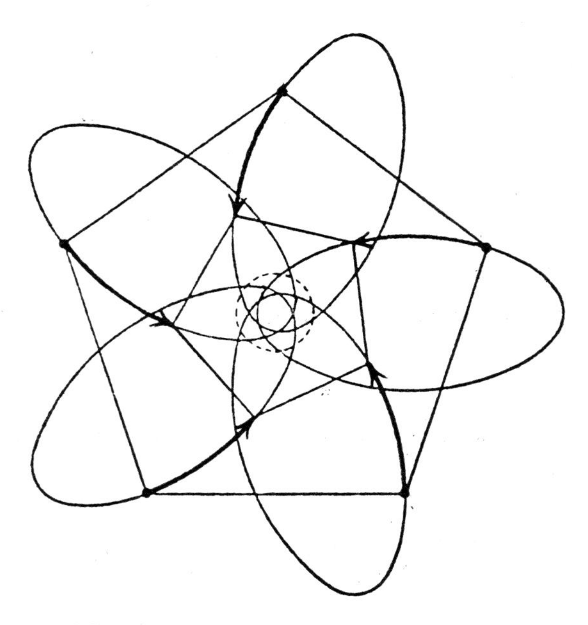

Fig. 6. – A 5-electron Ellipsenverein, from *Atombau*, ed. cit., p. 364.

while negotiating their private precessing orbits, always find themselves at the vertices of a regular polygon [148]. One will see how the scheme is to work from Fig. 7, which SOMMERFELD drew up for the edification of LANDÉ [149].

There was more behind Sommerfeld's assignment of two-quantum orbits to the normal atom. In particular, the separation of the $L$-levels, $W(2_2) - W(2_1)$, computed from (4.6), agreed extremely well with the differences in the frequencies of $K_\alpha$ and $K_{\alpha'}$, $L_\alpha$ and $L_{\alpha'}$, etc., which is precisely what Sommerfeld's theory caused him to expect. The agreement obtained for 22 elements from chromium to uranium, a remarkable fact when you consider that the doublets

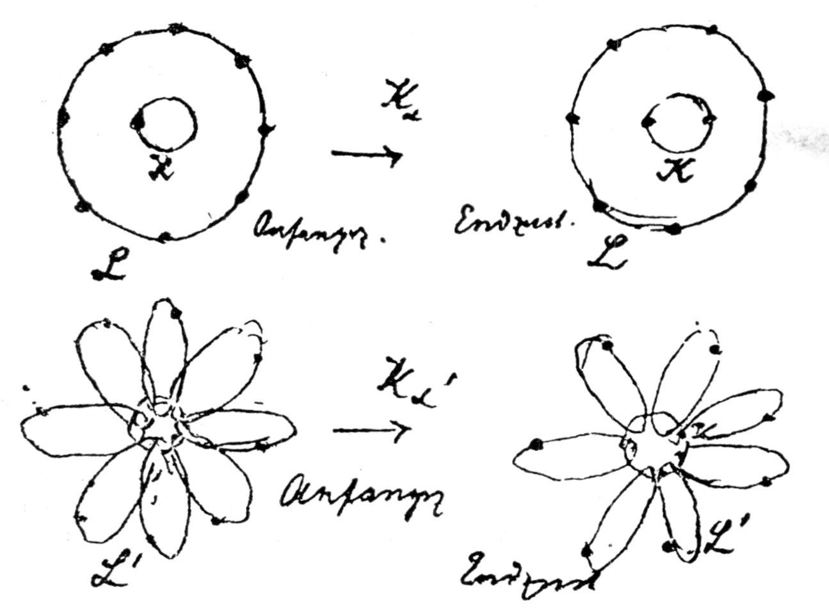


L. Landé!
Ich hoffe, Sie werden sich an Hand dieser Figuren u. der Litteratur resp. meines Buches überzeugen, dass Sie eine kolossale Confusion angerichtet haben. K-Zustand ist nicht gleich Endzustand! Sie dürfen hierüber wirklich nichts publicieren, ehe Sie sich über die ganz elementaren Verhältnisse der bisherigen (natürlich vermutlich falschen) Theorie klar sind. Born soll Schiedsrichter sein. Ich kann nichts mehr tun.

Beste Grüsse! Ihr A. Sommerfeld.

Fig. 7. – A note from SOMMERFELD to LANDÉ, c. May 1920 (AHQP), straightening out Landé's blunders in X-ray theory.

in question are not relativistic and that the formula allows but one disposable parameter, a screening constant appropriate to the $L$-region, to be subtracted from $Z$ in (4.6).

### 3. – The rejection of the one-quantum atom.

Two other lines of work on X-rays confirmed the double-quantum nature of the $L$-region and strongly suggested that the $M$ was three quantum, and the

$N$ four. The more persuasive line examined the absorption spectrum of certain metals for X-rays, and disclosed at least two edges in the $L$-region and three in the $M$. There was soon an *embarras de richesses*, for, by 1921, the three $L$ and the five $M$ edges had been found [150]. SOMMERFELD took these supernumerary levels in stride and worked out analogies, which need not concern us, to optical doublets [151].

The second line of work aimed to derive Moseley's formula

$$\nu_{K_\alpha}/R = \left(\frac{1}{1^2} - \frac{1}{2^2}\right)(Z-1)^2, \tag{4.7}$$

using Kossel's mechanism and adjustable parameters for the ring populations and their quanta. Let us say that $K_\alpha$ arises when one of the $q$ $x$-quantum electrons of the $L2_2$ circle falls into a vacancy among the remaining $p-1$ 1-quantum electrons of the $K$-circle. Then, since the energy of an $x$-quantum electron in a circular orbit subjected to an effective change $Z-s$ is $Rh(Z-s)^2/x^2$,

$$(4.8) \quad \nu_{K_\alpha}/R = \\ = \frac{p(Z-s_p)^2}{1^2} + \frac{(q-1)(Z-p-s_{q-1})^2}{x^2} - \frac{(p-1)(Z-s_{p-1})^2}{1^2} - \frac{q(Z-p+1-s_q)^2}{x^2},$$

where $s$ represents the centrifugal force exerted on a given electron by the other electrons in its ring.

The best fit required $x=2$, $p=2$ or 3, $q=8$ or 7 [152]. These numbers challenged not only Bohr's assignment of one quantum of angular momentum to every normal atomic electron, but also his belief that the $K$ and $L$ populations would continue to increase with $Z$. By the end of 1918 physicists concerned with the theory of X-ray emission had made explicit what SOMMERFELD had broadly suggested: the normal atom is $n$-quantum; the electron in a given region enjoys a multiple of angular momentum equal to the number of its region counting outwards from the nucleus [153]. As for the problem of size, it remained a difficulty. A few, like BORN and LANDÉ, minimized it by compromising the $n$-quantum principle, allowing the quantum number assigned to successive regions to *decrease* from some point in the atom outwards [154].

By 1918 two of the four special assumptions BOHR had introduced to support his attribution of electronic populations had become untenable: the atom was not one-quantum and its innermost ring had the same sparse population regardless of the total number of atomic electrons. Soon the Saturnian model itself, the pancake atom, was attacked from all sides. One blow fell from the subversive X-ray men, who, having failed to obtain precise agreement with their ever-improving measurements *via* formulae like (4.8), declared that the only possible salvation lay in admitting a three-dimensional atom [155]. The same conclusion came from studies of a quite different nature, namely calculations of the rigidity of crystals constructed of flat Bohr atoms. This line of work, quite untypical of the subjects cultivated at Copenhagen or at Munich, was pursued in Berlin by BORN and LANDÉ, who had been collected there with several other physicists to work on problems of interest to the Artillery [156]. But they also had, or took, the time for purer research, and came easily to the problem of the compressibility of crystals, on which BORN was a great expert. Their calculations revealed that pancake atoms gave crystals far weaker than Nature's. Spatial atoms, however, gave much better results. « Thereby », they concluded, with uncharacteristic understatement, « thereby a new problem arises for the quantum theory » [157].

LANDÉ tried to solve the problem. His suggestions, three-dimensional generalizations of the *Ellipsenverein*, include some curious pieces of atomic iconography. Figure 8, which pictures what LANDÉ called a *Würfelverband*,

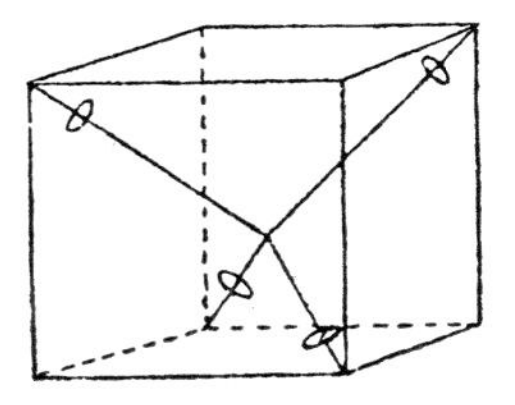

Fig. 8. – Würfelverband for small $Z$, from LANDÉ, *Verh. deut. phys. Ges.*, **21**, 651 (1919). The disks centered on the diagonals represent electron orbits.

is representative [158]. Each of the eight electrons, endowed with angular momentum $h/2\pi$, describes a small circle about a principal diameter of the cube, an arrangement which appears possible if the orbits are appropriately phased and if the effective nuclear charge $Z_{eff} = Z - s$ does not exceed 3. For large $Z_{eff}$, a more curious arrangement obtains (Fig. 9): each electron describes an equilateral spherical triangle occupying an octant and so phased

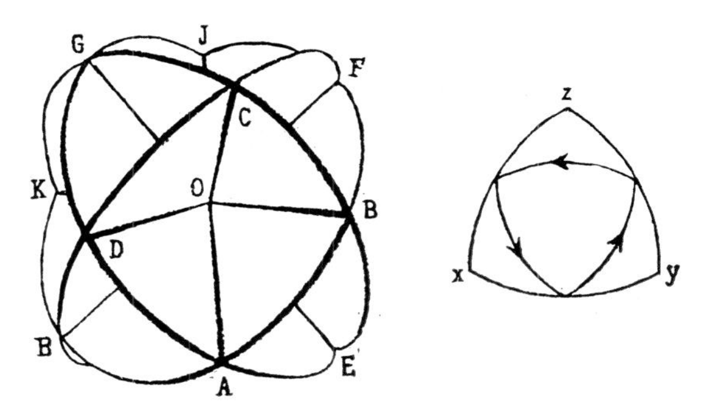


Fig. 9. – Würfelverband for large $Z$, from LANDÉ, *Verh. deut. phys. Ges.*, **21**, 651 (1919). The smaller diagram shows the orbit of a single electron in its octant; the larger gives the traces (ABD, DGC, CFB, BEA) of four orbits filling a hemisphere.

that as an electron completes each leg of its path it encounters a colleague moving in a contrary direction, which knocks it into the next segment of the orbit. You will be interested to know that these pictures excited great enthusiasm and applause on the part of both SOMMERFELD, who no doubt admired the harmonies, and BOHR, who saw in them a new purchase for the correspondence principle [159].

As they discredited the planar atom in their recondite ways, the physicists at last began to hear the chemists, who had known for generations that the simplest properties of carbon show the atom to be three-dimensional. They also thought that it was static, which did not recommend their views to the physicists, who in any case had little to do with them. RUTHERFORD, who, you recall, was a Nobel Laureate in chemistry, made no secret of the fact that he considered « damn fools » and « chemists » to be one and the same people [160]; while the chemists, or at least the British ones, deplored the invasion of their domain by mathematicians, radioactivists and speculative philosophers, who employed mysterious and powerful machines—Moseley's X-ray installation was meant—of « little chemical significance » [161]. But by 1918 or 1919 the chemists had succeeded in bringing their views, in the form recently given them

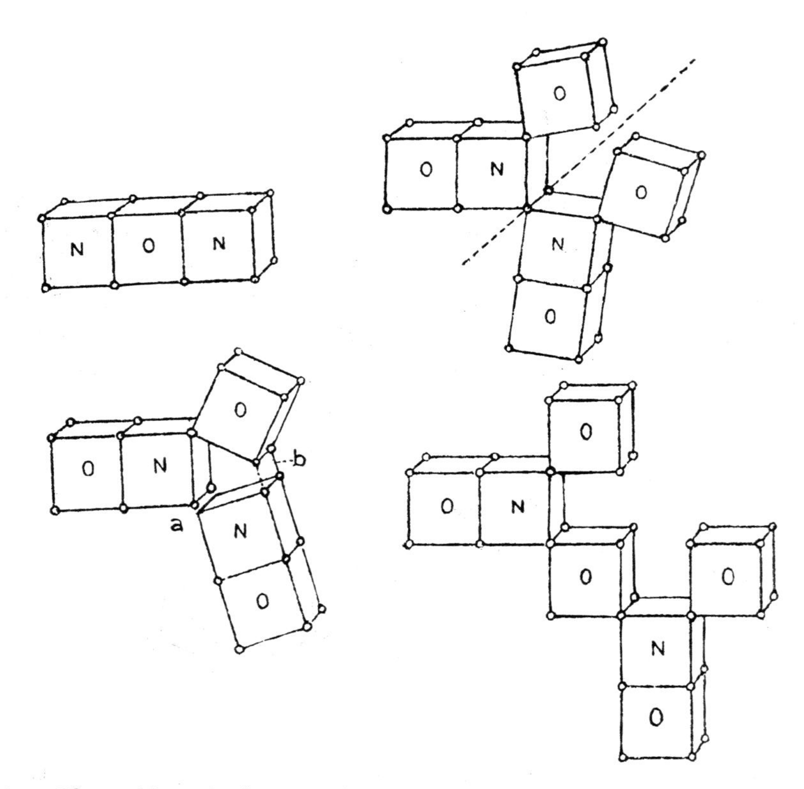


Fig. 10. – The oxides of nitrogen, from LANGMUIR, *Journ. Am. Chem. Soc.*, **41**:1, 898 (1919).

by LEWIS and LANGMUIR [162], to the attention of the physicists, who found themselves staring at the perplexing pictures of Fig. 10. Not knowing how electrons fitted into atoms they scarcely knew what to make of these diagrams, which fearlessly jumbled atoms together into molecules; but they perceived, without much study, that the chemical evidence, as interpreted by the chemists, also demanded a spatial atom.

By 1920, if not before, BOHR was well acquainted with the evidence undermining his atomic theory of 1913. In 1921, following a decisive intervention by SCHRÖDINGER, BOHR outlined a revised theory which penetrated far closer than the earlier to the heart of the atom.

## V. - THE SECOND BOHR THEORY.

In 1921, immediately after finishing with the problem of color vision, SCHRÖDINGER made a fast foray into the thickets of atomic physics. The difficulty that concerned him may be easily described. RYDBERG had distributed the line spectra of the alkalis (among other elements) into four series, each characterized by a running term of the form

$$\frac{R}{(m+\alpha)^2}.$$

Here $m$ is the running integer, which begins at 2 for the sharp and principal series, at 3 for the diffuse, and at 4 for the fundamental [163]; and $\alpha$ is a constant, different for each series, and small compared to unity except in the case of the sharp series, where $\alpha_s = s$ exceeds 0.5.

Now SOMMERFELD, in his masterful papers of 1915/6, had suggested that the orbits giving rise to the sharp series were characterized by $k=1$. This identification rested on a computation of the energy of a single electron describing a circular path of radius $r$ under the influence of the nucleus and an intervening ring of electrons assumed (for ease of calculation) to constitute a continuous ribbon of charge [164]. Since this ribbon, whose radius we shall call $a$, introduces a non-Coulomb potential $V(a, r)$, $k$ and $n$ enter separately into the energy formula. Expand $V(a, r)$ in powers of $a/r$ and keep only terms through $(a/r)^3$; the phase integrals yield

$$W = \frac{Rh}{(n+\text{const}/k^3)^2}.$$

SOMMERFELD proposed to identify the principal quantum number $n$ with the running integer $m$ of the Rydberg terms, and (because of the large value of $s$) to assign the sharp series to the smallest possible value of $k$. Since the alkali atom in its ground state absorbs the principal series, the ground state of the valence electron must be an $s$-level; and, since the first *excited* $s$-level was characterized by $m = 2$, it followed from Sommerfeld's identification of $n$ and $m$ that, in keeping with Bohr's original prescription, the outermost electron in a *normal* alkali atom described a circular orbit with $n = k = 1$.

## 1. – Penetrating orbits.

That was in 1915. Five years later, having discarded the one-quantum atom, physicists had lost the connection between the sharp spectra and the model. It was just this loss that SCHRÖDINGER undertook to repair. He looked first to Landé's model of the sodium atom: an inner core of two electrons each with angular momentum $1 \cdot h/2\pi$ enclosed by a *Würfelverband* of two-quantum electrons and (in Landé's version) a $2_1$ valence electron whose path, as SCHRÖDINGER observed, cannot be supposed to lie wholly outside the *Würfelverband* [165]. For LANDÉ had computed the radius of the shell circumscribing the *Würfelverband* to be $0.6a_0$, $a_0$ being the radius of the normal hydrogen atom; whereas SOMMERFELD had shown that the perihelion distances $u_{nk}$ of an $n_k$ electron bound by an effective nuclear charge of about unity is

$$u_{nk} = a_n(1 - \varepsilon_{nk}) = a_0 n^2 \left(1 - \sqrt{1 - k^2/n^2}\right) . \tag{5.1}$$

If $n = 2$ (as LANDÉ had thought), (5.1) makes the perihelion distance $0.54a_0$; the valence electron must therefore be presumed to *penetrate within* the shell (or Würfelverband) occupied by the two-quantum $L$ electrons [166].

SCHRÖDINGER had to quantize these penetrating orbits in order to be able to redo the argument from which SOMMERFELD had deduced the nature of the $s$ ground state [167]. He assumed that the path of the electron consisted of two parts (Fig. 11), an elliptical arc of eccentricity $\varepsilon$ described outside the

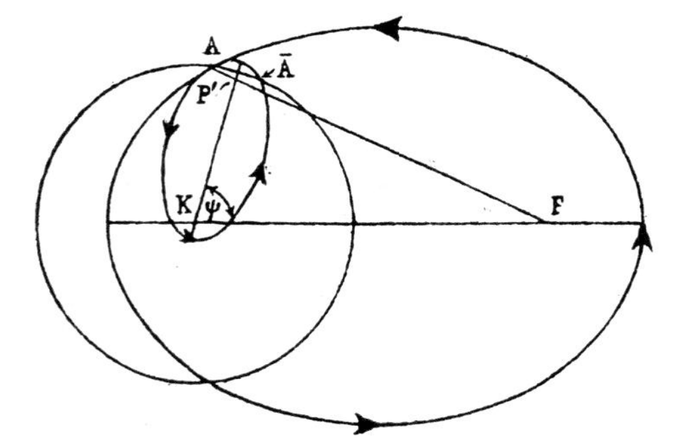


Fig. 11. – A penetrating orbit, from SCHRÖDINGER: *Zeits. Phys.*, **4**, 349 (1921). The circle represents the atomic core, the larger and smaller ellipses the path of the valence electron outside and within the core, respectively. $A$ and $A'$ are points of entry and exit, $F$ and $F'$ foci, $\psi$ the angle between the major axes of the ellipses.

$L$-shell under $Z_{\text{eff}} \sim 1$, and a second elliptical piece of eccentricity $\varepsilon'$ negotiated within the $L$-shell under $Z_{\text{eff}} = 9$. The action integrals offer no obstacle and, as usual, yield the allowed values of $\varepsilon$, $\varepsilon_{n1}$ (since $k = 1$ by hypothesis). For comparison with the Rydberg terms, SCHRÖDINGER wrote the energy for an $n_k$ path in the form

$$W = \frac{Rh}{k^2}(1 - \varepsilon^2),$$

and inserted for $\varepsilon$ the computed values of $\varepsilon_{n1}$.

The result may be expressed as

$$W = \frac{Rh}{n'^2},$$

where $n' = 1.26, 2.26, 3.26 \ldots$. Remember that this first value, $n' = 1.26$, corresponds to the ground state of the valence electron as supposed by LANDÉ, namely $n = 2$. Consequently, as appears by reference to the Rydberg terms, SOMMERFELD had erred in setting $n = m$, at least in the case of the sharp series: the quantum number $n$ does not fall short of the « effective quantum number » $n' = m + s$ by the amount $s$, but rather *exceeds* it by $1 - s$. In the case of sodium the quantum defect is really $2 - s$, for, contrary to the assumption of LANDÉ, the valence electron of sodium possesses a $3_1$ ground state, and penetrates within the $K$-shell. But neither this blemish, nor the fact that the computed value of $s$ ($\sim 0.26$) is only half that measured, should blind us to the importance of Schrödinger's result: the penetrating orbits and the quantum defect, by providing a plausible answer to the problem of atomic size, made it possible to resume the program of attributing to each atomic electron its appropriate quantum numbers.

BOHR had come to some of these conclusions before Schrödinger's paper appeared [168], and on March 24, 1921, he sent off a letter to *Nature* which traced the main difficulties in atom building to the tradition, to which he had himself succumbed, of conceiving atoms, like onions, to be built up of concentric layers or shells [169]. Whether one believed the atom to consist of rings and *Ellipsenvereine*, or of Landé's *Polyederverbände*, it was impossible—according to BOHR—to understand either the process of formation of the atom or its stability. For, he said, all electrons of a *symmetric* shell would have to be bound at once, which was scarcely plausible, and the resultant structure would not, in general, be able to rearrange itself in stable configurations after losing one of its members.

The last point is critical. Since any member of the group may be removed without destroying the whole, BOHR argued that the electrons must be so deployed that the removal of one causes no discontinuity in the motions of the others. It is the correspondence principle—that most mysterious and powerful of Bohr's *engins de guerre*—that requires continuity; and it is also the correspondence principle, according to BOHR, that reveals how the condition of continuity can be met by any given number of electrons.

## 2. – The correspondence principle and the periodic table.

The correspondence principle may for our purposes be formulated generally as follows: in the limit of high quantum numbers, where successive electronic orbits differ little one from another, one expects to obtain approximately the same results if one calculates according to the ordinary mechanics or according to the quantum theory. Note that the principle does not, as is often said, make the theories «equal» asymptotically; in particular, even in the limit classical electrons radiate continuously and modern electrons by quantum transitions. Perhaps the clearest application of the principle occurs in Bohr's first paper on the hydrogen spectra where it provides an alternative deduction of the Rydberg constant [170]. There the critical step compares the mechanical frequency of the $n$-th circular orbit, $\nu'_n$, with $\nu_n$, the frequency of the radiation produced in a transition from the $(n+1)$-st to the $n$-th orbit. Now

$$\nu'_n = \frac{1}{\pi e^2}\sqrt{\frac{2}{m}}\cdot W_n^{\frac{3}{2}},$$

and, in the region under consideration ($n \gg 1$),

$$\nu_n = R\left(\frac{1}{n^2} - \frac{1}{(n+1)^2}\right) \sim \frac{2R}{n^3}.$$

If one writes for $W_n$ the value implied by the Balmer formula as interpreted by BOHR, viz. $W_n = Rh/n^2$, then the correspondence principle, by requiring $\nu_n = \nu'_n$, gives an equation for $R$:

$$\frac{1}{\pi e^2}\sqrt{\frac{2}{m}}\left(\frac{Rh}{n^2}\right)^{\frac{3}{2}} = \frac{2R}{n^3}, \qquad \text{whence } R = \frac{2\pi^2 me^4}{h^3}.$$

This form of the derivation had superseded the others, at least in Bohr's practice, by the end of 1913 [171], and a few years later received an elegant formulation in terms of action and angle variables.

After BOHR returned to Copenhagen in 1916 he and KRAMERS applied to the intensities and polarization of spectral lines the same sort of correspondence that had worked so well with the frequencies. They made asymptotic connections between the Fourier coefficients of the classical electric moment of an electron describing a multiply periodic path and the number of emitted quanta. The detailed application of the correspondence principle plainly required wearisome, even awesome, calculations, for one had first to supply a complete solution to the classical problem. When KRAMERS managed to do so for a one-electron case, the Stark effect in hydrogen, his achievement was greeted, as it deserved to be, with great admiration and applause [172].

Atom building via the correspondence principle ought also to have been a staggering job. For the objective was to trace the classical orbit of each electron during its binding and, moreover, to examine the effects of different modes of capture on the motion of the electrons already bound. Only those end states of a new arrival could be allowed which might be reached by a classically acceptable path which did not cause discontinuities in the motion of any earlier arrival. In the letter to *Nature* of March, 1921, BOHR hinted that such calculations had been made, and that he had inferred from them (to use his words) that «for the atoms of the noble gases we must [!] expect the constructions indicated by the following symbols»:

helium $(2_1)$, krypton $(2_1 8_2 18_3 8_2)$,

neon $(2_1 8_2)$, xenon $(2_1 8_2 18_3 18_3 8_2)$,

argon $(2_1 8_2 8_2)$, niton $(2_1 8_2 18_3 32_4 18_3 8_2)$.

The large numbers are the electronic populations of successive shells moving outward from the nucleus; the subscripts indicate the principal quantum number $n$ of the shells. Of special interest are the attribution of $n=2$ to the second group of eight electrons in argon, the symmetrical arrangement of $n$, and the disposition of a two-quantum group at the atomic surface. One supposes that BOHR reached these curious arrangements not by following the correspondence principle, but by balancing two conflicting sorts of evidence: the X-ray data, favoring the many-quantum atom, and the Rydberg terms, suggesting the identification of $m$ and $n$.

Then came Schrödinger's analysis of the $s$-terms of the alkalis which severely diminished the force of the argument for identifying $m$ and $n$. After further consideration BOHR announced in a second letter to *Nature*, sent off in October, that his earlier attribution of principal quantum numbers to the electron groups of the noble-gas atoms had been in error [173]. There and in later work he employed the $n$-quantum atom, which—one is neither surprised nor reassured to learn—he says the correspondence principle demands.

These preliminary announcements made a great impression; some atomic physicists literally stopped work while awaiting further disclosures [174]. These came in 1922, in a German translation of a lengthy talk earlier given before the Danish Physical Society, and in one of the lectures in the important series BOHR gave at Göttingen in June [175]. In both presentations he proceeded straight through the periodic table, without missing an electron, calling each by its principal and azimuthal quantum number. Let us follow him.

Hydrogen evidently posed no problem: its single electron in its ground state circulates in a $1_1$ orbit. Helium required a touch of the correspondence principle. The ordinary mechanics, BOHR claimed, provided no way to add a second electron to the circle of the first without introducing discontinuities into the motion. Helium «therefore» contained two inclined $1_1$ orbits. Likewise the third electron, lithium's, cannot assume a $1_1$ orbit without interrupting the orbits already present; it therefore begins the second period, falling into a $2_1$ ellipse that just touches the $1_1$ circles (Fig. 12). The 4th, 5th and 6th elec-

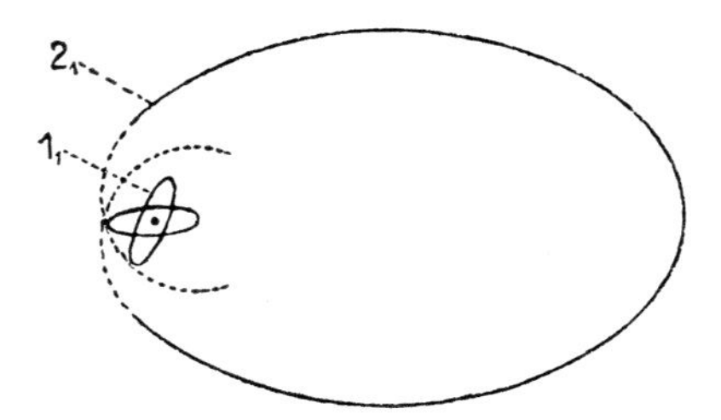


Fig. 12. – Lithium, from a chart BOHR had made to illustrate his lectures on atomic theory. From the reproduction in H. A. KRAMERS and H. HOLST: *Das Atom und die Bohrsche Theorie seines Baues* (Berlin, 1925).

trons also settle into $2_1$ paths, the third electron being unable to prevent their co-optation (as the first two electrons excluded it) because of the large eccentricity of the ellipses. All four $2_1$ electrons pierce the $1_1$ circles, thereby securing the necessary coupling, the «visits» being so phased that only one foreign particle is present at a time. Beyond carbon ($Z = 6$) the $2_2$ subgroup begins, to be completed with four members at neon. The eleventh electron, refused admission to the two-quantum level by the correspondence principle and the spectral-chemical data, assumes the single penetrating $3_1$ orbit characteristic of sodium. The further construction of the third period exactly parallels that of the second, so that argon possesses the structure $(1_1)^2(2_1)^4(2_2)^4(3_1)^4(3_2)^4$, the term $(n_k)^x$ indicating $x$ electrons of the type $n_k$ (Fig. 13). The fourth period

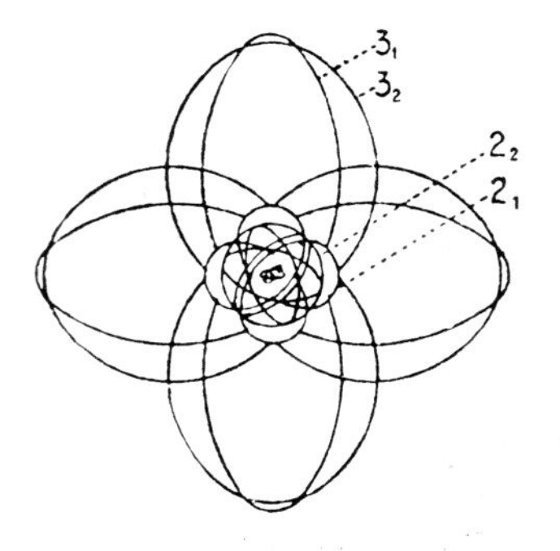


Fig. 13. – Argon.

opens with the addition of two $4_1$ electrons, to form potassium and calcium ($Z = 20$). So far the orbit of a newly-added electron has lain either wholly or (for penetrating paths) at least partly at the atomic surface, characterized by a value of $n$ equal to the number of the period in which the atom it completes is located. At scandium we meet the first of many deviations from this pattern. The twenty-first electron falls into a $3_3$ orbit because (according to BOHR) the nuclear charge has now become great enough to bind the $3_3$ circle more tightly than the highly penetrating $4_1$ ellipse. One might expect that, in analogy to the external structure of argon, only four $3_3$ electrons can be accomodated; but in fact with increasing $Z$ the $3_3$ circles tighten until they disturb the fine harmonic interplay of argon's eight three-quantum electrons. These groups then open until eighteen electrons in all are assimilated, in symmetric units of six, into what proves to be the definitive complement of the $3_1$, $3_2$ and $3_3$ subgroups. Krypton becomes $(1)^2(2)^8(3_1)^6(3_2)^6(3_3)^6(4_1)^4(4_2)^4$, as shown schematically in Fig. 14. (For brevity $(2)^8$ has been written for $(2_1)^4(2_2)^4$). The fifth period exactly repeats the fourth, for the same reasons, giving xenon (Fig. 15) the structure $(1)^2(2)^8(3)^{18}(4_1)^6(4_2)^6(4_3)^6(5_1)^4(5_2)^4$.

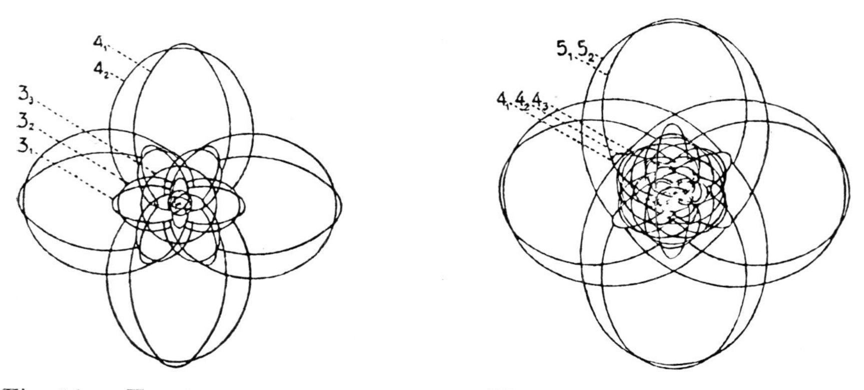


Fig. 14. – Krypton.

Fig. 15. – Xenon.

The sixth period starts with two $6_1$ electrons (cesium and barium) followed by the first representative of the $5_3$ subgroup, making up lanthanum ($Z = 57$) in precise analogy to the constitution of the third members of the preceding periods, scandium (21) and yttrium (39). Then the sequence is interrupted in a more extravagant manner: the nucleus has now become powerful enough to bind electrons into $4_4$ circles deep within the atom, and so to build up a family of elements of almost identical chemical properties. How many of these lanthanides (as we shall call lanthanum and its siblings) are there? The chemists could not answer unambiguously, as they had been forced to leave gaps for undiscovered elements precisely in the region of interest. BOHR supplied a characteristic argument. Since, he said, the two two-quantum groups each contain four electrons, and the three three-quantum groups six each, analogy suggests thirty-two four-quantum electrons, arranged in four groups of eight. Now eighteen four-quantum electrons already appear in xenon, and hence in lanthanum, leaving fourteen for subsequent elements. Assuming, as the correspondence principle (and the known chemical facts!) seemed to require, that all fourteen followed immediately on lanthanum, the lanthanides must stretch to lutecium ($Z = 71$), and no further. Thereafter, the five-quantum group is built out, taking us across the platinum family and up to mercury ($Z = 80$); then the six-quantum orbits are provisionally completed, ending the sixth period at radon.

Despite its numeralogical flavor, expressed particularly in the arbitrary opening and symmetrizing of the subgroups, Bohr's new survey of the periodic table did not disappoint the expectant atomic physicists. For one thing, it coordinated a quantity of information about the optical, X-ray and chemical properties of the elements. For another, it appeared to rest on vast, recondite calculations—none of which, I believe, has ever been found—consolidating vistas glimpsed in the pale light of the correspondence principle. Even the number and nature of the lanthanides received a natural explanation. « Indeed », as BOHR said in his Nobel Lecture in December, 1922, « indeed, it is scarcely an exaggeration to say that if the existence of the rare earths had not been established by direct experimental investigation, the occurrence of a family of elements of this character within the sixth period of the natural system of the elements might have been theoretically predicted » [176].

## 3. – The discovery of hafnium and the refinement of the orbits.

In the same address, in a postscript added immediately before delivery, BOHR was able to announce that his most hazardous speculation, the closure of the rare earths at lutecium, had just been confirmed in Copenhagen. This drama had not been of Bohr's making. About the time that he had reaffirmed his guess about element number 72 in his Göttingen lectures, URBAIN announced that a Parisian spectroscopist named Dauvillier had detected lines from 72 in Urbain's old preparation of rare earths, hopefully called « celtium », in which MOSELEY had found nothing some eight years before [177]. Now most chemists assumed that the space after lutecium belonged to one of several competing rare earths whose credentials remained to be established, the most popular competitors being celtium and a second thulium, the discovery of the Austrian industrialist the baron VON WELSBACH [178]. Since no one besides the discoverers possessed samples of these rare rare earths, one chose between them on—shall we say—extra-scientific grounds, and particularly on the basis of nationality. Urbain's revelation was immediately applauded by the French and English press, and received the sanction of DE BROGLIE, RUTHERFORD and the Chemical Society of London.

As you might imagine, it shook Bohr's confidence in the shaky symmetry arguments from which he had deduced the nature of 72. He continued to fret through the summer of 1922 without, however, starting a search for his own 72, which, according to his predictions, ought to have been closely associated with zirconium and thorium [179]. In the autumn, an expert X-ray man, COSTER, came to work in Copenhagen. Together he and HEVESY, who had come to Bohr's new institute in 1920, started a line of research which required HEVESY to know something of the techniques of X-ray spectroscopy. As a setting-up exercise he suggested that they look for Bohr's element 72; COSTER at first declined, expecting that, since the chemists had failed to find it, it must be extremely rare [180]. Finally, according to HEVESY, COSTER surrendered to his importunities and they sought 72 where BOHR had pointed, among ores of zirconium. It proved indecently easy. The new element showed up in all the so-called zirconium on display in nordic museums; it not only was not rare, it was as common as tin and 1 000 times more plentiful than gold [181]. COSTER and HEVESY called it hafnium.

The subsequent priority dispute between the physicists' hafnium and the chemists' celtium helped to advertise the strengths and the uncertainties of Bohr's new theory of the elements. In particular, it underscored the need for principles less arbitrary than Bohr's symmetries to account for the closure of the subgroups. We cannot follow here in any detail the line of thought which led to Pauli's exclusion principle of 1925. Suffice it to say that PAULI himself has traced his interest in the problem of closure to conversations with BOHR, whom he first met at Göttingen in 1922 [182].

It made a strong impression on me that BOHR at that time and in later discussions was looking for a *general* explanation that should hold for the closing of *every* electron shell and in which the number 2 was considered as essential as 8.

PAULI built particularly on the refinement of Bohr's assignments worked out by STONER, then a student at Cambridge, who concerned himself with the details of the X-ray spectra [183].

By 1924, when STONER wrote (and indeed already in 1922), one knew from examination of X-ray absorption edges that each principal shell $n$ must be associated not with $n$ chief sublevels $k = 1, 2, \dots n$, but with almost twice that number, viz. $2n - 1$. These supernumerary levels had been classified by SOMMERFELD in a purely phenomenological way with the help of a new quantity $j$, which, to avoid misleading identifications with the model, he noncommittedly called the «inner quantum number» [184]. Each $n_k$ level with the exception of $s$ terms ($k = 1$) became two levels $n_{kj}$, with $j$ equal either to $k$ or to $k + 1$. STONER transferred these attributions to the atomic shells and, moreover, estimated the population of each subshell $n_{kj}$ primarily from data on the relative intensities of the absorption edges. The scheme very much pleased SOMMERFELD, who endorsed it as more « trustworthy » than Bohr's. His reason is characteristic [185]:

It has (he said) an arithmetic rather than a geometrical-mechanical character; without assuming any symmetry of orbits, it exploits not some but all available data of X-ray spectroscopy.

Among other advantages of Stoner's scheme (Fig. 16), groups of electrons once closed remain so; there is no reopening of the type BOHR required for the symmetrizing of the three- and four-quantum groups.

| Element | Atomic number | Level ($n$) | Sublevel ($k, j$) | | | | | | |
|---|---|---|---|---|---|---|---|---|---|
| | | | I | II | III | IV | V | VI | VII |
| | | | 1, 1 | 2, 1 | 2, 2 | 3, 2 | 3, 3 | 4, 3 | 4, 4 |
| He | 2 | $K$ (1) | 2 | | | | | | |
| Ne | 10 | $L$ (2) | 2 | 2 | 4 | | | | |
| A | 18 | $M$ (3) | 2 | 2 | 4 | (4 | 6) | | |
| Kr | 36 | $N$ (4) | 2 | 2 | 4 | (4 | 6) | (6 | 8) |
| Xe | 54 | $O$ (5) | 2 | 2 | 4 | (4 | 6) | | |
| Nt | 86 | $P$ (6) | 2 | 2 | 4 | | | | |

Fig. 16. – Stoner's modification of Bohr's distribution, from *Phil. Mag.*, **48,** 722 (1924). The electronic complement of a given atom is indicated by all the numbers above and to the left of the heavy line drawn under the atom's symbol.

It is curious that the procedures used by STONER had been suggested before the war by THOMSON, who, in an address given at Oxford in 1914, called for measurements of X-ray absorption to establish the number and populations of the various atomic shells [186]. One might expect therefore that THOMSON would have vigorously applauded Stoner's analysis, especially after the discovery of the spin of the electron in 1925 made possible an easily visualized representation of $j$ as a total angular momentum.

THOMSON, however, had little sympathy for the new physics. He thought Bohr's theory of spectral emission « remarkable » but—and the comparison here with SOMMERFELD is most striking—« arithmetical rather than dynamical ». « If it is true (THOMSON said) it must be the result of forces whose existence has not been demonstrated » [187]. No theory for THOMSON could be adequate without some representation of the forces, without some visualization of « physical processes as the working of a model »; and during the war, in the time he could spare from heavy administrative burdens, he tried to rederive Bohr's results by allowing non-Coulombian forces between electrons and what he coyly called the positive centre of the atom. When he found it useful he would endow this center with magnetism: and indeed, with an appropriate distribution of the magnetic intensity and a bizarre variation of electrical forces with distance, he did succeed in deriving something which, with good will, one might regard as akin to Balmer's formula [188].

Unfortunately it appears that one cannot have models of Thomson's type and a quantitative fit with experiment. One must expand one's notion of allowable physical concepts or abandon the hope of a precise microphysics. This issue is not peculiar to quantum physics; it has been joined many times, perhaps most consequentially between NEWTON and the strict mechanists of his age, who insisted on tracing the causes of natural effects to mechanical motions alone; for otherwise, according to their leader HUYGENS, we must « renounce all hope of ever comprehending anything in physics » [189]. Physicists chose NEWTON and quantification over intelligibility, and found that, after a time, actions at a distance had become as intelligible (or the reverse) as elastic collisions.

The moral of the story is not that THOMSON failed, but rather that his program, which derived from the central concepts of classical physics, proved so fruitful, and achieved such secure results, that it could force physicists to seek—and find—a wider and richer set of concepts.

* * *

It is a pleasure to thank the Royal Society of London for permission to quote from the SCHUSTER and the LARMOR papers, and Professor A. BOHR and Dr.-Ing. E. SOMMERFELD for permission to quote from items in the Archive for History of Quantum Physics.

## REFERENCES

[1] E. WIECHERT: *Über das Wesen der Elektricität, Schriften phys.-ökon. Ges. Königsberg*, **37** (1896) [1-12]; *Experimentelles über die Kathodenstrahlen, ibid.* [12-16]. Cf. W. KAUFMANN: *Die Entwicklung des Elektronbegriffs, Verh. Ges. deut. Naturf. Ärzte*, **73**, 115-126 (1901); E. T. WHITTAKER: *A History of the Theories of Aether and Electricity*. - I: *The Classical Theories* (London, 1951²), p. 359-366; D. L. ANDERSON: *The Discovery of the Electron* (Princeton, 1964).

[2] J. J. THOMSON: *Recollections and Reflections* (London, 1936), p. 9; Lord RAYLEIGH: *The Life of Sir J. J. Thomson* (Cambridge, 1943), p. 18-20.

[3] See P. ZEEMAN: *Experimentelle Untersuchungen über Theile welche kleiner als Atome sind*, *Phys. Zeits.*, **1**, 562-565, 575-578 (1899-1900); H. A. LORENTZ: *Théorie des phénomènes magnéto-optiques récemment découverts*, *Rapports cong. int. phys.*, Vol. **3** (Paris, 1900), p. 1-33; J. BROOKES SPENCER: *On the varieties of nineteenth century magneto-optical discovery*, *Isis*, **61**, 34-51 (1970).

[4] N. BOHR: *Zeeman effect and theory of atomic constitution*, in *Pieter Zeeman 1865-1935. Verhandelingen op 25 Mei aangeboden aan Prof. Dr. Zeeman* (The Hague, 1935), p. 131-134; J. BROOKES SPENCER: *The historical basis for interactions between the Bohr theory of the atom and investigations of the Zeeman effect: 1913-1925*, *Actes XII. Cong. int. hist. sci.* [1968], Vol. V (Paris, 1971), p. 95-100.

[5] E. RUTHERFORD: *Uranium radiation and the electrical conduction produced by it*, *Phil. Mag.*, **47**, 109-163 (1899); *Collected Papers*, edited by J. CHADWICK, Vol. **1** (London, 1962), p. 169-215.

[6] See M. MALLEY: *The discovery of the beta particle*, *Am. Journ. Phys.*, **39**, 1454-1460 (1971).

[7] Cf. A. SCHUSTER: *The Progress of Physics 1875-1908* (Cambridge, 1908), p. 70-71.

[8] G. F. FITZGERALD: *On Ostwald's energetics* [1896], in *Scientific Writings*, edited by J. LARMOR (London and Dublin, 1902), p. 388.

[9] J. C. MAXWELL: *On the dynamical theory of gases*, *Phil. Trans.*, **157**, 49-88, on p. 60 (1867).

[10] *Scientific Writings*, p. 163-169.

[11] *Recent Researches* (Oxford, 1896), p. vi.

[12] Lord KELVIN: *Molecular Dynamics and the Wave Theory of Light*, papyrograph edition by A. S. HATHAWAY (Baltimore, 1884), p. 104-131. This edition contains several remarks about models omitted from the definitive version (London, 1904). Cf. A. ALIOTTA: *La reazione idealistica contro la scienza* (Palermo, 1912), p. 507-524; P. DUHEM: *The Aim and Structure of Physical Theory*, tr. P. P. WIENER (Princeton, 1954), p. 80-89.

[13] S. P. THOMPSON: *Address of the president*, *Proc. Phys. Soc. London* (1900/1), p. 21-22.

[14] O. LODGE: *Modern Views of Electricity* (London, 1892[2]), p. 5-6, 13.

[15] See P. FORMAN, J. L. HEILBRON and S. WEART: *Personnel, funding and productivity of physics circa 1900. A multinational statistical study*, *Hist. Stud. Phys. Sci.*, **4** (1974), Table A.6.

[16] See W. W. ROUSE BALL: *A History of the Study of Mathematics at Cambridge* (Cambridge, 1889); J. W. L. GLAISHER: *The mathematical tripos*, *Proc. London Math. Soc.*, **18**, 4-38 (1886); N. SHAW: *The Cavendish Laboratory as a factor in a counter revolution*, *Nature*, **118**, 885-887 (1926).

[17] H. ROSCOE to SCHUSTER, 19 May and 7 June 1881 (Schuster Papers, Royal Society of London). SCHUSTER won the chair; in his next encounter with THOMSON, over the Cavendish Professorship in 1884, he lost.

[18] H. HERTZ: *The Principles of Mechanics* (London, 1899), p. 18.

[19] H. VON HELMHOLTZ: *Preface*, to HERTZ [18]; H. POINCARÉ: *Électricité et optique*, Vol. **1** (Paris, 1890/1), p. viii; DUHEM [12], p. 85.

[20] G. KIRCHHOFF: *Vorlesungen über die Theorie der Wärme*, edited by M. PLANCK (Leipzig, 1884), p. 1-2.

[21] Cf. DUHEM [12]; R. DUGAS: *La théorie physique au sens de Boltzmann et ses prolongements modernes*, 5 (Neuchâtel, 1959).

[22] H. A. BUMSTEAD to J. LARMOR, 26 June 1908 (Larmor Papers, Royal Society of London).

[23] *E.g.*, DUGAS [21], p. 82; S. BRUSH: *Thermodynamics and history: science and culture in the nineteenth century, The Graduate Journ.*, **7**, 477-565 (1967).

[24] For the energeticist history of energetics: W. OSTWALD: *Elektrochemie, ihre Geschichte und Lehre* (Leipzig, 1896); G. HELM: *Die Energetik nach ihrer geschichtlichen Entwicklung* (Leipzig, 1898).

[25] See the account of the meeting in *Verh. Ges. deut. Naturf. Ärzte*, **67**, 28, 32-33, (1895); and HELM to his wife, 17 and 19 Sept. 1895, edited by H. G. KÖRBER, *Aus dem wissenschaftlichen Briefwechsel Wilhelm Ostwalds*, Vol. **1** (Berlin, 1961), p. 118-120.

[26] *Die Überwindung des wissenschaftlichen Materialismus, Verh. Ges. deut. Naturf. Ärzte*, **67**, 155-168 (1895). The speech appeared in French in *Revue générale des sciences*, **6**, 953-958 (1895).

[27] W. OSTWALD: *Lebenslinien*, Vol. **2** (Berlin, 1926/7), p. 197 ff.

[28] W. OSTWALD: *Grundriss der allgemeinen Chemie* (Leipzig, $1909^4$), p. iii-iv; J. PERRIN: *L'agitation moléculaire et le mouvement brownien, Compt. Rend.*, **146**, 967-970 (1908).

[29] J. J. THOMSON: *On the masses of the ions in gases at low pressures, Phil. Mag.*, **48**, 547-567 (1899), on p. 565; RAYLEIGH [2], p. 140-141. Cf. H. POINCARÉ: *Relation entre la physique expérimentale et la physique mathématique, Rapports cong. int. phys.*, Vol. **1** (Paris, 1900), p. 1-29, on p. 17; FITZGERALD: *Writings* [8], p. 524.

[30] J. LARMOR: *A dynamical theory of the electron and the luminiferous medium. II: Theory of electrons, Phil. Trans.*, **186** A:**2**, 741-742 (1895); for O. LODGE see J. J. THOMSON: *Indications relatives à la constitution de la matière*, in *Rapports* [29], Vol. **3**, p. 138-151, on p. 149.

[31] J. LARMOR: *Aether and Matter* (Cambridge, 1900), p. 3; G. F. FITZGERALD: *Writings*, p. 525-526; J. J. THOMSON: *Cathode rays, Phil. Mag.*, **44**, 293-316 (1897), on p. 312-313.

[32] A. SCHUSTER: *On harmonic ratios in the spectra of gases, Proc. Roy. Soc.*, **31**, 337-347 (1881); H. KAYSER: *Handbuch der Spectroscopie*, Vol. **2** (Leipzig, 1902), p. 498. See W. MCGUCKEN: *Nineteenth-Century Spectroscopy* (Baltimore, 1969), p. 103-156.

[33] J. LARMOR [31], p. 190-193.

[34] *Ibid.*, 225, and *On the theory of the magnetic influence on spectra, Phil. Mag.*, **44**, 503-512 (1897), on p. 512.

[35] J. J. THOMSON: *The magnetic properties of systems of corpuscles describing circular orbits, Phil. Mag.*, **6**, 673-693 (1903).

[35*a*] Lord RAYLEIGH: *On the propagation of waves along connected systems of similar bodies, Phil. Mag.*, **44**, 356-362 (1897).

[36] G. F. FITZGERALD: *Notes on Professor Ebert's estimate of the radiating power of an atom, with remarks on vibrating systems giving special series of overtones like those given out by some molecules, Brit. Ass. Report* (1893), p. 689-690.

[37] H. KAYSER [32], Vol. **2**, p. 279-280; cf. the review by CREW in *Astrophys. Journ.*, **7**, 150-155 (1898).

[38] J. J. THOMSON: *On a view of the constitution of a luminous gas suggested by Lorentz' theory of dispersion*, in *Recueil des travaux offerts à ... H. A. Lorentz* (The Hague, 1900), p. 642-643.

[39] H. KAYSER [32], Vol. **2**, p. 282. For Lockyer's views, MCGUCKEN [32], p. 76-100; A. J. MEADOWS: *Science and Controversy. A Biography of Sir Norman Lockyer* (Cambridge, Mass., 1972), p. 135-174.

[40] *Fortschritte der Physik*, **60**:**1**, 275-276 (1904); **67**:**2**, 148 (1911); *Beiblätter zu den Annalen der Physik*, **28**, 1270-1273 (1904); **36**, 54-56 (1912).

[41] Cf. the lectures sponsored by the Société française de physique, *Les idées modernes sur la constitution de la matière. Conférences faites en 1912* (by Mme CURIE, P. LANGEVIN, J. PERRIN, H. POINCARÉ and P. WEISS, among others) (Paris, 1913); and the theme of the second Solvay Congress, *La structure de la matière. Rapports et discussions du conseil de physique* (Solvay, 1913) (Paris, 1921), with THOMSON as key-note speaker.

[42] The numbers that follow come from the tables of [15] supplemented by CERN's *Report* for 1964 and 1966.

[43] A breakdown of the invited papers delivered at the Congrès will suggest the distribution of effort in fin-de-siècle physics:

| Subjects | Number | Percent |
|---|---|---|
| Questions générales. Unités. Mesures | 15 | 16 |
| Physique mécanique et moléculaire | 19 | 20 |
| Optique et thermodynamique | 14 | 15 |
| Électricité et magnétisme | 22 | 24 |
| Magnéto-optique, rayons cathodiques, uraniques | 8 | 9 |
| Physique cosmique (geo-physics) | 9 | 10 |
| Physique biologique | 5 | 6 |

[44] P. LANGEVIN: in *Le Collège de France 1530-1930* (Paris, 1932), p. 74; Lord RAYLEIGH: *J. J. Thomson* (Cambridge, 1942), p. 18; Cf. H. H. BELLOT: *University College, London, 1826-1926* (London, 1929), p. 312.

[45] J. TROWBRIDGE: in *Annual Reports of the Presidents and Treasurers of Harvard College* (1905/6) (Cambridge, Mass., 1906), p. 244-245.

[46] B. BRUNHÈS: *La physique dans les universités italiennes, Rev. int. enseign.*, **41**, 400-405 (1901).

[47] O. GLASSER: *W. C. Röntgen and the Early History of the Röntgen Rays* (Springfield, Ill., 1934), p. 97.

[48] J. JEANS: *The mechanism of radiation, Phil. Mag.*, **2**, 421-455 (1901).

[49] S. EARNSHAW: *On the nature of the molecular forces which regulate the constitution of the luminiferous ether, Trans. Camb. Phil. Soc.*, **7**, 97-114 (1842); see W.T. SCOTT: *Who was Earnshaw? Am. Journ. Phys.*, **27**, 418-419 (1959).

[50] Lord KELVIN: *Contributions of Lord Kelvin to the discussion of the nature of the emanations from radium, Phil. Mag.*, **7**, 220-222 (1904); *Plan of a combination of atoms to have the properties of polonium and radium, Phil. Mag.*, **8**, 518-534 (1904). Cf. A. S. EVE: *Rutherford* (New York, 1939), p. 140-142; S. P. THOMPSON: *Life of Sir William Thomson*, Vol. **2** (London, 1910), p. 1083, 1190.

[51] H. NAGAOKA: *Kinematics of a system of particles illustrating the line and band spectrum and the phenomena of radioactivity, Phil. Mag.*, **7**, 445-455 (1904); *Nature*, **69**, 392-393 (1904); *Proc. Tokyo Math. Phys. Soc.*, **2**, 92-107, 129-131 (1903/6). For an earlier Saturnian atom see J. PERRIN: *Les hypothèses moléculaires, Rev. Sci.*, **15**, 449-461 (1901). Cf. M. GLIOZZI: *I modelli atomici, Arch. Int. Hist. Sci.*, **12**, 329-339 (1957); C. E. BEHRENS: *Atomic theory from 1904 to 1913, Am. Journ. Phys.*, **11**, 60-66 (1943).

[52] Knott's predecessor was also a Scot trained by TAIT. *Anniversary Volume Dedicated to Professor Hantaro Nagaoka* (Tokyo, 1925), iii; (Notice of C. G. KNOTT), *Proc. Roy Soc.*, **102** A, xxvii-xxviii (1922/3); C. G. KNOTT: *Life and Scientific Work of P. G. Tait* (Cambridge, 1911).

[52a] Cf. J. J. THOMSON: *On the scattering of light by unsymmetrical atoms and molecules, Phil. Mag.*, **40**, 393-413 (1920).

[53] G. A. SCHOTT: *On the kinematics of a system of particles illustrating the line and band spectra, Phil. Mag.*, **8**, 384-387 (1904). NAGAOKA responded, precipitating an exchange in *Nature*, **70**, 124-125, 176 (1904).

[54] J. J. THOMSON: *Cathode rays., Phil. Mag.*, **44**, 313 (1897); A. M. MAYER: *On the morphological laws of the configurations formed by magnets floating vertically and subjected to the attraction of a superposed magnet ..., Am. Journ. Science*, **116**, 247-256 (1878); *Floating magnets, Nature*, **17**, 487-488 (1877); **18**, 258-260 (1878). THOMSON had once before brought Mayer's magnets into the service of atomic theory, in connection with the vortex atom (*Treatise on the Motion of Vortex Rings* (London, 1883), p. 107, inspired by Lord KELVIN: *Floating magnets, Nature*, **18**, 13-14 (1878)). Mayer's experiments were refined by R. W. WOOD: *On the equilibrium figures formed by floating magnets, Phil. Mag.*, **46**, 162-164 (1898), and adapted to electricity by J. MONCKMAN: *On the arrangement of electrified cylinders when attracted by an electrified sphere, Proc. Camb. Phil. Soc.*, **6**, 179-181 (1889).

[55] Lord KELVIN: *Aepinus atomized, Phil. Mag.*, **3**, 257-283 (1902); reprinted in Kelvin's *Baltimore Lectures on Molecular Dynamics and the Wave Theory of Light* (London, 1904), p. 541-568. Cf. J. J. THOMSON: *The structure of the atom, Proc. Roy. Inst.*, **18**, 49-63 (1905/7).

[56] J. J. THOMSON [35]. J. LARMOR and O. LODGE (*Nature*, **68**, 609 (1903)) and E. RUTHERFORD (*The succession of changes in radioactive bodies, Phil. Trans.*, **204** A, 169-219 (1904)) also connected the eventual radiation loss with radioactive transformation.

[57] J. J. THOMSON: *On the structure of the atom ..., Phil. Mag.*, **7**, 237-265 (1904); *Electricity and Matter* (New Haven, 1904), p. 90-139. The three-dimensional problem was later solved by a student of Hilbert's, L. FÖPPEL: *Stabile Anordnungen von Elektronen im Atom* (Berlin, 1912).

[57a] « This constitutes the first suggestion of anything worthy to be called a rational basis of Mendeléef's law, and its importance can scarcely be overestimated. It is, perhaps, the greatest contribution to theoretical physics during a decade ». W. S. F[RANKLIN]: *The electron theory, Science*, **19**, 896-899, on p. 896 (1904).

[58] J. J. THOMSON: *Die Korpuskulartheore der Materie* (Braunschweig, 1908); M. BORN: *Über das Thomsonsche Atommodell, Phys. Zeits.*, **10**, 1031-1034 (1909): « Although this comparison is no more than an analogy, yet it brings the remarkable periodicity of the elements much closer to the understanding. It seems to me a very happy idea of Thomson's to draw on the stability of equilibrium figures for the explanation of this deep law ... The boldest concrete ideas have often led to surprising consequences ». Cf. A. HERMANN: *Frühgeschichte der Quantentheorie 1899-1913* (Mosbach in Baden, 1969), p. 102; *The Genesis of the Quantum Theory* (Cambridge, Mass., 1971), p. 87; F. REICHE: *The Quantum Theory* (London, 1922), p. 172; K. HERZFELD: *Zur Statistik des Borschen Wasserstoffsatommodells, Ann. Phys.*, **51**, 261 (1916).

[59] J. J. THOMSON: *Treatise* [54], 107 ff. On Thomson's method see the article on THOMSON in *Dictionary of Scientific Biography*, and Lord RAYLEIGH: *Life of Sir J. J. Thomson* (Cambridge, 1943²), *passim*.

[60] In Bohr interview II, p. 6-7 (AHQP); see T. S. KUHN, J. L. HEILBRON, P. FORMAN and L. ALLEN: *Sources for History of Quantum Physics. An Inventory and Report* (Philadelphia, 1967), for description and location of this archive.

[61] J. J. THOMSON: *On the number of corpuscles in an atom*, *Phil. Mag.*, **11**, 769-781 (1906). Cf. J. BOSLER: *Sur le nombre des corpuscles dans l'atome*, *Compt. Rend.*, **146**, 686-687 (1908); J. L. HEILBRON: *The scattering of α and β particles and Rutherford's atom*, *Arch. Hist. Exact Sci.*, **4**, 269-274 (1968).

[62] P. LENARD: *Über die Absorption von Kathodenstrahlen verschiedener Geschwindigkeiten*, *Ann. Phys.*, **12**, 714-744 (1903).

[63] Cf. O. LODGE: *Electrons* (London, 1906), p. 151, 194; N. R. CAMPBELL: *Modern Electrical Theory* (Cambridge, 1907), p. 130.

[64] J. J. THOMSON: *On rays of positive electricity*, *Phil. Mag.*, **13**, 561-565 (1907); *Rays of Positive Electricity* (London, 1913); G. P. THOMSON: *J. J. Thomson* (New York, 1965), p. 93, 136-141.

[65] J. J. THOMSON: *Corpuscular Theory of Matter* (London, 1907), p. 157-160; G. DARWIN: *Periodic Orbits*, *Acta Math.*, **21**, 99-242 (1897).

[66] G. A. SCHOTT: *On the electron theory of matter and the explanation of fine spectrum lines and of gravitation*, *Phil. Mag.*, **12**, 21-29 (1906).

[67] J. J. THOMSON: *The scattering of rapidly moving electrified particles*, *Proc. Camb. Phil. Soc.*, **15**, 465-471 (1910); HEILBRON [61], p. 277-280.

[68] J. A. CROWTHER: *On the energy and distribution of scattered Röntgen radiation*, *Proc. Roy. Soc.*, **85** A, 29-43 (1911). BARKLA replied in a *Note on the energy of scattered X radiation*, *Phil. Mag.*, **21**, 648-652 (1911).

[69] See Rutherford's Nobel lecture, *The chemical nature of the α-particles from radioactive substances*, *Collected Papers* [5], Vol. **2**, p. 137-146.

[70] H. GEIGER and E. MARSDEN: *On the diffuse reflection of the α-particles*, *Proc. Roy. Soc.*, **82** A, 495-500 (1909); Lord RUTHERFORD: *History of the alpha particles from radioactive substances*, in *Lectures Delivered at... Clark University* (Clark University, 1912), p. 83-95.

[71] Lord RUTHERFORD [5], p. 169-215.

[72] M. MALLEY [6].

[73] Lord RUTHERFORD: *The magnetic and electric deviation of the easily absorbed rays from radium*, *Phil. Mag.*, **5**, 177-187 (1903); *Collected Papers*, Vol. **1**, p. 549-557.

[74] J. L. HEILBRON [61], p. 256.

[75] Lord RUTHERFORD and T. ROYDS: *The nature of the alpha particle from radioactive substances*, *Phil. Mag.*, **17**, 281-286 (1909); *Collected Papers*, Vol. **2**, p. 163-167.

[76] See C. G. DARWIN: *A theory of the absorption and scattering of the α rays*, *Phil. Mag.*, **23**, 901-920 (1912), on p. 909; J. L. HEILBRON and T. S. KUHN: *The genesis of the Bohr atom*, *Hist. Stud. Phys. Sci.*, **1**, 211-290 (1969), on p. 247.

[77] W. BRAGG: *The consequences of the corpuscular hypothesis of the γ and X rays*, *Phil. Mag.*, **20**, 385-416 (1910); J. L. HEILBRON [61], p. 283-297.

[78] J. L. HEILBRON [61], p. 303-304.

[79] A. VAN DEN BROEK: *Die Radioelemente, das periodische System, und die Konstitution des Atoms*, *Phys. Zeits.*, **14**, 32-41 (1913). Cf. T. HIROSIGE: *The van den Broek hypothesis*, *Jap. Stud. Hist. Sci.*, **10**, 143-162 (1971); J. L. HEILBRON: *H. G. J. Moseley. The Life and Letters of an English Physicist, 1887-1915* (Berkeley, 1974), p. 82-85.

[80] D. I. MENDELEEV: *Principles of Chemistry*, Vol. **2** (London, 1905[3]), p. 509-529; A. WERNER: *New Ideas in Inorganic Chemistry* (London, 1911), p. 6-9; J. R. RYDBERG: *Untersuchungen über das System der Grundstoffe*, *Årsskrift Lund Universitet*, **9**:**2**, 1-41 (1913).

[81] *The cause and nature of radioactivity*, *Phil. Mag.*, **4**, 370-396, 569-585 (1902); *Collected Papers*, Vol. **1**, p. 472-508. See A. ROMER: *Historical essay*, edited

by A. ROMER, *Radioactivity and the Discovery of Isotopes* (New York, 1970), p. 12-30; T. TRENN: *Rutherford and Soddy: from a search for radioactive constituents to the disintegration theory of radioactivity, Rete*, **1**, 51-70 (1971).

[82] *E.g.*, ionium and thorium: B. KEETMAN: *Über Ionium, Jahrb. Rad. Elek.*, **6**, 265-274 (1909) (English translation in ROMER [81], p. 164-173); F. SODDY: *Radioactivity, Ann. Reports Prog. Chem.*, 285-286 (1910); *Chemistry of the Radioelements* (London, 1911), p. 29-30. Cf. A. S. RUSSELL: *Lord Rutherford: Manchester, 1907-1919*, edited by J. B. BIRKS, *Rutherford at Manchester* (London, 1962), p. 87-101.

[83] Mme CURIE: in *La théorie du rayonnement et les quanta. Rapports et discussions* (du Conseil Solvay), edited by P. LANGEVIN and M. DE BROGLIE (Paris, 1912), p. 385; J. J. THOMSON: *Ionization by moving electrified particles, Phil. Mag.*, **23**, 449-455 (1912).

[84] A. S. RUSSELL: *The periodic system and the radioelements, Chem. News*, **107**, 49-53 (1913); F. SODDY: *Radio-elements and the periodic law, Chem. News*, **107**, 97-99 (1913); *Intra-atomic charge, Nature*, **92**, 399-400 (1913); G. VON HEVESY: *Die Valenz der Radioelemente, Phys. Zeits.*, **14**, 49-62 (1913); K. FAJANS: *Über eine Beziehung zwischen der Art einer radioaktiven Umwandlung und dem elektrochemischen Verhalten der betreffenden Radioelemente, Phys. Zeits.*, **14**, 131-136 (1913); *Die Stellung der Radioelemente im periodischen System, Phys. Zeits.*, **14**, 136-142 (1913) (both translated in ROMER [81], p. 198-218).

[85] *Supra*, p. 1-39.

[86] A. E. HAAS: *Über die elektrodynamische Bedeutung des Planckschen Strahlunggezetzes und über eine neue Bestimmung des elektrischen Elementarquantums und die Dimensionen des Wasserstoffsatoms, Sitzb. Wien Ak. Wiss.*, **119**:**2**a, 119-144 (1910); *Der Zusammenhang des Planckschen Elementarquantums mit der Grundgrössen der Elektronentheorie, Jahrb. Rad. Elek.*, **7**, 261-268 (1910). See HERMANN: *Frühgeschichte* [58], p. 102-117; *Genesis* [58], p. 87-101.

[87] See R. A. MILLIKAN: *The Electron* (Chicago, 1917), for details.

[88] *Théorie du rayonnement* [83], p. 124.

[89] N. BOHR: *Studier over metallernes elektrontheori* (Copenhagen, 1911).

[90] *Ibid.*, p. 5.

[91] *Ibid.*, p. 103, 108; P. LANGEVIN: *Magnétisme et la théorie des electrons, Ann. Chim. Phys.*, **5**, 73-97 (1905); HEILBRON and KUHN [76], p. 213-223.

[92] Interviews II, p. 6 (AHQP).

[93] Letter of 29 September 1911 in HEILBRON and KUHN [76], p. 223-224.

[94] *Ibid.*, p. 225-226.

[95] J. L. HEILBRON: *Moseley* [79], p. 43, 46-47.

[96] C. G. DARWIN: *A theory of the absorption and scattering of the α rays, Phil. Mag.*, **23**, 901-920 (1912); N. BOHR: *On the theory of the decrease of velocity of moving electrified particles on passing through matter, Phil. Mag.*, **25**, 10-31 (1913); THOMSON [83].

[97] HEILBRON and KUHN [76], p. 237-243.

[98] *On the Constitution of Atoms and Molecules* (Copenhagen, 1963), p. xxi-xxviii.

[99] *The spectrum of nebulium, Month. Not. Roy. Ast. Soc.*, **72**, 49-64 (1911/12); *The constitution of the ring nebula in Lyra, Month. Not. Roy. Ast. Soc.*, **72**, 176-177 (1911/12); *On the new nebular line at λ4353, Month. Not. Roy. Ast. Soc.*, **72**, 693 (1911/12); *The constitution of the solar corona, Month. Not. Roy. Ast. Soc.*, **72**, 139-150, 677-692, 729-739 (1911/12); See R. MCCORMMACH: *The atomic theory of John William Nicholson, Arch. Hist. Exact Sci.*, **2**, 160-184 (1966).

[100] NICHOLSON: *Constitution* [99], p. 729-730.
[101] ROSENFELD [98], p. xxxvi.
[102] *Ibid.*, xxxix; Interview III, p. 11 (AHQP).
[103] N. BOHR: *On the constitution of atoms and molecules, Phil. Mag.*, **26**, 1-25 (1913).
[104] The assumption that $n$ quanta are emitted in the transition probably came from PLANCK: *Ann. Phys.*, **31**, 758-768 (1910); **37**, 642-656 (1912). See T. HIROSIGE and S. NISIO: *Formation of Bohr's theory of atomic constitution, Jap. Stud. Hist. Sci.*, **3**, 6-28 (1964); *The genesis of the Bohr atom and Planck's theory of radiation, Jap. Stud. Hist. Sci.*, **9**, 35-47 (1970).
[105] N. BOHR: *On the constitution of atoms and molecules II, Phil. Mag.*, **26**, 476-502 (1913).
[106] *Infra* 83-84.
[107] *E.g.*, W. NICHOLSON: *The high-frequency spectra of the elements and the structure of the atom, Phil. Mag.*, **27**, 541-564 (1914); *Atomic structure and the spectrum of helium, Phil. Mag.*, **28**, 90-103 (1914).
[108] J. J. THOMSON: *On the structure of the atom, Phil. Mag.*, **26**, 792-799 (1913); *La structure de l'atome*, in *La structure de la matière* [41], p. 1-44.
[109] *Nature*, **92**, 305 (1913/14); A. CONWAY: *An electromagnetic hypothesis as to the origin of series spectra, Phil. Mag.*, **26**, 1010-1017 (1913); W. PEDDIE: *On the structure of the atom, Phil. Mag.*, **27**, 257-268 (1914); G. W. WALKER: *Approximately permanent orbits and the origin of spectral series, Proc. Roy. Soc.*, **91** A, 156-170 (1915).
[110] BOHR: Interviews II, p. 6 (AHQP).
[111] H. G. J. MOSELEY: *High-frequency spectra of the elements, Phil. Mag.*, **26**, 1025-1034 (1913); *Atomic models and X-ray spectra, Nature*, **92**, 554 (1913/4).
[112] H. G. J. MOSELEY: *High-frequency spectra of the elements* II, *Phil. Mag.*, **27**, 703-713 (1914); *Brit. Ass. Reports*, 299-300 (1914).
[113] W. NICHOLSON: *Atomic models and X-ray spectra, Nature*, **92**, 583-584 (1913/4); *The spectra of hydrogen and helium, Mont. Not. Roy. Ast. Soc.*, **74**, 425-442 (1914); and *op. cit.* [107]. Cf. W. M. HICKS: *High-frequency spectra and the periodic table, Phil. Mag.*, **28**, 139-142 (1914); J. R. RYDBERG: *The ordinals of the elements and the high-frequency spectra, Phil. Mag.*, **28**, 144-149 (1914); HEILBRON: *Moseley* [79], p. 103-107.
[114] *Op. cit.* [99].
[115] Interviews III, p. 9 (AHQP).
[116] *Ibid*, I, p. 4.
[117] See A. FOWLER: *Series lines in spark spectra, Phil. Trans.*, **214** A, 254-263 (1914).
[118] BOHR [103], p. 10-11. Cf. ROSENFELD [98], p. xxxix.
[119] E. J. EVANS: *The spectrum of hydrogen and helium, Nature*, **92**, 5 (1913/4); A. FOWLER: *The spectra of hydrogen and helium, Nature*, **92**, 95-96 (1913/4).
[120] N. BOHR: *The spectrum of hydrogen and helium, Nature*, **92**, 231-233 (1913/4); confirmed by E. J. EVANS: *The spectra of helium and hydrogen, Phil. Mag.*, **29**, 284-297 (1915), and J. STARK: *Bemerkung zum Bogen und Funkenspektrum des Heliums, Verh. Deut. Phys. Ges.*, **16**, 468-474 (1914). NICHOLSON had earlier employed the reduced mass in spectroscopic computation.
[121] *Messung der Ionisierungsspannung in verschiedenen Gasen, Verh. Deut. Phys. Ges.*, **15**, 34-44 (1913). The apparatus had been introduced by LENARD.
[122] *Über Zusammenstösse zwischen Elektronen und den Molekülen des Quecksilberdämpfes und die Ionisierungsspannung derselben, Verh. Deut. Phys. Ges.*, **16**, 457-467 (1914).
[123] *Über die Erregung der Quecksilberresonnanzlinie* 253,6 $\mu\mu$ *durch Elektronenstösse, Verh. Deut. Phys. Ges.*, **16**, 512-517 (1914); cf. H. RAU: *Über die Lichter-*

*regung durch langsame Kathodenstrahlen, Sitzb. phys. med. Ges. Würzburg,* 20-27 (1914). When they first found the mercury line FRANCK and HERTZ did not know of Bohr's theory; when they discovered it, they interpreted their experiments as disconfirmation (*Über Kinetik von Elektronen und Ionen in Gasen, Phys. Zeits.*, **17**, 409-416, 430-440 (1916)).

[124] N. BOHR: *On the quantum theory of radiation and the structure of the atom, Phil. Mag.*, **30**, 394-415 (1915), on p. 411.

[125] J. FRANCK and G. HERTZ: *Die Bestätigung der Borschen Atomtheorie im optischen Spektrum durch Untersuchungen der unelastischen Zusammentösse langsamer Elektronen mit Gasmolekülen, Phys. Zeits.*, **20**, 132-143 (1919).

[126] Cf. HERMANN: *Frühgeschichte*, p. 86, and *Genesis*, p. 72 [58].

[127] W. VOIGT: *Über das elektrische Analagon des Zeemaneffektes, Ann. Phys.*, **4**, 197-208 (1901).

[128] Cf. J. STARK: *Die Atomionen chemischer Elemente* (Berlin, 1913), p. 12-13; *Nobelvortrag, Les prix Nobel en 1919-1920* (Stockholm, 1920), p. 7.

[129] J. STARK: *Observation of the separation of spectral lines by an electric field, Nature*, **92**, 401 (1913/4); *Beobachtungen über den Effekt des elektrischen Feldes auf Spektrallinien, Sitzb. Berl. Ak.* (1913:2), p. 932-946. An Italian, A. LO SURDO, had found the effect before Stark's paper appeared, but—perhaps not fully appreciating its importance—put off publishing his results (*Sul fenomeno analogo a quello di Zeeman nel campo elettrico, Rend. Acc. Lincei*, **22**: **2**, 664-666 (1913)).

[130] N. BOHR: *On the effect of electric and magnetic fields on spectral lines, Phil. Mag.*, **27**, 506-524 (1914). For earlier attempts to obtain a quantum theory of the Stark effect see E. WARBURG: *Bemerkung zu der Aufspaltung der Spektrallinien im elektrischen Feld, Verh. deut. phys. Ges.*, **15**, 1259-1266 (1913); E. GEHRCKE: *Über ein Modell zur Erklärung der Lichtemission, Phys. Zeits.*, **15**, 123-127, 198-201 (1914); A. GARBASSO: *Sopra il fenomeno di Stark-Lo Surdo, Rend. Acc. Lincei*, **22**: **2**, 635-639 (1913).

[131] *E.g.*, *Elementarquantum der Energie. Modell der negativen und positiven Elektricität, Phys. Zeits.*, **8**, 881-884 (1907); *Beziehung des Döppler-Effektes bei Kanalstrahlen zur Planckschen Strahlungstheorie, Phys. Zeits*, **8**, 913-919 (1907)

[132] HERMANN: *loc. cit.* [126].

[133] *Supra* p. 64.

[134] ROSENFELD [98], p. lii.

[135] A. SOMMERFELD: *Autobiographische Skizze*, in *Geist und Gestalt. Biographische Beiträge zur Geschichte der Bayerischen Akademie der Wissenschaften*, Vol. **2** (Munich, 1959), p. 100-103.

[136] A. SOMMERFELD: *Mechanische Darstellung der elektromagnetischen Erscheinungen in ruhenden Körper, Ann. Phys.*, **46**, 139-151 (1892).

[137] A. SOMMERFELD [135], p. 104.

[138] *E.g.*, *Das Plancksche Wirkungsquantum und seine allgemeine Bedeutung für die Molekularphysik, Phys. Zeits.*, **12** (1911), p. 1057-1068; *Application de l'élément d'action aux phénomènes moléculaires non-périodiques*, in *Théorie du rayonnement* [83], p. 313-392.

[139] See K. M. MEYER-ABICH: *Korrespondenz, Individualität und Komplementarität* (Wiesbaden, 1965); M. JAMMER: *The Conceptual Development of Quantum Mechanics* (New York, 1966), p. 109-117.

[140] *Lectures* (London, 1926), p. 46. But cf. SOMMERFELD to EINSTEIN, 11 June 1922: « Ich kann nur die Technik der Quanten fördern, Sie müssen Ihre Philosophie machen », edited by A. HERMANN: *Einstein-Sommerfeld Briefwechsel* (Basel-Stuttgart, 1968), p. 97.

[141] KRAMERS to SOMMERFELD, 4 June 1924 (AHQP).
[142] SOMMERFELD to KRAMERS, 5 July 1924 (AHQP): «Jede Theorie muss doch schliesslich dogmatisch und deduktiv vorgehen. Vergleiche den Zustand der Maxwell'schen Theorie bei Maxwell, der die ,,physikalischen'' Gesichtspunkte voranstellte, (bei ihm die mechanischen Analogien), und den Zustand bei Hertz, der die Theorie formal-mathematisch entwickelte. Wir sind heute wohl einig, dass das letzere Verfahren das entgültige ist. Auch die Quantentheorie muss schliesslich auf ein einheitliches System formaler Regeln gebracht werden, aus der das übrige deduktiv folgt. Sie können natürlich sagen, dass wir hier noch lange nicht so weit sind, wie in der Elektrodynamik—was ich natürlich zugebe. Aber die Wirkungsintegrale $J$ sind doch diejenigen Grössen, die man für die vollständige Ausführung der Theorie nötig hat. Deshalb ist es mir sehr hart, sie nicht an den Anfang der Darstellung zu setzen, sondern zunächst mit dem Zeit-Integral der lebendigen Kraft zu beginnen». For Bohr's way see BOHR: *Theory of Spectra and Atomic Constitution* (Cambridge, 1922).
[143] A. SOMMERFELD: *Zur Theorie der Balmerschen Serien, Sitzb. Münch. Ak.* (1915), p. 425-458, expanded in *Ann. Phys.*, **51**, 1-94 (1916).
[144] *Die Feinstruktur der Wasserstoff- und Wasserstoffähnlichen-Linien, Sitzb. Münch. Ak.* (1915), p. 459-500.
[145] F. PASCHEN: *Bohrs Heliumlinien, Ann. Phys.*, **50**, 901-940 (1916).
[146] BOHR to SOMMERFELD, March 1916 (AHQP).
[147] W. KOSSEL: *Bemerkung zur Absorption der homogenen Röntgenstrahlen, Verh. deut. phys. Ges.*, **16**, 898-909, 953-963 (1914). Cf. J. L. HEILBRON: *The Kossel-Sommerfeld theory and the ring atom, Isis*, **58**, 462-466 (1967).
[148] A. SOMMERFELD: *Atombau und Röntgenspektren, Phys. Zeits.*, **19**, 297-307 (1918).
[149] SOMMERFELD to LANDÉ, *c.* May 1920 (AHQP).
[150] See R. LEDOUX-LEBARD and A. DAUVILLIER: *La physique des rayons X* (Paris, 1921), p. 381-382, 392.
[151] *E.g.*, the Λ doublets. A. SOMMERFELD: *Theorie der Röntgenspektren, Ann. Phys.*, **51**, 125-167 (1916).
[152] LEDOUX-LEBARD and DAUVILLIER [150], p. 385-386; D. COSTER: *Zur Systematik der Röntgenspektren, Zeits. Phys.*, **6**, 185-203 (1921); HEILBRON [147], p. 473-478.
[153] See SOMMERFELD [148]; L. VEGARD: *Bemerkungen zu den Arbeiten von M. Born und A. Landé über Kristallgitter und Bohrschen Atommodell, Verh. deut. phys. Ges.*, **21**, 383-385 (1919).
[154] *Kristallgitter und Bohrsches Atommodell, Verh. deut. phys. Ges.*, **20**, 202-209 (1918); *Über die Berechnung der Kompressibilität regulärer Kristalle, Verh. deut. phys. Ges.*, **20**, 210-216 (1918).
[155] HEILBRON [147], p. 477-479; A. SMEKAL: *Über die Erklärung der Röntgenspektren und die Konstitution der Atome, Phys. Zeits*, **22**, 400-402 (1921).
[156] A. LANDÉ: Interviews I and II (AHQP); M. BORN: *My Life and My Views* (New York, 1968), p. 9; *Physik im Wandel meiner Zeit* (Braunschweig, 1966[4]), p. 283-284.
[157] M. BORN and A. LANDÉ: *Berechnung der Kompressibilität* [154]; P. FORMAN: *Alfred Landé and the anomalous Zeeman effect, 1919-1921, Hist. Stud. Phys. Sci.*, **2**, 153-261 (1970), on p. 163.
[158] A. LANDÉ: *Dynamik der raumlichen Atomstruktur, Verh. deut. Phys. Ges.*, **21**, 2-12, 644-652, 653-663 (1919).
[159] P. FORMAN [157], p. 165.

[160] RUTHERFORD to BOLTWOOD, 5 Dec. 1905, edited by L. BADASH, *Rutherford and Boltwood. Letters on Radioactivity* (New Haven, 1969), p. 107.

[161] See *Discussion on the structure of atoms and molecules, Brit. Ass. Reports* (1914), p. 293-301.

[162] G. N. LEWIS: *The atom and the molecule, Journ. Am. Chem. Soc.*, **38**: **1**, 762-785 (1916); I. LANGMUIR: *The arrangement of electrons in atoms and molecules, Journ. Am. Chem. Soc.*, **41**: **1**, 868-934 (1919). See R. E. KOHLER, jr.: *The origin of G. N. Lewis' theory of the shared pair bond, Hist. Stud. Phys. Sci.*, **3**, 343-376 (1971).

[163] A. SOMMERFELD: *Atombau und Spektrallinien* (Braunschweig, 1921²), p. 222-239.

[164] A. SOMMERFELD [143].

[165] A. LANDÉ: *Bemerkung über die Grösse der Atome, Zeits. Phys.*, **2**, 87-89 (1920).

[166] A. SOMMERFELD [143].

[167] E. SCHRÖDINGER: *Versuch der modellmässigen Deutung des Terms der scharfen Nebenserien, Zeits. Phys.*, **4**, 347-354 (1921).

[168] BOHR to SCHRÖDINGER, 15 June 1921 (AHQP), acknowledging a reprint of *op. cit.* [167]: « Ich habe übrigens selbst vor längerer Zeit genau dieselbe Überlegung ausgestellt und entspechende Berechnungen durchgeführt ». SCHRÖDINGER replied, 28 June 1921 (AHQP), very pleased that his treatment agreed with Bohr's, « denn ich nehme wohl mit Recht an, dass Sie für die Wahrscheinlichkeit dieser Vorstellung viel allgemeinere Gründe haben als ich ». Bohr's double role as leader and magus of the atomic physicists comes through clearly.

[169] N. BOHR: *Atomic structure, Nature*, **107**, 104-107 (1921).

[170] N. BOHR [103].

[171] N. BOHR: *Om brintspektret, Fys. Tids.*, **12**, 1-18 (1913).

[172] H. A. KRAMERS: *Intensities of spectral lines, Kong. dansk. vidensk. selsk. math.-fys. medd.*, **8**: **3**, 287-385 (1919).

[173] N. BOHR: *Atomic structure, Nature*, **108**, 208-209 (1921).

[174] A. SMEKAL: *Über die Erklärung der Röntgenspektren und die Konstitution der Atome, Phys. Zeits.*, **22**, 400-402 (1921). Cf. SOMMERFELD to EINSTEIN, 14 March 21, in *Briefwechsel* [140], p. 79.

[175] N. BOHR: *Der Bau der Atome und die physikalischen und chemischen Eigenschaften der Elemente, Zeits. Phys.*, **9**, 1-67 (1922); an Ausarbeitung of the Göttingen lectures is preserved in AHQP.

[176] N. BOHR: *The structure of the atom, Nature*, **112**, 1-16 (7 July 1923: Supplement).

[177] G. URBAIN: *Les numéros atomiques du néo-ytterbium, du lutecium et du celtium, Compt. Rend.*, **174**, 1349-1351 (1922); cf. HEILBRON: *Moseley* [79], p. 131-139.

[178] A. VON WELSBACH: *Notiz über die Elemente des Thuliums, Sitzungsb. Wien Ak.*, **120**: **26**, 193-195 (1911).

[179] BOHR to COSTER, 3 July 1922 (AHQP).

[180] G. VON HEVESY: *Historical notes on the discovery of hafnium, Archiv för Kemi*, **3**, 543-548 (1951).

[181] *Ibid.*, and G. VON HEVESY: *Chemical Analysis by X-Rays and its Application* (New York, 1932), p. 278-279.

[182] W. PAULI: *Collected Scientific Papers*, edited by R. KRONIG and V. WEISSKOPF, Vol. **2** (New York, 1964), p. 1080-1096; JAMMER [139], p. 135-136.

[183] E. A. STONER: *The distribution of electrons among atomic levels, Phil. Mag.*, **48**, 719-736 (1924).

[184] A. SOMMERFELD: *Allgemeine spektroscopische Gesetze, insbesondere ein magneto-opticher Zerlegungssatz, Ann. Phys.*, **63**, 221-263 (1920), on p. 231-234.
[185] A. SOMMERFELD: *Zur Theorie des periodischen Systems, Phys. Zeits.*, **26**, 70-74 1925), quoted by JAMMER [139], p. 141.
[186] J. J. THOMSON: *The Atomic Theory* (Oxford, 1914), p. 24.
[187] J. J. THOMSON: *On the origin of spectra and Planck's law, Phil. Mag.*, **37**, 419-446 (1919), on p. 420.
[188] *Ibid.*
[189] C. HUYGENS: *Treatise on Light* (1690), tr. S. P. THOMPSON (Chicago, 1945), p. 3.

# Recollections of an Exciting Era.

P A. M. Dirac
*Florida State University - Tallahassee, Fla.*

## I.

I am very glad I came to Varenna to attend this summer school. I have learned a great deal here, not only individual facts about the history of science, which I have picked up from various lectures, but I have learned to appreciate the point of view of the historian of science. It is really a very different point of view from that of the research physicist. The research physicist, if he has made a discovery, is then concerned with standing on the new vantage point which he has gained and surveying the field in front of him. His question is, where do we go from here? What are the applications of this new discovery? How far will it go in elucidating the problems which are still before us? What will be the prime problems now facing us?

He wants to rather forget the way by which he attained this discovery. He proceeded along a tortuous path, followed various false trails, and he does not want to think of these. He feels perhaps a bit ashamed, disgusted with himself, that he took so long. He says to himself: What a lot of time I wasted following this particular track when I should have seen at once that it would lead nowhere. When a discovery has been made, it usually seems so obvious that one is surprised that no one had thought of it previously. With that point of view, one does not want to remember all the work that led up to the making of the discovery.

Now, that is just the opposite to what the historian of science wants. He wants to know the various influences at work, the various intermediate steps, and he may even have some interest in the false trails. These are quite contradictory points of view. Most of my life has been spent with the point of view of the research physicist, and that involves forgetting as quickly as possible the various intermediate steps.

However, with the understanding of what the historians of science are concerned with, I have tried to think over the past, and have done my best to remember the various incidents, things that happened 50 years ago. I have tried to figure out the influences, the effect of the various teachers that I had and the training that I received, to see how these things led me to the style of work which I followed in later life.

I have given as the title « Recollections of an Exciting Era », and this era should be counted as beginning in 1919. At that time, a wonderful thing happened. Relativity burst upon the world, with a tremendous impact. Suddenly everyone was talking about relativity. The newspapers were full of it. The magazines also contained articles written by various people on relativity, not always for relativity but sometimes against it. Relativity was understood in a very wide sense, and was taken up by philosophers and by people in all walks of life.

It is easy to see the reason for this tremendous impact. We had just been living through a terrible and very serious war. It was also in some ways a rather dull war. The overall picture did not change very much, except for heavy casualties which we were continually reading about. The position of the front line changed very little with the various attacks, just maybe advanced or receded a few hundred yards, and that was all.

Then this terrible war came to an end rather suddenly. The result was that everyone was sick and tired of the war. Everyone wanted to forget it. And then relativity came along as a wonderful idea leading to a new domain of thought. It was an escape from the war. The impact that relativity produced I think has never been equalled either before or since by any scientific idea catching the public mind.

This impact of relativity involved simultaneously the special theory and the general theory. Now, the special theory was actually very much older, dating from 1905, but no one knew anything about it except a few specialists in the universities. The ordinary person had never heard of EINSTEIN. Suddenly EINSTEIN was on everyone's lips. EINSTEIN was rather a remote figure in a foreign country. A person who was much more present was EDDINGTON. He was the leader of relativity in England at that time. He was the great authority whom everyone listened to with the greatest respect, and he was rather regarded as the chief exponent of relativity. EINSTEIN was in the remote background.

At this time I was a student of engineering at Bristol University, and of course I was caught up in this excitement produced by relativity. We discussed it very much. The students discussed it among themselves, but had very little accurate information to go on. Relativity was a subject that everybody felt himself competent to write about in a general philosophical way. The philosophers just put forward the view that everything had to be considered relatively to something else, and they rather claimed that they had known about relativity all along.

You can get a sample of what the various articles on relativity were like from what HOLTON told us (see p. 266). He actually read out an extract of some writings of Sir Oliver LODGE. Sir Oliver LODGE was rather critical of relativity. But you see, with writings of that style, there is nothing very precisely stated, so we engineering students were caught up in a discussion of a question about which we had very little hard and fast facts.

I first got some accurate information about relativity through attending a course of lectures by BROAD. Now, BROAD was a philosopher who looked at things from the philosophical point of view. He was a lecturer in Bristol University in philosophy at that time. He later became a professor in Cambridge and died just a few years ago. He gave a course of about ten or twelve lectures on relativity, discussing it from the point of view of philosophy. Several of the engineering students attended his course at the beginning, but they rather dropped out. I stayed on till the end. I tried very hard to understand philosophy. The engineering students with me had a very practical outlook and they said that these philosophical questions are of no use to an engineer, so they stopped coming. However, I thought there might be something in philosophy, and I did my best to understand the philosopher's point of view. I had also read a little about philosophy. I had read all through Mill's book on logic.

However, my attempts to appreciate philosophy were not very successful. I felt then that all the things that philosophers said were rather indefinite, and came to the conclusion eventually that I did not think philosophy could contribute anything to the advance of physics. I did not immediately have that point of view, but I came to it only after a lot of thought, and studying what philosophers said, in particular BROAD.

Now, I not only heard from BROAD what the general outlook of the philosopher was; he gave some exact information about special relativity, also about general relativity. I remember (I think it was in the second or third lecture) he wrote a formula on the blackboard

$$ds^2 = dx^2 + dy^2 + dz^2 - c^2 dt^2 . \tag{1}$$

Now, when I saw that minus sign, it produced a tremendous effect on me. I immediately saw that here was something new. Perhaps I can explain the reason for this big effect from the fact that previously as a schoolboy I had been much interested in the relations of space and time. I had thought about them a great deal, and it had become apparent to me that time was very much like another dimension, and the possibility had occurred to me that perhaps there was some connection between space and time, and that we ought to consider them from a general four-dimensional point of view.

However, at that time the only geometry that I knew was Euclidean geometry, and, if space and time were to be coupled in any way, they would have to be coupled with a plus sign here, and it was very easy to see that that would not work, and led to nonsense as soon as one tried to make any big change in one's time axis.

I could perhaps explain at this point that I was always very much interested in geometry. That was the branch of mathematics which fascinated me. You can divide all mathematicians into two classes, those whose main interest is geometry, those whose main interest is algebra. This division into two classes is also largely a racial division. People with European training, European background, tend to be interested in geometry, following from the old Greek school. Those whose main interest is in algebra tend to be Asiatic people, following from the discovery of algebra by the Arabs.

Now, a good mathematician has to be a master both of geometry and of algebra, and he has to be able to pass from one to the other quite freely according to the nature of the problem that he is working on. But still he will always keep his preference for one kind of thought, and my preference was strongly on the side of geometry, and has always remained so.

Well, this formula that BROAD wrote on the board of course gave me a new insight into geometry. I remember that previously, when I was at school, one of my mathematics masters had told me that I would probably be very interested in non-Euclidean geometry, and suggested books which I should read on the subject. However, I was not interested in it. The reason for that is that I was interested in the real physical world, and it seemed to me to be obvious that the real physical world was based on Euclidean geometry. There was therefore no need to consider any other kind of geometry. I was not interested just in logical developments, or in seeing what the possibilities are when one follows from a different set of axioms. That again is a trend which has stayed with me all through my life. I have been interested in the real physical world, and not in questions just of logic. This being interested in the real physical world was of course just what was supported by the engineering training which I received.

When I got this new outlook from the formula which BROAD wrote on the board, I was soon able to figure out by myself the basic relations of special relativity.

I completed my course in engineering, and I would like to try to explain the effect of this engineering training on me. I did not make any further use of the detailed applications of this work, but it did change my whole outlook to a very large extent. Previously, I was interested only in exact equations. It seemed to me that if one worked with approximations there was an intolerable ugliness in one's work, and I very much wanted to preserve mathematical beauty. Well, the engineering training which I received did teach me to tolerate approximations, and I was able to see that even theories based on approximations could sometimes have a considerable amount of beauty in them. A problem like arranging the windings in the rotor of a dynamo involved some mathematics. It was a mathematics of whole numbers, but there was quite a bit of beauty in it.

There was this whole change of outlook, and also another change of outlook which was perhaps brought on by the theory of relativity. I had started off believing that there were some exact laws of Nature and that all that we had to do was to work out the consequences of these exact laws. Typical of the exact laws were Newton's laws of motion. Now we learned that Newton's laws of motion were not exact, only approximations, and I began to infer from that that maybe all the laws of Nature are only approximations. I was quite prepared at that time to consider all our equations as only approximations representing our present state of knowledge, and to take it as a task to try to improve on them.

I think that if I had not had this engineering training, I should not have had any success with the kind of work that I did later on, because it was really necessary to get away from the point of view that one should deal only with exact equations, and that one should deal only with results which could be deduced logically from known exact laws which one accepted, in which one had implicit faith. Engineers were concerned only with getting equations which were useful for describing nature. They did not very much mind how the equations were obtained. Once they got them, they proceeded to use them with their slide rules, and get results which were necessary for their work.

And that led me of course to the view that this outlook was really the best outlook to have. We wanted a description of Nature. We wanted to find equations which would describe Nature, and the best we could hope for was usually approximate equations, and we would have to reconcile ourselves to an absence of strict logic and confine ourselves to trying to get the equations which really succeeded in describing Nature.

After my engineering course was finished, I continued at Bristol University for another two years studying mathematics. During this mathematical training, the man who influenced me most was FRASER. FRASER was a mathematician who never did any research, never published anything, but he was a wonderful teacher, able to inspire his students with real excitement about basic ideas in mathematics. I think he's pretty well unheard-of now, except that when he died HODGE wrote an obituary notice of him, *Journal London Math. Soc.* **34**, 111 (1959). He paid tribute to FRASER as being a really great teacher.

There were two things that I learnt from FRASER. One of them was rigorous mathematics. Previously I had been using only the nonrigorous mathematics which engineers were satisfied with. They just wanted to get practical results. They were not concerned with the exact definition of a limit, or how long to sum a series, and things like that. Well, FRASER taught that it was sometimes necessary to have strict logical ideas for dealing with such subjects.

However, I continued in my later work to use mostly the nonrigorous mathematics of the engineers, and I think that you will find that most of my later writings do involve nonrigorous mathematics. When I introduce a function, I do not stop to say whether it is continuous or differentiable or put down all the other conditions which a pure mathematician wants before he can make any statements about it. I just take the point of view that this is the sort of function that a physicist is interested in.

However, this kind of nonrigorous mathematics does not always work. There were a few occasions when I have been sharply stopped by the need to consider more accurately definitions, and possible sources of error, which the nonrigorous mathematics might lead one into.

The second thing I learned from FRASER was projective geometry. Now, that had a profound influence on me because of the mathematical beauty involved in it. There was also very great power in the methods employed. I think probably most physicists know very little about projective geometry. That I would say is a failing in their education. Projective geometry always deals with flat space, but it is a most powerful tool for dealing with flat space, and it provides you with methods, such as the method of one-to-one correspondences, which give you results apparently by magic; theorems in Euclidean geometry which you have been worrying about for a long time drop out by the simplest possible means when one uses the arguments of projective geometry.

Now, I was always much interested in the beauty of mathematics, and this introduction to me of projective geometry stimulated me very much and provided, I would say, a lifelong interest.

You might think that projective geometry is not of much interest to a physicist, but that is not so. Physicists nowadays are concerned very largely with Minkowski space. Now, if you want to picture relationships in Minkowski space, relationships between vectors and tensors, often the very best way to do it is by using the notions of projective geometry. I was continually using these ideas of projective geometry in my research work. When you want to discover how a particular quantity transforms under a Lorentz transformation, very often the best way of handling the problem is in terms of projective geometry.

It was a most useful tool for research, but I did not mention it in my published work. I do not think I have ever mentioned projective geometry in my published work (but I am not sure about that) because I felt that most physicists were not familiar with it. When I had obtained a particular result, I translated it into an analytic form and put down the argument in terms of equations. That was an argument which any physicist would be able to understand without having had this special training.

However, for the purposes of research, when one is entering into a new field and one does not know what lies in front of one, one wants very much to visualize the things which one is dealing with, and projective geometry does provide the best tool for this.

That applied also to my later work on spinors. One had quite a new kind of quantity to deal with; but for discussing the relationships between spinors, again, the ideas of projective geometry are very useful.

Well, I spent two years studying mathematics at Bristol, and then proceeded to Cambridge as a research student. Research students in Cambridge were each provided with a supervisor, someone who had to look over their work, suggest problems to them, and in general take an interest in their work and help them along with it.

My supervisor was R. H. FOWLER, another man who had a great influence on me. At first I was slightly disappointed when FOWLER was appointed my supervisor. The reason for that was that my main interest was on the geometrical side, and especially on relativity. Now, FOWLER was not concerned with relativity. There was CUNNINGHAM, who was an expert in special relativity. He had written a book about it in 1910. But he did not want to take on any more research students, and so I was passed on to FOWLER.

I soon found out that this shade of disappointment that I first felt was quite unjustified. FOWLER introduced me to quite a new field of interest, namely the atom of RUTHERFORD, BOHR and SOMMERFELD. Previously I had heard nothing about the Bohr theory. It was quite an eyeopener to me. I was very much surprised to see that one could make use of the equations of classical electrodynamics in the atom. The atoms were always considered as very hypothetical things by me, and here were people actually dealing with equations concerned with the structure of the atom.

I was very quickly plunged into the center of the problems concerning the explanation of atoms. The most profound problem of all was why are the electron orbits stable? Why do the electrons not just fall down into the nucleus like they should, according to classical mechanics?

I proceeded to think over these problems as hard as I could, and really spent my whole time on that kind of work, and also other mathematical work.

I kept up my interest in relativity. I studied Eddington's classical book, *Mathematical Theory of Relativity*, and found it rather tough at first but eventually mastered it. I was very happy to have EDDINGTON actually present, and sometimes I met him and had some discussion with him on the question of kinematic and dynamical velocity, which led to a little note of mine published in the *Philosophical Magazine*. It was really a wonderful thing for me to meet the man who was the fountainhead of relativity so far as England was concerned. EINSTEIN was just too remote to count.

Another mathematical activity that I had was provided by BAKER, the professor of geometry at Cambridge at that time. I did not go to any of Baker's lectures. I went to other lectures, by CUNNINGHAM and FOWLER, but I was not sufficiently concerned with geometry to go to Baker's lectures. However, on Saturday afternoons BAKER held a tea party to which I was invited, and at the end of the tea party someone would give a talk on some subject of geometry. It was always the geometry of flat space. It was always handled by the methods of projective geometry. Everyone considered then that projective geometry was the only kind of geometry worth studying. It was so much more powerful than the geometry in which one was confined to Euclid's axioms. They worked very often in a number of dimensions more than three, four, five or six, and studied the various figures which one could construct in these spaces of a higher number of dimensions, and I was much impressed by the power of their methods. By studying these figures in spaces of higher dimensions, one was often able to get quick proofs of results in the ordinary three-dimensional Euclidean space, results which would be very tedious to get by other methods.

These tea parties did very much to stimulate my interest in the beauty of mathematics. The all-important thing there was to strive to express the relationships in a beautiful form, and they were very successful. I did some work on projective geometry myself and gave one of the talks at one of the tea parties. This was the first lecture I ever gave, and so of course I remember it very well. I dealt with a new method for handling these projective problems.

Well, those were the influences which I had as my background. I was, effectively, very much in the world center for the development of atomic theory. I should say that BOHR was very friendly to RUTHERFORD all his life. He continually came to Cambridge and gave us lectures. Also FOWLER often went to Copenhagen and learned about the latest situation from there and of course kept me informed about it. In fact, FOWLER spent the winter of 1925, three months, in Copenhagen.

I also had the benefit of hearing the workers in the Cavendish speaking about their experimental work. There were RUTHERFORD, ASTON, WILSON, and so on, and I learnt to appreciate something of the problems of experimentalists.

I would like to say something about BOHR, when he came to lecture in Cambridge. He impressed everyone very much with his deep style of thinking. When he gave a lecture, he would always begin at the beginning, with his work leading to the explanation of the Balmer formula, and would proceed to develop his ideas on that basis. He spoke slowly, and of course it took him a long time to get to a more modern outlook which was the goal of his lecture. The result was that his lectures lasted usually for two hours and maybe more, but everybody listened with the greatest attention. People were pretty well spellbound by what BOHR said.

Of course, it was necessary to pay the greatest attention because he spoke in a quiet voice. Microphones were not in use at that time. One just had to strain one's hearing in order to find out what he said.

While I was very much impressed by what BOHR said, his arguments were mainly of a qualitative nature, and I was not able to really pinpoint the facts behind them. What I wanted was statements which could be expressed in terms of equations, and Bohr's work very seldom provided such statements. I am really not sure how much my later work was influenced by these lectures of Bohr's. That's something I cannot answer. He certainly did not have a direct influence, because he did not stimulate one to think of new equations.

At that time, I was just a research student with no duties apart from research, and I concentrated all my energy in trying to get a better understanding of the problems facing physicists at that time. I was not interested at all in politics, like most students nowadays. I confined myself entirely to the scientific work, and continued at it pretty well day after day, except on Sundays when I relaxed and, if the weather was fine, Itook a long solitary walk out in the country. The intention was to have a rest from the intense studies of the week, and perhaps to try and get a new outlook with which to approach the problem the following Monday. But the intention of these walks was mainly to relax, and I just had the problems maybe floating about in the back of my mind without consciously bringing them up.

That was the kind of life that I was leading. There was excitement produced from time to time. There was the excitement of the Bohr-Kramers-Slater theory. That provided a new outlook, and it seemed to me to be a very reasonable outlook. With the backing of BOHR behind it, it seemed to me that here was a theory that was certainly worth considering. It meant giving up detailed conservation of energy, but I did not especially mind that. Conservation of energy had only been proved statistically. Here was a way which did seem to provide an escape from some of the fundamental difficulties concerned with understanding radiation. Radiation was emitted continuously in waves, absorbed suddenly in the form of quanta—a picture which people had not previously thought of, but which did account for all the experimental evidence available at the time.

The satisfaction which we had with the theory of Bohr-Kramers-Slater was short lived, because within a year, GEIGER and BOTHE had made precise experiments with the scattering of X-rays by electrons, and had checked that conservation of energy did hold in detail. So that was just a passing interest which faded out.

I might mention that the Bohr-Kramer-Slater idea was revived in 1936 by SHANKLAND, who did experiments like the Geiger-Bothe experiments using gamma-rays instead of X-rays, and SHANKLAND reported that, in the case of gamma-rays, there was no detailed conservation of energy.

Well, at that time I had a great respect for experimenters and when they asserted something confidently, I had a tendency to believe them, and I rather accepted Shankland's point of view. I tried to think how it could be that conservation of energy broke down in the case of high energies, even though it held for lower energies, and I was even sufficiently interested in it to write an article in *Nature* about it.

However, within a year SHANKLAND repeated his experiments and he reported that his earlier results were inaccurate and there was detailed conservation of energy, so we were back again to the point of view of the precise quantum theory preserving accurately conservation of energy.

Another of these early ideas was provided by DE BROGLIE. He put forward a theory connecting particles with waves. Now, this was a very beautiful theory because the connection was relativistic, and the beauty appealed to me immediately. The connection is such that, when you make the rest mass of the particles go to zero, you get the relationship between light quanta and electromagnetic waves.

Although I appreciated very much the beauty of de Broglie's work, I could not take his waves seriously. I was so much imbedded in the Bohr theory, I took the Bohr orbits very literally—we had the electrons as real particles, and the de Broglie waves seemed to me to be just a mathematical fiction of no importance to physicists.

Of course, I was very wrong in this point of view. SCHRÖDINGER also read de Broglie's work. He had a different outlook. He also had a different training. He had a training involving very much eigenvalues and eigenvectors, which I knew nothing about. And SCHRÖDINGER with his different outlook was able to develop de Broglie's ideas and get to a brilliant result. He was able to extend the primitive ideas of DE BROGLIE, which applied only to free particles, to particles moving in an electromagnetic field, and that led to wave mechanics.

Well, that was one of the examples where I was seriously at fault.

You might wonder, what was the sort of work I was doing on my own trying to solve problems of physics? Well, I was very much concerned with understanding Hamiltonian dynamics. I had read Sommerfeld's book *Atomic Structure and Spectral Lines*. There was an English translation available, which was rather fortunate because I was not very fluent in German at that time, and that provided the basis for getting a working knowledge of atomic theory. There was an appendix to Sommerfeld's book dealing with Hamiltonian dynamics and its applications to quantum theory. I studied this Hamiltonian theory very closely, and I also did some other reading about Hamiltonian dynamics. I read up about the advanced transformation theory connected with Hamiltonian dynamics, and became familiar, or at any rate acquainted, with the general ideas of this theory.

There were many meetings among the students in Cambridge to discuss scientific problems, and among those there was the Kapitza Club. KAPITZA was a young physicist who had come from Russia. He was very talented and RUTHERFORD appreciated his talents and helped him to get established in Cambridge. KAPITZA also had a very dynamic personality and he established a club of physicists, theoretical as well as experimental. We would meet on Tuesday evenings after dinner when someone would read a paper on some recent development in physics.

That was not really a very convenient time for me because I was usually rather sleepy after dinner. I did my work mostly in the morning. Mornings I believe are the times when one's brain power is at its maximum, and towards the end of the day I was more or less dull, especially after dinner. I was not in the best frame of mind for taking in new information. But still it was well worth-while going to these meetings of the Kapitza Club.

In the Summer of 1925, HEISENBERG came to Cambridge, and he gave a talk to the Kapitza Club. The main subject of his talk was « Anomalies in the Zeeman Effect », and I followed most of it. Towards the end, though, he spoke about some new ideas of his. By that time I was just too exhausted to be able to follow what he said, and I just did not take it in. He was talking about the origins of his ideas of the new mechanics. But I completely failed to realize that he was really introducing something quite revolutionary. Later on I completely forgot what he had said concerning his new theory. I even felt rather convinced that he had not spoken about it at all, but other people who were present at this meeting of the Kapitza Club assured me that he had spoken about it. In particular FOWLER was quite positive, and I just have to accept that he really did speak about it and that I had failed to respond to it at all, and so missed a great opportunity of getting started on it.

It was a little later when I really got started on the new Heisenberg theory. Toward the end of August I returned to Bristol to be with my parents for part of the vacation, and HEISENBERG sent to FOWLER the proofs of his first paper on the new mechanics. FOWLER sent it on to me with a query, « What do you think of this? »

I received it either the end of August or the beginning of September, I am not sure of the date, and of course I read it. At first I was not very much impressed by it. It seemed to me to be too complicated. I just did not see the main point of it, and in particular his derivation of quantum conditions seemed to me to be too far fetched, so I just put it aside as being of no interest. However, a week or ten days later I returned to this paper of Heisenberg's and studied it more closely. And then I suddenly realized that it did provide the key to the whole solution of the difficulties which we were concerned with.

My previous work had been all concerned with studying individual states. If you want to know about the false trails of work that I had been engaged on, one of them was that I had examined the Hamiltonian theory of planetary perturbations and wondered whether such a theory could not be applied to the interaction of the various electrons in a Bohr atom. That work gave me a mastery of Hamiltonian methods, but of course it did not lead anywhere.

HEISENBERG brought out the quite new idea that one had to consider quantities associated with two stationary states rather than just one.

I think I had better stop at this point, and take it up later.

## II.

I told you yesterday about the background and training which I had for approaching the problems of the new mechanics. The background was essentially an intense interest in mathematical beauty which had been especially built up through studying projective geometry. Also I was very much interested in relativity, which was a new subject which had just recently appeared and was absorbing everyone's interest. And then there was quantum mechanics, which was full of problems and difficulties, which I had thought about intensively without making any real progress. I do not think I would ever have made any progress in studying atomic theory if it had not been for HEISENBERG. I was so much attached to the Bohr orbits. It needed quite a different kind of intelligence to be able to break away from just building up theories in terms of Bohr orbits.

My work during the first two years at Cambridge, which was before Heisenberg's theory appeared, was very much concerned with relativity. I did not point this out sufficiently yesterday. There was a sort of general problem which one could take, whenever one saw a bit of physics expressed in a nonrelativistic form, to transcribe it to make it fit in with special relativity. It was rather like a game, which I indulged in at every opportunity, and sometimes the result was sufficiently interesting for me to be able to write up a little paper about it.

That was the situation up to September 1925, when I had the opportunity of studying Heisenberg's first paper. As I said yesterday, my first reaction was unfavorable, and it needed about ten days or so before I was really able to master it. And I suddenly became convinced that this would provide the key to understanding atoms.

Now, what did I do under those conditions? You will perhaps be able to guess. I was dissatisfied with the nonrelativistic form of Heisenberg's work, and I wondered whether I could transcribe it so as to fit it in with special relativity by the same sort of arguments which I had used on various occasions previously.

The special feature of Heisenberg's work that jarred against relativity was that he had a number of matrix elements forming the building bricks of his theory, and each of them was associated with two energies. It was associated with two states, referring to two energy levels.

The obvious step to take from that was to suppose that, if these matrix elements each referred to two energies, they would each refer to two momentum values. Now, if you had these matrix elements referring quite generally to two energies and two momenta, that gave you a situation where you could not combine them very reasonably. The whole structure was too loose. So it occurred to me that you would have to put some restriction on the two momentum values. The natural restriction to take was that the difference between the two momentum values would be equal to the difference between the two energy values, divided by $c$, the velocity of light, and that the direction of this momentum difference would be the same for all the matrix elements.

That would mean that these matrix elements were all connected with light moving in one particular direction. This was a rather artificial situation, but still, from the relativistic point of view, it was less artificial than having all the matrix elements referring just to different energies in one particular Lorentz frame of reference and all referring to no momentum change.

Well, there was a definite idea which I could start to work on, and I proceeded to write it up, but I never got very far with it. I suppose I soon saw that this was not the essential problem, and this work I dropped.

These ideas did reappear about a year later in some work that I wrote on relativistic quantum mechanics with an application to Compton scattering. This contained the essence of these early ideas of matrix elements each referring to two energy levels and two momentum values, with a momentum difference equal to the energy difference divided by the velocity of light.

I was in Bristol at the time I started on Heisenberg's theory. I had returned home for the last part of the summer vacation, and I went back to Cambridge at the beginning of October, 1925, and resumed my previous style of life, intense thinking about these problems during the week and relaxing on Sunday, going for a long walk in the country alone. The main purpose of these long walks was to have a rest so that I would start refreshed on the following Monday.

The question that was bothering me was of course the noncommutation of the dynamical variables. HEISENBERG had set up a theory where the dynamical variables correspond to matrices, and they are such that, for two variables $u$ and $v$, if you multiply them in that order to get $uv$, the result is not the same as if you multiply them in the reverse order to get $vu$. There is a difference $uv - vu$ which was very hard to understand.

I heard later on that HEISENBERG himself was extremely worried when he first noticed that $uv$ was not the same as $vu$. He must have found it out pretty soon, and he was naturally very disturbed by it because it is a result that is completely foreign to physicists. They had been brought up to deal with physical problems on the basis of Newton's laws, and everything that followed from that, always assuming that the product of dynamical variables was commutative. When HEISENBERG first noticed this noncommutation, he felt afraid that this was a fatal objection to his theory and that he would have to abandon it. I think he needed quite a bit of support from his professor BORN to carry on in spite of this really frightening disturbance, coming essentially from the revolutionary character of the new ideas which he was bringing forward.

Of course, I did not have this fear of the whole theory collapsing which HEISENBERG must have had to begin with, so I was able to approach the whole question more boldly, and I realized pretty soon that the noncommutation was really the most important feature of the new theory. Just what disturbed HEISENBERG so much was really the main new feature of the theory, and it is what one had to try to understand. In trying to understand this, of course, I had as background the Hamiltonian form of dynamics with which I was familiar.

It was during one of the Sunday walks in October, 1925, when I was thinking very much about this $uv - vu$, in spite of my intention to relax, that I thought about Poisson brackets. I remembered something which I had read up previously in advanced books of dynamics about these strange quantities, Poisson brackets, and from what I could remember, there seemed to be a close similarity between a Poisson bracket of two quantities, $u$ and $v$, and the commutator $uv - vu$.

The idea first came in a flash, I suppose, and provided of course some excitement, and then of course came the reaction « No, this is probably wrong. »

I did not remember very well what a Poisson bracket was. I did not remember the precise formula for a Poisson bracket, and only had some vague recollections. But there were exciting possibilities there, and I thought that I might be getting on to some big new idea. It was really a very disturbing situation, and it became imperative for me to brush up my knowledge of Poisson brackets and in particular to find out just what is the definition of a Poisson bracket.

Of course, I could not do that when I was right out in the country. I just had to hurry home and see what I could then find about Poisson brackets. I looked through my notes, the notes that I had taken at various lectures, and there was no reference there anywhere to Poisson brackets. The textbooks which I had at home were all too elementary to mention them. There was just nothing I could do, because it was a Sunday evening then and the libraries were all closed. I just had to wait impatiently through that night without knowing whether this idea was really any good or not, but still I think that my confidence gradually grew during the course of the night. The next morning I hurried along to one of the libraries as soon as it was open, and then I looked up Poisson brackets in Whittaker's *Analytical Dynamics*, and I found that they were just what I needed. They provided the perfect analogy to the commutator.

The precise formula for a Poisson bracket is the following:

$$[u, v] = \sum_r \left( \frac{\partial u}{\partial q_r} \frac{\partial v}{\partial p_r} - \frac{\partial u}{\partial p_r} \frac{\partial v}{\partial q_r} \right) . \tag{2}$$

The $q$'s and the $p$'s there are a set of Hamiltonian variables describing the dynamical system, and the sum is taken over all the degrees of freedom. You see that it is a rather complicated formula, and I think I can be excused for not having it clearly in my mind during my walk, because there was no reason to believe that it was of any importance. It was just something which had appeared in the books when I was reading about the transformation theory of dynamics, and I had not bothered to learn the details by heart.

The situation was really rather confused by the fact that there was another kind of bracket expression also occurring in this theory, the Lagrange bracket. The Lagrange bracket is very similar to the Poisson bracket in general appearance, but has of course a different significance in the details. The way things have turned out, the Poisson bracket is of very great importance and the Lagrange bracket is of no importance at all, from the point of view of people studying quantum theory.

This idea of connecting Poisson brackets with commutators formed the beginning of my work on the new quantum mechanics. The connection between these two things, which look very different, is very close when one examines their properties. It is so close that one finds that one only needs to put in a suitable numerical coefficient, $ih/2\pi$, and then one can say this quantity $uv - vu$ is the analog in quantum mechanics for the quantity

$$\frac{ih}{2\pi}\sum_r\left(\frac{\partial u}{\partial q_r}\frac{\partial v}{\partial p_r}-\frac{\partial u}{\partial p_r}\frac{\partial v}{\partial q_r}\right) \tag{3}$$

in classical mechanics.

The importance of this step is that it provides us with some way of handling the dynamical variables in the quantum theory which takes the place of the partial differentiations which we have in classical mechanics. In classical mechanics we have our dynamical variables which we can add and multiply; we can do the same with our quantum variables. In classical mechanics we have differentiation with respect to the time $t$. We can take this process over directly into the quantum theory by supposing that the quantum variables like $u$ and $v$ are functions of a time parameter $t$. But there is no immediate way of taking over processes of partial differentiation into the quantum theory until one sets up this analogy. Then, whenever one has to do a partial differentiation with respect to some variable in classical mechanics, there is a corresponding process which one can do in the quantum theory of taking the commutator of the quantity with a certain other dynamical variable.

It was this result—that we have a process of differentiation of variables applicable to quantum mechanics—which appeared to me to be the main consequence of the work, and so I started to consider the problem again from the general point of view of just looking for a process of differentiation which we can apply to quantum variables.

Let us forget this formula connecting the commutator and the Poisson bracket and start again, and look for a process of differentiation which can be applied to quantum variables.

We have a quantum variable $x$. Let $v$ be some other quantum variable. These quantum variables are both represented by matrices in accordance with Heisenberg's ideas. What can one do to give a meaning to $\mathrm{d}x/\mathrm{d}v$?

Well, first of all one should require $\mathrm{d}x/\mathrm{d}v$ to be a linear function of $x$. This requires that the matrix elements of $\mathrm{d}x/\mathrm{d}v$ shall be linear functions of the matrix elements of $x$. Thus

$$(\mathrm{d}x/\mathrm{d}v)(nm) = \sum_{n'm'} a(nm;\, n'm')x(n'm'), \tag{4}$$

where the $a$'s are unknown coefficients, but they have to be independent of $x$.

Then we can go on to put down the requirement that we differentiate a product $xy$ according to the law

$$\mathrm{d}(xy)/\mathrm{d}v = (\mathrm{d}x/\mathrm{d}v)y + x(\mathrm{d}y/\mathrm{d}v)\,, \tag{5}$$

$y$ being another quantum variable, represented by a matrix.

We find that the conditions which are needed for this to hold are that $\mathrm{d}x/\mathrm{d}v$ must be of the form

$$\mathrm{d}x/\mathrm{d}v = xa - ax\,, \tag{6}$$

where $a$ is another quantum variable represented by a matrix. I then proceeded to examine this equation to see what it corresponds to in the case when we are dealing with large quantum numbers, and was led to the formula connecting the commutator and the Poisson bracket.

I wrote up my first paper on quantum mechanics following those lines. Most of the papers that I wrote followed the lines of presenting the ideas in the order in which they had occurred to me, but here I made an exception. I did not want to write up the paper based primarily on the idea of connecting the commutator and the Poisson bracket, which had come to me rather out of the blue (I could not really say how it came) and I preferred to set up the theory on this basis where there was some kind of logical justification for the various steps which one made. One had the need for getting a process of differentiation into the quantum theory, and this process of differentiation had to satisfy the linearity condition and also the product law (5). On that basis, one could set up an argument which leads to this formula (6), and connect it with the Poisson bracket.

I wrote up that paper. It was communicated by FOWLER to the Royal Society. The Royal Society appreciated the importance of this work and published it very quickly, much more quickly than papers usually took.

I also sent a copy of the paper to HEISENBERG, and I got a fairly prompt answer from him. I would like to tell you about this letter from HEISENBERG and some later letters. That is information that is not available from published work. HEISENBERG wrote to me in German. I knew enough German to be able to follow pretty well what he said. If one makes a rough translation of Heisenberg's letter, it would read like this (*):

« I read your beautiful work with great interest. There can be no doubt that all your results are correct, insofar as one believes in the new theory. » I think that phrase in Heisenberg's letter is a remarkable one, because it shows that HEISENBERG did not really have very much confidence in his theory, or at any rate professed to feel that there still might be doubts about it—he brings in this phrase, « insofar as one believes in the new theory ».

(*) HEISENBERG to DIRAC, 20 November 1925, 2 pages, in German.

« Your representation of the frequency condition and energy law are simpler and more beautiful than the proof of the equation $\nu = \partial H/\partial J$ in the classical theory »—that refers to the connection between frequency and energy which one has in classical Hamiltonian theory.

HEISENBERG goes on: « I hope you are not disturbed by the fact that part of your results have already been found here some time ago and are being published independently in two papers, one by BORN and JORDAN the other by BORN, JORDAN and me. Your results, in particular the general definition of differentiation and the connection of the quantum conditions with the Poisson bracket, go considerably further. I would like to refer to one point which is not important ». He refers now to the quantum condition which I gave in my paper. For a system of one degree of freedom the only quantum condition is

$$2\pi m(q\dot{q} - \dot{q}q) = ih \,. \tag{7}$$

Equating the constant part of the left-hand side to $ih$, one gets Heisenberg's quantum condition, but I said in my paper that, equating the remaining components of the left-hand side to zero, one gets further conditions which are not given by Heisenberg's theory.

That was what HEISENBERG objected to. He said that those further conditions are

$$\frac{\mathrm{d}}{\mathrm{d}t}(pq - qp) = 0 \,. \tag{8}$$

In special cases, this follows from the equations of motion, namely in those cases when $H$ is a function of $q$ plus a function of $p$. He gives details of that calculation, and he says that he believes this result can be proved generally provided $H$ is real.

He says again that this is not an important point. The letter goes on to say « In the work of BORN, JORDAN and me, we tried to give a fairly complete representation of the whole theory, with perturbation theory, degenerate cases, etc. » One can then deduce the Kramers-Heisenberg dispersion formula. He did not put it that way, he said « one can then deduce the Kramers dispersion formula », although everybody calls it the Kramers-Heisenberg dispersion formula.

He said finally that « PAULI has succeeded in getting the theory of the hydrogen atom and the Balmer formula on quantum mechanics. I would willingly send you proofs of this paper and would be glad to hear of your further progress ».

This was really a very kind letter that HEISENBERG wrote to me. I suppose I was a stranger to him at that time. I had seen him when he came to Cambridge and spoke at the Kapitza Club, but I was just a member of the audience. I do not know whether I was actually introduced to him. A lecturer would not normally pay any attention to just one member of the audience unless there was some special reason for it, and at that time there was no reason why I should be specially singled out to be introduced to HEISENBERG.

It was a very kind letter, because he did not want me to be disturbed by the fact that there were other people simultaneously working on these problems who had anticipated my results to some extent, and he was also full of praise for my own contributions.

I got a second letter from HEISENBERG just three days later (*). This second letter was sent off of course before he could have received any answer to his first letter. In his second letter he says « Since my last letter I have had considerable discussion of your work with Jordan, and some questions have come up which I would like you to explain. As a specially satisfying result, I find your general formula for differentiation—that is this formula (6)—and while I have no doubt about the correctness of the result, your proof is not clear to us. You conclude from the linearity condition this equation ». Equation (4). He says, « At first sight, this seems to imply that one could use the same argument for ordinary functions, which would lead to

$$\mathrm{d}x/\mathrm{d}v = ax\,, \qquad x = e^{av}\,, \tag{9}$$

which is not in general true. I do not write this as an objection, but I would like you to give your deduction more explicitly. I also have a wish with respect to a second point. It seems to me that the physical content of the theory is not sufficiently characterized when one says that the mathematical operations which are used to deduce physical results are different from the classical theory. I would rather believe that we really have to do with a change in the kinematics. This brings with it that there can be no question of the validity of classical mechanics. In spite of this, it is possible on the basis of the new kinematics to build up a mechanics which is largely analogous to the classical one, and which allows the energy law and frequency condition to be proved. I do not know if this difference appears to you unimportant, but it seems to me that an important physical point lies in it. »

Well, I cannot remember just what I wrote in answer to these letters of HEISENBERG. It was nearly 50 years ago. I suppose I answered something like this, that there was quite a considerable difference between his argument that led to $x = e^{av}$ and the argument in my paper. With regard to the second point I probably answered that I did think that the argument he gave was rather unimportant because it did not affect the equations.

The letters which I wrote to HEISENBERG on this and later occasions were kept by HEISENBERG until the end of the war, 1945. Then he had all of his important papers put together and they were removed by the American military authorities, and HEISENBERG has been unable to get them back. Presumably these papers are now lying somewhere among the secret files of the American Atomic Energy Commission. Still we must count them as lost for the present. Maybe at some future date they will be unearthed and historians of science

(*) HEISENBERG to DIRAC, 23 November 1925, 2 pages, in German.

will be able to get at them. But for the time being, we must do the best we can without them. I can only imagine what I would have answered to these letters, but of course, with the basis of much greater knowledge at the present time, the answers I imagine might be different from what I actually did answer then.

HEISENBERG wrote to me a third letter on the 1st of December, 1925, and he says (*) « Many thanks for your interesting letter. Unfortunately in my last letter I expressed myself unclearly in some points, and I would like to ask you again about it. My thoughts against the derivation of the formula (4) were intended so. If this formula follows generally from the condition of linearity

$$\frac{\mathrm{d}}{\mathrm{d}v}(x+y)=\frac{\mathrm{d}x}{\mathrm{d}v}+\frac{\mathrm{d}y}{\mathrm{d}v}, \tag{10}$$

then it must apply also when one has only a single stationary state. In fact, nowhere is the assumption made that the number of stationary states must be infinite. Thus for a single stationary state, when $n$ and $m$ and $n'$ and $m'$ are all equal to one, say, one would get the equations

$$\frac{\mathrm{d}x(11)}{\mathrm{d}v}=a(11,11)x(11), \qquad \frac{\mathrm{d}y(11)}{\mathrm{d}v}=a(11,11)y(11). \tag{4'}$$

Now let $x(11)=v^n$, $y(11)=v^m$. Your relations (10) (see above) and (5) are certainly fulfilled, but not (4′) which gives in the first case $a(11,11)=1/nv$ and in the second $a(11,11)=1/mv$, which seems to lead to a contradiction. Please do not think of this as a criticism, but only as a proof that it is difficult to understand your equations without further explanation.

A further point has given me much to think about in the course of the discussions. You write in particular the energy will be the same function of the action variables as in the classical theory. I find it hard to believe that this result is true in general. » And then he proceeds to give a counter-example of the anharmonic oscillator.

Well, the answer to this letter of Heisenberg's I suppose would run something like this: « In the case when there's only one stationary state, noncommutation is impossible, so the process of differentiation breaks down, and that is why the argument, trying to follow through with only one stationary state, leads to a contradiction. »

The second point which HEISENBERG raised in that letter was quite correct. I had been careless in saying that the energy is the same function of $J$ in the classical theory and in the quantum theory, and it was quite correct of HEISENBERG to point out this mistake to me. HEISENBERG actually said that one would not be happy if my statement were true, because then one would have no

---

(*) HEISENBERG to DIRAC, 1 December 1925, 3 pages, in German.

hope to understand the complicated spectra for which the classical calculation of the energy completely fails.

He says at the end of this letter « Please do not take these questions that I write to you as criticisms of your wonderful work. I must now write an article on the state of the theory for the *Mathematischen Annalen*, and still wonder about the mathematical simplicity with which you have overcome this problem. »

This letter was followed by a postcard written on the same day, and he says (*) « In my last letter which I sent to you this afternoon, I forgot to mention a considerable difficulty which has come up in connection with your equation (11). » That is the equation

$$xy - yx = \frac{ih}{2\pi}[x, y]\,. \tag{11}$$

« Let us take the case of one degree of freedom, and take the case when $x$ is equal to $p^2$ and $y$ is equal to $q^2$. Then your equation gives

$$xy - yx = p^2q^2 - q^2p^2 = \frac{ih}{2\pi}[p^2, q^2] = -\frac{ih}{2\pi}4qp\,. \tag{12}$$

But a simple calculation gives

$$p^2q^2 - q^2p^2 = p(pq^2) - (pq^2)p + p(q^2p) - (q^2p)p = \frac{h}{2\pi i}(2pq + 2qp)\,. \text{»} \tag{13}$$

This considerable difficulty that HEISENBERG refers to here is just the fact that the Poisson bracket as worked out in the quantum theory is equal to the classical Poisson bracket only in simple cases; in the more complicated cases, one has to use directly the result that one gets from the commutator rather than taking over the classical formula. In effect, that was the answer that I gave to HEISENBERG with regard to that difficulty.

That was the beginning of quantum mechanics so far as I was concerned. I might mention that, in writing up my paper, I gave careful consideration to the question of notation. I feel that people who are writing up work on a new subject should pay considerable attention to this question of notation, because they are starting something which will very likely get perpetuated, and if they perpetuate a bad notation they are really doing harm to the future development of the subject.

One of the questions of notation which I had to face was with regard to the Poisson bracket. I had got the information about Poisson brackets from Whittaker's *Analytical Dynamics*, and Whittaker's notation had round brackets for them. He used the square notation for the Lagrange brackets. Now, in quantum theory we do not want the Lagrange brackets at all, we only want

(*) HEISENBERG to DIRAC, 1 December 1925, in German.

the Poisson brackets, and it seemed to me that this notation was unsuitable. It makes one think of the scalar product which one has in vector analysis, and the scalar product is symmetrical between the two terms that are mentioned in it, while with the Poisson bracket we have something which is antisymmetrical between the two terms mentioned in it. So I boldly used the other type of bracket for the Poisson bracket, departing from Whittaker's notation. Everyone has copied me since then, and it is really quite suitable to have square brackets always referring to a quantity which is antisymmetrical between the two terms mentioned in it.

Another question of notation—it often happens that $uv = vu$ is a special case. When that is so, the mathematicians who had been handling noncommutative algebra said that $u$ permutes with $v$. It seemed to me that the word « permute » was not really very appropriate. One thinks of permutations as rearranging the order of several quantities, and here we are concerned only with two quantities. So I invented the word « commute. » I do not think it had been previously used in mathematics. I said that, when $uv = vu$, $u$ and $v$ commute with each other. That again is a notation which everyone has accepted since then.

Well, the situation which I had got into now was that I was dealing with these new variables, the quantum variables, and they seemed to me to be some very mysterious physical quantities, and I invented a new word to describe them. I called them $q$-numbers, and the ordinary variables of mathematics I called $c$-numbers, to distinguish them. The letter $q$ stands for quantum, or maybe queer, and the letter $c$ stands for classical or maybe commuting. Then I proceeded to build up a theory of these $q$-numbers; $c$-numbers can be looked on just as a special case of $q$-numbers, having the property that they commute with everything.

Now, I did not know anything about the real nature of $q$-numbers. Heisenberg's matrices I thought were just an example of $q$-numbers; maybe $q$-numbers were really something more general. All that one knew about $q$-numbers was that they obeyed an algebra satisfying the ordinary axioms except for the commutative axiom of multiplication.

I proceeded to develop a theory in which I felt free to make any assumptions I wanted to, unless they led immediately to an inconsistency. I did not bother at all about finding a precise mathematical nature for $q$-numbers, or about any kind of precision in dealing with them.

I think you can see here the effects of an engineering training. I just wanted to get results quickly, results which I felt one could have some confidence in, even though they did not follow from strict logic, and I was using the mathematics of engineers, rather than the rigorous mathematics which had been taught to me by FRASER.

It was perhaps the most suitable attitude to take for a quick development of the theory, but it did lead me to make mistakes. One of the mistakes that

I made was to assume that every $q$-number has a reciprocal. Another mistake that I made was to assume that if the product of two things, $A$ times $B$, is equal to zero, then one of the factors must be zero.

I got a letter from BRILLOUIN pointing out these mistakes to me. He wrote to me in March 1926, and he pointed out that the assumptions that I had made about $q$-numbers were not valid for matrices. It took me quite some time to get reconciled to the view that my $q$-numbers were not really more general than matrices, and had to have the same limitations that one could prove mathematically in the case of matrices.

Another assumption that I made was the following one.

I assumed that, if we had any two $q$-numbers, $u$ and $v$, one could always find a $q$-number $b$ such that $v = bub^{-1}$. On the basis of that assumption, I was able to set up a general theory of functions of $q$-numbers, which worked very nicely mathematically; but this assumption of course is not true. We have learned now that this assumption can be true only in the special case when $u$ and $v$ have the same eigenvalues.

However, these mathematical points did not bother me at the time, and I proceeded to develop my equations. I soon wrote a second paper in which I applied the methods of just handling these $q$-numbers in accordance with the algebraic rules to get a theory of the hydrogen spectrum and to deduce in particular the Balmer formula.

The method was just to take the equations of motion for the electron and count the dynamical variables as $q$-numbers, and then proceed to solve the equations. I worked in two dimensions only and that was adequate to get the result that I wanted. I had already heard from HEISENBERG that PAULI had succeeded in getting an application of quantum mechanics to the hydrogen atom, and I was really competing with him at this time.

I should mention that, while I was doing this work on $q$-numbers, a paper appeared by LANCZOS that involved transforming the Heisenberg matrices into functions of two continuous variables. I was not very much impressed by this paper because it seemed to me to be just a mathematical development which would not help forward the physics at all. I was really perfectly satisfied with my own method. But I was wrong in attaching so little importance to this work of LANCZOS which was really quite an important development, and paved the way for the later connection that was given between Heisenberg's matrices and the Schrödinger form of quantum mechanics.

When I wrote up my second paper containing the application to the hydrogen atom, I again sent a copy of it to HEISENBERG, and I got Heisenberg's answer, where he states the following (*):

« Since some days I am back in the world of physics, and I found your last work on the hydrogen atom. I congratulate you. I was quite excited as I

---

(*) HEISENBERG to DIRAC, 9 April 1926, 2 pages, in German.

read the work. Your division of the problem into two parts, calculation with $q$-numbers on the one side, physical interpretation of $q$-numbers on the other side, seems to me completely to correspond to the reality of the mathematical problem. With your treatment of the hydrogen atom, there seems to me a small step towards the calculation of the transition probabilities, to which you have certainly approached in the meantime. Now one can hope that everything is in the best order, and, if THOMAS is correct with the factor 2, one will soon be able to deal with all atom models. »

This reference to the Thomas factor 2 refers to the newly proposed spin of the electron. The idea of the spin of the electron had been introduced by GOUDSMIT and UHLENBECK and they applied it to describe the doublets which occur in the spectra of the alkali elements. The electron spin did explain the existence of the doublets, but it gave for the doublet separation twice the observed value. THOMAS showed that the factor 2 came from an error in the calculation, involving the use of a formula for the precession of the spin in a frame of reference in which the electron is at rest, whereas one should take into account the motion of the electron.

Heisenberg's letter continues: « The real reason for my letter is naturally that I want to ask you a few questions. A few weeks ago an article by SCHRÖDINGER appeared in the *Annalen der Physik* (vol. 79, p. 301), whose contents according to my opinion are closely connected with quantum mechanics. Have you considered how far this Schrödinger's treatment of the hydrogen atom is connected with quantum mechanics? These mathematical questions interest me especially because I believe that one can win a great deal for the physical significance of the theory. »

Well, my answer to that was that I had not considered Schrödinger's theory. I felt at first a bit hostile towards it. The reason was that I felt that we had already a perfectly good quantum mechanics, which I believed could be developed for handling all the problems of atomic theory. Why should one go back to the pre-Heisenberg stage when we did not have a quantum mechanics and try to build it up anew? I rather resented this idea of having to go back and perhaps give up all the progress that had been made recently on the basis of the new mechanics and start afresh. I definitely had a hostility to Schrödinger's ideas to begin with, which persisted for quite a while. I do not know in detail just what I answered to HEISENBERG about this, but I got his answer again, which was dated the 26th of May (*).

HEISENBERG begins with a detailed exposition of the connection between the Schrödinger theory and matrix mechanics. He went to the trouble of writing out in two or three pages the details of this connection. It was very helpful to me.

HEISENBERG goes on to say that he agrees with my criticism of Schrödinger's

(*) HEISENBERG to DIRAC, 26 May 1926, 4 pages, in English.

paper, that the wave theory of matter must be inconsistent just like the wave theory of light, but the real progress of Schrödinger's theory is that the same mathematical equations can be interpreted as point mechanics in a nonclassical kinematics, and as wave theory according to SCHRÖDINGER. HEISENBERG hoped that the solution of the paradoxes in the quantum theory would be found in this way. HEISENBERG asked for more information about what I had done with the Compton effect. He said that « People in Copenhagen have discussed the problem so much and are very interested in it ».

I should say that I had been developing my theory of $q$-numbers and had found a way of making the theory to some extent relativistic, following the lines of the very first ideas which I had when I read Heisenberg's paper in September 1925, and which I spoke to you about yesterday.

In March of 1926, SOMMERFELD visited Cambridge and I had a chance of meeting him. I was invited by EDDINGTON to a tea on March the 13th, where SOMMERFELD was also present. I was very happy to meet SOMMERFELD because I had learned so much from his book, and, during the course of the talk which we had then, I mentioned that I had worked out the problem of the Compton effect according to quantum mechanics. SOMMERFELD rather flared up at this point and said « Now, why haven't I heard about this? ». FOWLER, who was present also at the tea party, said that I had only just done this work, and soothed SOMMERFELD down.

I had worked out this theory of the Compton effect and published it soon afterwards, and this is what HEISENBERG was referring to in his letter. I wrote up all this work for my doctor's dissertation. I wrote it up in the Spring of 1926. At that time, there was a general strike going on in England. Everyone who was willing was called upon to do public service, like driving a train or a bus or something, to try to keep the essential services going, and a great many of my fellow students did leave their studies and go over to this kind of work. But I was too absorbed in writing up my thesis, and just stuck to it, and completed it in June 1926.

## III.

I wrote my doctor's dissertation in the Spring of 1926. I continued working at it steadily for some time, unperturbed by a general strike which was occurring in England at the time and which had disrupted very many people's activities. In this dissertation there were still some mistakes in my general ideas of $q$-numbers. Also there was no reference to Schrödinger's theory. I think I told you last time that when Schrödinger's theory first appeared, I rather resented it. I felt that we had, as a result of Heisenberg's work, quite a satisfactory foundation for quantum mechanics, and we could continue to develop that quite happily without there being any need for a further revision of the foundations.

However, one of the letters that I received from HEISENBERG explained in detail the connection between Schrödinger's theory and the matrix mechanics, and I saw from that that Schrödinger's theory would not require us to unlearn anything that we had learned from matrix mechanics; Schrödinger's theory, quite to the contrary, just supplemented the matrix mechanics and provided very powerful mathematical developments which fitted in perfectly with the ideas of matrix mechanics.

Of course, after that my ideas of the Schrödinger theory changed; maybe not immediately, it took a little while. Then I took up Schrödinger's theory with enthusiasm, learning all I could about it. I had to learn a new technique, the technique of eigenvalues and eigenvectors. It was a technique that SCHRÖDINGER had learned in his early training, but it was very little known in Cambridge.

After I had mastered this new technique, I considered how it could be made use of, and I was led to study the problem of an atomic system with many similar particles. I thought of the possibility of having a wave function which is symmetrical with reference to all the particles, or alternatively one that is antisymmetrical. These symmetry questions brought in the possibility of new laws of Nature. Examining their consequences I found that with the symmetrical wave functions we had the particles obeying a statistics which was precisely that which had been originally proposed by BOSE and somewhat corrected by EINSTEIN. This statistics was known as the Einstein-Bose statistics. It applied to photons and gave an explanation for Planck's law.

Then there were the antisymmetrical wave functions, which gave a new statistics. I worked out the basic relations for this new statistics, and I published this work.

Soon after publication I got a letter from FERMI pointing out that this statistics was not really a new one; he had proposed it some time earlier. He gave me a reference to where he had published this work. I looked up the reference and found that it was indeed as FERMI had said in his letter. He had considered the statistics which had the characteristic that there could not be more than one particle in any state.

When I looked through Fermi's paper, I remembered that I had seen it previously, but I had completely forgotten it. I am afraid it is a failing of mine that my memory is not very good and something is likely to slip out of my mind completely, if at the time I do not see its importance. At the time that I read Fermi's paper, I did not see how it could be important for any of the basic problems of quantum theory; it was so much a detached piece of work. It had completely slipped out of my mind, and when I wrote up my work on the antisymmetrical wave functions, I had no recollection of it at all.

I then wrote an apologetic letter to FERMI. I felt that FERMI had reason to be angry with me and that I should placate him. FERMI must have forgiven me, because he never wrote any further letter to me on the subject, and when we met in later life, he was most friendly. We never had any discussion about who was the author of the statistics. The statistics is now often connected with both our names. But the published records show quite clearly that it was first proposed by FERMI, and my later work showed how it could be fitted in with quantum mechanics, and is in fact a consequence of quantum mechanics, when one makes the further assumption that the wave functions have to be antisymmetrical.

After I had obtained my PhD degree, I was no longer confined to Cambridge, and I felt I would like to travel. The place which was most attractive to me was of course Göttingen, the birthplace of quantum mechanics. That was where HEISENBERG lived, and it contained also BORN and JORDAN, who had been extremely active in starting off matrix mechanics. However, when I talked about it with FOWLER, he suggested that I should go to Copenhagen. FOWLER himself had very close connections with Copenhagen. He had been there frequently. He told me how friendly the Institute in Copenhagen was, how BOHR was so friendly with everyone who visited his Institute, and I was therefore undecided over the question whether I should go to Copenhagen or to Göttingen. I decided eventually to divide my time during the coming year between these two places, going first to Copenhagen.

I went to Copenhagen in September of 1926, and I was very glad that I did so because I found, as FOWLER had said, that it was an extremely friendly place and that BOHR was especially friendly to me. I learned to become closely acquainted with BOHR, and we had long talks together, long talks in which BOHR did practically all of the talking.

BOHR had a habit, it seemed, of thinking aloud, doing all his very deep thinking aloud, and he liked to have an audience, maybe the audience of a lecture room or else the audience of just one or two people. Very often I was just his audience during this process of thinking aloud. I admired BOHR very much. He seemed to be the deepest thinker that I ever met. His thoughts were of a kind which were, I would say, rather philosophical. I did not understand them completely, although I struggled as hard as I could to understand them. My own line of thinking was really to put emphasis on thoughts which could be expressed in the form of equations, and much of Bohr's thoughts were of a more general character and rather remote from mathematics. But still I was very happy to have this close connection with BOHR and, as I mentioned once before, I am not sure to what extent hearing all these thoughts of BOHR influenced my own work.

Another person whom I met in Copenhagen, who had a very profound influence on everyone whom he came into contact with, was EHRENFEST. EHRENFEST would insist on absolute clarity in every detail in a discussion. He would never let a speaker get away with some fuzziness in his explanation.

He would go back and just stick to that point to get it absolutely clear before allowing the discussion to go on to further development. EHRENFEST was a most useful person to have in the audience whenever there was a lecture or a colloquium or anything like that. Not only would he jump up and insist on further clarification when the speaker had not expressed himself sufficiently clearly, he had other most valuable qualities.

Suppose the speaker was going into very great detail, elaborating some point, and the audience was getting a bit bored. Well, then ERHENFEST would get up, interrupt the speaker, but do it in a very polite and diplomatic way, so that the speaker would not be offended. He would say that « I'm quite certain this work is very important, but we would like to read about the details of this work later and we don't want to hear all the details now. Would the speaker please go on to discuss his conclusions and results? ». The speaker was pacified by this very diplomatic interruption, and would go on to the results, and everyone in the room would be grateful to EHRENFEST.

Then there were other occasions when the speaker was perhaps assuming too much, assuming something that many people in the audience did not know about. EHRENFEST would again interrupt and ask for further explanation of this matter. Again, many people would be grateful to EHRENFEST for doing this. Probably many people in the room also wanted further explanation of this point, but did not want to expose their ignorance by asking for it.

EHRENFEST would say on these occasions that he did not mind being laughed at. Occasionally he was laughed at when he had asked for an explanation of some point that needed only a quite elementary explanation. But EHRENFEST was not in the least perturbed by being laughed at. I never knew anyone who was so unperturbed at being laughed at. He would say « It doesn't matter in the least if I'm laughed at. The only important thing is that I should get to understand this point. »

If EHRENFEST was present in any audience, then of course one could be assured that the lecture would be a good one, that we should not waste time on unnecessary things, and the speaker would be confined to telling the audience just what they really wanted to hear.

There is another thing I might say about BOHR. During the course of the discussions that I had with him, he told me of the disagreement which he had much earlier with THOMSON. He said that he was a great admirer of THOMSON, and the last thing he wanted to do was to criticize THOMSON or upset him in any way. However, BOHR did want further explanation of some of the features of Thomson's atomic models, and he did not then know English very well and was not able to express himself in such a polite form as he would like to have done, and THOMSON did take his questions badly. THOMSON assumed that he was being criticized and got angry.

This distressed BOHR very much, and I think this incident made a lasting impression on him. It continued to distress him all his life, I should think.

He was very careful in later life that this sort of thing should not happen again. Whenever he was questioning any author about his work, he would keep on saying « This is not to criticize you, it is only to learn ». That became rather a stock phrase in Copenhagen: « This is not to criticize but only to learn ». It was often said in German « Nicht um zu kritisieren, nur um zu lernen ».

I believe HEISENBERG must have been affected by this phrase, because in the letters which he wrote to me, which I have been telling you about in my earlier talks, he is continually saying « I have no doubt that your results are right, but I would like to have further explanation of this point ». He was very careful not to say something which I might take as direct criticism and get offended with. It was really unnecessary for HEISENBERG to be so diplomatic. I was very honored in having these letters from HEISENBERG, and I would not have been offended if he had openly criticized me. But he was very careful not to do that.

Another person in Copenhagen who very much influenced the proceedings there was GAMOW. GAMOW was rather childlike, always wanting to play, and introducing a sort of light humor into all occasions. He was very fond of drawing pictures of Mickey Mouse. He added a lot to the entertainment that we had. He had some good ideas, applications which led to important developments in quantum theory, but I do not think he did any work which was very deep.

While I am on this question of giving my opinions of other physicists, I should also mention SCHRÖDINGER. I do not think I ever saw SCHRÖDINGER in Copenhagen. I do not remember any such occasion. But I met him frequently in later life, and, of all the physicists that I met, I think SCHRÖDINGER was the one that I felt to be most closely similar to myself. I found myself getting into agreement with SCHRÖDINGER more readily than with anyone else. I believe the reason for this is that SCHRÖDINGER and I both had a very strong appreciation of mathematical beauty, and this appreciation of mathematical beauty dominated all our work. It was a sort of act of faith with us that any equations which describe fundamental laws of Nature must have great mathematical beauty in them. It was like a religion with us. It was a very profitable religion to hold, and can be considered as the basis of much of our success.

Now, there is one point that you might wonder about when you read of Schrödinger's work. SCHRÖDINGER developed his quantum mechanics from de Broglie's wave equation. De Broglie's wave equation was relativistic, and SCHRÖDINGER of course was profoundly influenced by the beauty of relativity, and you may wonder why it is that his work, where he introduces the wave equation, is nonrelativistic. There is a contradiction there.

SCHRÖDINGER explained this matter to me many years later, I do not remember just when, around about 1940, when I had got to know him well. He said that he was working from the relativistic point of view inspired by DE BROGLIE, and he was led to a relativistic wave equation, which was a generalization of de Broglie's equation, bringing in the electromagnetic potentials. When he got this relativistic equation, his first concern was to apply it to the hydrogen atom to see what results it would give. The calculation gave results that were not in agreement with observation.

SCHRÖDINGER was extremely disappointed by that and thought that his wave equation was no good at all, and abandoned it. He gave it up for some months, then went back to it, and taking a second look at it, he noted that, if he used the equation with less accuracy in nonrelativistic approximation, the results that he got were in agreement with the experimental results, again with neglect of relativistic effects. So he was able to publish his wave equation in a nonrelativistic form, and in agreement with experiment.

Of course, the reason why Schrödinger's original equation, the relativistic one, did not agree with experiment was because it did not take into account the spin of the electron. The spin of the electron was a very new idea at the time, and possibly SCHRÖDINGER had never even heard of it. And SCHRÖDINGER then did not have the necessary boldness to publish an equation which definitely gave results in disagreement with observation.

The relativistic equation of SCHRÖDINGER was later on resurrected by KLEIN and GORDON and was published by them, and is known nowadays as the Klein-Gordon equation. It is considered as a good equation for use in a relativistic manner for a charged particle which does not have any spin. There was no such charged particle known at the time, and KLEIN and GORDON published this work purely as a mathematical development without any direct physical application. They had the boldness to publish an equation which was not connected with experimental results, while SCHRÖDINGER did not have that boldness.

Well, to return to my period in Copenhagen, in spite of meeting so many eminent physicists and having these discussions with them, I continued to work mainly on my own, following up my own ideas, and the problem that concerned me mainly was to get a general physical interpretation for quantum mechanics. One had the equations based on noncommuting things, $q$-numbers, as I called them, but one could use these equations to get results comparable with observation only by following various special rules. There was a great need for putting these rules together and getting some general method for physical interpretation. I worked on that for some time, and wrote a paper incorporating my results.

I would like to say that this work gave me more pleasure in carrying it through than any of the other papers which I have written on quantum mechanics either before or after. You may wonder why this is so. Many of my papers just were consequences of an idea that had come to me rather accidentally. The early work on Poisson brackets, for instance, and my later work on the relativistic wave equation were very definitely of this nature. They were consequences of an idea which had just come out of the blue. I could not very well say just how it had occurred to me. And I felt that work of this kind was a rather undeserved success. On the other hand, my work on the physical interpretation of quantum mechanics was a deserved success. There I was tackling a problem which was not too difficult to be solved by a direct approach. There were various stages in this problem which had to be disposed of one by one.

During the course of the work I was continually faced with the question of getting a suitable notation for writing down the equations that I was then dealing with. There was a frequent modification of notation. Everything proceeded step by step in a rather logical way, and led to a piece of work which laid the foundation of the general transformation theory of quantum mechanics, and also provided the essential features of a suitable notation.

With regard to this question of notation, I had to face the problem of writing down symbols which would contain an explicit reference to those factors which it was important to mention explicitly, and which left understood those quantities which it was safe to leave understood, to keep at the back of one's mind and not to write down explicitly. Well, this led to the notation which, with some slight modifications, has become the standard notation for use in quantum mechanics at the present time.

I had a very successful time in Copenhagen, because I developed this general physical interpretation which gave me so much pleasure, and I also started the quantum theory of radiation, and showed how it could be connected with the Bose-Einstein statistics, which follows from the use of wave functions which are symmetrical in the particles that they refer to.

While doing this work, I got one of those ideas out of the blue, namely to take the Schrödinger wave equation and apply a process of quantization to the wave function itself. The wave function was previously always considered as expressed by ordinary numbers, $c$-numbers. What would happen if you turned them into $q$-numbers, and assumed that they are noncommuting with their conjugates?

That led to a theory which was equivalent to the theory of radiation which I had been setting up, and provided an alternative way of introducing the subject. It gave rise to a process which has become known as second quantization.

It was towards the end of my stay in Copenhagen, probably in January 1927, that PAULI visited Copenhagen. I explained to him my work on the physical interpretation and transformation theory of quantum mechanics. We discussed how these ideas would apply to the spin of the electron. We were led to the introduction of the three $\sigma$-variables to describe the three components of spin. I believe I got these variables independently of PAULI, and possibly PAULI also got them independently of me.

Soon after PAULI left Copenhagen he wrote a paper (*Zeits. f. Physik*, **43**, p. 601) in which he incorporated the spin of the electron into the wave equation, in a nonrelativistic manner. SOMMERFELD in his book *Atombau und Spektrallinien II* refers to Pauli's paper (page 226) and says « The discovery of the

Pauli equation was an important step leading to the recognition of the true nature of the electron, *i.e.* the Dirac equation ».

This statement is not true, so far as I was concerned. I was not interested in bringing the spin of the electron into the wave equation, did not consider the question at all and did not make any use of Pauli's work. The reason for this is that my dominating interest was to get a relativistic theory agreeing with my general physical interpretation and transformation theory. I thought that this problem should first be solved in the simplest possible case, which was presumably the spinless particle, and only after that should one go on to consider how to bring in the spin. It was a great surprise to me when I later on discovered that the simplest possible case did involve a spin.

While on this subject of the relativistic wave equation I might mention that KRAMERS told me (some years after my equation had appeared) that he had independently obtained a second-order equation equivalent to my first-order equation. It is possible that KRAMERS worked from the Pauli equation. KRAMERS did not publish his work because it was superseded by mine.

In the beginning of February 1927, I moved from Copenhagen to Göttingen. I passed through Hamburg. At the time there was a meeting of the German Physical Society in Hamburg, which I attended for a few days. The work at this Physical Society meeting was largely concerned with a discussion of experimental results about spectra. Still, I did get to appreciate the way the German physicists worked. It seemed to me that they were very hard working, had long hours of lectures, and they did not seem to get tired from them. They had enormous energy.

I traveled from Hamburg to Göttingen in a fourth-class compartment of a train with a number of other physicists who had attended this Hamburg meeting and were returning to Göttingen. Among them was ROBERTSON, whom I got to know pretty well later in Princeton and who was concerned with cosmology, and I got my first interest in cosmological models of the universe from him.

I arrived in Göttingen and spent some months there. It was a rather more formal atmosphere than we had in Copenhagen. I increased my mathematical knowledge. I went to a course of lectures by WEYL on group theory. I met HEISENBERG and BORN on various occasions. I also met OPPENHEIMER and became a close friend of his, because we lived in the same pension and of course saw very much of each other.

I should say that, after leaving Cambridge, I continued my general mode of life, studying and calculating hard during the week and relaxing on Sundays and going for long walks in the country. In Copenhagen the walks were often not solitary walks. Sometimes I was accompanied by BOHR. He was also fond of walking and we had many enjoyable long walks together. Sometimes there was a whole party from the Institute all going together, providing a sort of excursion which refreshed us all.

I continued with these Sunday walks also in Göttingen. Sometimes I went with OPPENHEIMER. I remember in particular one long walk we had together on Easter Sunday in 1927, where we covered a great deal of ground.

I had received an invitation from EHRENFEST to visit his institute in Leyden. OPPENHEIMER was also invited, and we traveled together from Göttingen to Leyden in June of 1927. We spent some days with EHRENFEST at his institute, and we also visited KRAMERS in Utrecht for one day.

I went to the Solvay conference in Brussels in October of 1927, which was a great experience for me, meeting so many eminent physicists, among them EINSTEIN and LORENTZ. There are a few things that I remember clearly about this meeting. I had given a talk about my second-quantization method, and after this talk someone announced that there was a similar second-quantization method applicable with the Fermi statistics, a method which had been given by JORDAN and WIGNER.

At first I did not like this work of JORDAN and WIGNER, and I think I can attribute this dislike to my mind being essentially a geometrical one and not an algebraic one. In the case of the Bose statistics and the second quantization which was connected with it, one had a definite picture underlying the basic equations, namely the picture that the theory could be applied to an assembly of oscillators. There was no such picture available with the Fermi statistics, and I felt that that was a serious drawback. I did not appreciate therefore the importance of this other kind of second quantization.

Actually, of course, the importance lies in the very close connection that we have between these two kinds of second quantization, when looked at from a purely algebraic point of view. I will just put down the basic equations

$$\psi_n \psi_m - \psi_m \psi_n = 0\,, \qquad \bar{\psi}_n \psi_m - \psi_m \bar{\psi}_n = \delta_{nm}\,. \tag{14}$$

One gets equations like this when one quantizes the ordinary Schrödinger wave function, and these equations can be connected with those that describe harmonic oscillators, there being one oscillator for each state $\psi_n$.

With the other kind of second quantization, we have

$$\psi_n \psi_m + \psi_m \psi_n = 0\,, \qquad \bar{\psi}_n \psi_m + \psi_m \bar{\psi}_n = \delta_{nm}\,, \tag{15}$$

the same equations, except for a plus sign instead of a minus sign. There is this extremely close similarity between the two processes of second quantization when you look at them algebraically. If you try to get some pictures of the relations, then we have a picture in the Bose case and no picture in the Fermi case. But it is the algebraic connection which is important, and which has the effect that the second quantization for Fermi statistics is really just as important as the one for Bose statistics.

Another important question at this 1927 Solvay conference was the physical interpretation of quantum mechanics. Of course, there was a lot of discussion between those people who saw the need for indeterminacy in the results of quantum mechanics, and those who objected to any kind of indeterminacy appearing in fundamental natural processes. I gave my own point of view, which had been based on work on the general interpretation of quantum mechanics. This work led very directly to our being able to interpret the square of the modulus of the wave function as giving the probability of there being definite results for any observation applied to an atomic system. I should say that BORN had independently obtained the same result for use in connection with his scattering theory. With this probability coming into the interpretation, one had to accept that the results were not deterministic when one made an observation, and I expressed this situation by saying that, under these conditions, "Nature makes a choice". I think that that is perhaps still the best way of expressing the kind of indeterminacy which we have in atomic theory. There are occasions when we just have to admit that Nature makes a choice, and we cannot predict what this choice will be.

There is one incident that I remember about this Solvay conference. During the period before the lecture started on one occasion, BOHR came up to me and asked me « What are you working on now? ».

I said « I'm trying to get a relativistic theory of the electron ».

Then BOHR said « But KLEIN has already solved this problem ».

I was a bit taken aback by this. I began to explain that Klein's solution of the problem, based on the Klein-Gordon equation, was not satisfactory because it could not be fitted in with my general physical interpretation for quantum mechanics. However, I was not able to explain very much to BOHR before the start of the lecture interrupted our conversation, and I had to leave the question rather in the air.

This was a problem that was very much dominating me at the time: how could one get a satisfactory relativistic theory of the electron? I had the general physical interpretation for quantum mechanics which I felt sure was right, but it required one to work with a wave equation which was linear in the operator $d/dt$, giving $d\psi/dt$ equal to some definite function of $\psi$. Now, the Klein-Gordon equation involves $d\psi^2/dt^2$. This would not fit in with my general interpretation. If one tried to fit it in, one was led to a probability which could be sometimes negative, and that of course is physically nonsense.

KLEIN and GORDON had tried to get over this difficulty by saying that the quantity which I considered ought to be the probability was really the charge density. The equation should be applied to an assembly of particles, and an expression was proposed which gave the charge density, and of course the charge density could be either positive or negative, if one allowed the possibility of particles with negative charge as well as positive charge.

However, this was not good enough for me. It was not much use having a theory of several particles if one did not have first of all a theory of one particle. It could not be considered as a logical theory if it could not be applied to one particle, and so long as one was considering just one particle, it was necessary to be able to find probabilities for this particle, and the probabilities had to be positive, and that required that the wave equation should involve only $d\psi/dt$.

This was quite a problem for some months, and the solution came rather, I would say, out of the blue, one of my undeserved successes. It came from playing about with mathematics. I was playing about with the three components $\sigma_1$, $\sigma_2$, $\sigma_3$, which I had used to describe the spin of an electron, and I noticed that if you formed the expression $\sigma_1 p_1 + \sigma_2 p_2 + \sigma_3 p_3$ and squared it, $p_1$, $p_2$ and $p_3$ being the three components of momentum, you got just $p_1^2 + p_2^2 + p_3^2$, the square of the momentum. This was a very pretty mathematical result. I was quite excited over it. It seemed that it must be of importance. But it did not immediately answer the question of how one could get a satisfactory relativistic equation for the electron.

It provided effectively a method of taking the square root of the sum of three squares and getting it in a linear form. Now, if we are to have a relativistic theory of a particle, we would need to have the square root of the sum of four squares, and it was just impossible to use this method to get a square root for the sum of four squares. So it seemed that it was an interesting bit of mathematics, but just was not good enough to provide an answer to the problem.

It took me quite a while, studying over this dilemma, before I suddenly realized that there was no need to stick to quantities $\sigma$, which can be represented by matrices with just two rows and columns. Why not go to four rows and columns? Mathematically there was no objection to it at all. Replacing the $\sigma$-matrices by four-row-and-column matrices, one could easily take the square root of the sum of four squares, or even five squares if one wanted to.

Well, that led to a new wave equation for the electron, a wave equation which is linear in the four components of the relativistic four-vector of momentum and energy. It provided us with this wave equation

$$(p_0 - \alpha_1 p_1 - \alpha_2 p_2 - \alpha_3 p_3 - \alpha_4 mc)\psi = 0 \tag{16}$$

(I expect you are all familiar with it), a wave equation in which we have a wave function with four components, corresponding to the four rows and columns of the matrices, and every one of those components just satisfies the de Broglie equation.

Well, that was just an equation for one particle in the absence of any field of force. To get something interesting one had to bring in an electromagnetic field. There was the general problem of how to bring in an electromagnetic field when we had a theory for the particle in the absence of any field. I had met that problem some time previously. I think it was first in my work on the Compton effect. It was necessary there to bring the electromagnetic potentials into the description of the motion of the particle, and still to keep the equations in the Hamiltonian form.

When I first met this problem, I proceeded to solve it without bothering to look up the literature to see whether it had been solved previously. It was a problem of classical mechanics, to put the equations of motion of a charged particle, expressed relativistically, into the Hamiltonian form. I expect it was solved probably some time near the beginning of this century, but I never bothered to look up who first did it. That is a question for the historians of science. I proceeded to work it out for myself, which did not involve much difficulty, and I think was really simpler than looking up references.

Well, I used this same method again for the new wave equation, linear in the four $p^2$'s. It just involved replacing each of the four $p$'s by $p + (e/c)A$, $A$ being the corresponding electromagnetic potential.

Then I noticed that this was really a very successful equation. It led to the electron automatically having a spin of half a quantum, which is just what experiments required. It led also to the electron having a magnetic moment, and I applied it to the hydrogen atom in the first approximation, and got results in agreement with observation.

I wrote up this work and published it, keeping to the first approximation in my treatment of the hydrogen atom. You may wonder why I did not immediately go on to consider the higher approximations, but the reason is that I was really scared to do so. I was afraid that, in the higher approximations, the results might not come right, and I was so happy to have a theory that was correct in the first approximation that I wanted to consolidate this success by publishing it in that form, without going on to risk a failure in the higher approximations. The higher approximations were worked out later by DARWIN, who wrote and told me of his results, and I was very happy to hear that they agreed with observation.

The originator of a new idea is always rather scared that some development may happen which will kill it, while an independent person can proceed without this fear, and can venture more boldly into new domains.

There was then a wave equation for the electron which was satisfactory in many respects, but also had a serious failing, namely we had the matrices describing an internal motion, containing four rows and columns, whereas we need only matrices with two rows and columns to describe the two states of spin of the electron as it is observed. The result is that this equation gives you twice as many states as you want for describing the experimental situation. If you look into it more closely, you soon see that half of the states refer to negative energies for the electron, so you can say, well, just exclude these negative-energy states, which are unobservable. Let us confine our attention to the positive-energy states, and then we just have a theory giving us things which can be observed.

However, it is not so simple just to do that, because of the transitions which may occur between positive-energy states and negative-energy states. We have the negative-energy states occurring also in classical theory, but with the classical theory they can be ignored, because we do not then have any transitions from positive-energy states to negative-energy states. In the quantum theory these transitions cannot be ignored.

They occur rather seldom, if one is dealing with radiation which does not involve very high frequencies, and so one can get an approximate theory just by ignoring them. That is what we had to do for a time.

SCHRÖDINGER proposed a modification in which the transitions between positive- and negative-energy states were excluded, but he had to bring in a change in the wave equation which spoiled its relativistic character and spoiled its beauty, so it was not a satisfactory explanation.

The problem of the negative-energy states puzzled me for quite a while. The main method of attack to begin with was to try to find some way of avoiding the transitions to the negative-energy states, but then I approached the question from a different point of view. I was reconciled to the fact that the negative-energy states could not be excluded from the mathematical theory, and so I thought, let us try and find a physical explanation for them.

And that was not so difficult, when one remembered that electrons satisfy the Fermi statistics which does not allow more than one electron to be in any state. I was led to a picture in which we have a world with all the negative-energy states occupied, so that an electron in a positive-energy state cannot make a transition to a negative-energy state. Then, of course, we have to consider the possibility that some of the negative-energy states are not occupied; there are holes and these holes will appear as particles also but having a positive energy.

It was not really so hard to get this idea, once one had the proper understanding of what one needed, because there was a very close analogy provided by the chemical theory of valency. In this theory we have the inert gases, where all the electrons fill up closed shells. Then we get the alkali elements, where there are one or possibly two electrons outside the closed shells, and these are the chemically active electrons, and they are also the most active in producing spectra. Then we have to consider the possibility of there being a hole in a closed shell, which gives the halogen gases. The relationship between the holes and the electrons, which one gets from this chemical theory of atoms, could be taken over directly to the positive- and negative-energy states, so it did not need any great stretch of the imagination to set up this theory where we have nearly all the negative-energy states occupied.

Of course, as soon as I got this idea, it seemed to me that the negative-energy states would have to correspond to particles having a positive charge instead of the negative charge of the electron, and also having the same mass as the electron. Now, that was a serious difficulty. At that time, we had the electrons carrying negative charge, and we had the protons carrying positive charge, and everyone felt pretty sure that the electrons and the protons were the only elementary particles in Nature. It is true that RUTHERFORD had sometimes considered the possibility of there being a third particle, the neutron. He proposed the possibility of a neutron rather wistfully. He said it would be so useful for the experimenters if these neutrons did exist because they would provide ideal projectiles to shoot into atomic nuclei. They would not be disturbed at all by the electrons outside. People did not really have much faith in the existence of neutrons. It seemed to everyone self-evident that as there were just two kinds of electricity, there should be just two kinds of particles to carry them. People did not go beyond that.

Well, what was I to do with these holes? The best I could think of was that maybe the mass was not the same as the mass of the electron. After all, my primitive theory did ignore the Coulomb forces between the electrons. I did not know how to bring those into the picture, and it could be that in some obscure way these Coulomb forces would give rise to a difference in the masses.

Of course, it is very hard to understand how this difference could be so big. We wanted the mass of the proton to be nearly 2000 times the mass of the electron, an enormous difference, and it was very hard to understand how it could be connected with just a sort of perturbation effect coming from Coulomb forces between the electrons.

However, I did not want to abandon my theory altogether, and so I put it forward as a theory of electrons and protons. Of course I was very soon attacked on this question of the holes having different masses from the original electrons. I think the most definite attack came from WEYL, who pointed out that mathematically the holes would have to have the same mass as the electrons, and that came to be the accepted view.

OPPENHEIMER put forward a theory that the holes did have the same mass as the electrons, but there was some special reason in Nature why they were never observed. He could not say what this special reason was, but he just put it forward as some thing still to be explained. OPPENHEIMER was really close to the mark. These holes were particles with the same mass as the electron, and they had never been observed simply because the experimenters had never looked for them in the right place.

I remember that during my attendance at lectures given by experimenters in the Cavendish, there was one occasion, I am not quite sure whether it was 1926 or 1927, when, in the discussion after the lecture, the lecturer pointed out a rather curious fact which he had observed in his experiments. His experiments were done with tracks of particles in a Wilson chamber, in the presence of a magnetic field, and so they were all curved. Then if one knows the charge on a particle, one knows which way it is going. The remark was that it had often been observed that there were tracks leading into the source. He was assuming that the particles had to be electrons, and then the curvature of the tracks indicated that they were moving into the source.

It was just mentioned casually. Nobody thought of examining this point in greater detail, but if they had examined it they would have been led to an important discovery. What they thought were electrons going into the source were really positively charged particles with the same mass as the electron coming out from the source.

That just goes to show how an important discovery may be missed through people not attaching sufficient importance to something which they look upon as a curiosity and not worth further examination.

Well, I expect you all know the history after that. The positively charged particle with the same mass as the electron was discovered a few years later. It was actually first observed by BLACKETT. The first example was obtained by BLACKETT, but he was rather cautious and did not want to publish his result without confirmation. ANDERSON, getting a similar result, was bolder and published it and scooped the credit for being the first to observe a positron.

Well, that marked the beginning of the discovery of a whole lot of new particles. The neutron was discovered, then various kinds of mesons and a whole lot of new particles, and people are still going on discovering more.

It is very strange how the whole climate of opinion with regard to new particles has changed so drastically, from the late 1920's, when it was considered practically self-evident that there could not be any particles other than electrons and protons. People have gone over to the diametrically opposite point of view, and are willing to postulate a new particle on the flimsiest experimental or theoretical evidence. The number of particles which are considered as rather fundamental has gone up to several hundred now, instead of just two.

Well, this takes me to the end of what I counted as the exciting era. It was an era in which there was a rapid development of theoretical ideas on the foundations of our knowledge of atoms. Since then, physics, of course, has continued to develop strongly, but on rather different lines. Since then the experimental people have had it pretty much all their own way. They make experiments and report on the results of their observations. The theorists are not in a strong enough position to contradict those observations. They have to accept what the experimentalists say and do their best to construct theories to fit in with the observations, and most of their work consists in setting up theories to account for this host of new particles.

# Dirac's Quantum Electrodynamics and the Wave-Particle Equivalence (*).

J. BROMBERG
*Smithsonian Institution - Washington, D. C.*

Dirac's first paper on quantum electrodynamics was « The Quantum Theory of the Emission and Absorption of Radiation », submitted to the *Proceedings of the Royal Society* in the Winter of 1927 [1]. I have chosen it as my subject because of its connection with a fundamental historical problem: that of the way in which the two aspects of matter and light, wave and particle, came to be incorporated into quantum mechanics. Historians have paid some attention to the path which led from de Broglie and Einstein to Schrödinger's theory of matter waves. They have also paid some attention to the development of Bohr's complementarity interpretation and Heisenberg's uncertainty principle. On the other hand, quantum electrodynamics and quantum field theory have been little treated [2]. These parts of physics were of considerable importance, however. A version of the relation between wave and particle developed here which was different from those of SCHRÖDINGER, BOHR or HEISENBERG. My first purpose therefore is to exhibit the relation between the wave and particle aspects of light, as it is embodied in Dirac's « Quantum Theory of Radiation ».

At the same time, I wish to make some suggestions on the place of this paper in the history of matter theory. These are as follows: EINSTEIN published a well-known paper on the quantum theory of gases in 1924 and 1925 [3]. In it, he discussed the fundamental assumptions of statistical mechanics, and connected them with considerations on the wave and particle natures of light and matter. This paper, which was important for the development of wave mechanics [4], was also of central significance for the development of quantum electrodynamics. In particular, it was important for DIRAC. In the article « On the Theory of Quantum Mechanics » [5], submitted in August 1926, DIRAC, without taking up the wave-particle question explicitly, made use of Einstein's statistical arguments. In an unrelated part of the same paper, DIRAC also

(*) I am grateful to Prof. DIRAC for valuable comments on the talk which formed the basis for this paper, and for permission to consult the papers he has deposited at the Churchill College Archive, Cambridge.

discussed the emission and absorption of radiation. The « Quantum Theory of Radiation », of the following winter, brought together an attack on a problem left unsolved in the earlier discussion of emission, with a further development of the statistical considerations. And these developments, in turn, had implications for Einstein's considerations on waves and particles. They involved the necessity of altering the correspondence between light and matter, as EINSTEIN had given it. To develop the historical as well as the conceptual aspects of Dirac's paper, I must therefore begin with Einstein's work.

In his paper, EINSTEIN used the same statistics which BOSE had just applied to light quanta for a gas of material particles, and showed that this led to experimentally observed results. He gave a careful analysis of the difference between Bose's statistics and Boltzmann's. Emphasizing the noteworthy fact that both material particles and light obeyed Bose statistics, EINSTEIN remarked it implied a « deep, essential kinship » between the two kinds of matter [6]. He suggested deepening the kinship still further. He had already proposed in 1909 that light has both wave and particle aspects. Now he suggested juxtaposing to the duality between light quantum and light wave, a duality of matter-wave and matter-particle. He directed the reader's attention to the hypothesis of matter-waves that had just been advanced by DE BROGLIE.

KLEIN has shown how the latter remarks were important for Schrödinger's development of wave mechanics [4]. SCHRÖDINGER published his theory, together with a demonstration of its mathematical equivalence to matrix mechanics, in the first half of 1926. What DIRAC did in the summer may then be characterized as a bringing together of the wave mechanical formalism inspired by the comments in the last part of Einstein's paper, with the analysis of statistics in an earlier part. DIRAC showed that, if Schrödinger wave functions for a set of identical particles are constructed so as to be symmetric with respect to the interchange of the co-ordinates of any two particles, the set so described fulfills the conditions EINSTEIN had revealed as underlying Bose statistics. DIRAC went on, as we shall see, to investigate sets of particles whose wave functions are antisymmetric.

Dirac's argument made crucial use of the stricture given prominence by HEISENBERG. In his first paper on matrix mechanics, HEISENBERG had stipulated that the construction of quantum mechanics should be made in such a way that only observables would enter the equations [7]. The question raised in this way, of the proper relation between observations and the formalism, subsequently underwent a significant development. In Heisenberg's first paper, the restriction to observables was a rather crude means of inventing the theory. By mid-1926, the formalism had been immeasurably enriched. The problem had become the more subtle one of deciding which physical questions the formalism could and should permit one to handle.

In all its changing forms, this question was important for DIRAC, serving both as guide and as problem. Dirac's paper on « The Physical Interpretation of the Quantum Dynamics », for example, was a major contribution to the problem in its later phase. Its aim was to « show all the physical information that one can hope to get from the quantum dynamics, and [to] provide a general method for obtaining it » [8]. In the « Theory of Quantum Mechanics » DIRAC began his discussion of systems of identical particles by calling the reader's attention to the fact that Heisenberg's theory « enables one to calculate just those quantities that are of physical importance, and gives no information about quantities... that one can never hope to measure experimentally. We should expect », DIRAC continued, « this very satisfactory characteristic to persist in all future developments of the theory. »

« Consider now a system that contains two or more similar particles, say, for definiteness, an atom with two electrons. Denote by $(mn)$ that state of the atom in which one electron is in an orbit labelled $m$ and the other in the orbit $n$. The question arises whether the two states $(mn)$ and $(nm)$, which are physically indistinguishable as they differ only by the interchange of the two electrons, are to be counted as two different states or as only one state. » DIRAC concluded, « in order to keep the essential characteristic of the theory that it shall enable one to calculate only observable quantities, one must adopt the second alternative that $(mn)$ and $(nm)$ count as only one state » [9].

The consequence was that the Schrödinger wave function for an assembly of identical particles must be either symmetric or antisymmetric in the particle co-ordinates. DIRAC then made the usual assumption that every state of the assembly is equally probable. This led him directly to the insight that has been mentioned; an assembly described by symmetric eigenfunctions satisfies the conditions for a Bose-Einstein assembly. DIRAC appears to have taken these conditions directly from Einstein's paper. « The solution with symmetrical eigenfunctions must be the correct one when applied to light quanta, since it is known that the Einstein-Bose statistical mechanics leads to Planck's law of black-body radiation » [10].

At this point, Dirac's interest was not primarily in the symmetric functions, however, but in the antisymmetric ones. He recognized that they presented an ensemble satisfying the Pauli exclusion principle. Thus, he took them to be the correct wave function for electrons, and probably also for material gases. He saw that they demanded a statistics which was different from that EINSTEIN had discussed. DIRAC proceeded to the independent invention of what are now called the Fermi-Dirac statistics. He wrote down the new equations in such close correspondence with those of EINSTEIN that I think one can assume he either had Einstein's text before him, or had absorbed it completely [11].

He did not, as I have said, make use of the later sections of Einstein's paper. Nor did he comment on the wave-particle duality which EINSTEIN discussed there. Nevertheless, what he did had implications for that duality. DIRAC had demonstrated a distinction between two kinds of matter waves: symmetric and antisymmetric. Now, on the one hand, the pairing of an antisymmetric

Schrödinger wave with electrons and other material particles closely resembled Einstein's duality for matter. But, on the other hand, the pair of entities which DIRAC here associated with light consisted of a symmetric Schrödinger wave and an assembly of quanta. This was quite different from Einstein's duality of an electromagnetic wave and quanta. At this point, however, the change was still implicit.

The last section of « Theory of Quantum Mechanics » is devoted to a discussion of the emission and absorption of light [12]. Dirac began with a mathematical treatment of an ensemble of independent identical systems acted upon by an arbitrary, external perturbation. He then specified the perturbation to be due to electromagnetic radiation, representing the latter by a classical plane wave. The ensemble was specified to be an assembly of atoms. The perturbation energy was the classical term $\vec{\varkappa}/c \cdot \vec{\eta}$, where $\vec{\varkappa}$ is the vector potential of the wave, and $\vec{\eta}$ the electric polarization of the atom.

No attention was paid to the statistical peculiarities of the assembly. Thus, the considerations developed in the preceding sections for assemblies of matter or quanta were not yet being applied to an ensemble of matter in interaction with radiation. Instead, the ensemble was described in terms of one-particle eigenfunctions. The numbers of particles in each eigenstate—what we today call the occupation numbers—entered very simply here. The wave function for a particle at time $t$ was expanded in terms of the single-particle's unperturbed eigenfunctions. The squares of the amplitudes that appear in the expansion were then taken to be equal to the total numbers of atoms in each of the states, at time $t$. DIRAC then calculated the change in the numbers of atoms in a given state with respect to time. This gave him Einstein's $B$-coefficients for induced emission and absorption. It did not give a way to calculate Einstein's $A$-coefficient for spontaneous emission, so that this problem was left open.

It will be noticed that in the statistical sections of the paper DIRAC spoke of assemblies of light quanta, while in the final section he treated light classically. In his paper on Compton scattering, written a few months before the « Theory of Quantum Mechanics », he also treated light classically, using a model in which a free electron interacts with an electromagnetic wave [13]. What then was Dirac's view on the constitution of light? The answer, I think, is that at that time DIRAC would not have seen profit in elaborating such a view. The path he was following was that of continually expanding the mathematical formalism of the quantum theory, so as to render it applicable to a widening circle of physical problems. In my view, DIRAC held the expectation that the formalism thus achieved would ultimately enable one to read out answers to questions like the nature of light and matter. In this approach, DIRAC differentiated himself not only from the older generation of BOHR, EINSTEIN and SCHRÖDINGER, but also from his contemporary, Pascual JORDAN [14]. In the following winter, DIRAC succeeded in expanding the theory to give a quantum treatment of radiation. It is not surprising that it was in this and later papers that he expressly commented on the wave and particle aspects of light.

The introduction to the « Quantum Theory of Radiation » appears to have been written after reflection on the already completed body of the paper. It presented the construction of a quantum theory « of the emission of radiation and of the reaction of the radiation field on the emitting system » as the main object of the paper. This was to be accomplished by taking the Hamiltonian for an atom interacting with an electromagnetic field and converting the classical field quantities into quantum-mechanical operators. If the Hamiltonians were then written in terms of a closed system, that is, with terms representing the energies of the atom, the field and the interaction, a calculation of the coefficient of spontaneous emission, as well as of induced emission and absorption, would be possible, and thus the problem that had been left open would be solved. DIRAC also reported that he had discovered a way to write the Hamiltonian for matter and radiation in terms of the interaction of the atom with an assembly of light quanta. This gave him two equivalent expressions whose starting points were, respectively, waves and quanta. This equivalence, on the one hand, and the translation of the classical field variables into quantum operators, on the other, provided the formal basis for an understanding of the constitution of light.

In his interview for the *Archive of Quantum Physics*, DIRAC remembered the paper actually having originated in « playing about with [the Schrödinger equation]... seeing what happens when you make the wave function into a set of noncommuting quantities. ...I'm pretty sure that was the starting point of this work, and the application to radiation... turned out, without its being expected » [15]. It is this part, from which the Hamiltonian for atom and quanta results, which is the first and longest portion in the body of the paper. The derivation deserves to be described in some detail for its originality and beauty.

It rests upon the two independent parts of « The Theory of Quantum Mechanics » described above, the statistical mechanics and the calculation of the Einstein $B$-coefficients. They are here unified with the help of what became called « second quantization ». Thus, DIRAC began with the same treatment of an ensemble of systems under an external perturbation that he had given in the earlier paper. There he had specified the systems to be atoms and the perturbation to be due to radiation. Here he did the reverse. He specified the systems to be Bose particles, with an eye to the special case of light quanta. The perturbation was taken to be caused by an atom. Again, DIRAC introduced the set of eigenfunctions for a single system. Again he expanded the wave function for a particle in terms of these eigenfunctions. This, he wrote, « gives immediately the probable number of systems in (each) state at that time » for the whole assembly. This is the same connection between the occupation numbers and the coefficients of the expansion that he had used previously [16].

DIRAC now undertook to show that the equations for the change in occupation numbers can be put in Hamiltonian form. For this purpose, he introduced a set of phases that were canonically conjugate to the occupation numbers. « The development... which naturally suggests itself » is the conversion of the occupation numbers and the phases into operators. He therefore imposed suitable commutation relations on them [17]. This is the procedure that later received the name of « second quantization ». The $r$-th phase was intro duced into the Hamiltonian in such a way that it acted to increase (or diminish), by one, the number of particles in the $r$-th state [18]. It is this that became called a creation (or annihilation) operator. Finally, he wrote down the Schrödinger equation for a system with this Hamiltonian. The $\Psi$-wave in the equation was written as a function of the occupation numbers. Each different set of these arguments represents a particular distribution of the assembly's systems over the various possible states. The square of the $\Psi$ gives the probability for each such distribution. It therefore gives immediate expression to one of the central concepts of statistical mechanics.

It may be for this reason that DIRAC was led to turn next to the calculation based on the statistics section of the « Theory of Quantum Mechanics ». He took a wave function which represented a Bose-Einstein ensemble and which therefore was symmetric in the variables describing the particles, and derived the rules according to which these variables may be replaced by occupation numbers in the wave function and in the Hamiltonian for the ensemble [19]. When wave function and Hamiltonian were combined in a Schrödinger equation, the result was the same equation he had just derived. That is, the second quantization of the equation for a single particle led to the same result as did the ordinary quantum-mechanical equations for a Bose-Einstein system.

DIRAC now had a Hamiltonian for an atom and a set of Bose-Einstein particles. In order to apply it to quanta, he needed to define the single-quantum eigenstates. He did this by applying a result from the « Theory of Quantum Mechanics ». In that paper, he had defined the state of one of the mass particles in his gas by specifying its momentum and total energy. Now he specified the state of the quantum in the same way, by its three-momentum components [20]. This was a straightforward application of his treatment of mass-points to the case of zero mass. Nevertheless, it gave rise to a significant novelty, the concept of the state of zero energy. A quantum in this state would be unobservable. This is an interesting departure from the heuristic introduced by HEISENBERG. It had uses, however. DIRAC wrote: « When a light-quantum is absorbed it can be considered to jump into this zero state, and when one is emitted it can be considered to jump from the zero state to one in which it is physically in evidence, so that it appears to have been created » [21].

Adding a term for the energy of the perturbing atom to the energy of the quanta and the perturbation energy, DIRAC had a Hamiltonian for a closed system [22]. He was now in a position to complete his earlier work and to calculate the coefficient of spontaneous emission. A solution from this direction had not at all been anticipated previously [23]. However, the Hamiltonian in terms of the quanta did not in itself give sufficient information. It was at this point in the paper that DIRAC turned to the Hamiltonian in which light was represented as a continuous electromagnetic field. It may be that he was motivated to do so precisely by the desire to calculate the spontaneous emission. It would certainly be natural to return in this way to the equations he had already used for the $B$-coefficients, and to see what more could be done with them. If that were the case, however, Dirac's formulation of the wave-particle equivalence had its genesis in a concrete piece of problem solving.

The steps by which DIRAC developed his Hamiltonian were as follows: He wrote down an expression for the interaction of an atom and a field, which was almost identical with that he had used in the « Theory of Quantum Mechanics ». The classical vector potential is expanded in a Fourier series over all frequencies. The amplitude of each of the Fourier components is connected with the energy by a classical relation. Now DIRAC introduced the occupation numbers. He did this simply through the use of Einstein's relation: the energy of a given component of the radiation was set equal to the number of quanta associated with that component, multiplied by $h\nu$. This gave a classical quantum equation, so to speak. The final step was to interpret the numbers of quanta and the conjugated phases as operators and to rewrite the Hamiltonian in a manner appropriate to this interpretation. When one identifies this with the Hamiltonian for atom and quanta, one finds that the perturbation matrix of the latter, $v_{rs}$, is now specified as a function of the atom's polarization [24].

There were more precedents for this part of Dirac's paper. The interpretation of the energy of the field by assigning a number of quanta to each component, proportional to that component's energy, had been made by DEBYE and EHRENFEST before the World War [25]. And, as DIRAC pointed out in his introduction, transformations, similar to his, of field variables into operators had appeared in the matrix-mechanics papers of BORN and JORDAN, and BORN, JORDAN and HEISENBERG [26]. On the other hand, Dirac's final result, the demonstration of two equivalent Hamiltonians, was entirely new.

DIRAC pointed out the physical meaning of his result in the introduction to the paper. It expressed « a complete harmony between the wave and light-quantum descriptions of the interaction » between an atom and the field. The quantization of the electromagnetic field, as carried out in the construction of the second Hamiltonian for atoms and waves, by itself brought with it the introduction of light quanta. « (The assumption) that the variables $E_r$, $\Theta_r$ are $q$-numbers satisfying the standard quantum conditions... immediately gives light-quantum properties to the radiations » because $E_r/2\pi\nu_r$ becomes thereby an action variable and « can change only by integral multiples of the quantum $(2\pi h)\nu_r$ » [27]. Dirac's next paper « The Quantum Theory of Dispersion » [28] began with a re-statement of these ideas. His new quantum theory of radiation made it possible to discard the correspondence method of solving problems by inserting quantum quantities into classical formulae. The theory based itself on a quantum treatment of the entire interacting system. The result was « a complete formal reconciliation between the wave and light-quantum points of view » [29]. This « formal equivalence » emerged as Dirac's answer to the meaning of the proposition that light is both particle and wave.

We may compare this with Bohr's complementarity, on the one hand, and Jordan's interpretation, on the other. BOHR took the wave and particle models as both valid, but mutually limiting, or complementary. The two models did not both hold for the same situation. JORDAN did not accept the mutual limitation BOHR proposed. He based himself on second quantization. « We see », wrote JORDAN, « that it is not necessary to abandon or to limit the wave theory in favor of a representation by other models: but it is merely a matter of carrying the wave theory over into quantum mechanics ». Then the existence of quanta presents itself as a consequence of wave theory; the particle picture arises out of the wave picture when the waves are quantized [30]. Dirac's stand in 1927 was that of JORDAN, rather than that of BOHR [31].

In the « Quantum Theory of Radiation », DIRAC also drew explicit attention to the relations between particles and waves that his theory implied. « It should be observed », he wrote, « that there is a difference between a light-wave and the de Broglie or Schrödinger wave associated with the light-quanta ». In particular, their intensities are to be interpreted differently. A wave like a light wave only appears in connection with an Einstein-Bose assembly. « Thus there is no such wave associated with electrons » [32].

Einstein's scheme has therefore given way. Instead of the four terms: material particle, matter wave, light quantum, light wave, we now have an arrangement with five terms: material particle and antisymmetrical Schrödinger wave, on the one side, and light quantum, symmetric Schrödinger wave and light wave, on the other. Between the quanta and symmetric Schrödinger wave subsists the same kind of duality as between the electrons and antisymmetric waves. Between the quanta and light waves there is a second, distinct duality.

This pointed to a second, more fundamental, departure from Einstein's ideas, which was still only shadowed in Dirac's paper. In the new scheme Schrödinger waves were associated with both quanta and mass points. The fundamental distinction between the waves for the two cases was their symmetry, or, alternatively, the statistics of the particle they represented. This distinction has nothing to do with essential differences between matter and light. Indeed, EINSTEIN had already shown that material atoms could obey the same statistics as quanta. From 1905, EINSTEIN had been probing the properties of matter and light. His investigations of light's particulate aspect, of the energy-mass equivalence and of the identical statistical behaviour of certain ideal gases and radiation served to uncover similarities between these two categories of substance, and even to some extent to break down the division between them. Quantum electrodynamics, on the other hand, opened the way to emergence of entirely new categories which cut across the older division. Alongside of a classification of substance into matter and radiation, it was now also divided into Bose-Einstein systems, with their two associated waves, and Fermi-Dirac systems, with one wave. In 1927 light was still the only elementary member of the first class, and material particles of the second. But forthcoming discoveries and theories were soon to give these categories a larger content [34].

* * *

It is a pleasure to acknowledge the help of Dr. M. HOSKIN and Prof. S. S. SCHWEBER, and the critical reading of the first draft by Mr. M. LOWE and Prof. M. JAMMER.

## REFERENCES

[1] *Proc. Roy. Soc. London*, A, **114** (1927), p. 243, reprinted in J. SCHWINGER, ed.: *Selected Papers on Quantum Electrodynamics* (New York, 1958), p. 1.

[2] The exception is F. HUND: *Geschichte der Quantentheorie* (Mannheim, 1967). His work is the starting point for mine. Although I do not follow the particulars of his analysis, I am indebted to his emphasis on quantum field theory and its importance for concepts of matter.

[3] A. EINSTEIN: *Quantentheorie des einatomigen idealen Gases*, in *Berliner Berichte*: *Sitzung der physikalisch-mathematischen Klasse* (1924), p. 261; (1925), p. 3.

[4] M. J. KLEIN: *Einstein and the wave-particle duality*, in *The Natural Philosopher*, **3** (1964), p. 1.

[5] *Proc. Roy. Soc. London*, A, **112**, p. 661.

[6] *Quantentheorie des einatomigen idealen Gases*, note 3, Part II, p. 7 (« tiefe Wesensverwandtschaft »).

[7] W. HEISENBERG: *Quantum-theoretical re-interpretation of kinematic and mechanical relations*, in *Sources of Quantum Mechanics*, edited by B. L. VAN DER WAERDEN (Amsterdam, 1967), p. 262.

[8] *Proc. Roy. Soc. London*, A, **113**, p. 622.

[9] *Theory of Quantum Mechanics*, note 5, p. 666.

[10] *Ibid.*, p. 671-672.

[11] *Ibid.*, p. 672-673. Compare Dirac's equation (18) with Einstein's (2*a*), p. 5, and Dirac's p. 672 with Einstein's p. 5.

[12] *Ibid.*, p. 673.

[13] *Relativity Quantum Mechanics with an Application to Compton Scattering*, *Proc. Roy. Soc. London*, A, **111** (1926), p. 415. Compton's papers were published beginning May, 1923. See M. JAMMER: *The Conceptual Development of Quantum Mechanics* (New York, 1966).

[14] JORDAN, who worked in Göttingen with BORN, seems to have been influenced as much by EINSTEIN as by the Copenhagen physicists. He took from EINSTEIN the idea of the analogy of light and matter and made it a fundamental guide for his investigations. Dirac's paper on quantum electrodynamics had unique

importance for him. Combining it with the light-matter analogy, JORDAN saw Dirac's results on the relation between light wave and light particle as presenting a pattern for solving the « still deeper problem of the existence of the electron » (my translation). P. JORDAN: *Der gegenwärtige Stand der Quantenelektrodynamik*, *Phys. Zeits.*, **30** (1929), p. 702. See also the *Archive for History of Quantum Physics*, interview with P. JORDAN, June 18 and June 20, 1963.

[15] *Archive for the History of Quantum Physics*, interview with P. A. M. DIRAC, May 14, 1963, p. 19-21.

[16] Writing the function $\psi(\tau)_{\text{perturbed}}$ as $\sum_n a_n(\tau)\psi_{n\,\text{unperturbed}}$ he sets $N_n$, the number of particles in the $n$-th state, equal to $|a_n|^2$, p. 248. Compare *Theory of Quantum Mechanics*, p. 674. He then shifts to the more convenient variable $b_n = a_n \exp[-i\omega_n \tau/h]$. $N_n$ remains the same, but its canonically conjugate variable alters. If one calls the latter $\theta_n$, the Hamiltonian becomes $\sum_{r,s} H_{rs} N_r^{\frac{1}{2}} N_s^{\frac{1}{2}} \exp[i(\theta_r - \theta_s)/h]$, p. 249-250.

[17] The commutation relations are chosen so that $b_r b_r^* - b_r^* b_r = 1$, p. 251. These relations are analogous to those in use for dynamical systems at this time. P. JORDAN and E. WIGNER showed, in *Über das Paulische Äquivalenzverbot*, in *Zeits. Phys.*, **47** (1928), p. 631 (reprinted in SCHWINGER: *Selected Papers*, [1]), that, if anticommutation relations are chosen instead, Fermi-Dirac systems result.

[18] The $r$-th phase $\theta_r$ corresponds to $\partial/\partial N_r'$. It occurs in the Hamiltonian as $\exp[\pm i\theta_r/h]$. The latter operates on a function of the occupation numbers $N_r'$ to diminish/increase $N_r'$ by 1, p. 252.

[19] *Ibid.*, p. 253.

[20] *Ibid.*, p. 260. In addition, there is a variable depending on polarization. The energy is determined by the momentum, since the rest mass is zero.

[21] *Ibid.*, p. 260-261.

[22] *Ibid.*, p. 255 and 261.

[23] In August, Dirac thought that the spontaneous emission might depend on the relative positions of the atoms: *Theory of Quantum Mechanics*, p. 677.

[24] $v_{rs} = 0$ for $r, s \neq 0$, $v_r = h^{\frac{1}{2}} c^{-\frac{3}{2}} (\nu_r/\sigma_r)^{\frac{1}{2}} \dot{X}_r = v_r^*$, where $X_r$ is the polarization of the atom in the direction of the $r$-th component of the vector potential, $\sigma_r$ is the number of Fourier components with a given polarization, per unit frequency range and per unit solid angle of wave direction. $v_r \equiv v_{r0}(N_0 + 1)^{\frac{1}{2}} \exp[-i\theta_0/h]$, $v_r^* \equiv v_{0r} N_0^{\frac{1}{2}} \exp[i\theta_0/h]$ (p. 261).

[25] M. BORN, W. HEISENBERG and P. JORDAN: *On Quantum Mechanics, II*, in B. L. VAN DER WAERDEN: *Sources* [7], p. 376.

[26] *Quantum Theory of Radiation*, p. 245. SCHWINGER, on the contrary, picks out exactly this part of the paper as marking « the beginning of quantum electrodynamics ». This is worth pointing out because Schwinger's comments are likely to become a source for historical passages in physics texts. *Selected Papers* [1], p. vii-viii.

[27] *Quantum Theory of Radiation*, p. 244-245.

[28] *The Quantum Theory of Dispersion*, *Proc. Roy. Soc. London*, A, **114** (1927), p. 710.

[29] *Ibid.*, p. 711.

[30] *Gegenwärtiger Stand der Quantenelektrodynamik* [14], p. 702.

[31] I am indebted to Prof. P. A. M. DIRAC for pointing this out in his comments, and thus directing my attention to the relevant texts.

[32] *Quantum Theory of Radiation*, p. 247. A monochromatic light wave of frequency $f$ has an intensity proportional to the number of light quanta of that frequency. A monochromatic Schrödinger wave has an intensity proportional to the total number of quanta present at all frequencies.

[33] The most important particle to enter physics in the next few years, in this connection, was the neutrino. Among the important theoretical developments was the 1934 paper of PAULI and WEISSKOPF. See F. HUND: *Geschichte der Quantentheorie*, p. 187.

# Development of Solid-State Physics.

H. B. G. CASIMIR (*)

*Philips Research Laboratories - Eindhoven*

In one of his lectures KLEIN showed that the interpretation of the specific heat of solids at low temperatures played an important role in establishing quantum theory and in convincing the physical community of the existence of a quantum. Let me just recall briefly the main formulae. The average energy of a quantized oscillator is given by

$$\langle \varepsilon \rangle = \frac{h\nu}{\exp[h\nu/kT] - 1}.$$

In the original theory of EINSTEIN it was for simplicity supposed that there were as many oscillators as atoms, all with the same frequency, but, as was shown by NERNST, agreement with experiment can be improved by admitting the possibility of several frequencies. Of course the most general form for the internal energy would be

$$U = \int \frac{h\nu}{\exp[h\nu/kT] - 1} \varrho(\nu)\, \mathrm{d}\nu .$$

Evidently, if you take for $\varrho(\nu)$ a $\delta$-function, it brings you back to one single oscillator.

It was in 1912 that my countryman DEBYE calculated the distribution function $\varrho(\nu)$ by making a bold approximation. He assumed that the vibrations were acoustical vibrations. He started from long-wavelength acoustical vibrations in an elastic solid and that led him to a distribution

$$\varrho(\nu) = \frac{\text{const}}{v_{\text{sound}}^3} \nu^2 \, \mathrm{d}\nu , \qquad \nu < \nu_{\max} ,$$

---

(*) The author wishes to acknowledge the valuable assistance of J. KOOPS in preparing the manuscripts of his four Varenna lectures.

and

$$\varrho(\nu) = 0 \qquad \text{for } \nu > \nu_{\max}.$$

The cut-off frequency $\nu_{\max}$ is chosen in such a way that the number of vibrations is equal to the number of degrees of freedom of the system. This theory can be worked out, by taking into account that there is not one velocity of sound, $v_{\text{sound}}$, but that there are at least two, longitudinal and transversal.

This holds for an isotropic substance. In a crystal the situation is more complicated. For every direction of propagation there are three distinct velocities. For the specific heat, DEBYE found the formula

$$c_v = f\left(\frac{T}{\theta}\right),$$

where the (Debye) temperature $\theta$ is defined by

$$k\theta = h\nu_{\max}.$$

At high temperatures, this leads to the law of DULONG and PETIT and at low temperatures it leads to a specific heat that is proportional to $T^3$.

It is perhaps interesting, that NERNST, who did so much to get the quantum theory accepted, was, at first, not at all in agreement with this idea of DEBYE. As DEBYE once told me, he received as a reaction to his publication a nice letter from NERNST; it ran roughly as follows:

« Lieber Herr Debye,

Ihre Arbeit zeigt einmal wieder, dass Sie von der Quantentheorie weniger verstanden haben als mein schlechtester Schüler ».
(Your paper shows clearly, once more, that you have understood less of quantum theory than my worst pupil.)

NERNST had a rather special style! I should add, that later on he corrected his opinion; in his book the theory is treated and accepted. So far for the specific heat.

The physics of solids has been studied since the very beginning of physics: the determination of physical properties of matter, like thermodynamic, elastic and mechanical properties and even geometrical properties, was one of the current topics in physics.

In the 19th century there was an emphasis on optical properties—crystal optics—and also on elastic properties. A phenomenological theory was well developed, especially concerning the connection between crystal symmetries and physical phenomena, *e.g.* how many independent constants occur in an elastic tensor in an orthorhombic crystal.

This subject was almost forgotten but is now coming up again, occasionally, in connection with new problems.

Much of the older work is covered by the book of the German physicist VOIGT, *Kristallphysik*. He discusses in a very systematic way phenomena described by tensors of the second rank, by tensors of the fourth rank, phenomena depending on the relation between two vectors, or between a vector and a tensor and so on.

Of course, these things can be done more elegantly these days, using the formal methods of group theory.

Now, the main thesis of this lecture is that the later development of solid-state physics did not play an important role in formulating the basic ideas of quantum theory. It did in the early days of specific heat, it did not in later years. It was the other way around: the development of solid-state physics did not contribute to establish the ideas of quantum mechanics, but, as soon as the quantum theory had been developed to the extent of becoming a useful tool in calculating atomic phenomena, the great problems of solid-state physics were almost automatically solved one by one.

Automatically is not quite the word: it took a lot of mathematical technique and knowledge of new quantum mechanics to solve the solid-state problems, but it was not necessary to modify the essential ideas of quantum mechanics for that purpose. Nor did the successes in solid-state physics contribute in any very pronounced way to the question of acceptance of quantum mechanics and its interpretation.

So, from a very high-brow point of view one might say that even the basic study of solid-state physics is in a way a kind of technology.

From now on I shall deal mainly with the question of conduction of electricity in metals.

As soon as the idea of electrons was generally accepted, it became logical to look at the conduction of electricity in metals in terms of the movement of free or almost free electrons through matter. That was done by THOMSON, and in Germany by DRUDE. It was LORENTZ who, applying the ideas of Boltzmann's transport equation, worked out the theory in greater detail.

I shall not go through those details. I shall give only a very sketchy presentation of some of the results and point out the great difficulties that remained in that theory. Already the question of the number of free electrons gives difficulties. If there are $n$ free electrons per $cm^3$, then they should contribute, according to classical statistics, to the specific heat

$$C_{el} = \tfrac{3}{2}\, nk\, .$$

On the other hand we know that the law of DULONG and PETIT holds for most metals at room temperatures. And that means that every vibrational degree of freedom of the atom has an energy

$$\langle \varepsilon \rangle = kT\, .$$

So, if there were as many electrons as there are atoms, there should be a very sizable contribution to the specific heat. Yet Dulong and Petit's law holds. This puzzle became worse when one started to understand more about atomic models. For an atom like sodium, for instance, which can easily lose an electron, one would expect as many free electrons as atoms.

Next let us look at the conductivity. Let us write down a very sketchy formula which is sufficient to establish the main features

$$\sigma = \frac{nev_d}{E}.$$

The conductivity $\sigma$ will be equal to the number of electrons $n$, times the charge $e$, times the drift velocity $v_d$, divided by the electric field $E$. The drift velocity divided by the electric field is equal to

$$\frac{v_d}{E} = \frac{e}{m}\frac{l}{v},$$

because the charge times the electric field, multiplied by the time between collisions $l/v$, is the average momentum acquired, and, divided by $m$, the drift velocity. Therefore

$$\sigma = \frac{ne^2}{m}\frac{l}{v}.$$

More rigorous calculations on a well-defined model give essentially the same result.

This is a very nice and plausible formula, but a difficulty is the following one. The velocity $v$ will be proportional to the square root of the temperature.

We know that the conductivity is inversely proportional to the temperature. So the free path would have to be proportional to the square root

$$l = \frac{C}{\sqrt{T}},$$

and it is difficult to imagine any kind of model that would lead to that temperature dependence.

The electrons do not only contribute to the electrical conductivity but also to the thermal conductivity. That explains why metals are not only good electrical conductors, but also good thermal conductors. The thermal conductivity is, again in a sketchy way, equal to

$$k = \tfrac{1}{3}(nk)\, v_e\, l\,,$$

where $v_e$ is the average electron velocity. This is a very general formula; you can derive it from elementary kinetics or again, of course, from a detailed theory of transport phenomena along the lines of BOLTZMANN. One interesting point is that this gives

$$\frac{k}{T\sigma} = \text{const}\,\frac{k^2}{e^2}\,.$$

This explains the experimental law of Wiedemann-Franz, which states that there is a constant ratio between thermal conductivity and the product of the electrical conductivity $\sigma$ by the absolute temperature. It was one of the major successes of the early theory of conductivity that it could explain this experimental law, and that even the constant came out with a value that was roughly, say within 10 % or so, in agreement with experiment.

The theory could also explain the existence of the Hall effect, which is that when a current flows through a conductor in a tranverse magnetic field there exists a transverse potential difference. This is just an example of the Lorentz force: the electrons going through the metal have an average drift velocity and therefore they experience an average Lorentz force driving them in a transverse direction. One can easily write down what the transverse electric field will be:

$$E_t = \frac{v_{\text{drift}}}{c}\,H\,.$$

This formula for the Hall effect was alright as far as the order of magnitude was concerned. But then, one met with one of the major difficulties of this theory: the Hall effect is sometimes positive. The sign is for certain metals, for instance bismuth in certain crystal directions, such that it would appear that the charge carriers are positive, and not negative.

On the other hand, one assembled more and more overwhelming evidence that the electrons were negative and that the positive charges were heavy ions, and it was quite clear that, when a current moves through a metal, there is no ion transport.

This difficulty is emphasized in the writings of LORENTZ and I think this argument even induced NERNST to discard entirely the idea of free electrons; in the same way in which he discarded in the prequantum days the whole idea of statistical mechanics. He said, it is quite obvious that at low temperatures equipartition has no relation whatever to nature and then to develop a whole mathematical theory based essentially on the idea of equipartition is just a waste of time. Perhaps he also considered the mathematical work of LORENTZ a waste of time!

Let me summarize the situation: we do not understand the specific heat, we cannot understand the temperature dependence of the electrical conductivity, we do understand the Wiedemann-Franz law and we cannot understand the positive Hall effect. EHRENFEST always pointed out in his lectures that the occurrence of a positive Hall effect in metals was one of the major mysteries.

There were other serious difficulties in connection with temperature dependence of thermo-electrical quantities; those I shall not discuss today.

BOHR, in 1911 in his thesis, worked out Lorentz's theory in a more general form and relied far less on specific models than LORENTZ did. He found, roughly speaking, the above results but showed that they could be obtained with much more general assumptions about the interaction between electrons and obstacles. Therefore his thesis confirmed the weaknesses of this classical theory of electrons in metals. Incidentally, Bohr's thesis also contains an interesting discussion about the relation between thermo-electric quantities. KELVIN had derived certain relations between thermo-electric quantities by a kind of pseudo-thermodynamics.

This thermodynamic reasoning is unjustified. Yet BOHR was able to prove the relations in a rather general way. Let me just indicate the main lines of the argument. BOHR showed that Kelvin's relations are a consequence of a symmetry property of the integral equation for the transport phenomena. And that symmetry property he derives by inverting the time, by using time symmetry.

The later work of ONSAGER, who showed that in a very general way the invariance with respect to time reversal of the fundamental equations leads to certain relations for irreversible processes, was foreshadowed by a discussion of a model, a very general one, in Bohr's thesis. Well, this was just a little digression.

Now, what happened later on with quantum mechanics? First of all classical statistics was replaced by Fermi-Dirac statistics. As we know, the number of electrons in a state $\varepsilon$ is given by

$$n(\varepsilon) = \frac{1}{\exp[(\varepsilon - \zeta)/kT] + 1}\,.$$

The $\zeta$-level has to be chosen such that you get the right number of electrons at a given temperature.

It really means that, roughly speaking, up to a level $\zeta$ all states are filled; above that level they are not.

That is true at the absolute zero. At other temperatures states are filled almost up to the $\zeta$-level, but in a band of the width of $kT$ there are electrons missing below $\zeta$ and electrons running about above $\zeta$.

So you start by filling up to a level $\zeta$, then over an energy spread of about $kT$, exponentially decreasing on both sides, there are electrons jumping about above $\zeta$ and there are unfilled places, holes, just below the $\zeta$-level.

The first one to apply this Fermi-Dirac statistic to a typical solid-state problem was PAULI. That led to his famous phrase: « Ich mag diese Theorie des festen Körpers nicht, zwar habe ich den Anfang damit gemacht ». (I do not like the theory of solid-state physics, although I did start it.) He discussed the paramagnetism of an electron gas. As soon as one knew that the electron has a spin, this became a problem.

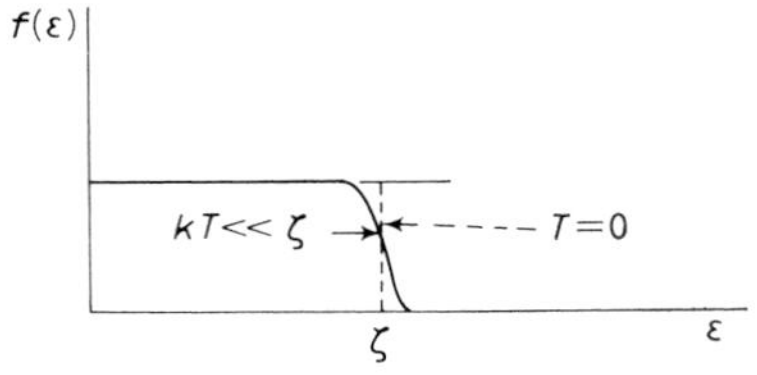

Fig. 1.

Applying the classical Langevin theory one finds for the magnetic susceptibility

$$\chi = n \frac{\mu^2}{kT},$$

where $n$ is the number of electrons and $\mu$ the magnetic moment of the electron. Taking as many electrons as atoms this gives an enormous paramagnetic susceptibility, especially at low temperatures which had not been observed.

PAULI pointed out that, for a Fermi distribution, an applied field does not orient all electrons. Below the $\zeta$-level they just remain with spin up and down. Only in the small $kT$ band are electrons able to change orientation in a magnetic field. As a result one finds a temperature-independent paramagnetism which is quite small.

This PAULI worked out and I think he was also the first to show that the notion of the grand ensemble introduced by GIBBS was ideally suited for treating this kind of statistics. As a matter of fact, applying the idea of a grand ensemble to Fermi-Dirac statistics makes the distribution almost a tautology.

However, PAULI did not look at other properties of the electron gas.

The next step was taken by SOMMERFELD. He rewrote the theory of LORENTZ, using Fermi-Dirac statistics instead of classical statistics. Now, what happened? First of all the specific heat becomes different. Fermi-Dirac statistics lead to a specific heat

$$C_v = \frac{3}{2} nk \left(\frac{T}{T_D}\right),$$

where $T_D$ is a temperature of degeneracy, which is within a small numerical factor given by

$$kT_D \approx \zeta .$$

Here $\zeta$ is again the energy to which the band is filled. If one takes into account the volume and the number of electrons, assumes one or two electrons per atom, according to valency, that energy is something in the order of electron-volts, $T_D$ something in the order of 10 000 °K. Therefore, at room temperature you do not get the Dulong and Petit term, but something which is a factor one hundred or so smaller. This solved the problem of specific heat.

At low temperatures this linear specific heat can be determined experimentally. You would say, why at low temperatures, because it gets even smaller. But remember that the specific heat of the lattice vibrations goes down with $T^3$. At about 4 °K, for most metals the linear term becomes comparable with the cubic term. This was found soon afterwards by KEESOM and his co-workers at Leyden. It provided a beautiful confirmation of this part of the theory.

What happens to the conduction properties? Well, you get essentially the same formula, only the average electron velocity has to be replaced by $v_F$, the velocity at the edge of the Fermi distribution:

$$\sigma = ne \frac{e}{mv_F} l .$$

Also for the thermal conductivity you have to take the velocity of electrons at the edge of the Fermi distribution and the specific heat $nk$ has to be multiplied by $T/T_D$:

$$k = \frac{1}{3} (nk) v_F l \frac{T}{T_D} .$$

But you still get the Wiedemann-Franz law, because a factor $mv_F^2$ cancels against $kT_D$. You get the Wiedemann-Franz law with a slightly different constant, which happens to be in somewhat better agreement with experiment than the original Lorentz-Bohr value.

But still one did not quite well understand the behaviour of the free path. The velocity at the edge of the Fermi distribution $v_F$ is essentially temperature independent, so now the free path would have to be inversely proportional to $T$. That looks plausible at least. It makes you think of equipartition of vibrations and what have you, but you could not derive it really from such a model because you could not find the magnitude. And still the positive Hall effect was not explained.

The next step was the theory of BLOCH. There was an intermediate phase; the American physicist HOUSTON tried to calculate the free path by looking at the scattering of electron waves, but it was BLOCH who in 1928 gave a more complete theory. BLOCH started to study the wave function of an electron moving in a purely periodic field of force. He showed that in that case the wave function $\psi$ is given by a plane wave multiplied by a periodic function, depending on the wave vector $k$ and the period of the lattice:

$$\psi = u_k(r) \exp [i\boldsymbol{k} \cdot \boldsymbol{r}] .$$

This means that in a perfectly periodic lattice these waves can propagate without any scattering, so in an ideal lattice you would have an infinite free path. This is a very fundamental result.

At once one understands why perfect metals have such a high conductivity. But thermal motion causes irregularities in the lattice and every irregularity will lead to scattering of these Bloch wave functions. That was already very satisfactory. A further refinement was the notion of energy bands. It soon turned out, by using simplified models or the elaborate theory of Mathieu functions, that in this case there are no discrete energy levels but that there exist zones of allowed energy and forbidden regions in between.

Often the energy is plotted *vs.* the wave vector in a certain direction. But in reality the energy bands have very complicated three-dimensional structures.

Often one also uses a picture where the energy bands are plotted against a co-ordinate. This is perhaps a bit misleading because a wave function corresponding to a definite energy is omnipresent. But if the situation changes slowly from place to place this picture is permissible. The picture of energy bands for electrons in crystals, with allowed zones and forbidden zones, with wave functions without scattering, is now well known and universally used.

PAULI was, as said, somewhat antagonistic to solid-state physics and may have felt that SOMMERFELD had not made much progress. It seems that BOHR felt differently and expected right away that this work would lead to further progress. In any case I remember seeing a reprint of Bloch's first paper, sent by PAULI to BOHR, on which he had written: « From this reprint you will see that your triumph over the world of physicists is complete ». I hope I am right in my recollection of this particular reprint with this inscription of PAULI. If not, let it be part of folklore, it will not do much harm.

Now the specific heat was explained and the long free path was understood, but BLOCH could do more. He found for the free path a linear dependence on the reciprocal of temperature, so that the resistance becomes linear with $T$ for higher temperatures. He also found that, at low temperatures where the specific heat goes down with $T^3$, the resistance is proportional to $T^5$. You might think, why not $T^3$. But that is because there are not only fewer acoustical quanta, but also their energy becomes too small to have a complete matching with the momenta of the electrons. In order to change the direction of an electron by 180° one collision is not sufficient; a number of collisions are required because the acoustical quanta are so small. This leads, when worked out in detail, to a $T^5$-law.

It is easily seen that impurities lead to a constant scattering and hence to a constant resistance. This again is in agreement with findings at low temperature; at low temperatures the resistance is equal to the residual resistance $R_0$ plus a term $AT^5$. The occurrence of a constant residual resistance near the absolute zero is again a beautiful confirmation of the theory.

Of course, there are refinements. In some cases, at very low temperatures $R_0$

goes up again, an effect first discovered by DE HAAS and VAN DEN BERG at Leyden and often referred to as Kondo effect, because much later KONDO gave a tentative explanation of this phenomenon.

There still remained the great puzzle of the positive Hall effect. The answer to that was given by PEIERLS in 1929.

PEIERLS worked in those days with HEISENBERG who had studied the question in some detail, which DIRAC also referred to in his lecture here when speaking about the positron, *viz.* the question of the behaviour of closed shells and almost closed shells. HEISENBERG had shown mathematically that if you have an almost closed shell with one electron missing, you can consider it, to a good approximation, as one particle which, however, would be a positively charged particle. Those ideas were certainly well known to PEIERLS when he started to work on the influence of a magnetic field on such energy bands. What did he find?

It had already been shown by BLOCH that the second derivative of the energy with respect to the wave number $k$, which is a pseudo-momentum, is a kind of effective mass. Now, BLOCH, and also PEIERLS in earlier work, had shown that at the edge of an energy band the curvature of $E$ *vs.* $k$ is such that this effective mass is negative. Which means that, if you would have an electron in such a state and you put on an electric field, it would move against the electric field, it would have a negative effective mass. Now suppose that such a band is almost filled, then you have missing particles in an almost full band; but in these unfilled states the particles would have a negative mass. Then these holes will behave as positive particles. This idea of holes behaving as positive particles occurs for the first time in Peierls' 1929 paper. And that explained the positive Hall effect. I hope that I have proved my point. By consequent and clever application of the ideas of quantum mechanics but without adding to its formal or epistemological contents, the main puzzles of specific heat, conduction of electricity, temperature dependence, positive Hall effect were solved.

I have a few additional remarks. One or two years later the idea of holes in a negatively charged particle distribution arose quite independently in connection with Dirac's equation for the electron and led to the theory of the positron. I do not think that at that time one was so much aware of the analogy between these two cases. Looking back and knowing more about semiconductors, it is rather amusing to work out the analogy. Looking from the point of view of semiconductors, one might say: empty space is a semiconductor with a band gap of a million volts.

In normally used semiconductors the band gap is of the order of one volt. In that case, as was shown by WILSON, when a band is entirely filled at absolute zero, then at higher temperatures you will get some holes in the valence band and some electrons in the conduction band. That is known now as intrinsic conductivity. The intrinsic conductivity of empty space is negligible, because $kT$ is usually quite small in comparison with one million volts. It is already very hot when $kT$ amounts to ten or one hundred volts, and a factor $e^{-100\,000}$ is a very small quantity indeed, so we do not have any intrinsic conductivity. You can also dope a semiconductor: if you put into a semiconductor an atom which easily loses its electrons, then there remains a positively charged stationary ion and an extra electron moving in the conduction band. You can also take an ion which likes to bind electrons. Then you get a state of affairs with an extra hole in the valence band. That is, what is called a *p-type* semiconductor. Well, our empty space is a *n-type* semiconductor. It has been doped by the presence of all the positive nuclei into having a surplus of negative electrons.

In a semiconductor you can have photoconductivity. By injecting light you can create a pair of an electron and a hole. You can also create an electron-positron pair. (Not in empty space, some nuclei are required to take up momentum.) Electron and positron recombine just as a hole and an electron can recombine. You can make an atom which does not exist very long, positronium, with a positive and a negative electron.

In a semiconductor there can exist a temporary structure of a hole and an electron, turning around one another; this is known in solid-state physics as an exciton.

Although there is a fargoing analogy between the two cases there are also differences. One difference is that because of the peculiar band structure there is no symmetry between holes and electrons in a semiconductor. The effective mass of the holes can be quite different from the effective mass of the electrons. Also, in a crystal, this effective mass may depend in a very anisotropic way on the direction of the momentum vector. Another major difference is that, in treating these phenomena in solid-state physics, it is not necessary to take into account relativity, because the velocities that occur are always of the order $\alpha c$, so relativistic corrections are very slight.

A further remark. We all know that later the properties of semiconductors became of great technological importance. But that was only after World War II, when in the Bell Laboratories an intense program was started with the technical goal in mind to make amplifiers that would function without thermionic cathodes. Therefore, they would presumably have an infinite lifetime and they could be switched on immediately without waiting time, which, of course, was very important for telephone amplifiers. For those who do not know, I might just mention that in the U.S.A. the telephone network is not operated like in most European countries by the government, but by private corporations, of which by far the largest, with about 80 % or so of the total coverage, is American Telegraph and Telephone (ATT). This corporation fully owns Western Electric, which manufactures all equipment, and it also controls the research and development corporation, the Bell Laboratories, which are the largest and, in many ways, the best industrial research laboratories existing today.

During the war the technology of germanium and silicon had been developed to a certain extent in connection with rectifiers, that were needed in radar, and then, with the above-mentioned theoretical background, an attack on this problem was launched. BARDEEN was one of the members of the team and also SHOCKLEY, who knew a lot about solid-state physics.

The basic discovery they made was that in $n$-type germanium (with a surplus of electrons) you can inject from the outside extra holes, like you might inject positrons into matter, and that these holes have a longer free path and a longer lifetime than was anticipated. Such that you can even inject them into a slice of *n-type* germanium and pull them out on the other side. This was the essential, physical discovery that was added to the basic framework of ideas of FERMI, DIRAC, PAULI, SOMMERFELD, BLOCH, PEIERLS and WILSON. This led to the transistor and was therefore, in a way, the beginning of modern electronics.

In this connection I should like to call attention to the fact that there was a rather considerable time lag between the formulation of the basic ideas of BLOCH and PEIERLS (which in themselves were an application of the new ideas of quantum mechanics) and the penetration and use of such ideas for technological, industrial purposes.

It will be one of the main theses of my last lecture that this is very characteristic of our time. Contrary to what can be read in newspapers and in many popular books, things are, from this point of view, not going more rapidly. What is going more rapidly is building up an industry. But the time lag between the basic thoughts and their applications is, during our century, remarkably constant or even increasing.

This is, I think, an important element to keep in mind in any discussion on the relation between science and its applications and in any discussion about the responsibility of scientists. This simple fact, that the time lag between the birth of basic ideas and their application is long and, if anything, is getting longer, is a factor which makes it in some ways perhaps easier, but in some ways more difficult to get a grip on the whole scientific and technological development.

# Superconductivity.

H. B. G. Casimir

*Philips Research Laboratories - Eindhoven*

In my previous lecture I tried to explain how quantum mechanics, once the appropriate formalism had been created, explained almost automatically some of the major problems of the solid state, more specifically, of the conduction of electrons. I hope this should not lead to misunderstanding. If I say almost automatically, it does not mean that it did not require a great amount of ingenuity and even new ideas to apply this quantum mechanics to the solid state, but, as I said, it did not require new points of view with respect to the really basic equations and basic theory; although it did require new mathematical developments of the theory.

However, one major phenomenon remained ununderstood, and that was superconductivity. Superconductivity, the property of some metals to lose every trace of electrical resistance below a certain temperature, had been discovered by Kamerlingh Onnes and his co-workers in 1911. I mention his co-workers, because the first observations on superconductivity were not done by Kamerlingh Onnes himself, although he prepared the ground by liquefying helium and by realizing that it was interesting to study the conduction of electricity at low temperatures. Should one say it was his co-workers who found superconductivity? After all, once Kamerlingh Onnes gave them liquid helium and told them to measure the resistance of metals, they could hardly fail to discover it.

That took place in 1911, about three years after the first liquefaction of helium. Now, why was it impossible to explain superconductivity in 1930 or thereabout? I would say for two reasons: first of all because the phenomenology of superconductivity was not well established, although the phenomenon had been known for almost twenty years. There were even some curious misconceptions—I will come back to that—about the nature of superconductivity. Therefore, one did not know *what* to explain, which made things more difficult. It was only after 1930 that one arrived at a more consistent picture of the macroscopic properties of superconductors. And, secondly, the theoretical explanation, as given about twenty years later, involved some further developments both of the mathematical formalism and also of its interpretation. We shall come back to that later, too.

Now, the early history of superconductivity is in a way easy to trace for the simple reason that the scene of low-temperature physics, that is of research at temperatures below the boiling point of liquid helium (about 4 °K), was concentrated in the Leyden physical laboratories.

For many years it was the only place where liquid helium was being made; then there were a few other centres, the National Bureau of Standards in the U.S.A., the Physikalisch-Technische Reichsanstalt in Germany; later in the twenties, a centre was created by MacLennan at Toronto. But the activity in those centres was not very extensive. Up to 1930 the major part of all work in low-temperature physics came from the Leyden laboratories. And because all the articles were always reprinted as Communications from the Physical Laboratories at Leyden, it is rather easy to get hold of the literature on the subject. You will find in the Leyden Communications also complete references to the few other papers that came from outside.

Of course, this is not true for work at liquid-nitrogen temperatures, not even quite true for work down to liquid-hydrogen temperatures, but it was certainly true for work at liquid-helium temperatures, and there the most spectacular phenomena occur.

This picture began to change in the thirties. Then a very active group started in Oxford under the direction of Simon, who had come from Germany with his collaborators, especially Mendelsohn and Kurti.

Also Kapitza had built a new type of helium liquefier, a very efficient one, and had started the Mond Laboratory in an inner court of the Cavendish Laboratory at Cambridge.

Kapitza then went back to Russia, in the early thirties, where he became the originator of the Russian school of low-temperature physics, which has contributed greatly both to experimental knowledge of the properties of liquid helium and superconductors and also to their theoretical interpretation.

The work at the Mond Laboratory was continued by Allen and Misener, who had come from Toronto where they had low-temperature experience, and later by Shoenberg and Pippard.

That was the extent of the activities in the thirties. It was only after the Second World War that helium liquefiers were produced commercially, mainly in the U.S.A., by an offshoot of A. D. Little Co., arising out of a design made by Collins at M.I.T., based on Kapitza's ideas. Then liquid helium became a current commodity in practically every self-respecting laboratory.

To summarize: up to the thirties we have practically only one laboratory, in the thirties we have a small group of laboratories and only after the Second World War liquid helium becomes a generally available commodity.

It may be interesting to say something about Kamerlingh Onnes. He had started at Leyden with work on low temperatures. Dewar, who was the first to liquefy hydrogen, had a head start of several years in this field. At first, as Klein said in one of his lectures here, it was almost a kind of sports event to liquefy all gases. Of course there was more to it; there was a theoretical guideline, *viz.* Van der Waals's equation of state

$$(p + a/V^2)(V - b) = RT\,.$$

Yet, I have it on good authority that Lorentz advised ONNES against concentrating all the efforts of his laboratory on low temperatures. LORENTZ, who was primarily a classical physicist, certainly in those days, did not see that very low temperatures would give anything very spectacular. Now, if there were no quantum of action, then low-temperature physics would not be so interesting. What makes low-temperature physics interesting is the fact that the states are quantized and that at low temperatures only the lower quantum states are occupied. This is also the basis of Nernst's theorem or postulate.

KAMERLINGH ONNES must have had a feeling of that. In any case, he concentrated all his energy in that field. And he was a great manager. In an era when physics was still very much in the string and sealing wax period, when in Germany physical laboratories were often known as « das physikalisches Kabinett », which suggests a collection of curiosities rather than equipment, he built what might be called an industrial organization.

I have mentioned already during a discussion that he had the idea of organizing a school—paid for by the Department of Education—of instrument makers and glass blowers. Glass blowing was not at a high level in Holland in those days, so he went to Jena and hired the very best glass blower he could find. This man, KESSELRING, became the beginning of a tradition of instrument glass blowing in the Netherlands and perhaps even in many other places, because quite a few laboratories, even in the U.S.A., hired glass blowers trained at Leyden. Not all his methods were equally agreeable: he got a number of positions for assistants and then, if a man appointed an assistant, this was such a privilege that he had to pay back his salary into a special fund, which was used for buying more equipment. In those days that could be done.

KARMERLINGH ONNES was rather authoritarian but with regard to his technicians and mechanics he was certainly a benevolent despot. He told them exactly what to do but he paid them better than was usual in those days. He saw to it that apprentices got good jobs, but I think he believed in a stratified society, with clear-cut social distinctions. His chief technician was a man called FLIM. His name hardly occurs in the literature, yet it is certain that the liquefaction of helium and the whole early development of low-temperature technology owes very much to FLIM, a man of little theoretical training, but of uncanny skill and of a great intuitive insight into what he was doing. Of course, in later years technological development and scientific development made his skill a little bit superfluous and then he may have had a retarding influence; many outstanding artisans become very conservative later in life. But in the early days he must have played a very important role. FLIM liked to speak in very short cut-off sentences. He usually started a sentence by « 'k zou zeggen », which means « I'd say », followed by a snapped-off statement. If, for instance, we wanted to have something done very rapidly or if we had some rather wild plans, he thought a moment about it and then said: « I'd say, madness ». That settled it.

There was a story in the Leyden laboratory that on one occasion FLIM really rebelled against KAMERLINGH ONNES. That was when he wanted to send his son to a secondary school preparing for university education. KAMERLINGH ONNES had a serious talk with him and said that this was inadvisable. FLIM should send the boy to a good technical school, train him in some technical trade, but he should not let him prepare for university education, as this did not correspond to his station in life. To which FLIM, according to the legend, replied in his characteristic way: « I'd say, my boy, my money, my decision ». And that was that.

In the same vein there is a story about Kamerlingh Onnes' funeral. He was living at the outskirts of Leyden and the family had a family vault at a nearby cemetery at a distance of four kilometers or so. It had been ordained by KAMERLINGH ONNES that his faithful technicians should follow the cortege on foot. It was a rather hot day and there they came in morning coats and top hats. As often in such cases, the procession was rather late in leaving the house and at the cemetery many officials were waiting. So, as soon as they were outside Leyden, the whip was put on the horses and they went at a rather brisk pace, with the technicians walking or trotting behind. They arrived at the cemetery sweating and puffing and then one of them looked at the others with a broad grin on his face and said: « That is just like the old man, even after his death he keeps you on the run ». Maybe this better characterized the man KAMERLINGH ONNES than the official funeral orations given later that day.

I do not think KAMERLINGH ONNES was always very generous when it came to his collaborators. Certainly, they got good opportunities to work and many physicists wrote a good thesis under his direction, but I think he did not like too much independent initiative by people working in his laboratory.

After the first World War, installations were considerably extended and the factory character became almost more pronounced.

EHRENFEST used to refer to the laboratory as the brewery, because of all the copper tubes and pumps.

Following the discovery of superconductivity in 1911, a number of measurements had been made, and these, together with some theoretical speculations, had led to some then generally accepted ideas, that happened to be erroneous. One was that superconductivity should be regarded as a curious property of the free path rather than as a special state of the electron gas. Why was this? Mainly because preliminary measurements had been done to ascertain whether there was a heat of transition at the transition point below which superconductivity appears, and there was not. Also, measurements had been done on the specific volume of the crystal and no change was found in the volume of the crystal lattice. There had even been done X-ray diffraction measurements which showed that nothing happened in the crystal lattice. Since there was no heat of transformation, one assumed that there was no phase transition and tried to explain a sudden change of the free path of the electrons. That was the first superstition.

The second one was even more curious. In 1919 the French physicist LIPPMANN pointed out that, if there is no resistance, that then there can be no change of magnetic field. Suppose we have a ring with a magnetic field passing through it. If that field would change, currents would be induced. Even a small $\boldsymbol{B}$ would give rise to an infinitely large current, so the thing must arrange itself in such a way that the flux through the ring does not change. If we cool a ring in an external magnetic field, then there is a magnetic flux passing through it. If we take away the outer magnetic field the total number of lines of force remains constant and a current is induced, which remains flowing for hours—we know now for years—without change. That is perfectly right.

Now it was thought that, if you would cool a solid body, for instance a sphere, in a magnetic field, the same would happen. The field would be unchanged at the transition, for one believed that only the free path was changing which would not give rise to any special magnetic phenomenon. Then, if you switch off the field, you would have a persisting current and a so-called frozen-in field. Rough measurements had been done on a hollow sphere or cylinder. They seemed to confirm this idea of a frozen-in field.

Let me repeat: there was the superstition that superconductivity was a change of the free path rather than a change of state of the electrons and there was the superstition of the frozen-in field even in singly connected bodies. That these superstitions could prevail so long may have had to do with peculiarities of the Dutch language. In Dutch « to measure » is called « meten » (messen in German) and to know is called « weten » (wissen in German). KAMERLINGH ONNES had created the slogan « Door meten tot weten » (by measurements to knowledge); because it happens to rhyme in Dutch it was not easily forgotten. This slogan had in some way a very bad influence on physics at Leyden. Not that one should not carry out measurements, of course one should; but it caused people to concentrate on measurements rather than on observations. It is very fine to carry out measurements of a property if you know what you are going to measure, but before you know that a phenomenon exists at all, you cannot measure. So there was a tendency not to make observations that might lead to something new, but rather to take well-known properties like specific heat, vapour tension and electrical resistance and then carry out long series of measurements under well-defined conditions. And that is not always the best way to discover new effects.

Another factor was the ritual surrounding liquid helium. That is difficult to understand for physicists these days, when liquid helium is always available. But in those days there was liquid helium only one day a week and you had to start very early with the necessary preparations; you had to apply for a helium run weeks in advance. In the morning your cryostat would be filled. Then, of course, you kept working with it, carrying out measurements as long as the helium lasted. If you were lucky, your cryostat would be filled once more towards the end of the afternoon; then you could continue until late in the evening. Anyone who was at that moment in some way connected with the measurements and who refused to stay was a traitor to the cause. This sort of ritual surrounding liquid helium made people rather reluctant to use it for qualitative observations. You felt you were almost obliged to use such a day for producing long columns of readings of well-defined instruments. Yet I must say that in a way this ritual of liquid helium was very nice. It created a team spirit that was very refreshing; I am missing this on the present scene of solid-state physics.

In the thirties the superstitions I mentioned were overcome. The first step was that DE HAAS and his collaborator VOOGD carried out measurements on the magnetic disturbance of superconductivity. It was known since the early days that an applied magnetic field destroys the phenomenon of superconductivity; then the metal behaves as any normal metal. Now it was shown that if you take single crystals, for instance of tin, and put them in a longitudinal field, you get a very sharply defined critical field. If you plot this critical field against temperature you get roughly a curve like Fig. 1. Such curves had been seen before, but not as sharply defined.

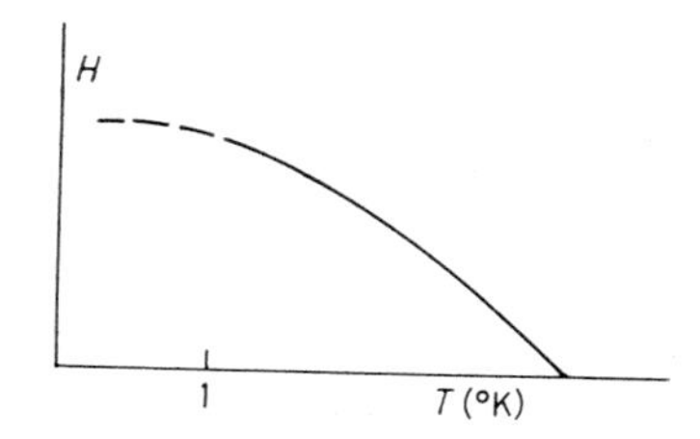


Fig. 1.

If, however, a transverse field is applied you get something quite different; you do not find a sharp transition of the resistance. The resistance begins to come back at a lower field—about half the longitudinal field—and there is an intermediate region where it behaves in a rather curious way.

On the other hand, the other successor to KAMERLINGH ONNES, KEESOM, carried out long and careful measurements on the specific heat, in the other section of the Leyden laboratories. The technique of calorimetric measurements at these very low temperatures had been much improved and one could do these measurements much more accurately than before.

The first thing KEESOM and his collaborators did was to measure the specific heat of normal metals and they found, as a matter of fact, at low temperatures a behaviour of the form

$$C_v = AT^3 + BT\,.$$

This is in agreement with the ideas of Fermi-Dirac statistics. The first term is the lattice specific heat, which is entirely determined by acoustical vibrations of the crystal lattice. The second term, the linear one, is the Fermi-Dirac specific heat of the electron gas. This is discussed in more detail in my lecture on solid-state physics.

Then they went on to measure the specific heat of a superconductor and they found a behaviour like Fig. 2, with a jump in the specific heat. This was a rather

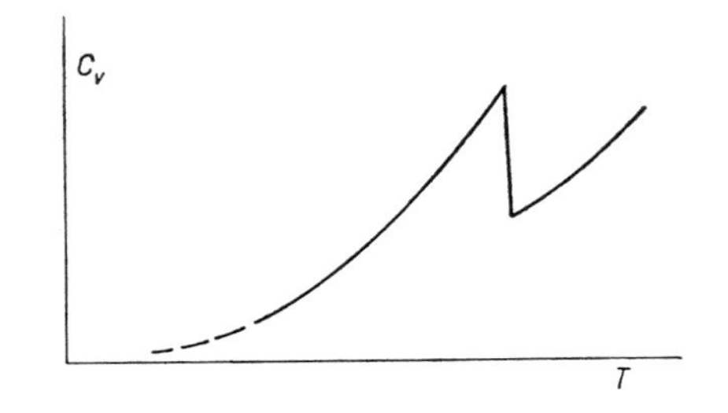


Fig. 2.

curious behaviour, but it so happens that just before KEESOM had also found a jump in the specific heat of liquid helium itself, at 2.1 °K. EHRENFEST had then pointed out that this was a novel type of phase transition; a transition in which there was no discontinuity in the internal energy, but a discontinuity in the specific heat, and he had worked out in some detail thermodynamic relations that would hold for such a transition.

One of Ehrenfest's students, RUTGERS, took a very bold step: he applied Ehrenfest's thermodynamics of transitions of the second kind to a superconductor. He argued as follows: if I cool a superconductor in a field zero and then I switch on a field, the field cannot penetrate because of eddy currents. If we apply thermodynamics, assuming that in a superconductor $B$ is zero, then one can derive the equation

$$\Delta C_v = \frac{1}{4\pi}\, T \left(\frac{\mathrm{d}H}{\mathrm{d}T}\right)^2,$$

where $\mathrm{d}H/\mathrm{d}T$ is the slope of the transition curve $H$ *vs.* $T$. The accurate measurements by DE HAAS and VOOGD of the transition curve and the measurements of the jump of the specific heat were in good agreement with this equation. Yet, that in itself was still not enough to let people believe there was something wrong with the idea of the frozen-in field, although according to this notion the argument of RUTGERS would not be justified if you would cool the body in an external field. The next step was made by VON LAUE, who gave an explanation of the difference between the effect of a longitudinal and of a transverse magnetic field. Suppose I have a superconducting cylinder and switch on a tranverse field, then there is no induction inside the cylinder because of eddy currents. You then get a pattern of lines of force like Fig. 3. The lines

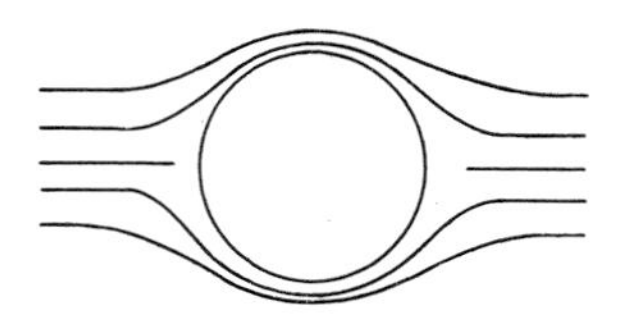

Fig. 3.

of force are compressed near the cylinder and the maximum tangential field will be twice the external field, as you can see from simple magnetostatics. So, VON LAUE said, the field near the cylinder will reach the critical value when the external field is only one half the critical value. Then it starts to penetrate and you will get some curious sort of transition. In a longitudinal field there is no such compression of lines of force.

DE HAAS did not want to believe VON LAUE and he repeated the measurement. He previously had found a behaviour roughly like Fig. 4

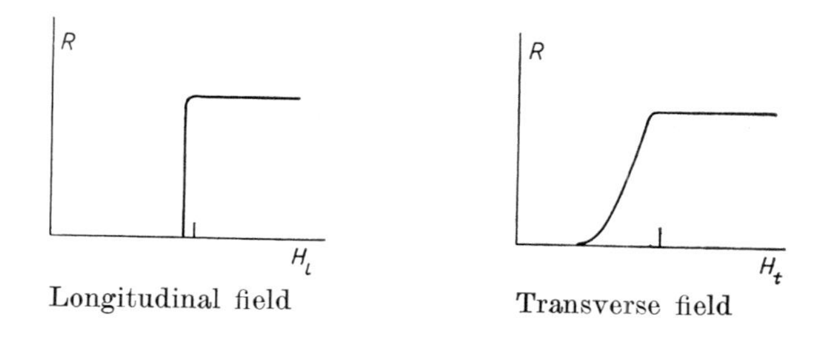


Fig. 4.

Well, DE HAAS said, to see whether VON LAUE is right I shall do a different experiment; I keep the field constant and I change temperature. Then there should not be any field expelled, the lines of force will go straight through the wire and it should make no difference whether the field is transversal or longitudinal. He did those measurements in February 1933 and they were reported at the Physiker Tagung at Leipzig (Leipziger Vorträge), a colloquium organized by DEBYE. DE HAAS found that you get essentially the same sort of behaviour, in a transverse field, whether you keep your field constant and change temperatures or you cool in a field zero and switch on the field.

The conclusion that one should draw from that seems obvious today. The magnetic-field distribution must be the same, whether you cool in a field zero or you cool in an external field.

But DE HAAS did not draw this conclusion. He was more inclined to say: these theoreticians do not understand anything at all, their arguments have nothing to do with reality. The thermodynamic evidence, plus this fact of cooling in a magnetic field being the same as cooling without a magnetic field, should have convinced people at Leyden that probably in all cases the $B$ in a superconductor is zero. But the belief in the frozen-in field was too strong.

By the fall of 1933, however, we of the younger generation began to doubt the dogma of frozen-in fields and experiments were set up to settle the point. At that moment there appeared a paper of MEISSNER and OCHSENFELD from Germany. They had really observed that, if you cool in an external field to below the transition point, the lines of force are expelled. Now the whole theory took a different appearance. If in a superconductor there is always the same behaviour, then there is some point in considering superconductivity as a special phase. There is a special state of the electrons, the superconducting state, separated in a $H$-$T$ diagram by a transition line from the normal state prevailing at higher magnetic fields. This superconducting state has the peculiar property that it does not support magnetic fields; for some reason there are always currents that screen off external magnetic fields.

It then became evident that you do not have to look for a property of the free path, but that you have to look for a new phase of the electron gas. It must be a property of the electron gas, because there are no changes in lattice parameters and all that. I must say that even somewhat earlier it had been found by DE HAAS and a number of collaborators that in a superconducting state the heat conduction is also different from what it is in a normal state. There is a pronounced difference between the two. This again confirms the idea that something is happening in the electron gas. And of course there is the difference in specific heat.

After this work was done, it was not very difficult to formulate the thermodynamics of this transition in more detail than RUTGERS had done. That was done by GORTER and myself and we arrived at a fairly clear-cut picture of the phenomenological behaviour of what are now called class I or soft superconductors.

Soon afterwards the first work on alloys was started. It was found that these may have very much higher critical fields. There is one critical field at which the field starts to penetrate the superconducting alloy and a much higher one at which all superconductivity starts to vanish. Tentative explanations were suggested. Also some attempts were made at explaining the thermodynamics of superconductors in terms of a statistical model. For instance, GORTER and I made a two-fluid model: we divided the electrons into superconducting electrons and normal ones and the ratio between them would change with temperature. There was also an interesting and striking attempt by WELKER. He postulated a gap in the distribution of Fermi states. If you do that, you can get the right specific heat and so on. You have to assume that the gap is temperature dependent; you can even reason that with such a gap you might get the Meissner effect. It would take too much time to go into it, but for the specialists I may remark that these reasonings are not gauge invariant and that the same trouble of not being gauge invariant persisted for a very long time in the more refined theories of superconductivity.

This was about the state of affairs by the mid-thirties. Then LONDON went further and proposed phenomenological equations for the current in a superconductor. The London theory has had an enormous influence both on experimentation and on the theory, for instance on BARDEEN. It is not very simple, and to understand it I might first refer to a little-known paper of three Germans, BECKER, HELLER and SAUTER. BECKER is especially known as the author of later editions of what started as Abraham-Föppl, a standard text on electromagnetic-field theory. HELLER was a young student who later emigrated to Holland; unfortunately he died during the war in a concentration camp.

SAUTER became a professor at Cologne; he retired last year.

They took up the argument that in a superconductor, or rather in a body without any resistance, you cannot have any change of magnetic field. They pointed out that because of the inertia of the electrons an electric field cannot lead immediately to an infinitely high current. Let us make the assumption that the $\dot{v}$ of an electron is given by

$$\dot{v} = \frac{eE}{m},$$

which is just the acceleration of a free electron. This leads to the equation

$$\frac{\mathrm{d}i}{\mathrm{d}t} = \frac{ne^2}{m} E .$$

With this expression you can easily work out the electrodynamics and arrive at the following equation:

$$\operatorname{curl} i = -\frac{ne^2}{mc} (B - B_0) .$$

In words, if you start with a field $\boldsymbol{B}_0$ and you change that field then curl $\boldsymbol{i}$ is under these conditions proportional to this change in magnetic induction. London's equations were closely related to this. You might explain his whole theory in the following way. You know from Meissner's experiments that, whether you cool in a magnetic field or in a field zero and switch on the field after cooling, you always arrive at the same situation; therefore we put the constant in the above equation equal to zero. This gives London's famous equation. It leads to a finite penetration depth of about $10^{-6}$ cm, which has been measured by a large variety of methods. It leads to a certain refinement of the thermodynamics of the transition and all that.

London's idea behind his equation was simple and beautiful. For electrons moving in a magnetic field the Hamiltonian is

$$H = \left(p - \frac{e}{c}A\right)^2.$$

We know there is a term with $A^2$, which leads to diamagnetism. Now, if there were no compensating term with $p \cdot A$, then the $A^2$-term would lead exactly to London's equation. But in a normal metal the second-order effects of the $(p \cdot A)$-term compensate almost entirely the first-order effect of the $A^2$-term. So in superconductivity the $p \cdot A$ effect (which is zero in first order) must be reduced. This can also be expressed by saying that the wave function must be stabilized in such a way that it does not change much when the field is applied. This had been a guiding thought both for experiments and theory.

What happened further? Well, work went on, with the Russian school, with the Cambridge school. The phenomenological theories of LONDON were developed to a higher extent. There was the Ginzburg-Landau theory. Alloys were studied in greater detail; better alloys were found with higher transition temperatures and with much higher critical fields, so it became possible to construct coils for producing magnetic fields and so on.

It was only in the fifties that a real theory was developed. One has to explain a phase transition, and that cannot be done by any perturbation theory; you can never really derive the properties of a new phase by perturbation calculus. You must make a guess what the new phase is. The essential idea of the Bardeen, Cooper, Schrieffer (BCS) theory is a very intelligent guess for a new state. A new state instead of the Fermi distribution accounts for superconductivity. In order to arrive at this wave function one had to be able to handle many electron systems; no explanation of superconductivity is possible on the basis of an independent electron picture. Further the theory is based on attraction between electrons. If we have a Fermi distribution with a very weak attraction between electrons, then there is at low temperatures a stable state where, in phase space, electrons with opposite direction of momentum form pairs and, in a way, condense out, introducing a gap in the energy spectrum. (This reminds us of the early work of WELKER.)

To work out this theory requires mathematical skill; it also calls for a lot of confidence, to believe in the attraction between electrons. After all, electrons show a very strong repulsion and HEISENBERG at one time even tried to make a theory of superconductivity based on the Coulomb repulsion between electrons. Later, through the work of people like PINES and BOHM and BARDEEN himself and LANDAU and others, one realized that there is not much repulsion between electrons in metals. That is crazy at first sight, electrons are negatively charged particles. But in a metal they are screened off by all the other electrons. It was really a miracle that the one-electron theory for conduction in metals worked so well. For from a more sophisticated point of view the « electrons » do not have one-electron states. We are dealing with pseudoparticles; every electron is, so to say, surrounded by a cloud of polarized electrons moving along with it. But the states can be labelled and the behaviour can be calculated as if we were dealing with independent particles without much interaction. That is in itself a curious thing. And then you have to believe that, on top of all that, there might even be a weak attraction. The attraction is caused by a second-order interaction with the lattice. The existence of an isotope effect has greatly contributed to arriving at this notion.

To explain this theory I should like to tell a little anecdote. Once the French theoretician NOZIÈRES visited Eindhoven and gave a lecture on the BCS theory. He was very familiar with this theory but still he found it astonishing that the total repulsion between electrons is removed and that even, through the lattice, there is an attraction. Someone then said, what these matrix elements in second order really mean is that one electron deforms the crystal lattice in such a way that another electron has a lower energy when it comes to that deformed place. One might compare it to a sagging bed with someone in the middle and some one else rolling towards him. To which NOZIÈRES, as a good Frenchman, at once replied « Oh yes, yes, certainly, but in that case there is no initial repulsion ».

The story of superconductivity is a curious story. Here we have a phenomenon discovered already in 1911, which is one of the most pronounced phenomena in physics. There are few quantities in physics that are as zero as the resistance of a superconductor; it is the zeroest quantity I know. It has been measured with an incredible accuracy; a current can circulate for a whole year with no change at all, even if we measure it by magnetic spin resonance to one part in $10^6$ or $10^7$. But it took some forty years before the theoretical explanation was found and applications are still around the corner. KAMERLINGH ONNES thought you could do terrific things with it, but he was disappointed when he saw there was a maximum current density and a maximum magnetic field.

At present superconductivity is used for producing high magnetic fields, but that, for the time being, is about all. The idea of making superconducting computers was taken up by IBM at one time, by General Electric at another time. RCA has been announcing during six or seven successive years that next year they would bring out a large superconducting memory, but they never did. For all I know the idea of using superconductors in logical circuits has now been entirely dropped.

The idea of transmitting power or information is often discussed in a science-fiction way; it has not been put into effect and so we still have to see whether this remarkable phenomenon will find widespread applications.

# Some Recollections.

H. B. G. CASIMIR

*Philips Research Laboratories - Eindhoven*

What I shall say now is not at all systematic. I do have some more systematic remarks to make concerning the relations of science and technology: those I shall present in a special lecture later on.

I should like to use the present opportunity to tell one or two anecdotes and personal recollections that have a bearing on the subject matter of this course.

To put chronology right, I started my studies in 1926. In 1928, after passing some preliminary examinations I began to work with EHRENFEST. In the spring of 1929 EHRENFEST took me along to a meeting at Bohr's Institute; that was the beginning of my first visit to Copenhagen. During the next two years I spent most of my time at Copenhagen, occasionally returning to Leyden for examinations.

I should like to mention some of the things which occupied the interest of BOHR and people at the Copenhagen Institute in those days.

I remember that at that meeting in 1929 BOHR was speaking about the spin of the electron. He had an argument to show that this spin was an essentially nonclassical feature and that you could never observe it by classical measurements on a free electron. This argument went roughly as follows. Suppose I want to make a Stern-Gerlach experiment; then I must have an inhomogeneous magnetic field $H$ and the force on a magnetic moment $e\hbar/2mc$ will be given by

$$F_x = \frac{e\hbar}{2mc}\frac{\partial H}{\partial x}.$$

The total momentum given to the particle is equal to that force, multiplied by the transit time, say $l$, the length of the slit divided by the velocity of the particle $v$, so

$$\delta p = \frac{e\hbar}{2mc}\frac{\partial H}{\partial x}\frac{l}{v}.$$

On the other hand, the particle will also experience a Lorentz force. You might arrange things in such a way that along the trajectory the magnetic field is zero. However, the Lorentz force is undetermined because you do not know exactly where the particle is:

$$\delta F_{\mathrm{L}} = \frac{ev}{c} \frac{\partial H}{\partial x} \delta x \, .$$

This would give a momentum

$$\delta p_{\mathrm{L}} = \frac{ev}{c} \frac{\partial H}{\partial x} \delta x \frac{l}{v} \, .$$

In order to measure the spin of the electron one should have

$$\frac{e\hbar}{2mc} \frac{\partial H}{\partial x} \frac{l}{v} > \frac{ev}{c} \frac{\partial H}{\partial x} \delta x \frac{l}{v} ,$$

and hence

$$\delta x < \frac{\hbar}{2mv} \, .$$

So you would have to have a wave packet which is smaller than $\hbar/2mv$ and that is impossible because a wave packet of that order of magnitude does disintegrate right away.

This, but much more refined, was the general type of argument to show that the spin of a free electron could not be measured. You could try to make ingenious arrangements with crossed electric and magnetic fields and one of the first things I did with BOHR was to spend quite some time going through some such proposals and studying qualitatively the quantum-mechanical diffraction in a variety of fields. Sure enough we always found that the classically calculated deviation of an electron would never be larger than the essential dimensions of the wave packet one was working with.

I do not think that these arguments had much follow-up later or were of very great importance for the further development of the spin theory. Of course it had at that time not yet been shown that scattering of an electron by a nucleus does lead to polarization. It took quite a number of years before this theoretically predicted polarization was confirmed experimentally, mainly because the experiments were almost always spoiled by double or multiple scattering. We know now that it is possible to make beams of polarized electrons, although until now they have not played a very important role in experimental physics.

Another problem was the conservation of energy. BOHR was very much impressed by the fact that in β-decay you have a continuous spectrum, although the initial and final states seemed to be well determined by the mass defects. All attempts to detect energy, escaping in some form, had so far failed and this induced BOHR to consider once more the idea that there might be a breakdown of the principle of conservation of energy. We know that these same phenomena prompted PAULI to formulate his theory of the neutrino. There is a story attached to that, which may be partly a folkloristic legend. I know that in those days PAULI wrote a letter to BOHR, which BOHR did not show to me, which contained a number of penetrating questions. BOHR did not feel that he could answer that letter easily. He was working on other things, so he asked Mrs. BOHR to write PAULI a nice letter with some family gossip and to explain that he himself would write on Monday. Three or four weeks later there came a kind letter from PAULI to Mrs. BOHR, thanking her for the letter and saying that her husband had been very wise in saying that he would write on Monday without specifying on which Monday. But, PAULI added, « Er soll sich keineswegs an Montag gebunden fühlen, ein Brief am irgendeinem anderen Tag geschrieben wäre genau so willkommen ». (He should not in any way feel tied to write on Monday, a letter written on any other day would be equally welcome.) I have some reason to believe, but I have not been able to ascertain with a hundred percent certainty, that that letter of PAULI which BOHR was going to answer on Monday was a letter containing the first suggestion of the neutrino as a means to solve the problem of apparent nonconservation of energy.

Here the following question arises: suppose you postulate a particle which takes away the missing energy and momentum and suppose that that particle would really be undiscoverable and unobservable, like the neutrino was for a very long time. If such a particle with its energy, angular momentum and so on just disappears forever into space is that very different from saying there is no conservation of energy? Well, it is a slightly philosophical question and I think it almost becomes a question of words.

However, it is a fruitful idea that the loss of energy is connected with a simultaneous loss of momentum and with a simultaneous change in spin and statistics. We know that by taking up that idea FERMI was able to successfully develop a description of β-disintegration later on.

In those days GAMOW was writing his book on the nucleus. He was writing some of the chapters at Copenhagen, and the Cambridge University Press had provided someone to translate his English into English. So there was Miss SWIRLES—later Mrs. JEFFREYS—sitting in the library transcribing Gamow's texts with an eternal grin on her face because of Gamow's grammatical and other extravagancies.

At that time there was not only the question of conservation of energy, there was also the problem that it was hard to see how an electron with its long de Broglie wavelength could be contained in a nucleus. Later this was solved by saying that there is no electron in the nucleus; it is created when a neutron changes into a proton. But in those days you wanted to enclose an electron in the nucleus. However, its mass is so small that this does not work very well: you would get enormous energies. Therefore GAMOW marked all the passages in his book dealing with β-rays with a skull and cross-bones. He

Schrödinger, Dirac, Born and so on he wrote his name in bold capitals; then, taking out a penknife, he underlined it with a deep scratch.

I was at that time Ehrenfest's assistant and I thought that, after all, one should not put the boy in even greater difficulties than he quite obviously was, so I carefully erased his name. Being in those days less of a handyman than one has been forced to become of late, I did not make any attempts to fill up the plaster. The scratch remained. I never heard any comments by Ehrenfest about it and I think it is still visible. It must be said for the young man that he later played a courageous role in the resistance movement in Amsterdam and eventually was caught by the Germans and shot. And if, at present, at the Leyden colloquium you still see a rather mysterious scratch between all the famous names, it is in a way a monument for an unknown man in the Dutch resistance during the war years.

## From Bruno to Kepler: Man's Position in the Cosmos (*).

P. Rossi

*Istituto di Filosofia dell'Università - Firenze*

1. – The thesis of the noncentrality of the Earth, the enlargement of the traditional boundaries of the Universe, the affirmation of the plurality of worlds and the infinity of the cosmos—these opinions provoked in European culture not only a sense of exaltation and enthusiasm, but also, as is well known, a feeling of astonishment and bewilderment, of beginning as well as ending. The seventeenth century seemed to Mersenne to be at the start of radical changes which went well beyond the confines of astronomy. In March 1644 he asked Peiresc what he thought of these upheavals. Did they not perhaps give one the feeling of the end of the world? [1]

The impression that the destruction of the old order of the world would be tantamount to the establishment of disorder, and that such disorder could spill over into the establishment of moral and religious values, was an impression shared by many. Naudé, a lucid witness to his times, had considered clearly the revolutionary import of the new cosmology and its implicit perils:

« I fear that the old theological heresies are nothing compared to the new ones that the astronomers want to introduce with their worlds, or rather lunar and celestial earths. For the consequence of these latter heresies will be much more dangerous than that of the preceding ones and will introduce yet stranger revolutions » [2].

In the face of many reactions to Copernicus, as in the face of many uncertainties about the new astronomy, it is never to be forgotten that Bruno had placed the world of Copernicus and its innumerable worlds within an infinite and homogeneous space « which we may freely call the void ».

*Omne movetur aut e vacuo, aut ad vacuum, aut in vacuo.* Atoms and worlds move in the void or infinite space, similar to atoms dispersed in the heavens. The infinite void of the Lucretian and atomistic vision of the world could really seem a kind of « natural location » for the Copernican solar system and

(*) In a more enlarged form this paper will be published in *Science, Medicine and Society in the Renaissance, Essays to honor Walter Pagel*, edited by A. G. Debus, New York, Neale Watson Academic Publications.

for a plurality of such systems [3]. That reference to atomism present in Donne's verses certainly was not casual. The Sun and the Earth were « lost » in a crumbled universe.

Of course in COPERNICUS there was no negation of the circularity and regularity of celestial motions, whereas BRUNO had insisted at length upon this negation, interpreting spheres and epicycles as « poultices and prescription for doctoring nature... to the service of Master ARISTOTLE ». Certainly the text of *De Revolutionibus* could not make one think about the dissolution of many world « systems » or of the shattering of the Universe. It was BRUNO, in the text of *Cena*, who had refuted the idea of every « continuous and regular » celestial motion « around the centre », who had affirmed the impossibility of perfect motions and perfect forms in the physical universe, given that every movement of natural bodies differs « from the simply circular movement around a center ». It was BRUNO who saw in the laws of motion of the celestial bodies something that is characteristic of the individual stars and planets; who had entrusted the path which the heavenly bodies should take in the heavens to their « very soul » and their « intrinsic principle ». It was BRUNO who had conceived the heavenly bodies as animated beings in free movement and had insisted on the impossibility of constructing a harmonic picture of the Universe within which one could effect precise calculations [4].

BRUNO had made a clear distinction between the Universe and the worlds. To speak of a world system does not mean, in his vision of the cosmos, to speak of a system of the Universe. Astronomy, as the science of heavenly bodies, is legitimate in so far as it is the science of the worlds which fall into the range of our sensible perception. But beyond those worlds extends an infinite universe which contains an infinite plurality of worlds as well as all those « grand animals » which we call stars. It is a universe that has neither dimension nor measure, neither form nor figure. Both uniform and formless, it cannot yield a « system » [5].

**2.** – COPERNICUS, KEPLER and GALILEO, beyond the differences, the affinities and the divergences, maintain the firm image of a universe as a unitary system. They see in the world the expression of a divine order, the manifestation of principles or mathematical-geometrical archetypes. Their « geometric » astronomy contrasts on the whole, from this point of view, with that which has been not inappropriately called Bruno's « astrobiology ». GALILEO keeps away from Bruno's perspectives, notwithstanding their common enthusiasm for COPERNICUS and their common negation of the interpretation of the Copernican doctrine as a hypothesis. GALILEO repeatedly insists on the world as a « perfect body », a well-ordered composition of parts arranged according to the best possible order. Like COPERNICUS he places at the base of his consideration of the cosmos two premises of a general nature, the order of the Universe and the pre-eminence of circular motion [6].

That singular mixture of Lucretian, Copernican, Neoplatonic and Hermetic themes was characteristic of many authors—among them CUSANO, PALINGENIUS and BRUNO—who were quite distant from the rigour and coherence of the scientific thought of GALILEO. And yet in this mixture one may discern those five innovative ideas of « revolutionary cosmographical theses » which LOVEJOY delineated, more than thirty years ago, as characteristic of a changed world view:

1) the assertion that other planets of our solar system are inhabited by sentient and rational beings;

2) the destruction of the external walls of the medieval universe, whether they be identified with the extreme crystalline sphere or with a definite region of the fixed stars, and the dispersion of these stars within vast and irregular spaces;

3) the idea of the fixed stars as suns similar to ours all or nearly all surrounded by their own systems of planets;

4) the hypothesis that the planets of these other worlds are inhabited by rational beings; and

5) the assertion of the actual infinity of the physical universe in space and of the number of solar systems contained therein [7].

LOVEJOY was correct in affirming that none of the five theses was present in COPERNICUS, and that both the doctrine of infinity and that of the plurality of worlds were rejected, in various ways, by the three great astronomers of Bruno's time and the succeeding generation: BRAHE, KEPLER and GALILEO.

**3.** – If there are more worlds, as would be asked at the end of the seventeenth century, is CHRIST to have redeemed all those worlds? And is this not in contrast to the Scripture which calls him the Savior of a single world? And how can we know that specifically ours was so-favoured and not some other one of which we know nothing?

DREYER and KOYRÉ, among others, were attentive to the « exceptional » position which COPERNICUS attributed to the Earth, with the planets rotating around the centre of the Earth's orbit [8]. But it is actually in KEPLER, who first transported the Sun to the centre of planetary motions and the Universe, that we find the vivid consciousness of a radical contrast: the theses of the infinity of the Universe and the plurality of worlds are irreconcilable with the affirmation of the centrality of man in the Universe. If BRUNO is right, KEPLER suggests, if the Universe no longer has a centre and is no longer enclosed within external limits, if there is no longer a limit consisting of that « skin or shirt of the Universe » (*mundi cutis sive tunica*, in which KEPLER firmly believed) which is similar to a lantern protecting the flame of the Sun from the winds and reflecting the light everywhere like an opaque and illuminated wall, if it is true that every point could be the centre, if there are as many worlds as there are fixed stars, if the solar system could appear to a hypothetical inhabitant of the constellation Canus just as the fixed stars appear to us, if the Sun is only one of the infinite stars dispersed in an infinite space and if other planets rotate around those suns, then the biological and moral status of the Earth and the solar system is no longer unique and the image of a universe constructed for man had really been destroyed. The privileged position of the solar system, situated equidistantly from the fixed stars, hereby falls; the image of man as master and dominator of creation must be abandoned [9].

KEPLER, as is well known, resolutely opposed the Brunian infinitization of the Universe, decisively rejecting the assimilation of the Sun to the fixed stars. He firmly maintained the uniqueness of the Sun and of the solar system, contrasting it to the immobile profusion of fixed stars. Kepler's *Dissertatio* has often been used to narrow the gap between the positions of GALILEO and BRUNO, without paying attention to Kepler's principal preoccupation in that text; he seeks to clearly distinguish between precisely these two positions showing that Galileo's astronomical discoveries do not in any way constitute a proof of the validity of Bruno's infinitist cosmology.

The discovery of new planets rotating around one of the fixed stars or around the Sun would pose grave difficulties for his vision of the world, proving correct the theses of BRUNO and WACKHER VON WACKHENFELS, the enthusiastic follower of Bruno's theses [10]. In the same way, and for the same reasons, it was perfectly possible for KEPLER to conceive of the existence of inhabited planets inside the solar system; but the idea of the plurality of worlds or of systems in which life would be present had to be firmly rejected. Our world, the solar system, constitutes a *unicum* in the Universe. It was created for man to serve his needs, his requirements and his hopes.

**4.** – The beginning of the *Dissertatio cum Nuncio Sidereo* [11] gives the precise meaning of a discussion where the foundations are riddled with a series of doubts and uncertainties which KEPLER saw in a dramatic way. KEPLER begins his account at the time when he had withdrawn to his home for some time for a brief vacation. Around the middle of March the news reached Germany that GALILEO had discovered four previously unknown planets. WACKHER, from his carriage in front of the house, brought KEPLER the incredible news:

« In considering it I was struck with such amazement and my soul was so moved that—he for joy and I for blushing, both for laughter and disoriented by the news—he could hardly speak and I could hardly listen » [12].

Waiting to see the text of *Sidereus Nuncius* « with extraordinary greed to read its contents », KEPLER and VON WACKHENFELS gave it two different interpretations. According to KEPLER, just as the Earth has its Moon which turns around it, so might GALILEO have seen four other tiny moons rotating in very narrow turns around the small masses of Saturn, Mars, Jupiter or Venus:

« To WACKHER on the other hand it seemed certain that these new planets spun around some fixed star. If four planets had remained hidden up until now, what was to keep us from believing that innumerable others might be subsequently discovered? And that therefore this world is itself infinite, as was suggested by MELISSUS and by the Englishman William GILBERT, author of magnetic philosophy? And that there are infinitely many other worlds similar to ours (or, as BRUNO says, infinitely many other Earths), as was held by DEMOCRITUS and LEUCIPPUS and, among more recent philosophers, BRUNO and Edmund BRUCE » [13].

The reading of the Galilean text proved KEPLER correct, and he emerged encouraged. Bruno's exile into the infinite seemed averted:

« If you had discovered planets rotating around one of the fixed stars, the stock and prison were already ready for me by way of Bruno's innumerability, or rather exile into that infinity. For the moment (*in presens*) you have therefore freed me from the great fear which had welled in me at the first news of your book, given that you affirm that these four planets do not rotate around one of the fixed stars but around the planet Jupiter » [14].

KEPLER has one principal preoccupation. He seeks to demonstrate not only:

« That this system of planets, in one of which we men find ourselves, is found in the principal spot in the Universe, around the heart of the Universe which is the Sun; but also in particular that we men find ourselves on the globe which is most suited to the most important and most noble rational creature among all physical bodies » [15].

For KEPLER, then, the Earth occupies a unique place in the structure of the solar system and the Universe. On the Earth lives the « contemplative creature » created in the image and likeness of God. This creature is in a position to rationally reconstruct that perfect architecture in which the grandeur of God is expressed; he is even in a position to reconstruct those « archetypal laws » which, in God, presided over the creation of the world. The Universe was created to serve this contemplative creature and the laws of the Divine Mathematician were enacted for him. For KEPLER, man and his home remained at the centre of the cosmic drama of creation and redemption.

The propositions expressed in the *Dissertatio* certainly were not momentary or isolated positions. One need only open Kepler's *Epitome astronomiae copernicanae*, published at Linz and Frankfort between 1618 and 1621, to find the Earth's exceptional position in the Universe confirmed with unequivocal emphasis, even within a more strictly technical argument:

« Where do you feel the investigation on the proportions of heavenly bodies should begin? From the Earth, because it is the home of the contemplating creature that was made in the image of God the Creator ... because the earthly orb is the figurative middle among the planets ... and the proportional middle between the limits of the superior and the inferior planets. The order of these proportions, finally, proclaims that the Creator ... began from Earth as his first measure ... The Earth in fact had to be the home of the contemplative creature, whom the Universe was created to grace .... Since the Earth was destined to become the home of a measuring creature, it is clear that it had to become the measure of the heavenly bodies with its body, and the measure of lines or distances with its radius, in so far as it is linear » [16].

The sense of secret, hidden horror that KEPLER had manifested in 1606 for Bruno's vision would never be repaired:

« That thought bears with it I know not what secret, hidden horror: we feel lost in that immensity to which limits and the centre are denied, to which every determinate place is consequently denied » [17].

**5.** – In another area, historically very complicated but in any case different from that within which the major astronomers of the seventeenth century had been moving, several notions had to be reinforced and developed to full maturity. These ideas included the negation of anthropomorphism; the image of a nature infinitely vaster and more powerful than man. Within this perspective, reference to the Democritean and Lucretian tradition operated with determining force, and, at its hands, that « terrestrial » vision of the cosmos, which had persisted even in the new astronomy, was also destined to topple [18].

MONTAIGNE had already asked, in the *Apologie de Raymond Sebond*, how man could assume that the wonderous movement of the heavenly vault, the eternal light of the flames which turn at its peak and the awesome movements of the sea all could have been determined and then could have continued to exist for his utility and his advantage. Was it not ridiculous that a miserable creature, riveted to the worst and most putrid part of the Universe, incapable of dominating even himself, would call himself owner or master of the world and presume to set himself up as the only being that was in a position to recognize the architecture of the cosmos? [19].

At mid-century, and quite far from this toning-down, stands Cyrano DE BERGERAC, disciple of the doctrine of an organic universe comparable to a gigantic living being. BERGERAC, linked to the thought of CAMPANELLA, GASSENDI and LA MOTHE LE VAYER, mixes together themes from hermetic platonism, Cabala, Democritean and Epicurean atomism, the Averroistic tradition and the new cosmology of COPERNICUS, GALILEO and KEPLER. In a quite beautiful page he protests « the insufferable pride of humans », « the insolence of these brutes » who conceive of the Universe as *serving* the needs of man.

« Like a man whose boat sails near dry land, who has the impression of standing still while the coast is moving, so men, spinning around the sky with the Earth, held that the sky itself was turning around them. Add to this the insufferable pride of humans, who persuade themselves that Nature was created expressly for them, as if it were probable that the Sun ... was lit to make their medlar-trees mature and to till their cabbages. As for myself, far from subscribing to the impudence of these brutes, I believe that the planets are worlds around the Sun, and that the fixed stars are also suns with planets around them, *i.e.* worlds which we cannot see from here ... For how can one imagine in good faith that these immense globes are only great deserted earths, while ours, precisely because we crawl about there, a dozen arrogant rogues, would have been created to command everyone? What ! Because the Sun measured our days and our years, does this mean that it was made to keep us from banging our heads against walls? No! No! This visible God illuminates man purely by accident, just as the flame of the king accidentally illuminates the teamster who passes in the street » [20].

To man's viewpoint, MONTAIGNE had opposed that of the duckling and the crane; CYRANO had spoken of medlars and cabbages. With less broad-mindedness but with much greater theoretical force, DESCARTES had also rejected the legitimacy of making use of an anthropocentric point of view, at least in physics. In the *Principia* of 1644 he said:

« It is in no way probable that all things were created for us in such way that God had no other reason for creating them » [21].

In 1686, the limpid, brilliant argument of FONTENELLE would come forth:

« Our own particular folly also consists of the belief that all of Nature, without exception, is intended for our uses. And when our philosophers are asked the use of that prodigious number of fixed stars, a part of which would be sufficient to do that which all of them do, they respond coldly: they serve to enliven the view » [22].

You are not, exclaimed BOREL nearly half a century earlier, like those farmers who, never having seen the big cities, do not understand throughout their lives that there may be cities bigger and more beautiful than their village.

The Earth had now taken form as a lost village and a province, as had happened at the beginning of the century for the Mediterranean and the whole Western world, in the face of geographical discoveries, voyages to unknown lands and distant peoples, and in the face of *Terra australis incognita*:

« What are we doubting? What are we terrified of? Of shadows? Of ourselves? In that place there is a sky, there is an earth, no doubt there are men, perhaps much more civilized than ourselves. Who would ever have suspected such intelligence and such good judgement in the Chinese? Such a great number of arts? Such a vast and various science of all things? While we continue to believe that all the Muses reside in that humble dwelling which is our Western world (*in hoc occidentali gurgustiolo*), they smile. And not without reason » [23].

## REFERENCES

[1] Cf. R. LENOBLE: *Mersenne ou la naissance du mécanisme* (Paris, 1943), p. 342. Many historians have analysed the characteristics and the meaning of this crisis: A. O. LOVEJOY: *The Great Chain of Being* (Cambridge, Mass., 1936); E. M. TILLYARD: *The Elizabethan World Picture* (London, 1943); V. HARRIS: *All Coherence Gone* (Chicago, 1949); M. H. NICOLSON: *The Breaking of the Circle* (Evanston, Ill., 1950); A. KOYRÉ: *From the Closed World to the Infinite Universe* (Baltimore, Md., 1957).
[2] G. NAUDÉ: letter to Ismäel BOUILLIAU on August 15, 1640, published in R. PINTARD: *Le libertinage erudit* (Paris, 1943), p. 472.
[3] Cf. M. H. NICOLSON: *Science and Imagination* (Ithaca, N. Y., 1962), p. 30.
[4] G. BRUNO: *La cena de le ceneri*, edited by G. AQUILECCHIA (Torino, 1965), p. 165; *De immenso*, in *Opere*, I, 1, p. 369; *De l'infinito universo e mondi*, in *Opere*, I, p. 340.
[5] On the irrelevance of the term *system* to the Brunian vocabulary, on the counter-position of an astrobiology to a cosmic geometry cf. P. MICHEL: *La cosmologie de G. Bruno* (Paris, 1962), p. 194, 232-33.
[6] G. GALILEO: *Opere*, VII, p. 55-56.
[7] A. O. LOVEJOY: *The Great Chain of Being*, p. 108. The quotation which follows is drawn from *Athenian Mercury* (London, 1961), Suppl. 2, p. 13.
[8] A. KOYRÉ: *La révolution astronomique* (Paris, 1961), p. 69, 113; J. L. DREYER: *A History of Astronomy from Thales to Kepler* (New York, N. Y., 1953), p. 343.
[9] Cf. KEPLER: *De stella nova in pede serpentarii*, in *Opere*, II, p. 688, and for the previously quoted expression, *Epitome*, in *Werke*, II, p. 259.
[10] Cf. A. KOYRÉ: *From the Closed World*, p. 74.
[11] The text of the *Dissertation* is contained in Vol. III of the *Opere* by GALILEO, as well as in the edition of the works by KEPLER (*Gesammelte Werke* (München, 1938-1959), IV, p. 281-311). Recent translations include those of F. HAMMER (Gräfelfing, 1964) and E. ROSEN (New York, and London, 1965).
[12] KEPLER: *Dissertatio*, in *Werke*, IV, p. 288.
[13] *Ibid.*, p. 289.
[14] *Ibid.*, p. 304.
[15] *Ibid.*, p. 307-308.
[16] KEPLER: *Epitome*, in *Werke*, VII, p. 276-79.
[17] KEPLER: *De stella nova*, in *Opere*, II, p. 688.
[18] On these themes G. McCOLLEY: *The seventeenth-century doctrine of a plurality of worlds*, in *Annales of Science* (1936), p. 385-430 (but see also the specifications concerning COPERNICUS in A. KOYRÉ: *From the Closed World*, p. 36). Barely usable is C. FLAMMARION: *La pluralité des mondes habités* (Paris, 1862). McColley's work concerning the classical world is integrated by F. M. CORNFORD: *Innumerable worlds in the pre-Socratic philosophy*, in *Classical Quarterly* (1934), p. 1-16. A large number of textual references may be found in M. NICOLSON: *Voyages to the Moon* (New York, N. Y., 1960).
[19] M. DE MONTAIGNE: *Essais*, II, 12.
[20] *Les états et empires de la Lune*, written in 1649, was published in Paris in 1656. On the relationship among CYRANO and DESCARTES and GASSENDI, cf. Chapter 3 in J. S. SPINK: *French Free-Thought from Gassendi to Voltaire* (London, 1960), p. 48-66.

[21] *Oeuvres de Descartes*, edited by ADAM et TANNERY, VIII, 1, p. 180-81.
[22] B. DE FONTENELLE: *Entretiens sur la pluralité des mondes* (Amsterdam, 1719), p. 23. The first edition dates from 1686.
[23] MERCURIUS BRITANNICUS (HALL, JOSEPH): *Mundus alter et idem sive Terra Australis ante hac semper incognita* (Hanoviae, 1607), p. 9. In the edition of 1643 the texts of Campanella's *Città del Sole* and Bacon's *New Atlantis* were added. A modern edition of the English translation of 1620 (*The Discovery of a New World or a Description of the South Indies*) was edited by H. BROWN (Cambridge, Mass., 1937).

# The Historical Roots of Modern Physics.

Y. ELKANA

*The Hebrew University of Jerusalem, The Van Leer Jerusalem Foundation - Jerusalem*

## 1. – The growth of knowledge.

1·1. « *The past* » *vs.* « *history* ». – The historian PLUMB drew a sharp distinction recently between the past and history: « Man, from the earliest days of recorded time, has used the past in a variety of ways: to explain the origins and purpose of human life, to sanctify institutions of government, to give validity to class structure, to provide moral example, to verify his cultural and educational processes, to interpret the future, to invest both the individual human life or a nation's with a sense of destiny » (1). « History », on the other hand, « like science, is an intellectual process. Like science, too, it requires imagination, creativity and empathy as well as observation as accurate as a scholar can make it. History, like science, has grown intellectually out of all recognition with its ancient self in the last three hundred years in Western societies ».

What is said here about history in general is eminently true for history of science in particular. The chief difference lies in their timetables, for science, like history in some sense, is a product of the last three hundred years; but history of science is a product of the last fifty years. Until nearly the end of the 19th century, science had only a story of its past, and even today at least half the books catalogued as history of science are more like stories of the past than like history in Plumb's sense. The difference is apparently important. For general history, as against « the past », is an attempt to objectivize and intellectualize the subject and thus free it from the ideological burden that makes the past so eminently useful to those who wish to use it. Opposed to that is a great deal of literature telling the story of science as if it were an objective, ideology-free, culture-free, international, empirical enterprise wherein truth is slowly accumulated and mistakes are eradicated. Because this purported body of true scientific knowledge is growing, a systematic splitting —specialization—becomes necessary.

---

(1) J. H. PLUMB: *The Death of the Past* (Boston, 1970), p. 11.

But this tale of an objective body of knowledge is not true history: it is only the past used to sanctify the present institutions of science and give validity to its meritocratic structure—not to say, also, to provide moral examples of the making of science [2], to unify the educational process that claims to be able to teach the method of science, to interpret the future development of science and to invest the individual scientist or science itself with a sense of destiny. In real history of science, however, we begin to realize that science does not grow by simple accumulation of true data and that empirical findings do not always determine the rejection or acceptance of theories; we begin to doubt whether science is a purely intellectual product of disembodied ideas and to ask why some discoveries were made at a given place and time by one genius and not another, and—even more tantalizing—why some discoveries were delayed although all the intellectual tools (so far as can be seen by hindsight of the body of scientific knowledge in the field) were available. Or, further, to question whether simultaneous discoveries occur so often as is believed.

Let us recall a few of these cherished so-called « truths » from the past of science:

1) science is growing so rapidly that no human brain can contain all that has been found out, and therefore specialization is inescapable;

2) specialization is not only inevitable but is an asset—because it also serves to sharpen analytical tools, for greater precision in more and more limited specializations;

3) specialization necessitates true co-operation between experts, and therefore teamwork becomes a beneficial reality;

4) the present institutions of science make such specialization and teamwork possible;

5) the leaders of the scientific community—elected democratically, although by the criterion of excellence—exercise severe control of quality, and thus in good conscience they can represent the scientific community in its demands on society;

6) the scientific community supervises and actively participates in science education, thereby assuring the transmission of the latest achievement—ac-

---

(2) A famous example comes to mind: both BACON and DESCARTES (much as they were otherwise in opposition) agreed that learning their method would enable anybody to do science creatively. This is a typical *moral* to be drawn from standard histories of science. Those who wish to draw the other moral, namely the « great-genius theory » (science grows by the unpredictable strokes of genius of the chosen few) generally reserve the great steps to the genius and the small steps to the « correct method » which is in everybody's reach.

companied by insight into the scientific method of clarity, rationality, precision, empirical certainty and excision of all methaphysics and prejudices.

That the above state of affairs indeed assures progress can be seen from the story of the past. Of which, again, let us recall a few examples:

1) Kepler's painstaking observations and measurements finally showed him that the planets were moving in elliptical orbits and not in circles.

2) GALILEO, one of the first true scientists, after years of patient experimentation with inclined planes—and visits to the leaning tower of Pisa—finally established the first true law of science, the law of free fall of bodies.

3) With the clarification of the theory that heat is motion, the law of conservation of energy was found simultaneously by several great scientists, *e.g.* JOULE, HELMHOLTZ, MAYER.

4) When EINSTEIN read and interpreted correctly the Michelson-Morley experiment, he then drew the necessary conclusion of the constancy of the speed of light; this led him to the discovery of the theory of relativity.

5) Heisenberg's formulation of the uncertainty principle proved, once and for all, that Nature is not fully deterministic.

6) Quantum theory was born out of the failure to account for the observed energy distribution in the continuous spectrum of black-body radiation.

The above views have immediate implications for the teaching of science, though most of these examples have been treated in great detail by historians of science and found to be purely mythical. But all of them represent the Victorian image of science and its growth, of which I shall treat below. Here, the important point is that implicit in both lists of examples is a whole theory of knowledge and of the growth of knowledge, which must be analysed critically. My thesis is that the theory implied here hinders progress. I shall suggest a different one. But before that, we must clarify the need for such a theory.

**1·2.** *Why do we need a theory of the growth of knowledge?* – Growth of knowledge is a topic on which everybody has a theory, though not always a conscious or a verbalized one. And, in some cases, even when it is both conscious and verbalized, it is not fully developed and coherent. But the theory is there, in some form. Whatever statement we make on science or scientists, on validity of a theory or on criteria of truth, we carry with that statement the implied burden of a whole aggregation of other statements on the genesis of knowledge. If we reread carefully the two groups of statements listed in the previous Subsection, it will inevitably be clear that, even pictured at random, they imply that knowledge grows by accumulation and, generally, that knowledge consists of clearly established facts (positive contributions to knowledge) independent of any theoretical frameworks and free of unwarranted presuppositions.

It then follows that this growing mass of data creates a severe problem (« information explosion ») and therefore makes strict professionalization inevitable. It is also implies that, with this exploding growth, the frontiers of knowledge shift: *e.g.*, some areas become clarified and are left behind, all in good order and with their problems solved. Thus, if a scientist wishes to contribute anything original, he must first of all read up the subject in order to reach the frontiers of the field; then and only then can he see farther and develop the field. All of which presupposes, although without so stating, that at least the facts so far established are eternal (even if the theories are changeable); therefore these facts constitute the only correct basis for further discoveries.

If, then, knowledge advances all the time, and all the time along more and more narrowly delimited frontiers—which must be reached before one can get on—analytical tools become more and more sophisticated, and therefore the initial investment in the quality of the tools (mathematical or experimental) becomes longer. What follows from that state of affairs is a very large emotional investment in the importance of these tools—and thus more and more emotional opposition to a readiness to go down to fundamentals from time to time; this activity, which needs an ability to over-simplify—to think away the tools so to speak—does not promise any certain success and tends to encounter no institutionalized encouragement.

The organized scepticism of scientists, it would appear, relates only to the quality of the tools, not to the basic theory. Let us recall three fairly recent developments.

1) The Ph.D. dissertation, which used to be considered an « original contribution to human knowledge », has changed its meaning. In today's scientific community it is considered immoral for a supervisor to allow a student to choose a problem for which the supervisor does not at least vaguely see a solution, because, if the student fails to solve the problem and succeeds only in analysing it or showing in what ways it should not be approached (I am not referring to a *positive* experiment showing that something does not work, which is acceptable), then his career is ruined. Since all this not only constitutes a heavy burden on the supervisor's conscience but is also a waste of society's investment, which is huge, in each Ph.D. in science, a redefinition of the term « original contribution » has been compelled into being. Another corollary of this situation is that only rarely does a student choose a topic: he is generally *given* one, which is usually in his supervisor's field of immediate interest, and a rather enforced form of teamwork ensues. But my chief criticism is that, at an age when scientists are admittedly at their best, they are thus given problems instead of encouragement in original or speculative probing. That in this situation the independent mind must clash with his superior's is clear. If he succeeds, we all admire him for having gone against the trend and triumphed; but if he fails, it is not recorded that he was a daring genius.

2) Referees of scientific journals (especially in biochemistry) tend to reject papers if their experiments have not been executed on the most recent equipment; this irrespective of whether the theory described depends in any way on the precision of the results.

3) Experimental and sometimes even theoretical scientists, when asked what they are working on, will respond not by specifying a problem but by giving the name of the technique or of the machinery they are using. Naturally, this could be no more than just a way of talking; but very often the expression reflects accurately the real situation of a man who, having invested so many years of study, is now most efficiently used as a higher form of tool, *i.e.*, in the concentration on technique, his knowledge is drafted for work on those problems that depend for their solution on his technique. If the problem is interesting and chosen by a good scientist, then the whole question of whether such men are scientists or research technicians is a mere matter of labels. But if these research technicians are monarchs of their kingdoms, they will wield vast financial power in a search for their own problems—with the technique they want to use as their only criterion for locating the problems. The result is a huge percentage of shoddy papers reporting totally unnecessary results. When this state of affairs is discussed with open-eyed leading scientists, the consensus generally is that the high percentage of meaningless scientific work is a necessary by-product of the scientific enterprise (« It has always been so ») and thus is the price that has to be paid.

I shall return to this point later. Let me here only run ahead with the historian's remark that all this has resulted from the 20th-century image of knowledge and it has *not* always been so. In the 17th, 18th and even 19th centuries a much higher percentage of results was meaningful, and *not* because then there was more to be discovered (3). As things stand today, at no stage in the process of education, apprenticeship, or even of high-level consultation is a scientist encouraged to subject the very foundations to critical questioning, or to engage in such intellectual exercises as verifying which « eternal » facts would become irrelevant (if not wrong) if the theoretical framework were changed. On the other hand, it is widely admitted that nobody can influence the Einsteins of this world, who would go their way anyway. But the presupposition there is that science's accumulative course suddenly changes via big

(3) Though dealing here with the problems of progress in science, one should not forget that whether this wasteful system advances (as usually considered) or hinders (as I hold) that progress, that social pressures apparently will not allow it for very long. With spreading popular scientific education, more and more *enlightened* criticism of the scientific profession is to be expected.

leaps produced by individual geniuses who alter the course of development (4). Since they are almost always, as if by a law of development, in radical opposition to their peers anyway, why bother to remove the hobbles from students in general? To recapitulate: all these views, however contradictory in part, when taken together constitute a theory of the growth of knowledge.

According to the present image of knowledge, theories are believed to be accepted or abandoned on experimental evidence, irrespective of the price of abandoning a theory or accepting a new one. We all suspect that *ad hoc* hypotheses are very often invented to save theories, only to be tested on their own experimentally. This is more or less true. The problem is—and here we must act differently from scientists in, let us say, the 18th century— that, if we succeed in formulating an *ad hoc* hypothesis that can be tested experimentally, our experimentation is vindicated and the scientific community rewards us. Since this process of experimentally testable *ad hoc* hypotheses is a practically infinite one, an ingenious scientist can spend a lifetime on that—the *image* of knowledge—without social pressure from the consensus-minded scientific community, which does not push first towards investigation of the meaning of an undesirable experimental result with respect to either fundamentals or other scientific theories. In other words, under the ruling image of knowledge, we think in terms of single theories, not networks of theories. I shall return to this when I deal with scientific research programmes in the next Section.

One last aspect of our current social image of knowledge: there is a widespread view that great discoveries are often made simultaneously in different places. Since science is universal (we believe) and, since results are communicated immediately, this notion is not surprising. I admit that because of good and speedy communication simultaneous discoveries are much more profitable in the 20th century than previously, but I claim that this is only a seeming phenomenon, occurring because we see results rather than problems and thus reconstruct developments by hindsight. Hindsight in historical context can be very illuminating; but when it is used in a conceptual framework of science as an accumulation of eternally true, theory-independent data, then hindsight is very misleading. What is actually happening in most cases is that different problems are being asked and different answers given; these different answers then supplement each other and become one theory, which is read back into both results.

For a recent instance of alleged simultaneity, let us take the case of POINCARÉ and EINSTEIN. It is often stated that because POINCARÉ and EINSTEIN simultaneously and independently (I ignore here, as both uninteresting and

(4) Here, I am merely describing the current view that knowledge grows by accumulation. I shall show later that the details of this theory are explained by the presently accepted image of knowledge and do not relate to history at all. It will also be explained why the Kuhnian picture is so well received by scientists.

false, the unworthy and unproved claim of some writers (5) that there might have been conscious copying on Einstein's part) developed identical transformation equations, POINCARÉ is thus a co-discoverer of at least part of the theory of relativity. Generally, historians then ask why POINCARÉ did not go the whole way: why did he not discover the whole of the theory? For, as it is pointed out, there was no *information* available to EINSTEIN that was not available to POINCARÉ! But this point is true not only for POINCARÉ—so far as information goes, the theory could have been discovered at least twenty years earlier. GOLDBERG in two papers showed that POINCARÉ and EINSTEIN were actually working on two different problems:

> The fact that LORENTZ and EINSTEIN had arrived at the same transformation equations does not mean that their theories are the same. Nor is it significant that POINCARÉ realized that the Lorentz equations form a group or that the equations imply that the velocity of light is the ultimate velocity. POINCARÉ (and LORENTZ) and EINSTEIN were doing different things, working on different theories from different points of view (6).

A different apparent simultaneity, which is more recent (again, resulting from speedy communication), is the iterationlike process by which a result is taken up and read somewhat differently from the way it was written (see my discussion on EINSTEIN and the monkey in the next Section); this leads to a new result somewhat different in direction, and so on. Finally the two individuals, or the two teams, reach the same result by two different processes that rely heavily on each other (the Crick-Watson story illustrates a similar process) (7).

---

(5) G. H. KESWANI: *Origin and concept of relativity*, I., *Brit. Journ. Phil. Sci.*, **15**, 286-306 (1965); and II., *Brit. Journ. Phil. Sci.*, **16**, 19-32 (1965).

(6) S. GOLDBERG: *Poincaré's silence and Einstein's relativity: the role of theory and experiment in Poincaré's physics*, *Brit. Journ. Hist. Sci.*, **5**, 82 (1970); and his previous, *Henri Poincaré and Einstein's theory of relativity*, *Amer. Journ. Phys.*, **35**, 934-944 (1967).

(7) Here are some sources on the simultaneity argument:

W. F. OGBURN and D. S. THOMAS: *Are inventions inevitable?*, *Political Science Quarterly*, **37**, 83 (1922). (The authors collected some 150 cases of what they consider cases of independent discovery, or, as MERTON calls them, « multiples ».)

R. MERTON: *Priorities in scientific discovery*, in B. BARBER and W. HIRSCH (eds.): *The Sociology of Science* (New York, 1962), p. 447-485. Another important essay by MERTON on essentially the same problems, *Singletons and multiples in scientific discovery: a chapter in the sociology of science*, *Proc. Amer. Phil. Soc.*, **105**, 420-486 (Oct. 1961). Also, *Resistance to the systematic study of multiple discoveries in science*, *European Journal of Sociology*, **4**, 237-282 (1963).

T. KUHN: *Energy conservation as an example of simultaneous discovery*, in M. CLAGETT (ed.): *Critical Problems in the History of Science* (Madison, 1958), p. 321-356.

Y. ELKANA: *The conservation of energy: a case of simultaneous discovery?*, in *Archives*

To round out the present discussion with a positive remark, we must create an image of knowledge according to which scientists are encouraged to think fundamentals, and to differ from each other (theoretical pluralism); as it is now, discoveries are being made by those who have intellectual gifts plus the daring to dissent [8]. If the present image changes, intellectual gifts will be sufficient, and the group will not be delimited by the additional demand of daring.

What I have described thus far is a randomly illustrated theory of the growth of knowledge. Since my claim is that everybody has such a theory (albeit sometimes inarticulate), I argue that it was important to articulate the one that most scientists share and to juxtapose to it other theories, including the one I hold to be true. In describing the present theory I relied on the concept of the image of knowledge which is an integral part of the theory of the growth of knowledge I shall endorse below.

1·3. *Theories of the growth of knowledge.* – Philosophy of science is many things to many people. At least three different meanings are current. For some it is the logical analysis of established scientific theories, which are first artificially closed down (no open problems) and then an analysis of their foundations and of their coherence is supplied. This approach, mainly represented by the logical positivists, is held to be a very respectable quasi-scientific activity. On the one hand, scientists respect such philosophers as colleagues working in a scientific field different from their own. On the other hand, the philosopher of science's results are held to be as relevant to the elementary-particle physicist as those of the molecular biologist (the solid-state physicist and the molecular biologist are more relevant to each other).

For others, philosophy of science is the story and the history of cosmologies, *i.e.* great philosophical world-views; this group considers it much less of a science, but interesting bedside reading for the scientist. If philosophy of science is directly relevant to the scientist's work, that is unnoticed and unadmitted: who can tell whether a new idea springs from reading WHEWELL or WORDSWORTH?

For a growing number of historians, sociologists and philosophers of science, philosophy of science is a study of the theories of the growth of knowledge and the theories of rationality and scientific epistemology that go with them. This is the view I represent here, and this activity, according to my view, is highly relevant to science. I shall devote Sect. 4 to discussion of this relevance.

---

*Internationales d'Histoire des Sciences*, 90-91, Janvier-Juin 1970.
For the genetic code, see J. WATSON: *The Double Helix*; and G. STENT: *DNA*, and ROBERT OLBY: *Francis Crick, DNA, and the central dogma*, both in *Daedalus*, Fall, 1970.

(8) Very rarely can one read moderate criticism of the scientific enterprise written by insiders. The only recent source is J. Ravetz's fascinating book: *Scientific Knowledge and Its Social Problems* (Oxford, 1971).

Serious discussion of scientific rationality, scientific epistemology and the growth of knowledge started with the various writings of Sir K. POPPER. The chief difference between him and the other leading school of philosophy of science, the logical positivists, was one of emphasis: their chief interest lay in determining how to justify, validate, or corroborate an established scientific result; Popper's main purpose was always to establish a theory of how science grows. According to his view, science grows by a method of trial and error (which, he notes, is typical of all learning processes in animals and man) whereby scientific theories are replaced by better ones. The criterion of progress is higher empirical content of a theory and thus lower probability, *i.e.* higher falsificability. Therefore, to advance science one must make an attempt to falsify ([9]). In order to support his argument, POPPER builds a time-independent theory of how science grows (always)—« science » being disembodied ideas that are very little influenced by any development other than pure ideas. He illustrates this by a few great ideas in the history of the physical sciences, using historical case studies to illustrate inductively a time-independent, culture-independent process, which is rooted first in biological traits and then in anthropological universals.

What I would like to save of Popper's theory is his great insight that knowledge grows through conflict between competing theories, *i.e.* by critical dialogue between competing conceptual frameworks. I chose the expression « conceptual framework » so as to be free to decide later on whether these are theories, metaphysics, paradigms, scientific research programmes, or whatever. This insight is fundamental to any history-conscious attempt to understand the process of growth, despite the fact that POPPER was looking for a universal, *i.e.* unhistorical, process.

Philosophers of science have invested most of their energy in recent years in developing a theory of the growth of knowledge: that was the case with AGASSI, TOULMIN, FEYERABEND, KUHN, LAKATOS and many others. I shall refer here first to KUHN—because his theory is the only really historically-conscious one, has been more or less accepted by the scientific community, and best represents its own self-image (I shall try to show in Sect. **4** why this is so)—and then to LAKATOS, because scientific programmes occupy a central position in his theory and I wish to construct my theory on his insight of the centrality of the programmes ([10]).

---

([9]) POPPER has elaborated his theories in many publications and they are so well known that there is no need for full reference here. I mention only one: *Truth, rationality and the growth of knowledge* which is Chapter 10 of his *Conjectures and Refutations* (London, 1963).

([10]) AGASSI has already remarked, in his *Towards a Historiography of Science* (Middletown, 1967), on the fruitfulness of this concept; but it was LAKATOS who made it the cornerstone of his theory.

Kuhn's *The Structure of Scientific Revolutions* is so well known that it scarcely needs describing; it is one of the very few books about science that those working scientists, who do not make a virtue of illiteracy, would read ([11]). In rough outline Kuhn's theory is that knowledge grows by sudden revolutions which entail changes in world-view (paradigm changes) in an otherwise calm, precise and methodical activity of normal science carried out by average scientists—all of whom do puzzle-solving on a basis of their shared paradigm. What brings about such changes, nobody can tell; but to do research on that one must deal with a socio-psychological phenomenon that is irrational and thus external to science. Critical dialogues have a place only during the short unsettled period when the revolution is taking place; after that, a new, generally hard paradigm change occurs and a new normal science is at work. And in periods of normal science textbooks are written to modify the knowledge achieved at any given moment.

Kuhn's defence of his theory is sophisticated and rich in historical perspective but its main characteristic is that it eminently suits the scientist's own image of science and its growth, and it reassures him that what he is doing is the best thing to ensure progress ([12]).

Lakatos' theory is stated in two recent publications ([13]). According to LAKATOS, knowledge grows by a critical dialogue between whole scientific research programmes (not between single theories) that include a whole net of theories, more or less coherent, at the core of which is a scientific metaphysics, *i.e.* a belief in a given aspect of the structure of the world that is not experimentally testable but is defended by a whole research programme. (Sections **2** and **3** of this discussion will be devoted to description of these). In order to understand such a programme LAKATOS suggests a rational recon-

---

([11]) The second edition includes a postscript which is at once a philosopher's delight and food for controversy (Chicago, 1970). A whole book is dedicated to critical analysis of Kuhn's ideas. LAKATOS and MUSGRAVE (eds.): *Criticism and Growth of Knowledge* (Cambridge, 1971). This is the proceedings of a conference focusing on a discussion between KUHN and POPPER on the growth of knowledge.

([12]) A very illuminating sociological paper appeared recently: H. MARTINS: *The Kuhnian « revolution » and its implications for sociology*, in NOSBITER *et al.* (eds.): *Imagination and Precision in the Social Sciences* (London, 1971), p. 13-58. He points to the important conclusion, which I fully endorse, that in order to do sociology of science intelligently, one cannot separate it anymore from a general theory of sociology of knowledge.

([13]) I. LAKATOS: *Falsification and the methodology of scientific research programmes*, in LAKATOS and MUSGRAVE (eds.): *op. cit.*, p. 91-195; and *History of science and its rational reconstruction*, in *Boston Studies Phil. Sci.*, **8**, 91-136 (New York, 1972). It is also reprinted in Y. ELKANA (ed.): *Chapters in the Interaction Between Science and Philosophy* (New York, 1973), and it is followed by a detailed criticism in my own *Boltzmann's Scientific Research Programme and its Alternatives*, Hebrew Univ. and Van Leer Foundation (Jerusalem, 1971), p. 30.

struction (which is logically criticizable) instead of a « true » historical narrative, the truth of which can never be ascertained. The research programme with its metaphysical core and the critical dialogues between research programmes are « internal history » (*i.e.* of disembodied ideas). Why some such research programmes are accepted and others are rejected, one cannot tell —surely irrational « external » elements are active; but, at least by rationally reconstructing, one could say why they *should have been* rejected or accepted. In any case, for understanding science and its growth only such a reconstruction is necessary: true history with its external factors is, though fascinating, irrelevant. Thus in Lakatos' theory of the growth of science, the present dictates the past by hindsight (for rational reconstruction is just that, institutionalized hindsight); presupposing that the present is the highest stage of science, it serves as a basis for the future.

Changes, according to LAKATOS, occur in the body of knowledge by way of mutations, which he calls « progressive problem shifts », in which new problem areas are uncovered while some of the previous ones are being solved; the mutations survive via degenerative problem shifts that will lead finally to the slow disappearance of a theory.

I shall attempt now to outline my theory of the growth of knowledge, and I shall try to do so in a critical dialogue with LAKATOS.

Knowledge grows by the interaction of three factors, which can be distinguished only if time is stopped and a socio-cultural situation is, so to speak, photographed. (In other words, I am idealizing for the sake of clarity.) These three factors are

*a*) the body of knowledge,

*b*) the socially determined image of knowledge, and

*c*) other factors that do not directly depend on knowledge.

*The body of knowledge.* At any given moment there is a state of knowledge with its methods, solutions, open problems, net of theories and—at its core—a scientific metaphysics: which is to say, at every stage there is a field of scientific research programmes, shared by some parts of the scientific community, while others share another scientific research programme with a different scientific metaphysics at its core. These two, or more, scientific research programmes conduct a critical dialogue. Depending on the stage of the science and on the time, the place and the culture, there will be one or more scientific research programmes and the dominant ethos among the scientists will be either consensus or critical dialogue with disembodied ideas—objective scientific knowledge— as the object of the discussions. However, whether there is consensus or debate has nothing to do with the body of knowledge; it is a function of the socially determined image of knowledge.

*The socially determined image of knowledge.* If the body of knowledge is the sum total of statements on Nature, then the socially determined image of knowledge is the sum total of statements on knowledge of science, scientists, truth, etc. Rationally held beliefs about the task of science (*e.g.* understanding, prediction), about the nature of truth (certain, probable, attainable, etc.) and about knowledge (by revelation, by ratiocination, by experiments through the senses) are all part of the time-dependent image of science. It is this image that decides which problems are chosen, out of the infinity of open problems suggested by the body of knowledge; their scale of importance is fixed by socially formulated criteria, as is the determination of the frontier of science. Theories of explanation too, in the final account, boil down to what is socially convincing as an explanation. The whole body of *method* is part of the time-dependent, culture-dependent, social image of knowledge.

*Other factors, not directly dependent on knowledge.* These are the true irrational elements— the ideologies, political considerations and social pressures that either give or withdraw support for an institution or a research programme. These are the factors that put a scholar into science or keep him out by causing him to choose more rewarding fields; they determine how a career looks and what it consists of. All the economic materialistic factors that underline the vulgar Marxist ([14]) theory of the growth of knowledge belong in this category. As has been emphasized, the distinctions among the above three factors are artificial and can be drawn only if time is stopped. On a time scale, the three interact and cannot be disentangled. According to LAKATOS, his theory is good because this kind of reconstruction can be, and is, nearest to the historically true story (true in the primitive sense in which it is clear that BOHR did not discover the spin, NEWTON had no energy concept, etc.). I shall try to show that the demand for rational reconstruction as against historical narrative is unnecessary, and that the internal-external dichotomy itself constitutes a degenerating problem shift.

Even if we leave the factors of speed, locality, etc., of historical events to nonrational explanation, much more than just the growth of objective scientific knowledge remains to be explained rationally. I fully agree that « whether an experiment is crucial or not, whether an hypothesis is highly probable ... whether a problem shift is progressive or not is not dependent in the slightest on the scientist's beliefs, personality or authority » ([15]). But it is dependent on what the scientist sees as his own role and the role of science; on what he believes to be basic concepts, theorizing in terms of which constitutes legitimate scientific thought; on what he thinks is the accepted limit of speculation for

---

([14]) The distinction here is between vulgar Marxism (like *Science at the Crossroads*, Cass Reprints, 1972) and the much more sophisticated latest books by J. D. BERNAL.
([15]) I. LAKATOS: *op. cit.*

advancement in terms of grants or position in the scientific community; on how he defines the connections between the animate and the inanimate worlds; on whether he considers it rational to explain nature on the basis of conservation laws, or symmetry rules, or fundamental entities; and on whether, in his context, daring speculation or solid minute theoretical connections are thought of as a hallmark of science.

Those are the rational considerations, the cognitive considerations I shall call the « image of science »: it is the sum total of *thoughts* on what science is and should be, and it has a major rational influence on the scientific programmes of individuals, schools and communities. If this be psychology, it is cognitive psychology and not motivational. As to its source—the image of science is created in part by the philosophy of science-teaching in a given scientific community; this philosophy in turn is influenced by a mixture of considerations, some of which LAKATOS would call internal and others external. The rational influences are the objective state of knowledge and the accepted theory of the growth of knowledge (for science teaching is and always has been at least openly normative), as well as the accepted epistemology of the time.

The nonrational factors are the web of beliefs and the social and economic factors that all contribute to the creation of what is called the « spirit » of an age. All in all, one can with good reason abhor Lakatos' rational reconstructions without becoming a classical inductivist. One can view the growth of knowledge as the result of sharp, slow-moving critical dialogues between competing research programmes with degenerating as well as progressive problem shifts, and realize that these critical dialogues ensue because of differences in the three interacting areas cited above. Moreover, one must take into account also the competing images of science, a critical dialogue which in itself is perennial.

The hard-core scientific metaphysics of a scientific research programme is heavily influenced by socio-political irrational factors and by a previous generation's image of science. On the other hand, the next generation's image of science depends to an even larger extent on achievements in the body of knowledge and on the scientific metaphysics at core.

One of the most important consequences of dealing with scientific research programmes is the emphasis on a network of theories rather than one or another single theory. Instead of checking the validity of a theory, one considers the fruitfulness of a scientific research programme that relies on a whole network of theories, some of which may be less valid than others. GONSETH, in his description of Einstein's 1912 lectures in Zurich, emphasized this point:

> What EINSTEIN mentioned in very summary fashion and with never a thought of analysing it was not the validity of this or that train of reasoning, the result of this or that measurement, or the value of this or that interpretation, but rather a whole web of reasoning, results and successful applications at the centre of which Maxwell's equations assumed a special significance. Now, the solidity of such a web and the

security it gives to a scientist can be so compelling that, when it comes into conflict with some of the most firmly anchored proofs, it is opting for the proofs which lay him open to the risk of very grave error. In such a situation, theoretical reason is no longer sole arbiter; but neither can the practical reasons be on their own. Here is a complex of circumstances and consequences to be appreciated and evaluated. What must be perceived is what is best retained or discarded, taken or left. This can be made into a principle, that of the *greatest suitability*, which I prefer to call the principle of the best fit ([16]).

That LAKATOS draws our attention to this is one of his greatest insights. It frees us also of, among other things, the sterile all-or-nothing methodological debates about whether theories are confirmed or falsified. For they are both accepted and rejected out of a whole complex of considerations from the body of knowledge and from the image of science—including, always, the attempt to remain in the framework of a whole scientific research programme. In the net of theories around a metaphysical core, which constitutes the scientific research programme, a delicate balance obtains wherein some theories must be falsified by the confirmation of some others.

One more issue must be discussed, however cursory this survey may be. Very often, the notion of the objective culture-free character of science is supported by pointing to the impersonal scientific result as written up in the scientific paper. If that were indeed an accurate reflection of the nature of science, to understand science would require only acquisition of the body of knowledge. However, that is not the case.

The question before us is: « Is the scientific content of an objectively understandable printed article independent of the mind that wrote it and thus certain to carry the same message to every reader, *i.e.* to be independent of the individual mind that reads it? » To this my answer is an unqualified « no » ([17]).

---

([16]) In *Science and synthesis, An International Colloquium Organized by UNESCO on the Tenth Anniversary of the Death of Albert Einstein and Teilhard de Chardin* (New York, 1971), p. 5.

([17]) On this point, see the discussion between LAKATOS and ELKANA, following papers by MENDELSOHN and AGASSI in Y. ELKANA (ed.): *Chapters in the Interactions between Science and Philosophy* (New York, 1973).

There, LAKATOS said: « ... For instance, if I want to judge the merits of Einstein's 1905 paper, it does not matter at all that it was written by EINSTEIN. It would have had exactly the same value if typed out by a monkey. (One can easily calculate the probability with which a monkey, once taught to type, would type out Einstein's 1905 paper.) Thus one can, and I think one should, have "rationality" in *methodology*, while giving way to "irrationality" in discovery, in the creative process. In the latter I agree with POPPER, POLANYI and KUHN. (Few people have appreciated that POPPER mentions BERGSON with approval in his *Logik der Forschung*, while, say, RUSSELL abhors BERGSON as being irrationalist.) »

« Now let me add a footnote to all this. There is much less probability that the monkey should type out, say, Newton's *Principia* than just the inverse square law—and this is why NEWTON is superior to HOOKE. This is why it is more rational to take as a unit

A written statement never amounts to more than a tenth of its meaning. It carries all its meaning only if writer and reader share scientific metaphysics, problem situation, technical know-how and a social image of knowledge—but, if they share all this, then we are indeed dealing with a very limited minority. And if all the members of this small group also both share a network of accepted theories and select identical theories for rejection, then in all probability the communication will be trivial.

The fruitful clash—the critical dialogue—occurs when for the reader (or the writer) the argument or the experiment becomes a crucial argument or experiment between two theories. There is no crucial experiment or proof which is crucial as such: it can be crucial (*i.e.* classically at the « cross-roads », as BACON put it) only in a given context. The context that the writer or reader brings to the theoretical or the experimental case is his scientific research programme, *i.e.* a whole network of diverse theories that are accepted, others that are rejected, and inconsistencies that are tolerated although others are not. The context is also the social image of knowledge, which affects the kind of explanation that is acceptable at a given place at a given time, and the respect or scepticism with which an experimental, or for that matter a mathematical, proof is treated. To see this effect, it is enough to follow the debates between experimentalists (as a group) and theoreticians (as a group) when widely divergent interpretations are inevitable—which occurs often in elementary-particle research. It was also typically visible in Berlin in the 1840's, when *any* theoretical argument was rejected by experimentalists because of the prevalent image of scientific knowledge. There was also a time, in the 1930's, when, under the influence of Einstein's general theory of relativity and because there was little hope of successful experimentation on the burning issues, all theoretical speculations were given a high degree of receptivity. Finally, in

---

of appraisal a whole *research programme* rather than the simple idea formulated in the hard core. Or in the case of EINSTEIN, a monkey, with relatively high probability, can type out the sentence: *Time and space are somehow interrelated.* But it is *much less* probable that the monkey should type out the entire special, let alone the general relativity theory. »

ELKANA said: « ... Our most important historical task is still that of finding the connection between what goes into a piece of scientific work and what comes out as a finished product. The moment we do not have to take into account in the output what the input was, the way we try to understand how science is being done stops being interesting. It is in this sense that it is important whether a paper was written by EINSTEIN or a monkey. In other words, what we can do is to try to rationalize scientific creativity. This will not give us a complete account of how the creation was made, of the act of creation, but we can collect quite a wide range of not sufficient but necessary conditions. We can ask what things have to have happened rationally in the intellectual make-up of the scientist for the creation to have taken place. This then becomes very important for understanding the final product, and this you would lose the moment you claimed that it is not important who or what wrote it. »

the middle of the 18th century—when the critical dialogue centred around such matters as « Is the essential quality of matter "impenetrability" (Euler's view) or "repulsive force" (BOSCOVICH)? »—the sole criterion of the quality of an explanation was its logical coherence. No experimental proof was even hinted at ([18]).

Add to this that the problem-choice itself is determined not by the body of knowledge but by the social image of knowledge, as we have seen, and a reader's context can emerge in which a written scientific paper is not even accepted as « science » and its results are disregarded. All this comes from *context* and is not mentioned in the written work. Precisely because of awareness of this state of affairs, it has become part of the *rules* that a scientific paper must be written impersonally, omitting any narration of the history of the work; mistakes and irreproducible results are not to be communicated—and nowadays, even the broader problem situation is not to be stated. In short, the rules dictate saying as little as possible, so that the message can be absorbed by as many contexts as possible—which, if it is *universalism*, is certainly so by default.

Sections **2** and **3** will be devoted to a bird's-eye view of the history of the last three centuries of scientific growth, seen on the basis of the above theory of the growth of knowledge.

## 2. – From the scientific revolution of the 17th century to the Einsteinian revolution of the 20th century-(A).

2·1. *Was it revolution?* – If it is true—and I think it is—that knowledge grows by a continuous critical dialogue between competing scientific research programmes and especially between the competing scientific metaphysics at the core of these programmes, and also by a continuous critical dialogue between competing social images of knowledge, then there are no revolutions ([19]). However, hindsight is inescapable: the knowledge considered true and scientific in the 19th and in the 20th century, the conceptions prevalent today, and even the different protagonists occupying those of us today who see a continuous critical dialogue—all those have their source in the 17th century. It is very probable that, with the elimination of the objectivist instrumentalistic and methodological fallacies (see discussion in Sect. **4**), the 17th century will seem a direct continuation of the 15th and 16th centuries, and the Victorian 19th century

---

([18]) See Y. ELKANA: *Scientific and metaphysical problems*: *Euler and Kant, Boston Studies in the Philosophy of Science,* to be published.

([19]) The evolutionary view, or rather the view that knowledge grows by a continuous series of microrevolutions, has been forcefully championed by S. TOULMIN, see his *Conceptual revolution in science,* in R. S. COHEN and MARX W. WARTOFSKY (eds.): *Boston Studies in the Philosophy of Science,* Vol. **3** (New York, 1968), pp. 331-337, and his recent book, *Human Understanding,* I (Oxford, 1972).

will become the « Dark Ages ». The historiographic changes undergone by the concept of the Middle Ages or of the Industrial Revolution, under the pressure of changing images of the past as a reflection of changing self-images of the present, are well known ([20]).

In other words, a revolution is a revolution only with respect to a course of events that is considered normal and is generally of long standing. We consider the 17th century revolutionary because it is our view that knowledge —true, scientific knowledge— all started in Greece but failed to develop further for lack of experimental method, and that in the Dark Ages humanity retreated into the obscurantism of superstition with a spark of light preserved only in the monasteries, where it soon degenerated into the ash of empty quibblings of scholastic philosophy. Such analyses apply to all intellectual revolutions (it is different with social revolutions, which are very real indeed but serve as wrongly chosen models for intellectual history). When we look at them in the correct historical context, we shall find continuous critical dialogue between the competing scientific research programmes and the competing images. Some trends will appear to be more closely connected with the past and some with the future; from our (20th-century) point of view, some will seem successes and others failures.

For the 17th century, two important kinds of historical research had to be done in order to dismiss the myth of the revolution (though the label is retained, for convenience). One kind of research was initiated by DUHEM, who showed that our favourite characteristics of « good » science— *e.g.* the experimental method, the critical attitude, the heliocentric model—all have a history of many hundreds of years before the 17th century ([21]). The second approach

---

([20]) The English never underrated the glorious 13th, 14th and 15th centuries. For them it was a time of relative peace and prosperity, when the Continent was swamped by wars, famine, plague, and the repeated enthusiastic millenarian movements. It is enough to compare a book like G. G. Coulton's *Medieval Panorama* (Oxford), and N. Cohn's *The Pursuit of the Millenium* (London, 1970). Also Johan Huizinga's remarks on the discrepancy of images in his *The Waning of the Middle Ages* (London, 1965).

([21]) In addition to DUHEM, the great rediscoverers of pre-17th century science were George SARTON, Lynn THORNDIKE (in his majestic 10-volume *History of Magic and Experimental Science*), Anneliese MAIER and Marshall CLAGETT and A. C. CROMBIE. CROMBIE, on p. 1 of his *Robert Grosseteste and the Origin of Experimental Science: 1100-1700* (Oxford, 1953) says: « The history of science shows that the most striking changes are nearly always brought about by new conceptions of scientific procedure. The task demanding real genius is the revision of the questions asked, the types of explanation looked for, the criteria for accepting one explanation and not another. Underlying the conception of scientific explanation accepted, for example, by GALILEO, HARVEY and NEWTON, was the theory of formal proof developed by the Greek geometers and logicians. The distinctive feature of scientific method in the seventeenth century, as compared with that in ancient Greece, was its conception of how to relate a theory to the observed facts it explained, the set of logical procedures it contained for con-

was developed fairly recently by the historians who showed the « irrational », « unscientific » traditions that were predominant on the 17th-century scene, *e.g.* the Paracelsian, Hermetic, Rosicrucian, and other scientific research programmes ([22]).

If one adds that such important attributes of modern science as conservation laws were completely alien to the leading scientific research programmes of the 17th century (except to the synthesis-minded ones like Leibniz's—see Subsect. **2·6**) and were introduced only in the 18th and 19th centuries, then the 17th century is seen to be no more revolutionary than the 13th or the 19th. What we must admit, however, is that most of the scientific theories held today as part of our commonsense science (not by working scientists) were formulated in the 17th century. The « scientific revolution » is sheer hindsight. However, since we do tend to see the world from our own vantage point, let me choose the 17th century as a simple starting point for my general survey.

---

structing theories and for submitting them to experimental tests. Modern science owes most of its success to the use of these inductive and experimental procedures, constituting what is often called "the experimental method". The thesis of this book is that the modern, systematic understanding of at least the qualitative aspects of this method was created by the philosophers of the West in the thirteenth century. It was they who transformed the Greek geometrical method into the experimental science of the modern world. »
But in order to accept such a continuity theory one need not see the 17th century's experimental methods as the chief value of its science. The great A. KOYRÉ, in his *Origin of Modern Science, A New Interpretation* says: « As for myself, I do not believe in the explanation of the birth and development of modern science by the human mind turning away from theory to *praxis*. I have always felt that it did not fit the real development of scientific thought, even in the seventeenth century; it seems to me to fit even less that of the thirteenth and fourteenth. I do not deny, of course, that in spite of their alleged—and often real—"other-worldliness", the Middle Ages, or to be more exact, a certain, and even a rather large number of people during the Middle Ages, were interested in techniques; nor that they gave to mankind a certain number of highly important inventions. Some of them, had they been made by the Ancients, would probably have saved the Ancient World from collapse and destruction by the predatory Barbarians. Yet, as a matter of fact, the invention of the plough, of the horse harness, of the crank, and of the stern rudder had nothing to do with scientific development; even such technical marvels as the Gothic arch, stained glass, the foliot or the fusee of late medieval clocks and watches did not depend on, nor result in, any progress in corresponding scientific theories. Strange as it may seem, even such a revolutionary discovery as that of firearms has had no more scientific effect than it had scientific bases. Bullets and cannon balls brought down feudalism and medieval castles, but medieval dynamics resisted the impact. Indeed, if practical interest were the necessary and sufficient precondition of experimental science—in our sense of the word—this science would have been created a thousand years at least, before Robert GROSSETESTE, by the engineers of the Roman Empire, if not by those of the Roman Republic ».

([22]) This is the contribution of W. PAGEL, A. DEBUS, CH. WEBSTER, P. RATTANSI, T. MCGUIRE and P. HEIMANN.

2·2. *Where did it come from?* – During the bitter fight between Reformation and Counter-Reformation a sceptical crisis arose, challenging the basic principles of theology, humanistic studies moral studies and the sciences. Man's confidence in discovering truth by the use of human reason—a confidence so typical of both Scholastics and Renaissance naturalists— was undermined by a devastating attack from a group whose chief spokesman was the philosopher-essayist DE MONTAIGNE (1533-1592). But total scepticism could not last long. The « pyrrhonics », as they were called, dwindled, and the generation after MONTAIGNE emerged optimistic. It gave three different answers to its teacher's scathing criticism. One was the religious answer of Father CHARRON, for whom any scientific theory is actually blasphemy because every theory presumes to limit God's power and ability to that which man can understand. Thus the very foundation of science—the presupposition of Nature's comprehensibility—is anathema, for certain knowledge can be gained only by revelation. As Koyré put it, « ... to the uncertainty of natural reason CHARRON opposes the supernatural certainty of faith ». Here *Pistis* once again replaces *Gnosis*.

Francis BACON offered experience as a remedy: « ... the sterile uncertainty of reason left to itself the fruitful certitude of well-ordered experience » ([23]). Bacon's programme failed scientifically but succeeded socially: his remedy became part of the ethos of science. From the 17th century till today, our image of certain knowledge relies on experience.

The third response, reason, came from DESCARTES. Going beyond common sense and classification (which BACON aimed at just as intently as ARISTOTLE) he followed the order of ideas, not of things.

Common to all three answers—faith, reason, experience—is their polarity. All are analytic in nature and claim dogmatically to be capable of reducing the whole world to faith, reason, or experience. This polarity seems the cause of certainty: whatever knowledge there is, on the basis of faith or reason or experience, is certain knowledge—the rest is excluded, is not considered knowledge at all. Thus the image of knowledge is that the only possible knowledge is certain knowledge. The opposite of this is the vague, metaphysical synthesis that accepts all kinds of knowledge and believes in the comprehensibility of Nature, in the existence of a pre-established harmony, or in a biologically conditioned fitness between the human mind and the world it is to comprehend; this kind of synthesis creates great systems of thought. It is broad, tolerant and much less certain than revelation, clear and distinct ideas, or induction.

---

([23]) A. KOYRÉ: *Introduction* to E. ANSCOMBE and P. GEACH (eds.): *Descartes Philosophical Writings* (London, 1970), p. xiii.
See also R. POPKIN: *History of Scepticism from Erasmus to Descartes* (New York, 1964) and H. G. VAN LEEUWEN: *Problem of Certainty in English Thought 1630-1690* (The Hague, 1970).

The greatest representatives of this tradition are LEIBNIZ, EULER, FARADAY, HELMHOLZ, BOLTZMANN and EINSTEIN—all of whom are nonpolar, construct a weight-bearing web of scientific metaphysics, and are physical realists. Their contributions to the body of knowledge are well known; however, their image of knowledge was suppressed. From the 17th century onward, scientific knowledge claimed certainty, and it claimed to have found it in either reason or experience—mostly, in experience. The polar reactions to the 16th-century scepticism can be characterized as *constructive scepticism*, which became the hallmark of science first in the 17th century and then again more forcefully in the 19th century. The great natural philosphers of the 17th century, *e.g.* GALILEO, DESCARTES, GASSENDI, NEWTON and BOYLE, were all constructive sceptics.

However, the difference between the two views did not result in differences in the quality of science: it would be a mistake to brush this distinction aside with a casual « well, what you mean is simply that some were scientists and the others were, at best, philosophers ». I am referring only to the best minds in the 17th century, and all of them were simply natural philosophers. Moreover, their *positive* contributions to present-day knowledge (that criterion for excellence belongs not to me but to those positivists with whom I am debating here) are comparable. True, it is NEWTON who formulated the general law of universal gravitation; but it is LEIBNITZ from whom all conservation ideas stem.

Attempts to characterize these two kinds of scientist-philosophers have been numerous. NIETZSCHE distinguished between Dionysian and Apollonian artists, with APOLLO symbolizing the art of sculpture and DIONYSOS the art of music; attic tragedy synthetized the two. Ostwald's distinction between the classical and the romantic scientist is well known, as is Matthew Arnold's. Recently, the biologist SZENT-GYÖRGYI remarked on it in a letter to *Science*, where he pleaded for allocating to intuition (according to him, the Dionysian element) its righful place in science ([24]). But that dichotomy is much too simple. The growth of knowledge cannot be viewed only as a fierce battle between the rational and the irrational, or between the clearly defined and the vague. Neither can it be seen on the social level as a battle between elitist and populist. (« APOLLO always moved in the best society, whereas DIONYSOS was much more the god of the common man ») ([25]).

---

([24]) Letter to the editor, *Science*, **176**, 966 (1972); OSTWALD expounded his theory in *Grosse Männer* (Leipzig, 1909); M. ARNOLD stated his in his *Pagan and Christian Religious Sentiment* and in his juxtaposition of « Hellenism » to « Hebraism ».

([25]) Quoted by Antony ANDREWS in *Greek Society* (London, 1971), p. 262. The tension between the Apollonian and Dionysian penetrates Greek culture in all its manifestations. HOMER tried to exclude the irrational. Even when he (or some other Homeric poet) writes a hymn to Demeter, the goddess becomes almost sober, not accompanied by the usual wild music and ritual. On the other hand, ANDREWS says of Euripides'

All these are attempts at classifying individuals according to their genius, *i.e.* according to inborn qualities, whereas my attempt is to distinguish not different kinds of individuals but different kinds of scientific research programmes. In my opinion, one can find all the various kinds of genius and all the various psychological characteristics among the representatives of both kinds of research programmes. Moreover, I do not claim that there are only two such general trends. What I do claim is that because all change in knowledge occurs dialectically, every question or problem creates its own dichotomy. Very many different questions occupy people's minds simultaneously, and it is only in a given context that a dichotomy seems so important. The context I chose here was the scepticism of the end of the 16th century, in relation to which problem the two trends, the dogmatic and the synthetic, emerged. One could concentrate on religious, social, or economic issues that would point to other critical dialogues ([26]).

The important thing about a critical dialogue is that it is much more subtle than a mere list of differences in view. Critical dialogue occurs between those *a*) who are ready to tolerate clear-cut discrepancies between well-formulated theories and securely established experimental results in order to save one metaphysical view of the world, and others *b*) who cannot tolerate such discrepancy and prefer to restrict themselves to one single theory (not a network of theories), thus ignoring the broad metaphysical world-view. The scientific elite in both groups at all times formulated their theories as clearly as the times in which they lived dictated, and performed their experiments as accurately as their times expected of them. Their methods were rational, *i.e.* attentive to a conscientiously acknowledged purpose; their clinging to either a metaphysical world-view or to logical coherence of a partial world-view can be seen as irrational, if we wish to see it so. In terms of the image of knowledge, we can detect a critical dialogue between those who wanted to explain all phenomena by reducing them to essences (essentialists) and those who thought it was the task of science to quantify all phenomena (mathematical physicists). Or the critical dialogue is between the realists, who claim that science must make us understand the world, and the instrumentalists, who claim that the task of science is merely to predict. Or the critical dialogue is between those who say Nature is not infinitely comprehensible (« *ignorabimus* ») and therefore concentrate on the visibly solvable problems, and those who believe Nature is comprehensible to the last (at most, « *ignoramus* ») and therefore attack all problems not yet solved. At all ages both sides are represented in each dia-

---

*Bacchae*: « Whatever else the play may contain, it has to be read as a sermon on the danger of trying to suppress the irrational altogether, a danger to which Greeks were liable » (p. 258).

([26]) An example of this occurs in my discussion of Newtonianism in the 19th century (see Section **4**).

logue; yet most ages are characterized by one side, with a dissenting majority representing the other. This we shall see below.

The answers given by the constructive sceptics and by the great synthesizers (LEIBNIZ, the Helmontians, the Paracelsians) established the source of the major themes in the 17th century. I shall deal with some of them now, in each instance taking up a critical dialogue centering on a given theme.

One of the central problems in modern physics is the tangled issue of determinism, a probabilistic account of physical states. Here scientific considerations *per se* become indistinguishable from moral-religious issues, and even from a deep psychological craving for certainty that seems inherent in the human being. Is it indeed so? Where does our quest for certainty come from? Is there anything specifically scientific in our preference for a deductive proof over a conclusion inductively arrived at? Is a Laplacian deterministic world indeed an unreachable ideal, by comparison with which all our present theories are only second-best?

We have just seen the 17th-century sources of the quest for certainty, which the three polar answers promised. Religious certainty by revelation separated out of the world of knowledge and became a different branch of human consciousness; it will be of little concern for the next two hundred years. The Cartesian programme of certainty by ratiocination failed as a scientific research programme aimed at understanding the physical world, and in the mid-18th century it yielded to the Newtonian research programme that claimed to be following BACON directly. The Cartesians' social image of science—*i.e.* the aim of science is the mathematization of all fields of enquiry—was taken up by the Newtonians and became the dominant force in science for a hundred and fifty years. But this is true of only one of the Newtonian traditions, for the Newtonian research programme split into two parts: the one based on the *Principia* was mathematical and reductionistic; the other, based on the *Opticks*, represented a programme to develop a theory of matter that would explain chemistry and biology and fit the 17th-century image of God. The hard-core metaphysics of the former was that the world consists of discrete particles between which central forces are acting at a distance; at the core of the latter programme was the nutshell theory of matter ([27]). This theory was taken up by natural philosophers—materialists who believed in an all-pervasive aether, which they saw as a vehicle of transmission of various forces.

From the point of view of the history of critical dialogues it is interesting that the Cartesian scientific research programme failed although the Cartesian image of science prevailed. The two Newtonian traditions fare very differently:

---

([27]) See A. THACKRAY: *Matter in a nut-shell: Newton's Opticks and 18th century chemistry*, *Ambix*, **15**, 28-53 (1968). Also his *Quantified chemistry—The Newtonian dream*, in D. S. L. CARDWELL (ed.): *John Dalton and the Progress of Science* (Manchester, 1968), p. 92-108.

the mechanistic tradition ruled supreme and, on having accomplished its task by 1800, simply merged with another theory to create the field concept. The other Newtonian tradition became a participant in all 18th-century critical dialogues on theory of matter, and finally merged in the 19th century with the *Naturphilosophie*, yielding the theory of energy. (I shall return to these in Subsect. **2**˙6 and Sect. **3**.)

However, the nonpolar, nondogmatic answers to 16th-century scepticism did not take up the banner of certainty. Their contributions were of a different kind. Though all these nonpolar trends were connected, for clarity's sake it is easier to talk separately of the Hermetic philosophers, the Paracelsians and LEIBNITZ.

**2**˙3. *17th-Century themes and problems.* – The best single-volume historical introduction to a view of the 17th century, one that sees the growth of knowlege as a series of critical dialogues, is Trevor-Roper's *The crisis of the Seventeenth Century* [28]. Though its emphasis is not on sciences, and not even on the history of ideas, it gives a superb picture of intellectual environmnent. I shall presuppose it here as general background. The other major themes that should be dealt with here are:

1) The critical dialogue between the Royal Society natural philosophers and the different groups of intellectuals who were engaged in medicine, iatrochemistry, education and social utopias, *viz.* the « Helmontians », the « Rosicrucians » and above all the Paracelsians. This critical dialogue was concerned with both their views of knowledge and their views of Nature (a mystical quest for revelation and a basic « *ignorabimus* » against rational natural philosophers, *i.e.* theologians who saw God's glory in man's understanding of creation). The controversies expressed themselves most strongly in various pamphlets and antipamphlets on the Royal Society, and in clashing views on university reform.

2) The problem of the origins of the Royal Society and the intellectual environment it created.

3) The Académie Royale des Sciences in Paris and the relations between this institution and the governments in Paris and in London.

4) The metaphysics and image of science of *a*) NEWTON and *b*) LEIBNIZ, and their critical dialogue.

5) Galileo's view of the tasks of science and his influence on European science.

---

[28] New York, 1956.

6) The interactions of science and technology and the image of science ([29]).

I shall be able to take up only a very few of these themes.

A problem of central importance for those of us who, from the vantage point of the 20th century, see something special—a starting-point—in the 17th century is this: who were the people who represented that growth of knowledge in which our 20th-century ideas originate? Is our science a descendant of the kind of late-Aristotelian scholars who filled the universities of Europe until the middle, and in some places almost to the end, of the 17th century? Does our science originate in the Paracelsian research programme or in the alchemical dreams of the late Renaissance? Has it anything to do with the Cartesian metaphysical heritage or the Baconian inductivist method? Is it a product of the members of the Royal Society or of the university reformers?

As is well known, one theory claims that our science—unique, experimental, nonmetaphysical, objective science—was born among people who were English Puritans and built the Royal Society on the intellectual foundations of Baconian inductivism and the Puritan values of hard work, thrifty utilitarianism, sceptical religious dissent, hard empiricism and a concept of universal brotherly love that, in stressing equality, denied any need of an intermediary class of interpreters between God and his two « books », Nature and Scriptures ([30]). This approach poses two severe problems—whether to consider the evidence amassed in its favour convincing, and whether to accept the conception of science supposed to have originated in the intellectual-cultural milieu of the English Puritans. Since, in most of the studies, the conception of science underlying the Weber-Merton theses was accepted—that is, science was seen as an accumulation of positive data collected by a purely experimental method—the question was whether this activity was indeed concentrated among the 17th-century Puritans. A great many historians then did some more or less sophisticated counting of scientists. Interestingly, even on the conception of science the evidence became confused recently; just as one example, Catholics were found as numerous among the scientists as Protestants.

As will be remembered from Sect. **1**, I do not accept the conception of science and the history of its growth as represented by the Weber-Merton

---

([29]) It is no use to give a detailed bibliography here. I shall again point out one book of central importance: P. Rossi: *Philosophy, Technology and the Arts in the Early Modern Era* (New York, 1970). (Original in Italian: *I Filosofi e le macchine* (Milano, 1962).)

([30]) The literature on this topic is enormous, starting with Max Weber's theory of capitalism and the Protestant ethic, and continuing with Merton's now-classic *Science, technology and society in seventeenth century England*, *Osiris*, **4**, part 2, 360-630 (1938) (where Weber's thesis was extended to science and the Puritan ethic) and hundreds of critical and admiring essays in historical, sociological and philosophical journals. For an enlightening sample covering the whole spectrum of views I suggest the volumes of *Past and Present Quarterly*, which have been appearing since 1952.

theses. That conception is mainly the Victorian image of science, generalized into a time-independent truth that, according to their thesis, was born in the 17th century.

In my view the 17th century is an unhappy age of religious strife and mutual killing and a time when, after 200 years of religious controversy, the intellectuals were seeking a noncontroversial field of intellectual activity and everybody was for consensus. As I have shown in Subsect. 2·2 they hoped to find a noncontroversial answer in, variously, revelation, fact and ratiocination. It is true that the believers in each of these answers disagreed among themselves; but *a*) these did represent, at least, three great consensus groups, and *b*) the discussion among them was rational dialogue and not scholastic *disputatio*, or, what is worse, religious partisanship. The Baconians and the Cartesians disagreed in salons and not in university *aulas* among cheering wild crowds, *disputatio* style, or, like the religious partisans, on the battlefield. Actually, the distinction we, with our Victorian image of knowledge, try to see as experiment-*vs.*-theory, did not exist in the 17th century. Even BACON, in spite of the interpretation of the 19th-century vulgar Baconians, never imagined a science without a theory ([31]); and DESCARTES did not preach a science without experiments.

The real enemy of both was the *disputatio*: it was against the *disputatio* that BACON wanted the inductive method and DESCARTES the extreme rationalization and mathematization of all fields of knowledge. The *disputatio* embodied the schoolmen's image of knowledge, and it was between these two competing images of knowledge, *i.e.* between the schoolmen and the representatives of the « new science », that critical dialogue existed. It is in this light that the mythical story about the « obscurantist » Aristotelians who refuse to look through the telescope of GALILEO must be understood. They were not afraid of seeing the evidence which would damn their theory, but they refused to accept the evidence of the senses as a criterion for solving a dispute ([32]): to them, whatever was seen at the other end (or perhaps inside) was, though true, not relevant to the solution of the problem (whether any heavenly body but the Earth can have satellites).

Science was a happy solution for the many 17th-century gentlemen of leisure who were tired of religious strife. They grouped around their masters and spent their leisure deciphering the book of Nature—some by collecting facts, some by looking inward, and the synthesizers and mystics among them

---

([31]) See P. ROSSI: *Bacon* (Chicago, 1968).

([32]) See the articles by P. FEYERABEND: *Problems of Empiricism*, Parts I, II, in R. G. COLODNY (ed.): *Beyond the Edge of Certainty* (New York, 1965); and COLODNY (ed.): *Nature and Function of Scientific Theories* (Pittsburgh, 1970); also *Realism and instrumentalism*, in M. BUNGE (ed.): *The Critical Approach to Science* (New York, 1964). The story in its modern good-science-*vs.*-« Aristotelian fools » interpretation serves as a central dramatic scene in Brecht's *Galileo*.

by looking for great general principles of Nature to be discovered or revealed. What united all of them was that they were tolerant people: our science was born of that latitudinarian spirit.

This thesis is best defended for the English scene by SHAPIRO [33]:

> It is not difficult to see why scientific activities attracted so many during this period of religious and political upheaval. Science provided a respite, a noncontroversial topic of conversation, where men might have the satisfaction of breathing a free air, and of conversing in quiet one with another, without being engaged in the passions and madness of that dismal Age. As SPRAT wrote: « For such a candid, and unpassionate company, ... and for such a gloomy season, what could have been a fitter Subject to pitch upon that Natural Philosophy? To have been always tossing about some Theological question, would have been, to have made that their private diversion, the excess of which they themselves disliked in the publick: ... It was Nature alone, which could pleasantly entertain them in that estate. The contemplating of that, draws our minds off from past or present misfortune, ... that never separates us into mortal Factions; that gives us room to differ, without animosity; and permits us to raise contrary imaginations upon it, without any danger of a Civil War ».

and also:

> The alliance between latitudinarianism and science, however, went far deeper than a common core of practitioners and a mutual distaste for dogmatism. For the two movements also shared a common theory of knowledge, and members of both became the principal proponents of a rationalized religion and natural theology. In their respective areas both scientists and theologians sought a *via media* between scepticism and dogmatism. On the scientific side this search resulted in an emphasis on hypothesis and a science without overt metaphysics. In spiritual matters it led to an emphasis on broad fundamentals and the eschewing of any detailed, orthodox theology claiming infallibility.

This view also relieves the tension between those who see 17th-century science as mainly an intellectual activity and those who see in it a utilitarian problem-solving activity to help industry, navigation, mining, or agriculture. For the gentlemen of leisure who were indulging in these noncontroversial topics did not distinguish so sharply between experiment and theory, nor did they want either to understand Nature *or* to rule it. Unlike the 19th-century vulgarizers, they read their Bacon very carefully and found as much sense in performing *experimenta lucifera* as *experimenta fructifera*. All problems amused them and interested them, and they discovered God's greatness in Nature in all its wonderful manifestations: enlightenment could come from the artisan as well as from the astronomer.

---

([33]) Both quotations are from her *Latitudinarianism and science, Past and Present*, No. 40, July 1969, p. 16-41, and her intellectual biography *John Wilkins 1614-1672* (San Francisco, 1969). The whole problem of the Founders of the Royal Society was taken up by Marjorie PURVERM: *The Royal Society, Concept and Creation* (Cambridge, Mass., 1967), and J. AGASSI: *The origins of the Royal Society, Organon*, **7** (1970).

The Cartesians were mostly Catholics (but many were Huguenots) and religious tolerants. They found their anchor in philosophical introspection—not, however, by eliminating experience, but by fixing the criteria of relevance and of truth in the results of such introspection rather than in reliance on the evidence of the senses. Much less is known about the Cartesian scientific research programme than, let us say, about the Baconian or the Newtonian, for in the Cartesian-Newtonian critical dialogue NEWTON had the upper hand: with the 19th-century image of science as a continuous success story, DESCARTES and the Cartesians were exorcised ([34]).

This is not the place to elaborate on the great synthesizing movements of the Hermetic philosphers, the Rosicrucians, Paracelsians and Helmontians with their iatrochemistry, alchemical dreams and universal social and educational reforms. I shall confine myself to quoting these two passages from Van Helmont's *Oriatribe* ([35]):

« I praise my bountiful God, who hath called me into the Art of the fire, out of the dregs of other profession. For truly Chymistry, hath its principles not gotten by discourses, but those which are known by nature and evident by the fire: and it prepares the understanding to pierce the secrets of nature, and causeth a further searching out in nature, than all other Sciences being put together: and it pierceth even unto the utmost depths of real truth ... .

And all those things, not indeed by a naked description of discourse, but by handicraft demonstration of the fire. For truly nature measureth her works by distilling, moystening, drying, calcining, resolving, plainly by the same means, whereby glasses do accomplish those same operations. And so the Artificer, by changing the operations of nature, obtains the properties and knowledge of the same. For however natural a wit, and sharpness of judgement the Philosopher may have, yet he is never admitted to the Root, or radical knowledge of natural things, without the fire. And so every one is deluded with a thousand thoughts or doubts, the which he unfoldeth not to himself, but by the help of the fire. Therefore I confess, nothing doth more full bring a man that is greedy of knowing, to the knowledge of all things knowable, than the fire. Therefore a young man at length, returning out of these Schooles, truly it is a wonder to see how much he shall ascend above the Phylosophers of the University, and the vain reasoning of the Schooles ».

DEBUS summarizes the intellectual passions of the time thus:

In short, the Scientific Revolution was not simply the forward march of a new experimental method coupled with the powerful tool of mathematical abstraction. For some the two were incompatible and the growing predominance of mathematical abstraction could be interpreted as a step backward—a step away from a truly experimental study of nature. For these men the purely experimental studies of chemistry seemed the best answer to the training of the logic-ridden universities. We have seen this in the glorification

---

([34]) Nowadays this is repaired: see A. I. Sabra's admirable *Theories of Light from Descartes to Newton* (London, 1967).

([35]) ALLAN G. DEBUS: *The Chemical Dream of the Renaissance* (Cambridge, 1968).

of PARACELSUS in the Rosicrucian texts and in the chemically orientated laboratory science of the utopian Christianopolis. And with the two chemists we have touched on, FLUDD and VAN HELMONT, we note a similar hope for educational change—a hope dominated by their belief that chemistry was the true key to nature ([36]).

The Newtonian research programme is too well known to require description here. I shall return later to the main elements in the 18th-century critical dialogues about it. However, I turn now to the much less known scientific research programme of LEIBNIZ, which is also less generally admitted to be relevant to the history of physics.

**2·4.** *The scientific research programme of Leibniz.* – The intent of Leibniz's programme was to unify body and mind, man and Nature and God, and to establish such overall principles as would point to such unity. In Leibniz's world there is no reduction of all explanations to faith *or* experience *or* ratiocination: his pre-established harmony is supported by all three, and he sees no incoherence in that. Then the scientific metaphysics and the core of his research programme is the unity of the world, focused in one fundamental entity identified first as a concept of force, and then generalized as the monad. To this world conservation laws apply.

Leibniz's image of knowledge is shared by the Paracelsians, Rosicrucians, Hermetic philosophers, and the great humanists of his age: it is to help man to understand, to feel, to empathise with God in all-pervading harmony. The aim of knowledge is not explanation or prediction or certainty, but empathy with God. The important aspect of Leibniz's work is that it contributed at least as much to what we accept today as positive knowledge as did the dogmatic polar reductionists. The critical dialogue between these two world-views (the reductionists and the synthesizers) and their different images of knowledge kept science progressing steadily.

There is no serious historical treatment of Leibniz's monad as a scientific concept, attempting to understand it as a generalization of his concept of force: yet LEIBNIZ made it very clear that this generalization was the connection between the two concepts. LEIBNIZ himself says, in his letters to DE VOLDER ([37]), « ... you are already tacitly assuming what matter would be except through monads, since it would always be an aggregate, or rather the result of a plurality of phenomena, until we arrived at these simple beings » and « ... It is essential to substance that its present state involves its future states and *vice versa*. And there is nowhere else that force is to be found or a basis for the transition to new perceptions ... It is also obvious that in actual bodies

([36]) *Ibid.*, p. 32.

([37]) L. LOEMKER (ed.): *G. E. Leibniz: Philosophical Papers and Letters* (Dordrecht, 1969), p. 539.

there is only a discrete quantity, that is a multitude of monads or of simple substances, though in any sensible aggregate or one corresponding to phenomena, this may be greater than any given number ».

In his « Reply to the Thoughts on the System of Pre-Established Harmony » (contained in the second edition of *Mr. Bayle's Critical Dictionary*) *Article Rorarius* LEIBNIZ repeats: « And I do in fact regard souls, or rather monads as the atoms of substance since there are no material atoms in nature according to my view and the smallest particle of matter still has parts. Since an atom such as EPICURUS imagined has a moving force which gives it a certain direction, it will carry out this direction without hindrance and uniformily if it encounters no other atom » (p. 579). Finally in *The Monadology* (1714) he says, « It follows from what I have said that the natural changes in monads come from an *internal principle*, since an external cause could not influence (*influer dans*) their interior » (pp. 643-644). LOEMKER himself also draws attention to the close relationship between monad and force: « All monads are temporal series of active force and passive content, representative of the universe and striving toward the purposes defined in the individualized law from which they proceed » (Introduction, *op. cit.*, p. 45) and in his annotation to « matter » he says « ... secondary matter, as matter is phenomenal, and only the aggregated monads whose passive force is expressed as resistance or inertia are substantial » (*op. cit.*, p. 508, note 12).

By pointing clearly to the connection between his concept of force and his concept of the monad, LEIBNIZ required us to presuppose coherence among the various aspects of man's thought. Though it may be that the choice of the term « coherence » is unfortunate, for this coherence is not the strict logical kind ([38]): it resembles the coherence between different works of art by the same artist. There is a sense in which the man who composed the « Razumowsky » quartets must be the same man who wrote *Fidelio*. The question is how to translate such a vague insight into practical terms—into fruitful historical research. But clearly, a commitment to such a belief in coherence implies that the works of the philosopher-scientists must be treated as a whole.

---

([38]) Anthropologists have written extensively on coherence which is not of a logical kind. LÉVY-BRUHL in his *Les carnets de L. Lévy-Bruhl* (Paris, 1948), writes that « mystical thought is organized into a coherent system with a logic of its own » (p. 61). According to EVANS-PRITCHARD in a review essay on LÉVY-BRUHL (E. E. EVANS-PRITCHARD: *Lévy-Bruhl's theory of primitive mentality*, *Bull. of the Faculty of Arts*, **2**, 1-36 (1934)) mystical thought is « intellectually consistent even if it is not logically consistent » and « primitive thought is eminently coherent, perhaps over-coherent. Beliefs are co-ordinated with other beliefs and behaviour into an organized system ». Both quotations are brought out by Steven LUKES in his paper *Some problems about rationality* which appeared in BRYAN R. WILSON (ed.): *Rationality* (Oxford, 1970), p. 202. See also G. HOLTON: *Science and new styles of thought*, *The Graduate Journal*, **7**, 399-422 (1967). The motto of this article is « not in logic alone ».

A unifying principle goes through the work of LEIBNIZ like a *Leitmotif*. In his early dynamical works ([39]) he spoke of conservation of force, a fundamental principle. It is nothing like the Newtonian force vector, and it is certainly not energy (as some modern commentators would like to translate it) ([40]). What LEIBNIZ himself actually tells us is, first, that force has an effect $mv^2$, *or* $mv$, *or* the height reached by a body thrown upward; then, in later works, he says it is « a metaphysical entity », « the essence of matter », or « the main attribute of a monad ... ».

The monad serves as a final generalization of Leibniz's concept of force, now uniting in it not only all the physical effects of this fundamental entity, which is conserved in Nature, but also the physical and the spiritual, *i.e.* mind and matter. To us, all this may sound rather confused; but for creating science the importance of vague concepts like Leibniz's force or monad is enormous ([41]). Like most scientific concepts at the early stages of their evolution (as well as that of the theories in which they occur), the concept of the monad defies any attempt at exact definition. Historical research on the evolution of such vague concepts and half-formulated theories will contribute to our understanding of those elements in science which, by contemporary standards, look irrational. If we do not follow the Leibnizian scientific research programme through to the mid-19th century, we can easily be puzzled by the obscure origin of conservative ideas that burst « suddenly » on the scene from so many quarters ([42]).

2·5. *18th-century themes*: *Newtonianism, Euler, Kant.* – The role of the 18th century for the development of physics in the 20th century cannot be summarized in a few pages. I shall choose several themes which seem to me at once of dominant importance and generally neglected so far. This selection will result in the following omissions.

1) The 18th century saw the slow completion and perfection of the mathematical programme of the Newtonians—a strictly deterministic mechanistic world picture, described in mathematical language. This task was accomplished

---

([39]) *The New Physical Hypothesis* (1671), which appeared in two parts:
1) *The theory of abstract motion*, and
2) *The theory of concrete motion.*

([40]) On the problem of translation of terms into modern terminology see my *Helmholtz' Kraft*, *Historical Studies in the Physical Sciences*, Vol. **2** (Philadelphia, 1970), p. 263-298.

([41]) A famous quotation by H. A. KRAMERS is so poignant that it is worth repeating here: « In the world of human thought generally and in physical science particularly, the most fruitful concepts are those to which it is impossible to attach a well-defined meaning ».

([42]) See also J. AGASSI: *Leibniz's place in the history of physics*, *Journ. Hist. Ideas*, **30**, 331-344 (1969).

by the BERNOULLIS, EULER, MCLAURIN, D'ALEMBERT, LAGRANGE, LAPLACE, and the whole school of French mathematical physicists. Without them, the development of 19th- and 20th-century physics would have been impossible. Their work is well documented and their contribution well known ([43]).

2) 18th-century rationalism, embodied in the Enlightenment, helped the institutionalization of science and the almost total separation between science and religion. This is the age of the great Academics, the scientific expeditions, the first great systematizations. Again, this aspect of the 18th century is well documented and well known ([44]).

The topics I wish to deal with are the controversy around the various theories of matter and the influence of the great synthetic schools of EULER and KANT. But before I review matter theory, let me dissect the myth that the 18th century was a direct fulfilment of the Newtonian research programme. According to this myth, the two views competing in the 18th century were Newtonianism and anti-Newtonianism, and in the ensuing controversy Newtonianism won, thus opening the road to progress ([45]).

What is to be called « Newtonian » and what « anti-Newtonian »? This is not a rhetorical question but a historiographic problem of fundamental importance, any reasonable answer to which would presuppose a comprehensive knowledge of the growth of science from the 17th century onwards. To describe the world in terms of discrete particles between which central forces are acting at a distance is certainly Newtonian: that is the metapysical core of the *Principia.* The programme of mathematization of mechanics, as perfected by the French school of rational mechanics, is also Newtonian. The various matter theories of the 18th century, first presented in the « Queries » and in the letters to BENTLEY, are also characteristically Newtonian. The idea of chemical affinities is Newtonian; but so was the Daltonian revolution that rejected the affinities. LAVOISIER was a Newtonian of sorts and so was PRIESTLEY, whom LAVOISIER rejected; so was Humphrey DAVY, who refuted Lavoisier's central doctrine, that all « elements » contain oxygen. Some of

---

([43]) R. DUGAS: *A History of Mechanics* (Paris, 1955); E. J. DIJKSTERHUIS: *The Mechanization of the World Picture* (Oxford, 1961); CH. TRUESDELL: *Essays in the History of Mechanics* (New York, 1968); and T. L. HANKINS: *Jean d'Alembert: Science and the Enlightenment* (Oxford, 1970).

([44]) Though science plays almost no role in Peter Gay's book on the 18th century intellectual environment, the great schools of thought are admirably depicted. *The Enlightenment,* Vols. I and II (New York, 1967 and 1969).

([45]) See two recent scholarly monographs: ROBERT E. SCHOFIELD: *Mechanism and Materialism: British Natural Philosophy in the Age of Reason* (Princeton, 1970), and A. THACKRAY: *Atoms and Powers: An Essay on Newtonian Matter Theory and the Development of Chemistry* (Harvard, 1970). Also, my review of them: Y. ELKANA: *Newtonianism in the 18th Century, Brit. Journ. Phil. Sci.*, **22**, 297-306 (1971).

these great natural philosophers called themselves Newtonians because they adhered to a world-view where the most important force was gravitation-acting-at-a-distance; some others considered themselves Newtonians because they accepted a material substratum, the ether, which transmits all physical action; still others made the same claim because they believed themselves to be doing scientific work in the hypothetico-deductive way, which they considered the hallmark of Newtonianism. Needless to say there is little similarity between Newton's thoughts and speculations and the conceptual framework they thought of as Newtonian.

In addition to those who believed themselves *bona fide* Newtonians, others used the label politically, for legitimization of their theories; an example is YOUNG, a disciple of EULER and HUYGENS, who introduced his famous paper on interferences of light by attributing the main idea to NEWTON. Finally there were the Continental natural philosophers, all of whom accepted Newtonian mechanics and attempted to blend it into their Cartesian or Leibnizian conceptual frameworks; later historians in their positivistic whitewashing exercises classified as Newtonians—to name only a few, BOSCOVICH, EULER and KANT.

But, allowing for the moment that we could sufficiently refine our notion of Newtonianism, it still seems to me that to attempt to describe the main lines of the development of 18th-century science (even if we restrict ourselves to England, and to matter theory or chemistry) in terms of the struggle between the Newtonian and anti-Newtonian traditions is to put ourselves into a conceptual straitjacket.

In my opinion at least three great traditions or scientific research programmes competed for primacy in 18th-century science. These are the Cartesian, Newtonian and Leibnizian research programmes. The critical dialogues among these three were usually conducted in pairs—Newtonianism *vs.* Leibnizianism, Newtonianism *vs.* Cartesianism, Leibnizianism *vs.* Cartesianism—but, rarely, two joined forces against the third. To lump all general explanatory hypotheses that are not Newtonian under the heading « anti-Newtonianism » is an over-simplification. Those conceptual frameworks which can justly be labelled anti-Newtonian focus their opposition either on Newtonian science or Newtonian methodology.

However, anti-Newtonians proper and Newtonians share a fundamental problem-situation: *a*) Should one, or could one, describe the Universe in terms of discrete particles with forces acting between them? *b*) Can force act through a vacuum? *c*) Are forces essential properties of matter? On the other hand the 18th-century Leibnizians and the two different brands of Cartesians (which separated out of the original Cartesian framework at the turn of the century) were required to face different problem-situations, and each had a scientific research programme different from the 18th-century Newtonians. The two Cartesian groups were *a*) the mathematical rationalists like D'ALEMBERT,

DIDEROT, and later LAGRANGE; and *b*) the matter theorists like MAUPERTUIS, EULER and Johann BERNOULLI. Cartesian mathematical rationalism developed a programme aimed at subsuming all phenomena under mathematically formulated laws. There was no discussion of fundamental concepts and no search for underlying principles, and the criterion of truth was rarely empirical; instead, mathematical formalizability and elegance became signs of truth. These Cartesians were occupied in developing mechanics as a branch of mathematics; they concentrated on attacking the Leibnizians rather than the Newtonians.

However, the main argument between the matter-theorist Cartesian and the Newtonians centred on the primacy of the concept of force. These Cartesians also accepted Newton's results—that is, the laws of mechanics and the law of gravitation—but they insisted that there are essential qualities of bodies to which forces can be reduced. If forces were introduced into the Cartesian programme, they were thought of as mathematical abstractions useful for smooth calculations, an attitude somewhat similar to Hertz's 150 years later.

The mind-body dichotomy was part of the Cartesian tradition, but it played only a very minor role in the controversy with the Newtonians. This problem was, however, the core of the Cartesian-Leibnizian critical dialogue. The Cartesians separated mind and body, and scientific metaphysics (that is, those views on the structure and genesis of the physical world which are on principle untestable but form the core of their research programme) from theology. On the other hand, both the Newtonians and the Leibnizians attempted to justify their scientific metaphysics by their theology. This justification became one of the foci of the Newtonian-Leibnizian critical dialogue, as exemplified in the Leibniz-Clarke correspondence and as continued by EULER in the « Letters to a German Princess » written in the 1770's.

The central Newtonian conception is force, whether acting at a distance or at short-range contact. Newtonian physics, astronomy, chemistry and physiology all involve forces. Whether the forces are inherent in matter or reducible to their relational properties is another focus of the dialogue between Newtonians and Leibnizians. On the other hand, the concept of force is as foreign to the Cartesian as it is inseparable from both the Newtonian and the Leibnizian research programmes.

Another difference between Newtonians and Leibnizians lies in conservation principles which are alien to the former but fundamental to the latter. Even though an anticonservation-principles attitude is not explicit in Newton's writings, it seems to me one of his deep-seated anti-Cartesian biases. Unlike DESCARTES, he refuses to address himself to such questions as « What are fundamental entities? » and « Are they conserved? ». He takes for granted four « fundamental notions »—space, time, mass and force—and operates with them. For LEIBNIZ, too, the concept of force is fundamental; but it is the *conservation* of force that is at the core of his scientific metaphysics. The idea of conservation of force served LEIBNIZ in doing away with the Cartesian mind-body dualism and helped him develop his monistic theory.

In short, in order to gain any reliable picture of 18th-century science one must explore at least three competing traditions. All three left indelible marks on the developments of science in the 19th, and even in the 20th century, and each has had the upper hand at times in the long critical dialogues. Newtonianism is the paradigm of success in terms of positive scientific results; the positivistic attitude does not find a place for either the Cartesians or the Leibnizians in the history of science. Thus « Newtonian » *vs.* « anti-Newtonian » covers the ground adequately only if we judge the development of science with the presupposition that science grows by accumulation. If we view the growth of knowledge as a result of a dialogue between competing research programmes, we must think in terms of at least the above-mentioned three traditions.

2·6. *The 18th-century image of science.* – The image of science outlined here emerged from the critical dialogues of the Newtonians, Cartesians and Leibnizians.

The conclusion drawn from Newton's unprecedented success was that certainty of nonrevealed knowledge is possible. This was the starting point of the quest for certainty that has marked our science, philosophy, literature and politics ever since the 17th century; since FREUD, it has been described as an innate need (*i.e.* noncognitive and non environment influenced) shared by all human beings (if not indeed by all live creatures). Because of the Baconian influence this certainty centred around the concept of «fact» in England; in France (because of DESCARTES!?), it looked to mathematics. Germany, after LEIBNIZ, sought certainty in consensus, *i.e.* a metaphysical synthesis that would harmonize the differing schools and discover underlying unitary principles (conservation laws in Nature) that apply to everything. This became the basic task of the scientist-philosopher, and no problem seemed worth working on, if it did not contribute directly to the broad metaphysical view. This image of science and this view of the role of the scientist kept alive the Leibnizian tradition through EULER, KANT, HEGEL, FICHTE, HELMHOLTZ, FECHER, FREUD, MARX, EINSTEIN and BOHR. It is the tradition of the critical dialogue between the competing Newtonian and Cartesian traditions.

In the Introduction of Euler's *De Curvis Elasticitis* ([46]) we find a passage, most typical of this approach to physical problems, which also casts some light on his image of science and methodology:

([46]) Additamentum I to his *Methodus Inveniendi Lineas Curvas Maximi Minimive Proprietate Gaudentes.* In this work EULER solves the isoperimetric problem and invents the calculus of variations that became so famous during the quarrel between Jacob and Johann BERNOULLI (Lausanne and Geneva, 1774).

Wherefore there is absolutely no doubt that every effect in the universe can be explained as satisfactorily from final causes as it can be by the aid of the method of maxima and minima, from the effective causes themselves. Now there exist on every hand such notable instances of this fact, that, in order to prove its truth, we have no need at all of a number of examples; nay rather one's task should be this, namely, in any field of Natural Science whatsoever to study that quantity which takes on a maximum or a minimum value, an occupation that seems to belong to philosophy rather than to mathematics. Since, therefore, two methods of studying effects in Nature lie open to us, one by means of effective causes, which is commonly called the direct method, the other by means of final causes, the mathematician uses each with equal success. Of course, when the effective causes are too obscure, but the final causes are more readily ascertained, the problem is commonly solved by the indirect method; on the contrary, however, the direct method is employed whenever it is possible to determine the effect from the effective causes. But one ought to make a special effort to see that both ways of approach to the solution of the problem be laid open; for thus not only is one solution greatly strengthened by the other, but, more than that, from the agreement between the two solutions we secure the very highest satisfaction. Thus the curvature of a rope or of a chain in suspension has been discovered by both methods; first, *a priori*, from the attractions of gravity; and second, by the method of maxima and minima, since it was recognized that a rope of that kind ought to assume a curvature whose centre of gravity was at the lowest point ([47]).

Thus Euler's methodological demand is for theoretical proliferation; moreover, his image of science is such that whatever we do in our mathematics is a way of revealing actual reality in Nature—and, by necessity, whether we proceed by the direct method (*i.e.* the method of minima and maxima from efficient causes) or by the indirect method of proceeding from final causes (*i.e.* from *a priori* principles), we must reach the same conclusions. This mutual reinforcement is the real satisfaction for the natural philospher. The sharp distinction between final, *a priori* causes and effective causes is Leibnizian rather than Newtonian; in this KANT turns out to be, a few decades later, Euler's most faithful disciple.

These views on the role of science and metaphysics combined with the strongly Puritan view of German Pietism, which stressed the habit of self-renouncing labour, of singleness of purpose. This was a *sine qua non* of developing for scientific metaphysics the unshakable foundations needed to build a whole system of science, ethics and social theory and an anthropology.

In addition, EULER—and, later, KANT—developed a theory of growth of knowledge that rejects the view of knowledge-by-accumulation and comes

([47]) The first sentence of this passage—taken from the annotated translation by W. A. OLDFATHER, C. A. ELLIS and D. M. BROWN: *Isis*, **20**, 72-160 (1933)—has been amended to correct a curious mistake. Their translation reads: « ... every effect in the universe can be explained as satisfactorily from final causes, by the aid of the method of maxima and minima, as it can from the effective causes themselves ».

much nearer to the view that knowledge grows by a continuous critical dialogue between competing metaphysics. The connecting 19th-century link in this tradition was WHEWELL. Its chief 20th-century representatives are POPPER, AGASSI and LAKATOS and their followers.

Euler's realistic attitude and his opposition to the growth-by-accumulation theory of the growth of scientific knowledge is beautifully illustrated by his systematic, detailed description of trials and experiments that failed; it allows us to gain a true insight into his method of discovery. In all the literature on EULER, the only source I know that calls attention to this at all, although without drawing any conclusions as to Euler's theory of the growth of knowlege, *is* a 1948 essay by FUCHTER (48). Euler's method is in complete harmony with the above-mentioned theoretical pluralism.

The Eulerian-Kantian view of science and the resulting scientific metaphysics determined for the next generation a choice of problems dealing with general principles in Nature. Their work complemented the Newtonian achievements and the French mathematization of mechanics; the principle of conservation of energy; the continuum approach, which gave birth to the field concept; the new, specifically biological approach, both in its vitalist formulation and in the « hard-approach » cell-theory-reductionism—all these grew on Kantian soil.

I have not entered here into the problems of the influence of the general intellectual environment on the image of knowledge. But some of the questions that must be raised follow.

What was the impact of the fact that the Berlin Academy consisted mainly of French, Swiss, Russians and other non-Germans? How did Frederick the Great's German nationalism (of which he was the originator as well as the charismatic leader) harmonize with the cosmopolitan « Enlightenment » attitude towards science and culture in general? Why did EULER write his *Anleitung zur Naturlehre* in German (instead of the usual French) (49) and why was it never published? What was the institutional difference between the Berlin and St. Petersburg academies, and how did this influence the work of EULER, who was in contact with both? Why was Germany the centre of growth of the strong synthesizing and reconciling tradition best represented by LEIBNIZ, EULER and KANT? How did KANT deal with his own German nationalism, his pro-French-Revolution attitude, and the changing Prussian-Russian occupancy of his beloved Konigsberg? Finally, how was the social and polit-

(48) R. FUCHTER: *Leonard Euler* (Basel, 1948), p. 1-24. On page 14 it is noticed that Euler's style of topics starts with a thorough analysis of the problem at hand. This characterizes his mathematical, physical and epistemological-philosophical work.

(49) According to the directives of Frederick the Great, all publications of the Academy were to appear in French; parallel to the French translation, the work could be printed in any other language the author desired.

ical Rousseau-influenced philosophy of KANT (constructed on his scientific metaphysics) influenced by the absolute autocratic rule of Frederick the Great's successor?

## 3. – From the scientific revolution of the 17th century to the Einsteinian revolution of the 20th century-(B).

3·1. *The problem of science, technology and the industrial revolution.* – The 18th century brought the industrialization of Europe, along with the accompanying development in technology and growing social unrest and dissatisfaction, the relative betterment of human destiny. For our understanding of the growth of knowledge we must face the problem of the kind of interaction that took place between science and technology. Are those who claimed that the steam engine did more for thermodynamics than *vice versa* correct, or are their violent opponents in the right? The literature on the question is huge, and the controversy was vociferous. I shall not go into this problem here in detail but only suggest a new way of looking at the question.

Those who deal with this problem ([50]) generally presuppose that only well-formulated scientific ideas in the body of knowledge can influence technical development, before they go on to ask whether these specific ideas did in fact do so. Also, it is generally presupposed that pure science, applied science and technology form a continuous spectrum. In my opinion, both presuppositions are mistaken.

To begin with, the cross-fertilization between science and technology does not need to result in specific ideas. Whether it does depends on the image of knowledge accepted in a society at a given stage of its cultural development —on whether knowledge is open speculation, revelation, results of disputation, or merely any idea which can be expressed in palpable form. In the 18th century that last became the accepted form of knowledge for those who had only the idea, as well as for those who had only the palpable results. At least in England these two had a common conception of knowledge and interacted fruitfully: these were the people who inquired into the nature of heat, the nature of light, the nature of various imponderable fluids, the nature of minerals,

---

([50]) To mention only a few recent publications: LANDES: *The Unbound Prometheus* (Cambridge, 1969); MATHIAS: *The First Industrial Nation* (London, 1971); MATHIAS (ed.): *Science and Society 1600-1900* (Cambridge, 1972); HOBSBOWM: *Industry and Empire* (London, 1969); CARDWELL (ed.): *From Watt to Clausius* (New York, 1971); HILL: *Reformation to Industrial Revolution* (London, 1969); and MUSSON (ed.): *Science, Technology and Economic Growth in the Eighteenth Century* (London, 1972).

the nature of plants, etc. Some others inquired into the nature of machines and technical devices and—irrespective of what they were engaged upon personally—all accepted this wide range of activities as relevant to what they considered knowledge. Whether in this general mood there were specific influences of some specific ideas is much less important than the fact of that acceptance. In other words, the first group was Baconian, with equal respect for *experimenta fructifera* and *lucifera* because what counted was that they were both *experimenta*. The critical dialogue in which these « scientists » and « technologists » were involved conflicted, however, with the images of knowledge of the German Kantians and the Scottish metaphysicians or French mathematicians; and the interaction between these two outlook-groups was exclusively one of mutual disapproval.

The other presupposition is connected with the following point: it is that technological innovation does not result from the direct extrapolation of applied science, but generally from a rethinking of the most basic science with a different aim and thus a different angle of view. Whether the fundamentals that must be reviewed are seen in the « primitive » commonsensical terms of a well-established basic scientific truth, or in terms of the most sophisticated and latest scientific results that have stemmed from them, is unimportant. Thus all technological innovation presupposes knowledge of some level of fundamental science, but not necessarily of what is called « science ». The maser and the laser will, I think, illustrate my point.

The Industrial Revolution is the result of the interaction between various well-known socio-economic factors and the image of knowledge, which gave equal importance to experimental science and technology without distinguishing sharply between the two.

**3·2.** *Helmholtz and the conservation of energy* ([51]). – The central concepts in the physics of NEWTON were space, time, mass and force. By the end of the 19th century the central concepts in physics were space, time, mass and energy.

The general concept of energy became meaningful only through the establishment of the principle of conservation of energy in all its generality; thus the story of the emergence of the energy concept and the story of the establishment of the conservation law are difficult to disentangle. The man who first formulated the principle mathematically, in all its generality, was HELMHOLTZ.

---

([51]) This subject was developed by me in two articles: *Helmholtz's Kraft: an illustration of concepts in flux*, *Historical Studies in the Physical Sciences*, Vol. **2** (Philadelphia, 1970), pp. 263-298; and *The conservation of energy: a case of simultaneous discovery?*, in *Archives Internationales d'Histoire des Sciences*, 23me 90-91, Janvier-Juin 1970, pp. 31-60. It is also discussed in my book *The Discovery of the Conservation of Energy* (London, 1973).

His was a towering scientific personality, and his life's work has left its mark on all branches of 19th-century science, from theoretical mechanics to applied physiology. The concept of energy as we know it today (« today » meaning classical, pre-relativity physics) has emerged from Helmholtz's 1847 paper *Über die Erhaltung der Kraft*; until that time, nobody, including HELMHOLTZ himself, had a clearly defined concept of energy.

But the problem is not a purely scientific one. In view of the prerequisites for the establishment of the principle of conservation of energy, this final step had to take place in 19th-century Germany; it could not possibly have happened in England or France. In German universities, the student, whatever he studied, could not have avoided facing sooner or later the great metaphysical problems posed by the various *Weltanschauungen*; although in England « science » was pursued, or in the traditional spirit « natural philosophy » was taught, German *Philosophie* covered the whole of the human intellectual enterprise. Speculation was encouraged; even today, after the great late 19th-century battle to erase the last remnants of an influence of the *Naturphilosophie*—or rather, of a degenerated, ridiculously trimmed-down version of it—« speculation » does not provoke the contempt in German that it does in English. In Germany the « schools » or « laboratories » of the great scientists represented a complete philosophical system, and every student was required to take a stand. One could not work in Weber's physical laboratory, or in Liebig's laboratory, without taking a considered philosophical approach to Kant's epistemology or to the mechanistic-vitalistic controversy. In that atmosphere, then, one could not readily separate experimental data from highly speculative hypotheses.

That atmosphere, which was detrimental for many other scientific problems (*e.g.* for the first formulations of electrodynamics—where all German theories failed, while the English, French and the isolated ØRSTED did the work), was indispensable for the establishment of the conservation-of-energy principle, *i.e.* the emergence of the concept of energy. In those German universities the « schools » of LIEBIG, WOEHLER, J. MUELLER, WEBER and MAGNUS were founded. And there the further prerequisite—namely, the cross-fertilization between physical and physiological thought—was made possible. Moreover, it was there that biology was born. Men like SCHWANN, BRUECKE, DU BOIS-RAYMOND and HELMHOLTZ matured in these universities. Their shared background consisted of a vast reading in the philosophers and a readiness to face their questions, an awareness of a deep-seated connection between all the sciences, and a tendency to search for coherence between their philosophies and their scientific knowledge. Helmholtz's father was a close friend of Fichte's son, who was a professor of philosophy; and HELMHOLTZ himself, according to his own testimony, was deeply influenced by both KANT and FICHTE. These influences, and the fact that Helmholtz's education in physics and his mathematical ability made him the ideal man for the task awaiting him, were all crucial. That time and place, those circumstances, constituted the background against which the physical concepts in use in the early 19th century would be revealed in all their aspects.

The confusion between « force » and « energy » (as we use these terms) in the works of HELMHOLTZ and some of his contemporaries, was not only not merely a verbal one, as most of the commentators on this topic tend to assume, but rather a necessary prerequisite for the final clarification of the concepts. Only an undefined entity could have been the subject of a general belief in principles of conservation in Nature, and such a belief was one of the major factors in the actual establishment of the conservation-of-energy principle in its final, mathematical—that is, correct and well-defined—form.

One of the most cherished beliefs of inductivist historians of science is that the principle of conservation of energy grew directly out of the realization of the impossibility of a perpetual-motion machine. That realization was indeed arrived at inductively, and it dates back at least to the 17th century. Certainly STEVIN had already drawn physical conclusions from it; and the French Academy's 1775 decision not to consider any longer suggestions for the construction of such a machine covered not only mechanics but all the branches of physics. Thus, at least seventy-five years before the establishment of the conservation principle, the alleged « intellectual father » had been identified beyond doubt.

It is also often assumed that the principle of conservation of energy was a direct generalization of the law of conservation of mechanical energy, as formulated, for example in Lagrange's *Mécanique Analitique*. But here, the use of the modern name of mechanical or kinetic and potential energy leads us into errors of hindsight: the assumed implication is that the concept of energy was extant, and that the workers in this field thought in these terms and worked with this notion. But actually it is only now that we view all sorts of work—*vis viva*, *Spannkraft* and the many others—as instances of the all-embracing concept of energy. At the time when, in mechanics, the sum of *vis viva* and « potential function » (under that or any other name) was found to be conserved, nobody thought of the necessity or possibility of generalizing any further; that was general enough. The notion of energy as something so general that all the special forms are instances of it was not discovered until 1847, and then only in precise mathematical language. Before that, it was a vaguely understood entity, which was conserved and related to mechanical energy, and even served very fruitfully as a working concept. It was used also by FARADAY, or MAYER, or others of the twenty-odd « simultaneous discoverers » of the principle. The proof that, owing to exact mathematical and dimensional considerations, the conserved entity must be related to mechanical energy simply by being reducible to it is the work of HELMHOLTZ.

In the historical or even in the physical literature, one sees often that at the beginning of the 19th century two theories of the nature of heat were still in vogue—the mechanical, or, rather, dynamical theory (the name « mechanical » is really justified only after the work of CLAUSIUS) and the material-caloric theory. It is implied, or sometimes explicitly stated, that CARNOT had a clear conservation law in mind and was misled only by his use of the caloric theory, and that the mechanical theory had to be established so that the principle of conservation could be finally enunciated. Actually, the connection between the actual development of the early thermodynamics and the theory of the nature of heat was very weak. Even a year after Helmholtz's proof, some scientists held to the caloric theory of heat (for example, CLAUSIUS, although he knew of Helmholtz's work), and the conversion processes—available because of the work of many physicists, from CARNOT to JOULE— did not really point to a general conservation law. Again, that seems to us so natural only because we already view heat as one of those many instances of a general concept of energy. Moreover, the famous supporters of the mechanical theory, RUMFORD and DAVY, did not entertain any conservation ideas. Indeed, what RUMFORD showed was the exact opposite: his argument is that because the heat generated is inexhaustible (that is, clearly not obeying any conservation law), it cannot be material (for material substances do obey a conservation law). DAVY, if his experiment teaches anything at all (which was cast in doubt by ANDREADE), has nothing to do with conservation. In short, the historical development was again the only logically possible one: first the establishment of the principle of conservation of energy leading to the emergence of the concept of energy, and then the formulation and separation of the two laws of thermodynamics and the mathematical formulation of a true, mechanical theory of heat. It was only after these developments of thermodynamics that the actual processes of nature took an important place among the possible ones, and a new interest arose in those extremum principles that then became an integral part of the new energy-centred mechanics.

The factors that constituted, in my view, a solid basis for the enunciation of the conservation principle are

1) an *a priori* belief in general conservation principles in Nature;

2) a realization that two mathematically equivalent formulations of mechanics, the vectorial Newtonian and scalar-analytical Lagrangian, must also be conceptually correlated;

3) an awareness of the physiological problem of « animal heat », or more generally of « vital forces », and a belief that these are reducible to the laws of inanimate Nature;

4) a mathematician's certainty that whatever is the entity conserved in Nature, it must be expressible in mathematical terms; and a mathematician's skill to perform the task of expression.

At a very early age, HELMHOLTZ had read the works of NEWTON, EULER, D'ALEMBERT and LAGRANGE (though not HAMILTON); he was aware of the double tradition of mechanics and knew that something had to be done about it. It was clear to him that the central concept of Newtonian-vectorial mechanics was the concept of force, and at the same time he saw that the only quantity conserved in scalar-Lagrangian mechanics was the sum of *vis viva* and the « potential function ». By temper and intellectual heritage he was a disciple of KANT and thus committed to a belief in the great unifying laws of Nature; this took the form of conservation laws, and naturally the conserved entity had to be that vaguely defined entity *Kraft* or « force » (in the Faraday sense). All this was in complete harmony with Helmholtz's mechanistic philosophy, a belief that all phenomena of Nature are reducible to the laws of mechanics. A physician by training, he spent several years in the laboratory of the famous physiologist J. MUELLER. There HELMHOLTZ came to confront the problem of « vital forces », and especially that of animal heat, and his first works were in this field. Again his approach was that « vital forces are like other forces, conserved in Nature and, like all phenomena, reducible to mechanics—so « vital forces » must be reducible to mechanical forces. But HELMHOLTZ was also a mathematician of the first rank: he saw clearly that if *Kraft* is conserved in Nature, and mechanical energy is conserved in mechanics, then all *Kraft* must have the same physical dimension as mechanical energy and moreover must be reducible to it. That is exactly what he did in his 1847 paper.

What this story teaches us is the way we use of the past in the service of the 19th-century image of science distorts history. The chapter on the conservation of energy can be viewed as an important critical dialogue betwen the competing traditions of the Newtonian and the Leibnizian science research programmes and the greatest failure of the Newtonians—or, if one wishes, the greatest success of the Leibnizians. It will be interesting to see how this success was again jeopardized with the 1924 Bohr-Kramers-Slater paper.

3·3. *The Victorian image of science.* – German *Naturphilosophie*, with its search for underlying unitary principles and great (mystically potent) conservation laws, was the greatest antithesis of the English matter-of-fact, commonsense, inductivist philosophy of Nature. The former was speculative, idealistic, broad and very often vague; the latter was empiricist, materialistic, narrow and often so precise as to descend to meaningless pettiness.

The two competing images of knowledge thus described dominated the 19th century. Both had great influence on scientific problem-choice in their cultural environments: all over Europe, there were representatives of both conceptions. The influence of the German *Naturphilosophie*—an influence that combined the Hegel-Fichte-Schelling romanticism with Kantian metaphysics and the search for underlying principles—spread to England through the works of COLERIDGE and T. YOUNG. It introduced the acceptability of the Eulerian wave theory as a legitimate area of research, and in the hands of YOUNG and

FRESNEL it triumphed. Though YOUNG tried to appear Newtonian (52), his was the first direct proof that NEWTON had been wrong in the critical dialogue between the Newtonian and the Leibniz-Euler-Kant scientific research programmes.

On the other hand, the German scientists, physicists and biologists alike, who turned away from *Naturphilosophie* with disgust, instituted what amounted to a reign of terror on behalf of empiricism. Scientific papers that contained even the slightest attempt at a theory not yet proved by hard experiment were rejected (Helmholtz's famous paper of 1847, rejected by POGGENDOFF, is a case in point). Their influence was such that in 1905 Einstein's relativity paper was nearly rejected; MAX VON LAUE, who happened to be assistant editor in charge, saved it.

The controversy around the scientific research programmes was no less violent than that about the images of knowledge. The Newtonian scientific metaphysics, according to which the world consists of discrete particles with *central forces* (and only such) acting between them at a distance, was crumbling in light of the aether of the later Newtonians and of the wave theory. Yet it was dominant enough for many years to prevent Ørsted's discovery (1820) of the magnetic lines of forces of a current-carrying wire because he was looking only for lines of force radially pointing out of the wire. But Ørsted's Newtonian research programme did not disturb his *Naturphilosophie*-influenced image of knowledge—romantic, mystical and speculative (53).

---

(52) Thomas YOUNG: *On the Theory of Light and Colours* (*Phil. Trans.*, 1802, p. 12) says: « A further consideration of the colours of these plates, as they are described in the second book of Newton's optics, has converted that prepossession which I before entertained for the undulatory system of light, into a very strong conviction of its truth and sufficiency; a conviction which has been since most strikingly confirmed by an analysis of the colours of striated substances. The phenomena of these plates are indeed so singular, that their general complexion is not without great difficulty reconcilable to any theory, however complicated, that has hitherto been applied to them; and some of the principal circumstances have never been explained by the most gratuitous assumptions; but it will appear, that the minutest particulars of these phenomena are not only perfectly consistent with the theory which will be detailed, but that they are all the necessary consequences of that theory, without any auxiliary suppositions; and this by inferences so simple, that they become particular corollaries, which scarcely require a distinct enumeration.
A more extensive examination of Newton's various writings has shown me that he was in reality the first that suggested such a theory as I shall endeavour to maintain; that his own opinions varied less from his theory than is now almost universally supposed; and that a variety of arguments have been advanced, as if to confute him, which may be found nearly in a similar form in his own works; and this by no less a mathematician than Leonard EULER, whose system of light, as far as it is worthy of notice, either was, or might have been, wholly borrowed from NEWTON, HOOKE, HUYGENS, and MALEBRANCHE ».

(53) Ørsted's great literary work, *The Soul in Nature* (London, 1966), is a revealing document of his image of knowledge.

Two observations should be noted here: it is often claimed that the *Naturphilosophen* were anti-scientific and even anti-intellectual in their approach, and it is also often implied—if not stated directly—that they knew no science. These again are myths, nourished on the inductionist historiography of the Victorian image of knowledge. It is enough to read Hegel's works on natural philosophy, or Schelling's 1832 appraisal of the state of electromagnetism, to realize that these views have no basis. We must get used to the idea that, in spite of their full awareness of the great achievements of science, their image of science was unlike ours and unlike that of our Victorian grandfathers.

The Victorian image of knowledge was a vulgarized Baconianism of great extremity and power ([54]). The image is of knowledge triumphant—certain, inductively collected, down-to-earth knowledge. And the Victorians, in their thoroughness, documented their view of knowledge by rewriting the history of the accumulation of scientific knowledge ([55]). This huge undertaking was then supplemented by creating a tradition of typical 19th-century biographers of great scientists, all of whom had been ardent experimentalists and hard workers from early childhood.

It is the 19th-century Victorian image of knowledge that also created the myth of the Protestant ethic as a cause of the development of modern science. The self-image of the 17th century, as we have seen, was of the theologically nonpartisan, latitudinarian gentlemen of leisure as the source of all new ideas. This image was now systematically replaced by the hard-working, preferably lower-middle-class boy, who lived up to God's expectations for the genius with which he had been endowed. The biographies started with Macauley's Bacon and went on through Brewster's Newton, Pouso's Davy, Peacock's Young and countless others.

The Victorian image of science was sustained by another new tradition, the writing of textbooks—no longer treatises written from the personal point of view of the authors, but impersonal reports of the organized and well-ordered progress of scientific success. The need for textbooks arose from teaching science on a vast scale in the newly institutionalized scientific profession. But its result was the final establishement of the Victorian image of knowledge.

The first such typical textbook, of enormous influence, was Thomson and Tait's *Principles of Mechanics and Dynamics*, which first appeared, under the title *Treatise on Natural Phylosophy*, in 1856. The greatest influence in spreading the vulgar-Bacon image of science was Sir John Herschel's *A Preliminary Discourse on the Study of Natural Phylosophy* ([56]).

---

([54]) On this see A. Thackray: *The industrial revolution and the image of science*, in A. Thackray and E. Mendelsohn (eds.): *Science and Values* (New York, 1974).

([55]) See H. Butterfield: *The Whig Interpretation of History* (London, 1960).

([56]) New York and London, 1966. See also J. Agassi's *Sir J. Herschel's philosophy of success*, *Historical Studies in the Physical Sciences*, Vol. **1** (Philadelphia, 1969), p. 1-36.

3·4. *Some 20th-century critical dialogues.* – The developments in 20th-century physics offer a rich field, richer than any other period, for investigation of any theory of knowledge. We can distinguish several reasons for this.

1) The general speeding-up of the process of change has rendered visible developments that formerly unfurled slowly and thus tended to go unnoticed.

2) The success of the 19th-century professionalization of science and the quantitative growth of the scientific enterprise caused developments in various sciences that brought about mutual interaction between sciences.

3) The political developments of the last fifty years made science grow, or at least helped its establishment among non-Western cultures in various stages of development; this has led to the clash between scientific metaphysics and, even more, to clashing images of knowledge. One can follow the growth of knowledge by following these developments: it is a rich field indeed ([57]).

The interaction between the different sciences is interesting on its own, for one can see how incommensurable research programmes introduce changes into each other. One can also most easily detect there the shift of emphasis to new problem situations, and thus the periodic redivision of the spectrum of knowledge into new disciplines. I shall not deal with this phenomenon here ([58]).

The growth of science in non-Western cultures is another rich subject not discussed here ([59]). All my examples here will deal with the mutually interacting and rapidly changing critical dialogues in Western physics.

EINSTEIN is at the centre of most of the critical dialogues in 20th-century physics ([60]). With his work, the long atomistic-*vs.*-continuous-matter picture of the world controversy ended and two new critical dialogues were inaugurated: *a*) the critical dialogue in the body of knowledge, *i.e.* is the world strictly deterministic or is it fundamentally statistical in character; and *b*) the critical

([57]) Most of the lectures, seminars and talks at this meeting deal with this area and this period. What I shall try to do below is to outline briefly the main dialogues as I see them in the 20th century, and refer to the lectures of my colleagues here as evidence for my points. If I had to refer to a limited bibliography, I would choose unhesitatingly the three volumes of McCormmach's *Historical Studies in the Physical Sciences* (herein after abbreviated « HSPS ») (Philadelphia, 1969, 1970, 1971).

([58]) See also N. R. HANSON: *The contributions of other disciplines to 19th century physics*, in S. TOULMIN and H. WOOLF (eds.): *N. R. Hanson*: « *What I do Not Believe and Other Essays* » (Dordrecht, 1971).

([59]) See my « Theory and practice of cross-cultural contacts: some philosophical queries ».

([60]) The historic-philosophical literature about EINSTEIN is already impressive: for the past ten years it has been no longer anecdotal but serious, critical and enlightening. With the new project of publishing all his works and correspondence the field will grow enormously.

dialogue between the competing images of knowledge as a guide to understanding and knowledge as a tool of prediction—the 20th-century version of the realistic-instrumentalistic dialogue.

On the plane of experience and experimental evidence in the process of scientific discovery, the problem of Einstein's reliance on the null result of the Michelson-Morley experiment is a paradigm case. That this reliance is a myth—for EINSTEIN scarcely remembered having learned about the Michelson-Morley results before 1905 (in Lorentz' work)—and how and why the myth was established are all beautifully demonstrated by Holton in his « Einstein, Michelson and the "crucial" experiment » ([61]). For the study of critical dialogues the following two main points are especially interesting.

1) It is often presupposed that the null result of the Michelson-Morley experiment dealt the death-blow to the aether hypothesis, and that EINSTEIN had to absorb this result before he could possibly discover the theory of relativity. HOLTON shows that for anyone in an intellectual framework of the special theory of relativity the whole problem of the aether was insignificant, and the null result of the Michelson experiment was taken for granted. In other words, if, according to the scientific metaphysics at the core of a scientific research programme, there is an aether which must be dealt with ([62]) (whether positively or negatively), then it is a totally different scientific research programme, the core of which is to formulate a theory « whose object is the totality of all physical appearances » ([63]). To quote from HOLTON once more:

> For example, in foregoing for the second time the opportunity to mention the Michelson experiment, EINSTEIN is only facing the fact that, from the point of view of relativistic physics, *nothing important happens at all* in the experiment. The result is « natural », fully expected, and trivially true. The abandonment of the aether and the acceptance of the transformation equations meant the disappearance of both the objective and the very vocabulary for discussing the aether-theoreticians' interests in the null result and in the possible « causes » of the « contradiction ». The two views of the experiment were thus different to the point of being unbridgeable, which accounts for the inconclusiveness of debates between the two factions long after the paper was published, for example, at the 1927 meeting on the Michelson experiment. The relativists simply could not see the complex problems that were seriously evident to the aether theorists, to whom, in Dugas' happy phrase, the aether still formed « the substratum of thought in physics » ([64]).

---

([61]) *Isis*, **60**, 133-197 (1969).
([62]) « Nothing better could probably be done as long as the aether was kept at the head of physics; and that secured to LORENTZ, as to POINCARÉ, MICHELSON and many others, a thematic necessity to the end of his life ». HOLTON: *op. cit.*, p. 175.
([63]) From EINSTEIN'S *Autobiographical Notes*, in SCHILPP: *A. Einstein, Philosopher, Scientist* (London, 1960), p. 23 (as quoted by HOLTON).
([64]) *Ibid.*, p. 175.

The two views were unbridgeable—which is exactly how a critical dialogue between competing scientific research programmes generally operates. Each of the two protagonists successfully drives the other to a point where no bridge exists—and thus fundamental insights are obtained [65].

2) The second very interesting lesson from Holton's paper is the portrait of a generation of historians and physicists attempting to tune the whole scientific enterprise to its prevailing image of science, *i.e.* science is purely experimental and all new theories are simply « deduced from the phenomena ». The way EINSTEIN conducts the critical dialogue is in itself very intersting: he simply ignores other images of knowledge and from time to time repeats his own [66].

Einstein's image of knowledge is beautifully formulated in the abstract of the Einstein-Podolsky-Rosen paper [67]:

In a complete theory there is an element corresponding to each element of reality. A sufficient condition for the reality of a physical quantity is the possibility of predicting it with certainty, without disturbing the system. In quantum mechanics, in the case of two physical quantities described by noncommuting operators, the knowledge of one precludes the knowledge of the other. Then either (1) the description of reality given by the wave function in quantum mechanics is not complete, or (2) these two quantities cannot have simultaneous reality. Consideration of the problem of making predictions concerning a system on the basis of measurements made on another system that had previously interacted with it leads to the result that if (1) is false then (2) is also false. One is thus led to conclude that the description of reality given by a wave function is not complete.

And, from the last page of the abstract:

One could object to this conclusion on the grounds that our criterion of reality is not sufficiently restrictive. Indeed, one would not arrive at our conclusion if one insisted that two or more physical quantities can be regarded as simultaneous elements of reality *only when they can be simultaneously measured or predicted.* On this point

---

(65) On the whole problem see: Loyd S. SWENSON: *The Michelson-Morley experiment before and after 1905, Journ. Hist. Astronomy*, **1**, 56 (1970); and his book, *The Ethereal Aether* (Houston, 1972).
For the part played by LORENTZ, see: K. F. SCHAFFNER: *The Lorentz electron theory of relativity, Amer. Journ. Phys.*, **37**, 498-513 (1969); T. HIROSIGE: *Origins of Lorentz' theory of electrons and the concept of the electromagnetic field, HSPS*, **1** (1969); R. Mc CORMMACH: *H. A. Lorentz and the electromagnetic view of nature, Isis*, **61**, 458-487 (1970); and *Einstein, Lorentz and the electron theory, HSPS*, **2**, 41-87 (1970).

(66) « The principal theme of Bohr's essay was the series of critical attacks that EINSTEIN directed over the years against quantum mechanics, and the successive defeat of each of these attacks by BOHR and his collaborators, with each exchange leading to a new and deepened understanding of the fundamentals of the new physics. » M. J. KLEIN: *The first phase of the Bohr-Einstein dialogue*, in *HSPS*, **2**, 1 (1970).

(67) A. EINSTEIN, B. PODOLSKY and N. ROSEN: *Can quantum-mechanical description of physical reality be considered complete?, Phys. Rev.*, **47**, 777-780 (1935).

of view, since either one or the other, but not both simultaneously, of the quantities $P$ and $Q$ can be predicted, they are not simultaneously real. This makes the reality of $P$ and $Q$ depend upon the process of measurement carried out on the first system, which does not disturb the second system in any way. No reasonable definition of reality could be expected to permit this.

The important statement is that last, « No reasonable definition of reality could be expected to permit this », which is a superb example of the social determination of what is acceptable as knowledge. The Einstein-Podolsky-Rosen paradox was not solved—it just faded away. Why? Not because changes occurred in the body of knowledge that made their point obsolete. What happened is that nowadays a « reasonable definition of reality » permits that the reality of $P$ and $Q$ depends upon the process of measurement.

Einstein's views on science and his scientific research programmes were of the greatest influence on the development of physics through his continuous critical dialogue with BOHR. The meaning of their disagreement has as yet only been touched upon, but it is already clear that in the history of thought this critical dialogue will occupy the same fundamental place as the Leibniz-Newton controversy two hundred years earlier. EINSTEIN and BOHR differed in their view of the world, in their scientific temper and in their image of knowledge. As is well known, EINSTEIN never reconciled himself to the seemingly essential feature of quantum mechanics according to which our description of Nature is in terms of probabilities only. For EINSTEIN, understanding Nature in terms of statistical averages was no understanding; probabilistic knowledge was no knowledge. BOHR, on the other hand, felt very much at home in the uncertain world of these averages. But he was also deeply committed to the great achievements of 19th-century physics and could tolerate no new idea that was not at least a continuation or extension of 19th-century ideas. Thus we have here a fascinating double dialogue: EINSTEIN, a revolutionary in scientific metaphysics, held an extremely conservative image of knowledge; whereas BOHR, who held an extremely conservative view of the world, was a revolutionary in his image of knowledge ([68]).

This double clash is best seen in the controversy on the 1924 Bohr-Kramers-Slater paper ([69]). The essence of their paper is this: in order to escape the necessity of admitting light quanta, they were willing to abandon the strict law of conservation of enery and momentum. To put it in the terminology introduced

---

([68]) For the details of the controversy, see first N. BOHR: *Discussion with Einstein on epistemological problems*, in *Atomic Physics*, and Einstein's reply—both in P. A. SCHILPP (ed.), *op. cit.*, then the above mentioned paper by M. J. KLEIN on the Bohr-Einstein dialogue; and K. M. ABICH: *Korrespondenz, Individualität und Komplementaritat* (Wiesbaden, 1965).

([69]) N. BOHR, H. A. KRAMERS and J. SLATER: *The quantum theory of radiation*, *Phil. Mag.*, **47**, 785 (1924).

here, for EINSTEIN knowledge was certain knowledge, which must describe a nature that could be determined absolutely and must deal with the sum total of phenomena—never partial theories that yielded partial successes. This is a strictly conservative 19th-century image of knowledge [70] (without the Baconian element). As to what the world looked like, EINSTEIN was bound to no loyalties: if the continuum interpretation of Maxwell's theory needed modification, and quanta of light had to be introduced, he saw no difficulty in that. In his scientific metaphysics EINSTEIN indeed carried on the Greek tradition of unfettered imagination [71]. Another interesting point that should be made here is this: EINSTEIN was the grand master of statistical considerations [72], which, for him, had nothing to do with probabilistic physics. Unlike us, he regarded the two thought-worlds as totally different.

BOHR, on the other hand, was not fettered by 19th-century ideas on the role and task of knowledge. He was ready to abandon any general principle

---

(70) « If one has studied the development of scientific theories one notes here a familiar theme: The so-called scientific "revolution" turns out to be at bottom an effort to return to a classical purity. » (From G. HOLTON: *On the origins of the special theory*, *Amer. Journ. Phys.*, **28**, 630 (1960).)

(71) EINSTEIN had some specific things to say in a letter to EHRENFEST, as can be seen in a previously unpublished letter to P. EHRENFEST, dated 31st May, 1924, quoted by M. J. KLEIN: *op. cit.*, p. 32-33.

« I reviewed the BOHR, KRAMERS, SLATER paper at our colloquium the other day. This idea is an old acquaintance of mine, but I do not consider it the real thing. Principal reasons:

(1) Nature seems to adhere strictly to the conservation laws (Franck-Hertz, Stoke's rule). Why should action at a distance be an exception?

(2) A box with reflecting walls containing radiation, in empty space that is free of radiation, would have to carry out an ever increasing Brownian motion.

(3) A final abandonment of strict causality is very hard for me to tolerate.

(4) One would also almost have to require the existence of a *virtual* acoustic (elastic) radiation field for solids. For it is not easy to believe that quantum *mechanics* necessarily requires an electrical theory of matter as its foundation.

(5) The occurrence of ordinary scattering (not at the proper frequency of the molecules), which is above all standard for the optical behaviour of bodies, fits badly into the scheme .... »

(72) « By 1909 Einstein's ceaseless preoccupation with the incredibly important and difficult question of the constitution of radiation had led him to a deeper insight into the theoretical situation. He was now convinced that the next phase of the development of theoretical physics will bring us a theory of light that can be interpreted as a kind of fusion of the wave and emission (particle) theories. EINSTEIN had reasons for holding this opinion, and he explained them at a meeting in Salzburg in September of that year. He had analysed the implications of Planck's law for the black-body radiation spectrum, using an approach that was peculiarly his own: the study of fluctuations. » (M. J. KLEIN: *op. cit.*, p. 4.) See also M. J. KLEIN: *Einstein and the wave-particle duality*, *The Natural Philosopher*, **3**, 1 (1964): « The methods were not so much those of the "old quantum theory" as those of statistical mechanics. And the presiding genius and principal guide was not BOHR, but EINSTEIN ».

(or what looked to him like a principle) dealing with the nature of knowledge rather than with the nature of the world. Conservation of energy, in the strict sense, seemed to him the 19th-century demand to save the deterministic aim of science. He was ready to give that up; what he wanted to save there were the achievements of 19th-century physics, and he thus looked for continuity with them. Yet BOHR was no Neo-Kantian, and was not even averse to mystical depths in his thinking: was not the conservation principle one of the deepest truths about Nature? By the early 1900's, it was not. Planck's 1887 prize essay on the principle, and the Poincaré explanation that it was tantamount to a definition of energy, demythologized the conservation principle. It had become a simple methodological device, part of the image of science. BOHR was a revolutionary in this respect. But, on the other hand, in his scientific research programme he was striving for great mystical depths, which he thought to have found in the principle of complementarity.

And finally, there was a difference in temper between EINSTEIN and BOHR: EINSTEIN favoured simplicity, and BOHR searched for complexity ([73]).

The socially interesting question is this: Why did the majority of physicists tend to accept the Bohrian abandonment of the conservation law? (PAULI was a notable exception.) What was there in the intellectual environment of Europe, and especially of Germany, that freed the scientists' image of knowledge from previous commitments ([74])? This question still awaits the historian's answer.

The case of BOHR ([75]) illustrates the way the image of science can influence the creative thinking process itself. In the body of knowledge in 1912, nothing was missing that could be needed as a basis for introducing that hypothetical entity, the electron spin. Why was it that BOHR missed this « obvious » completion of his theory? LAKATOS sees in this problem a proof that rational reconstruction gives an improved version of history: « The historian, describing with hindsight the Bohrian programme, should include electron spin in it, since electron spin fits naturally in the original outline of the programme. BOHR might have referred to it in 1913. Why he did not do so is an interesting problem which deserves to be indicated in a footnote (such problems might then be solved either internally, by pointing to rational reasons in the growth

---

([73]) « Bohr's favourite aphorism was Schiller's *Nur die Fülle führt zur Klarheit.* Unlike the situation in earlier periods, clarity does not reside in simplification and reduction to a single, directly comprehensible model, but in the exhaustive overlay of different descriptions that incorporate apparently contradictory notions. » G. HOLTON: *op. cit.*

([74]) On the problem of Europe, Germany, France, England, etc., see S. GOLDBERG: *In defence of aether: the British response to Einstein's special theory of relativity, 1905-1911*, *HSPS*, **2**, 89-125 (1970).

([75]) See J. L. HEILBRON and T. S. KUHN: *The genesis of the Bohr atom*, *HSPS*, **1**, 211-290 (1969).

of objective impersonal knowledge, or externally, by pointing to psychological causes in the development of Bohr's personal beliefs) » ([76]).

Now what does it mean to « fit naturally » or to claim that BOHR « referred » to « spin »? Does not the concept of energy « fit naturally » into Newton's programme, and might he not have referred to it? Does not the concept of field fit into Leibniz's dynamics? Or, to become more absurd, does not the idea of relativity fit into the physics of the Stoa? By considering the spin part of the Bohrian research programme, or energy part of the Newtonian research programme, we miss the most fascinating chapters in the history of critical dialogues between major research programmes.

We saw the mid-19th-century debates about the place of energy in the Newtonian research programme, and we saw that the introduction of energy was Helmholtz's anti-Newtonian programme; here, by including spin in Bohr's research programme we could miss the fundamental early controversies causing the problem-shift that ended in four quantum numbers. In addition, though we should pay that price, we should not even gain any insight into the growth of knowledge. By introducing the so-called « improvements » into a rational reconstruction, we could easily cause degenerative problem-shifts in our historiographical research programmes. The practice invites improvements like Tait's letter to THOMSON asking him to look for the concept of energy in Newton's *Principia*—the argument being that NEWTON was such a great man that he must have invented such an important concept ([77]).

In other words, such improvement can easily lead to hero-worship in the Victorian style of vulgar-Baconianism. If, on the other hand, we introduce mathematical compatibility as a criterion for what « might have » been referred to, we end with the inductivist picture of knowledge by accretion, where Einstein's theory does not refute Newton's but only serves as an extension of it. Thinking in terms of research programmes helps to eliminate precisely that kind of approach. By looking for competing scientific research programmes and competing images of science, we can find rational reasons why BOHR did not think of the spin; if we do, then it becomes an important topic which certainly should not be relegated to a footnote ([78]).

In all probability, the reasons for Bohr's not having discovered the spin prove to be rational, but of a new kind that I have been advocating. Because it belongs to the image of science, it is ignored by all internalists and only partly treated by externalists who wish to consider all psychological factors motivational rather than cognitive. To be specific: BOHR in the 1920's had

---

([76]) *Op. cit.*, f.n. 69.

([77]) I am referring to a well-known story, about which see my forthcoming book, *The Emergence of the Energy Concept*, now in press.

([78]) If we find that the reasons for not discovering the spin are purely psychological (*e.g.*, caused only by Bohr's personal beliefs—a notion I do not believe), then they are also out of place in Lakatos' research programme, even for a footnote.

an image of science according to which the main role of science was to describe the world in general atomistic terms (this included the Einsteinian view of quantized energy), and he could not see the task of science in speculating about hidden properties of elementary particles. For such a new view, Boltzmannian methodology, which ruled from 1900 till around 1920, had to be replaced by a revised version of phenomenology—a development influenced at least as much by the Einsteinian revolution in science as by the general disenchantment of the post-World-War-I period and the Vienna Circle of logical positivism.

The best case study I know of, documenting how socio-political factors influence the social determination of knowledge, and how in turn this image of knowledge interferes with the choice of problems in science and even alters the scientific metaphysics (thus changing a whole scientific research programme) is Paul Forman's « Weimar culture » ([79]). Part III of Forman's paper is actually called « Adaptation of knowledge to the intellectual environment ». One can see from this work how useful it is to distinguish (artificially, by stopping time) the views on Nature from the views on science, and again the views on science from the general intellectual environment. Without these distinctions, the interaction on a time-scale could not be followed up.

If the story could now be carried further, we would see how the prevailing image of science (instrumentalist-positivist) is again upheld with difficulty in a new « hostile intellectual environment », which reacts with an anti-intellectual, antirational image of knowledge that clashes seriously with the established order. What is « reasonable » is again at stake.

These examples could be multiplied by the hundreds. The few given here will suffice to show that the concept of growth by accumulation has no chance of survival. It may be hoped that, from the examples brought up here and from the many seminars and lectures held here at Varenna, enough insight will be gained to be persuasive of the fruitfulness of viewing the growth of knowledge as a continuous critical dialogue between competing scientific research programmes and competing images of knowledge.

Now to the relevance of all this for scientists.

## 4. – Lesson from history: today's perspectives.

4'1. *The irreducible value of progress.* – The past is dying and history is beginning to occupy its proper place. Again, Plumb says: « The present weakness of the past springs from deeper causes that penetrate far into the nature of industrial society. Industrial society, unlike the commercial, craft and agrarian

---

([79]) P. Forman: *Weimar culture, causality and quantum theory 1918-1927: adaptations by German physicists and mathematicians to a hostile intellectual environment*, *HSPS*, **3**, 1-115 (1971).

societies which it replaces, does not need the past. Its intellectual and emotional orientation is toward change rather than conservation » [80]. The past was used mainly to conserve; it was the handmaiden of tradition. What we are interested in is change. This brings us face to face with the perennial question of the relevance of history. Can we draw lessons from history? From the past we cannot, because the past only sanctifies.

Drawing lessons from history is the same as learning from our mistakes [81]. That means we must always ask for the hidden presupposition behind any question and look for the connection between the rationale for a problem and its solution; thus it means refusing to view history as a chain of questions out of the blue, each with its successful answer.

For me, drawing lessons from the past means fixing our priorities (or irreducible values) and then studying history to find out what can work against, and what can support, those values. In all this it is understood that there are no absolutely irreducible values: it is always in a given context that some values seem most fundamental. I do not pretend to believe in the possibility of a rigid coherent ethical system which, once fixed, is absolute and independent of its context.

But history does help us to fix our priorities. History can show us how change takes place and how the continuous tension between tradition and change expresses itself. After we have fixed our priorities independently of history, we can look for supporting trends that, if amplified, could help us in the light of our chosen value or values.

In the present intellectual-cultural context I shall choose as my irreducible value the growth of knowledge in general and of scientific knowledge in particular. In other words, my value is scientific progress [82].

I presuppose that all of us—scientists, historians of science, philosophers of science, teachers of science—in our various activities are motivated, *ceteris*

---

(80) J. H. Plumb: *op. cit.*, p. 14.

(81) K. R. Popper: in *Conjectures and Refutations* (London, 1963): « The history of science, like the history of all human ideas, is one of the very few human activities—perhaps the only one—in which errors are systematically criticized and fairly often, in time, corrected. This is why we can say that, in science, we often learn from our mistakes, and why we can speak clearly and sensibly about making progress there ». See also, J. Agassi: in *Can we learn from history?, XII Congrès International d'Histoire des Sciences* (Paris, 1970): « But learning from history means learning from our mistakes—as Popper has suggested—and hence to retain our stock of knowledge is not to reject our mistakes but to retain them ».

(82) Naturally, one must deal also with the question of what progress is. The framework of the theory of growth of knowledge outlined here relies heavily on Lakatos' theory of scientific research programmes. There the concept is clear « positive problem shifts ».

*paribus*, by the growth of knowledge as a value ([83]). If so, the central problem posed will be: how can we advance scientific knowledge? I shall deal with the problem under two headings, *the relevance of history in doing science* and *the relevance of history in teaching science.* The distinction here is made for the sake of clarity only; it does not reflect a belief in a difference between science for research and science for teaching.

4·2. *The relevance of history in doing science.* – One of the most important lessons, for me, from the history of science is that the body of knowledge offers a far broader spectrum of open problems and points to a far greater number of avenues of approach than are actually taken up by science. The actual scientific research programme at any given stage is the result of a long process of elimination whereby the number of possibilities is systematically reduced to one. This elimination, this choice of problems, owing to an especially dominant problem-situation, is the result of the cognitive process that stems from the social image of knowledge. In other words, our lessons from history enable us to see our activity in context, to see the body of knowledge against its social image. Is this important for solving any specific scientific problem? Definitely not. The solution of a well-formulated scientific problem belongs to synchronics, not to diachronics ([84]).

However, what must be done before one arrives at a well-formulated scientific problem depends heavily on considerations not within the body of contemporary knowledge. Clearly, the awareness of what problem-situations are important is socially conditioned. The demarcation of open problems in the context of a given problem-situation, and the location of the open problems in the matrix of theories acceptable according to a given scientific research prog-

([83]) This is not a naive remark, blind to the various economic and status pressures of the scientist's career. What I mean is that the norm from which we sometimes deviate, the value for which, « everything else being equal », we work is the growth of knowledge. The status of this value is not like that of the famous Mertonian « four sets of institutional imperatives: universalism, communism, disinterestedness, organized scepticism » (*Science and democratic social structure,* Chapt. XVIII of Merton's *Social Theory and Social Structure* (New York, 1968), p. 605) by which he characterizes the ethos of science. Those four depend on the current social image of science: sometimes they indeed represent what the scientist stands for, or at least accepts as a norm, and at other times they do not even serve as lip-service imperatives. At times, when science is expected to give positive answers and to counterbalance growing scepticism, organized scepticism is not encouraged, it is fruitful credulity which is socially supported. Indeed, as we have seen above, this characterized the 17th century Hermetic and Paracelsian chemists or the *Naturphilosophen* in early 19th-century Germany. These cases, naturally, can always be dealt with by calling them « non science » and « pseudo science ». Also, at those times when theory is in disrepute, and extreme care is taken with the description of experimental set-ups, universalism is not a value.

([84]) This distinction is taken from linguistics. In a diachronic approach, every problem is viewed in the context of its genesis; the synchronic approach is strictly a-historical.

ramme, depend on *a*) the criteria of satisfactory explanation, *b*) the techniques and mathematical apparatus considered relevant and *c*) the parameters within which a problem is judged solvable. This last point is especially important: to demand exact formulation of a problem as a condition for its solubility is a serious decision. Thus the presently accepted norm—that open problems are those that are not yet solved but are exactly delimited and well formulated—is a temporary norm, socially defined. We speak of the experimental detection of particles as a well-defined open problem, but we would not call the creation of a unified field theory an open problem. The same applies in other fields.

Finally, the choice of problems from among the many open problems again has nothing to do with the body of knowledge. The considerations there are either purely external (techniques available, money and time necessary, number of people working on the problem and ferocity of competition) or social, in terms of the image of knowledge determining which problem is considered nearer the « frontiers of knowledge » and so more important.

All these considerations, if they are not seen in a historical context, serve as rigid constraints on the freedom of the scientist. In most cases he is not even aware of them, for he is not made aware of the fact that one could decide otherwise, or that at other times other scientists acted differently in the framework of different scientific research programmes.

One could pause here and ask how it is possible that scientists, admittedly of high intellectual ability, are not aware of all this—especially since not only do philosophers and historians write about such problems, but the greatest scientists, when they are no longer doing science and have begun to reminisce of philosophize, also talk about the historical context of their own and their colleagues' work ([85]). Here again the social image of knowledge gives the answer: it is part of today's image that one does not philosophize as long as one can do science. It is accepted that one cannot engage in both intellectual activities simultaneously, and even that diachronic thinking is detrimental to problem-solving. Moreover, because progress is admitted, it is generally believed that, in order to say something about the future of science, you must start at the absolute verge; thus it is believed that only those who are at the very frontier can see where we are heading.

Those great scientists who reach the age of philosophizing and historicizing science are no longer at the very frontier; therefore, what they are saying is

---

([85]) As Bentley GLASS put it in his AAAS Presidential Address, *Science: Endless Horizons or Golden Age?* (*Science*, **171**, No. 3966 (1971)): « Scientists themselves often seem little concerned about the matter. They are concerned about the quantitative measurement of all sorts of phenomena except that of the growth, status, or decline of their own collective activity ... With advancing years one commences to wonder about the relative position of one's own work in a broader, a more extended perspective. Thus one is led to try to evaluate present scientific knowledge not only in comparison with the past but also the future of man's understanding of Nature ».

not considered relevant by their younger colleagues. This is a major change in the image of knowledge: the speeches, dinner-table-talk, inaugural lectures, *Rektorats-reden*, even the Nobel Prize speeches of the great do not serve as intellectual milestones for the profession. These are food for the historian and the philosopher of science and belong to the ritual of science, but in the 18th century, and in the first half of the 19th century such summaries also served as guidelines for the young professionals. This recent change explains why the communication between the doers and the reviewers of science is negligible (86).

Behind all this—though of course without general awareness—still lurks the growth-by-accumulation theory of progress. Only because of this theory is absolute up-to-dateness a requirement for legitimizing any prediction on progress. This proves again that, whether we are aware of it or not, the social image of knowledge is always part of the whole theory of the growth of knowledge (87).

Neither is « progress » a time-independent and culture-independent concept: the basic attitude towards the idea of progress, and towards the future of science, is again determined socially. We have seen the typical 17th-, 18th- and 19th-century attitudes. But the problem is with us again: whether we continue to search for a unified field theory, whether we are thorough reductionists in biology, whether we find psychophysics a worth-while problem (as MERTON would call it, a « strategic site ») all depends on whether we believe the world, including human beings, is ultimately comprehensible. And whether we believe that depends on the balance we strike between our puzzling, bewildering, irrational, social life and our astounding successes in science. More or less we

---

(86) One could add another consideration here: it is more and more often stated by psychologists and educators that « transfer » from one universe of discourse to another is a matter of attitudes rather than one of the objective measures of intelligence. The politicians and rhetoricians of the ancient world knew this very well.

(87) The enormous success of T. S. Kuhn's *Structure of Scientific Revolution* cannot be explained solely by the intellectual greatness of Kuhn's theory of the growth of knowledge, although it is undoubtedly very important. The majority of the working scientists never showed any interest in any theory of the growth of knowledge, so why should they enthusiastically read this one and quote it everywhere? The reason is that Kuhn's theory coincides with the modern scientist's self-image: he is relieved to know that his puzzle-solving activity, the textbooks he writes, the « universalism, communism, disinterestedness, organized scepticism » in which he believes, all contribute to the growth of knowledge and to progress, to which he is fundamentally committed. All the external conditions that bring about sudden changes are social-psychological: they « happen to him », he does not need to invest mental energy, for he could not influence them anyhow. The individual scientist needs only to do exactly what he is doing, and progress will be automatically taken care of—or so he finds it comforting to believe.

live in an age where the successful, competent scientists all claim, at most, *ignoramus* (88).

Finally, what history can teach us is that at least three dogmas of modern science are nothing but aspects of the present social image and are thus life- and culture-dependent. If they are indeed changeable, and if we accept growth of knowledge as an irreducible value and as the basis for the theory of growth of knowledge propounded here, we are forced to ask ourselves whether these three aspects of the social image of knowledge contribute to the growth of knowledge, hinder it, or are neutral with respect to it. The three dogmas I am referring to assert that *a*) science is objective, *b*) there is a method of science and *c*) scientific theories do not describe the world as it is but seek only to predict its behaviour. Prejudging the issue, I shall call them the objectivist fallacy, the « language of science » fallacy and the instrumentalist fallacy, respectively.

*The objectivist fallacy.* The 17th-century mood among English scholars, the founders of the Royal Society and others, was marked by weariness with internal strife and disillusionment with the sterility of religious partisanships. For most of them the scholarly world became a haven from prejudices, religious commitments, and dispute. As we have seen, it was in this intellectual climate that freedom from controverial presuppositions became part of the *ethos* of science. It was then that the Baconian solution—relying on observable evidence and drawing inductive conclusions from the amassed data—became « the new science »; these were principles on which the scholars could agree, free from source material for the hated disputations of the schools, so science became objective. But objectivity did not mean for them the elimination of fruitful dialogue on basically different world views—as can be seen in the minutes of the meetings of the Royal Society or the treatises of the natural philosophers which are all critical dialogues of a sort (89).

---

(88) The reference is to Du Bois-Raymond's *Die Sieben Weltraetsel*, mentioned above. Recently, Lawrence BADASH noted some of these views (*The completeness of 19th century science*, *Isis*, **63**, 48-58 (1972)). He says: « The physicist Eugene WIGNER, for example is on record as believing that our scientific understanding will not likely become much deeper than it is today, while Robert HUTCHINS, former President of the University of Chicago, asserted not long ago in his syndicated newspaper column that the conquest of Nature is almost complete. We can have every confidence that science will unlock her last secrets and that technology can speedily put the new knowledge at the service of mankind ».

(89) The literature is very rich and amply documented. Here I will refer only to the Ph. D. dissertation (soon to be published in book form) of Zeev BECHLER (Hebrew University of Jerusalem, 1972), who discussed the style of the dialogues in detail, and to his paper *Newton's 1672 optical controversies: a study in the grammar of scientific dissent*, in Y. ELKANA (ed.): *Chapters in the Interaction between Science and Philosophy.*

It was, again, 19th-century Victorian philosophy that introduced the value of consensus and interpreted scientific objectivity as the elimination of all metaphysics, that is, all critical dialogue. This is the objectivist fallacy, which sees science starting from fully agreed-upon experimental results (raw facts that are theory independent) and growing by accumulation of such results. It is on this foundation also that the late 19th- and early 20th-century image of science is built.

But there is another level on which science is considered objective, and on this level objectivity has lately become an issue: *i.e.* the view that scientific activity produces results independent of the use that people make of them. This is the oft-quoted statement that science is value free and culture independent. It, too, seems to me quite wrong, but in a very special sense. It is accurate, but trivial to say that, having built an atomic bomb, I am free to use it or not. However, any scientific theory has a hard-core metaphysics—that is, a view of the world that in its very formation is not testable—and if we remember that concepts have a meaning only within the framework of a given theory, then there are no theoretical results which are independent of the theory's metaphysics: Newton's laws do not have the same meaning in two cultures if « time », « space » and « force » have different meanings. Thus, the first sense of the objectivist fallacy yields to the evidence that science is not culture free. But even in the second sense, there is a nontrivial level on which science is not objective: if we agree that the choice of problems is determined by the social image of knowledge—which is, historically, interacting with the socio-political and cultural factors—then research is influenced by cultural factors. And a scientific result, once established, still must be interpreted in a given cultural context that in turn can be used by the people as they choose, and as they interpret it in their cultural context.

If we wish to advance science, instead of viewing it as objective, we should articulate the underlying metaphysics and the social image of knowledge that determined the problem choice.

*The « language of science » fallacy.* We must again go back to the 17th century. In order to prevent the Royal Society from becoming a junk-store of curiosities or a depository of bestiaries (as we have seen, collections of such used to be the task of the naturalist of the 15th and 16th centuries, according to the then-current image of knowledge), the founding fathers—and especially Robert BOYLE and John WILKINS—created rules for doing science and writing up the results. It is here that the origins of our cherished experimental method are to be found. BOYLE insisted that irreproducible results should not be recorded and should form no part of the body of knowledge. A rule was established that all factors possibly relevant to an experiment should be carefully noted down. But it was also established that the lengthy speculations and personal stories that had led an enquirer to a given experiment should not be included as part of the scientific report—which meant, incidentally, that the theory or metaphysical view according to which it was decided what factors were relevant to an experiment was suppressed. This also created an impression of the existence of « raw facts ». But there was as yet no sign that the 17th-century natural philosophers believed in a unique scientific method as a method of discovery. For them, scientific method was still only a manual for experimental procedures, as it is for all good scientists today.

However, parallel to the above development, Galileo's famous saying—that Nature is a book written in mathematical language and the task of the natural philosophers is to decipher that language—took hold of the imagination of the natural philosophers. In addition, ever since LUTHER, and especially among the English Hebraists in the 17th century, an interpretation of the Bible—a deciphering of God's code—had become the basis of religion, and the emphasis on language became very strong. (The widespread attempts to create a universal language was an interesting side-result (90). Thus the image of a language of Nature was born.)

With the professionalization of science in the 19th century, when more and more people were enlisted in the service of science, and with 19th-century Victorian ethics reinterpreting science and creating an image of experimental handiwork on data as the sole truly scientific activity, came the conflation of the method of science and the language of Nature into the « language of science ». There was only one method of discovery; and if you learned it well, you were on the sure path to the making of discoveries. This I call the « language of science » fallacy.

Although it is shared by very few scientists, it is an integral part of science-teaching on all levels (91). Needless to say, principles of science-teaching change with the changing image of knowledge; thus we are confronted with a changing « method of science »—always taught as the one and only certain method. That the present instrumentalistic image of knowledge is detrimental to progress will be shown below. But in principle it seems to me that the « language

(90) « When John WILKINS lay dying in the fall of 1672, a constant stream of friends and associates brought counsel and advice ... His visitors reported that the Bishop of Chester (*i.e.* John WILKINS) regretted only that he would leave unfinished his darling as he called the artificial language. » From Barbara J. Shapiro's fascinating biography, *John Wilkins* (Berkeley and Los Angeles, 1969), p. 1.

(91) The new science-teaching methods (P.S.S.C., E.P.S., Chem. Studies, B.S.C.S., with the exception of the Harvard Project Physics, which is historical), which emphasize the method of enquiry or method of discovery, make this fallacy a basis of their philosophy: the idea behind them is that one can teach the student a correct method of scientific discovery. On that, see my forthcoming *The theory and practice of cross-cultural contacts in science: queries and pre-suppositions.*

of science » fallacy can be overcome in only one way—by methodological pluralism ([92]) supported by historical case studies.

*The instrumentalist fallacy.* We have seen in Sect. **2** and **3** how the relativistic attitude towards Nature waxed and waned with the changing image of knowledge through the ages. With the Einsteinian revolution in the 20th century, confidence in the magnificent structure of 19th-century physics disappeared and creeping doubt about various aspects of it began to appear. One of the strongest elements of that bold and proud structure was its adamant realism: man knew that the world behaved, even *was*, as science described it. Even the famous controversy on the reality of atoms between the atomists (chiefly BOLTZMANN) and the energeticists and phenomenologists (OSTWALD, MACH, KIRCHHOFF) was couched in terms of hard realism. The world was either this way or that, depending on what experimental evidence each party chose to trust; the evidence was very controversial ([93]), but the choices were exclusive. When all that solidity crumbled, the most pervasive doubt centred around the question of whether physics had to do with the world « as it is » at all. It is my firm belief that this sceptic mood played a role in formulating the Copenhagen interpretation of quantum mechanics, though others see the instrumentalism of modern physics as an outcome of the Copenhagen interpretation ([94]). (But, firm as my belief may be, it needs research and documentation which have yet to be done.)

Whether instrumentalism is cause or effect, one of the most often quoted conclusions of quantum mechanics is that there is no longer any way to « picture » elementary processes and particles, and that scientists must learn to think in terms of probabilities and distribution functions. Though the proposition is rarely formulated in extreme instrumentalistic terms—*i.e.* it is rarely admitted that scientific theories serve as sheer instruments of pre-

---

([92]) This view has been propounded and defended in repeated publications of LAKATOS, TOULMIN, AGASSI, BUNGE, and others, but its main protagonist is Paul FEYERABEND. See especially his *On the improvement of the sciences and arts and the possible identity of the two,* in R. S. COHEN and M. WARTOFSKY (eds.): *Boston Studies in the Philosophy of Science,* Vol. **3** (New York, 1968), p. 387. See also Arne NESS: *A plea for pluralism in philosophy and physics,* in W. YOURGRAU and A. D. BRECK (eds.): *Physics, Logic and History* (New York, 1970), p. 129.

([93]) See my *Boltzmann's scientific research program and its alternatives,* in Y. ELKANA (ed.): *Chapters in the Interaction between Science and Philosophy.*

([94]) Paul Forman's *Weimar culture, causality and quantum theory 1918-1927: adaptation by German physicists and mathematicians to a hostile intellectual environment,* in R. McCORMMACH (ed.): *HSPS,* **3** (Philadelphia, 1971), p. 1-115, is a superb example of studying the interaction between the environment and the social image of knowledge, which then dictated the problem-choice and influenced the hard-core metaphysics of the scientific research programmes. There are also other studies, on Vienna, to which I referred above.

diction and not as pictures of reality—it is now an important part of physics education to teach students to manipulate mathematics without making an attempt to form pictures corresponding to each mathematical result. The acceptability of a result (ideally) depends either on mathematical reasoning or on experimental confirmation or refutation of predicted results. The interpretation of the mathematical prediction in terms of experimental results is itself rarely unambiguous and often must rely on past experience and not on picturing; it is rarely admitted, and never encouraged, that acceptability should be decided on the basis of whether a description makes sense. This is instrumentalism *par excellence*.

The fallacy here, in my opinion, lies in the belief that this is dictated by the state of knowledge in physics and therefore is the only way to ensure progress. The scientific metaphysics that serves as hard core to the present scientific research programme is the instrumentalistic Copenhagen interpretation of quantum mechanics, and according to its current « consensus » image of knowledge, a critical dialogue with competing science research programmes is discouraged. To give one example: the Einstein-Bohr controversy, which has not been settled—and which would serve as a basis for studying modern physics if the accepted theory of the growth of knowledge had been that knowledge grows by critical dialogue—has been swept under the rug ([95]).

The view that the present instrumentalistic approach is detrimental to progress relies on these three arguments, in ascending order of weight:

1) I claim that there has been no great step forward in conceptual thinking in physics ever since realistic picture-forming has been discouraged and controversies silenced. Admittedly, this is my weakest argument, since one can prove neither that the progress of the last twenty years is insufficient nor, even if it is, that the cause is instrumentalism.

2) It can be shown that, in the past, all those discoveries we admit were steps in progress were made by realists and never by instrumentalists. This applies to the 20th century, too, with the possible exception of HEISENBERG. The argument can be fruitful if one takes case history by case history.

3) If one accepts the theory of growth of knowledge propounded here, namely that knowledge grows by critical dialogue, then no stage at which such dialogues are suppressed can contribute to progress.

4·3. *The relevance of history for teaching science.* – We aim at progress, and I believe that progress can be achieved by introducing continuously progressive problem-shifts in our scientific research programmes while continuing the crit-

---

([95]) See the original discussion in the two volumes of A. SCHLIPP (ed.): *A. Einstein, Philosopher-Scientist* (London, 1959); and M. J. KLEIN: *The first phase of the Bohr-Einstein dialogue*, in R. MCCORMMACH (ed.): *HSPS*, Vol. **2** (Philadelphia, 1970), pp. 1-39.

ical dialogue between competing research programmes and competing images of knowledge. In order to do that, we must teach science as a continuous critical dialogue. To this end we must teach the ever-changing scientific research programmes with their ever-changing hard-core metaphysics, and thus follow the social image of knowledge in history ([96]).

Here the basic tenets of an historical approach will be outlined and a guide presented for a course in modern physics, taught historically. I shall presuppose the following theses, most of which I have referred to above.

1) The purpose of physics-teaching, or of science-teaching in general, is to advance the understanding of the growth of human knowledge—in treating science both as part of general education and as professional scientific education on any level—and not merely to impart facts and techniques dealing with logical reconstruction of pseudo-finished theories.

2) The fundamental interest in the dynamics of the growth of knowledge is basically different from the reconstruction of knowledge. The positivistic view of philosophy of science is mainly occupied with reconstruction of science and a search for a language of science or a method of science, using logic as its main tool (the synchronic view). Philosophy of science which is interested in the growth of science (the diachronic view) relies at least as much on the history of science as on logic.

3) Science grows by a continuous critical dialogue between scientific metaphysics, *i.e.* fundamental conceptual frameworks dealing with the structure and mode of operation of the physical world. These views, which are rooted in the problem-situation of every age, are metaphysical in that they cannot be directly tested by any method whatsoever, and scientific because they deal with questions about the physical world and because a rational dialogue between different metaphysics is possible. A rational dialogue between different metaphysics means that arguments brought forth in defence of the one against the other, although they are not conducive or final, can contribute to conviction or change in scientific metaphysics. Such changes are of definitive importance for scientific research programmes.

4) The scientific questions formulated, by various ways, can then be checked, answered and refuted or confirmed—experimentally in the laboratory, or on paper if they are mathematically formulated questions.

5) The connection between metaphysical frameworks and testably formulated scientific questions is obscure. Here, however, it is assumed that such a connection exists, and that it is causal in character. With this assumption it becomes of definitive importance to spell out the connection as a prerequisite

---

([96]) I have written on this topic elsewhere; here I wish only to emphasize the main points.

for any insight into the process of growth of scientific knowledge. Reconstructionist attempts to explain how science grows all started with the well-formulated scientific questions and ended by giving a logically correct, though historically wrong picture of absolutely no relevance to either advancing the growth of knowledge or teaching science. In addition to logic, there are two ways to approach the problem: *a*) by attempting to develop the psychology of scientific discovery, and *b*) by listing numerous case histories. The latter does not include a clearly formulated theory on how one learns from history, however. But despite the fact that no conclusive theory of the psychology of discovery, or a conclusive argument as to how one learns from history, has yet been propounded, it is presupposed here that these last two modes are infinitely more fruitful for an understanding of the growth of science than is the logical reconstruction of any given scientific field. The lessons to be drawn for science-teaching out of these presuppositions are that science can best be taught in the way that allows the student to participate in a critical dialogue between competing metaphysical frameworks and competing images of science; thus he is shown the problem-situations out of which these metaphysical frameworks grow. By rational arguments, the critical dialogue between the metaphysical frameworks is repeated; and an attempt is made to follow by many historical examples the different ways and methods of science by which various scientists advanced from metaphysical problems to scientifically testable questions.

6) The theory of the growth of knowledge expounded in Sect. 1 is also presupposed here.

7) At every stage of his development every human being has a model, perhaps not articulated or not conscious, of the physical world. This is so because everybody has a scientific metaphysics, be it primitive or complex, undeveloped or developed, « true » or, by now, clearly false. Thus the best way to teach anything new is to discover the conceptual framework or scientific metaphysics of the student (I repeat: his age and previous studies are irrelevant in this context) and, by way of a critical dialogue between his metaphysics and the one to be communicated to him, introduce the new ideas ([97]).

---

([97]) At this point, however, two serious limitations to this teaching programme appear: first, because of shortage of time, it is impossible for the teacher to analyse the metaphysical framework of each individual student in order to introduce at this point his new idea, and to bring about a dialogue with the competing metaphysics to be taught. Second, it is accepted as true that the more concise the treatment of any scientific field, the more compact its formulation, and the more it resembles a logical reconstruction rather than a historically true narrative, the easier it is to study it. Thus the place of conventional science-teaching (it being a logical reconstruction) is secure. I am only suggesting that it should be put in a completely different framework, whereby the

In the following pages I shall sketch a blueprint for a possible approach to a course in modern physics. It is generally accepted that quantum mechanics began with the problem of black-body radiation, and even a historical work like Jammer's begins with the statement: « Quantum theory in its earliest formulation had its origin in the ability of classical physics to account for experimentally observed energy distribution in the continuous spectrum of black-body radiation » ([98]). In survey courses which involve less depth, one usually gives a list of eight to ten formulae dealing with radiation, beginning with Kirchhoff's, law Stefan's law, and the Boltzmann proof of it (called the « Stefan-Boltzmann law »), continues with the Wien displacement law and the Wien distribution law and then, in rapid succession, Rayleigh's or Rayleigh-Jeans' law and concludes with Planck's law. In very rare cases several corrections to Wien's distribution law are mentioned, but generally Planck's radiation law is the pinnacle of the list.

Rarely does a whole trimester of a one-year course on quantum mechanics deal with the black-body radiation, so that whatever must be said about this problem from the vantage point of 1960's physics is then stated; but the late Giulio RACAH used to teach his course in Jerusalem that way. Thest udent was left with the impression that in 1900 scarcely any other problem had occupied physicists, who were all running madly after the definitive formula of black-body radiation. In reality, that was far from the case, and a course in modern physics should begin, ideally, with a penetrating description of the problem of the 1890's. (KLEIN has already hinted at this, in his first paper on PLANCK ([99]).)

With even a perfunctory look at the titles of papers in the important journals (*e.g. Annalen der Physik* or the *Philosophical Magazine*) during the last decade

---

other side of the story, that is the historical narrative and the problem of the growth of knowledge, is not less emphasized than logical reconstruction.

These two severe limitations can be overcome in the following manner: first, as we do not have the time to find out the metaphysics of each individual student, we can at least eliminate the erroneous image of « one correct method » of science by examining enough historical case studies to show the variety of the methods of science. Second, a very sensitive balance must be developed (here several years of experimentation in the classroom are needed) between that part of science-teaching which is conventional so far as reconstruction of science goes and the historical part that would emphasize the diverging methods of science. Additional problems are psychological educational in character, *e.g.* how do we know that authoritative teaching at some stage is not a prerequisite for attributing some certainty to knowledge? But this, important as it may be, is not for discussion here.

([98]) M. JAMMER: *The Conceptual Development of Quantum Mechanics* (New York, 1966), p. 1.

([99]) M. J. KLEIN: *M. Planck and the beginning of the quantum theory*, *Arch. Hist. Exact Science*, **1**, 459 (1962); and *Planck, entropy and quanta, 1901-1906*, *The Natural Philosopher*, **1**, 83 (1963).

of the 19th century, it can be seen immediately that the fundamental issues were still problems of energetics, the generality of the laws of thermodynamics, ideas on the kinetic theory of gases and—above all, in the last few years of the century—the revolutionary ideas of Roentgen rays, radioactivity, the structure of the atoms and the various consequences of Hertz's proof that light is indeed an electromagnetic phenomenon. Such famous textbooks as Drude's *Optics* or Chwolson's *Theoretical Physics* put no emphasis whatever on the problem of or even on the description of black-body radiation. DRUDE does not even mention it as an open problem in physics. Drude's book appeared early in 1900, several months before Rayleigh's paper in June 1900 or Planck's corrections in October and December.

The main problems in the 1860's, 70's, and even 80's were the result of the works of HELMHOLTZ, CLAUSIUS, JOULE and KELVIN. And it is no accident that in 1887, when PLANCK wrote his very important book on the principle of the conservation of energy, he was actually answering the call by the Faculty of Philosophy at Göttingen to find out whether the principle formulated by HELMHOLTZ in 1847 (as « conservation of force ») and what was by 1887 called the « principle of conservation of energy » were one and the same ([100]). Also, it was not until the 1860's that RANKINE, and others after him, realized that what the English school had found to be the final proof for the dynamical theory of heat and what the German school had found to be the proof for the conservation of energy could really be pooled together into one comprehensive theory of energy, with the corollary that heat was motion—and that this meant the final downfall of the caloric theory of heat.

PLANCK himself was mostly interested in what he called the « generality » of the laws of Nature, and he formulated his program of science accordingly ([101]). His whole program depended on whether the laws of thermodynamics were sufficient to describe this unified world picture: if they were not, the additional assumptions needed would then involve an atomistic view that PLANCK had vehemently opposed at first. Here, a beautiful example of critical dialogue on the highest level can be shown: I refer to the debate on atomism between PLANCK and his followers and BOLTZMANN (who had worked completely alone for many years in Vienna). PLANCK hoped to show that the second law of thermodynamics, together with classical electrodynamics, would make any atomistic or statistical consideration superfluous. At the same time BOLTZMANN emphasized that those presuppositions were lacking, and that a further hypothesis must be added that would describe a tendency in Nature to go always from less-

---

([100]) See my *Helmholtz*: *Kraft*: *an illustration of concepts in flux*, in *HSPS*, **2**, 263-298 (1970); and my book *The Discovery of the Conservation of Energy*.

([101]) PLANCK defined progress as a process by which a unity of the physical-world picture is attained.

probable to more-probable states. This is the basic content of Boltzmann's definition of entropy, which PLANCK and ZERMELO had rejected for several years ([102]). In those important years, when PLANCK was trying to describe and explain the true nature of radiation (between the years 1897 and 1900), he worked on purely classical lines.

Thus, for students, a description of the problem-situation would be followed by a description of the critical dialogue ([103]). A serious analysis of the problem-situation facing PLANCK and BOLTZMANN would naturally introduce all those different elements that are missing in most introductory courses: the basic ideas of thermodynamics, the problem of the generality of its laws, their place in the programme of modern science and the higly important point that one cannot deal with the foundation of quantum mechanics and of the black-body radiation by simply saying that one can take from statistical mechanics the Boltzmann formula for entropy and from it derive the relevant laws. According to BOLTZMANN, entropy is equal to the constant $k$ times the natural logarithm of the probability of states (that is, $S = k \ln W$); it is completely meaningless if its conceptual framework remains unanalysed, yet that happens very often in conventional courses on modern physics. Also, it must be explained to students why PLANCK so long opposed the use of this formula, to what extent he changed his attitude when he accepted the Boltzmann considerations, and where he differed from BOLTZMANN. Here I refer to the idea mentioned by KLEIN, and stated clearly by PLANCK himself (in his Leyden lecture in 1908, « On the unity of the physical world picture »), that he had accepted Boltzmann's view except for one point: although BOLTZMANN also allowed for processes which were entropy-decreasing and as such were not irreversible in character, PLANCK could not accept the possibility of such cases in Nature. If the student is not shown this, he cannot possibly understand why PLANCK opposed for at least ten years the really revolutionary Einsteinian interpretation of the quantum of action, or really understand the meaning of the laws of thermodynamics and of the concept of entropy that PLANCK tried to defeat.

After such a broad treatment of the problem-situation and of the metaphysical debates on atomism, we reach a point where a systematic treatment (conventional, formalized, mathematical) of all radiation laws is needed. I shall not describe this in detail, as any good textbook would serve the purpose. But even in the formal treatment, some lesser problems must be dealt with in the course: some of them have been raised already, and others, to the best of my

---

([102]) It is also the idea in Boltzmann's $H$-theorem, which is actually a hypothesis of molecular chaos in gas theory. PLANCK finally was forced to admit its equivalent for harmonic oscillators radiating energy.

([103]) The students should read Boltzmann's highly important paper on the *Methods of theoretical physics in recent times*, written in 1899, passages from Planck's five radiation papers, written in 1897 and 1898 and his introduction to his summary, which appeared in 1900 in Vol. I of the *Annalen der Physik*.

knowledge, either have never been asked, or were partly answered and partly forgotten. Let me list a few.

What were the problem-situations of KIRCHHOFF, STEFAN and WIEN? What were the presuppositions on which WIEN built his theoretical treatment in 1896, when he developed his distribution law? It is accepted that Wien's treatment is theoretical. If so, why was Wien's distribution law not considered completely satisfactory theoretically, and why did PLANCK find it necessary to spend years on an attempt to put it on a firm theoretical basis, before he succumbed to the need to throw it away and find a new distribution law? WIEN himself has referred to his own work as somewhat conjectural, but what he had in mind was only that he did not provide a mechanism by which radiation is emitted by harmonic oscillators. PLANCK did not give a mechanism either; yet, in the introduction to his five important papers on irreversible radiation processes, he stressed that his main motivation was to put Wien's work on a firm theoretical basis.

On the other hand, if Wien's formula was not considered theoretically conclusive, why was it accepted anyhow as absolutely in accordance with experimental facts? How is it possible that Rayleigh's radiation formula, published in June 1900 and mentioned by the two experimentalists RUBENS and KURLBAUM (who finally proved the failure of the Wien distribution law), is never even mentioned by PLANCK? What was the status of the equipartition theorem on which RAYLEIGH built his conclusions, and why did PLANCK refuse even to refer to it? That PLANCK must have known of the existence of the Rayleigh formula is almost certain. As is well known, the Rayleigh formula is valid exactly in those wavelength ranges where the Wien formula fails: it describes the spectrum of radiation correctly in the long-wave region, though it completely fails in the short-wavelength range. But, if PLANCK did not even find it important enough to mention, when did this formula become so important that its failure to fit the experimental data was called by EHRENFEST in 1911 the « ultraviolet catastrophe »? Then again, what was Jeans' part in the Rayleigh-Jeans formula? From 1905 onwards JEANS was trying to eliminate the Einsteinian interpretation of the radiation formula and save the validity of the classical picture. What were the aspects of the failure of classical physics on which he finally concentrated—and which, against his own intentions, proved to everybody that the Einsteinian interpretation was indeed the correct one?

It is only when we come to ask these questions (even if we do not know the answers to some of them) that we can succeed in giving our students a better understanding of the ideas of classical physics, and of what a breakdown of classical physics actually meant, than does the superficial picture given nowadays. Actually, it was in the work of EINSTEIN that the radiation problem, its various solutions, and its different interpretations became a central feature of physics. The usually accepted picture gives the impression that EINSTEIN in 1905 enlarged or extended the Planck treatment of his quantum of action by stating that *a*) not only is energy in the oscillators quantized, but light itself consists of quanta of energy, and *b*) after that the photoelectric effect, and its experimental proof by FRANCK and HERTZ, solves the whole problem. This again is a fundamentally wrong historical picture, built up by hindsight, that once more closes off any possibility of understanding what really went on—what the problems were and why physics developed as it did, and, even more important, why EINSTEIN and some others never agreed to the completely instrumentalistic view that developed in physics.

It should be brought to the student's attention that EINSTEIN ([104]) wrote important papers year after year, attempting to re-interpret and understand the true meaning of radiation. These papers are very rarely mentioned, and yet they have the greatest importance for the development of physics and for our present world picture. In 1906 and 1907-8, EINSTEIN repeatedly searched for lacunae in Planck's theory of radiation and attempted to give a broader interpretation to his own 1905 results. He was far from satisfied. Only after the 1911 Solvay Congress did a consensus emerge on radiation problems and on the meaning of the quantum of action. It seems to me that only such a consensus could have paved the way for a fruitful connection of the quantum of action with atomic theory; in other words, these discussions served as a necessary prologue to the Bohr atom. Along the same line, our introductory course in modern physics should show clearly the connection between the photoelectric effect and Einstein's treatment of the specific heat of solids; such a connection emerges clearly when the problem-situation that occupied EINSTEIN and is eminent in all his papers from 1905 is examined thoroughly. Such an analysis would also explain why in 1905, and in spite of what we generally think nowadays, EINSTEIN did not rely solely on the Planck radiation law but instead used the Wien distribution law as a basis for his theoretical statement.

A formal treatment of Einstein's work on radiation should be followed by a survey of Planck's changing views on the place of classical physics in our world picture. Included should be *a*) the interpretation of the laws of thermodynamics, beginning with Clausius' formulation of the second law; *b*) Planck's own correction of this formulation and its generalization for the Boltzmann picture of the second law, and later the unification of this picture with Planck's original attempt to connect thermodynamics and electrodynamics; *c*) his opposition to any interpretation of the radiation formula that would do away with classical physics, and his changing views on the interpretation of his own formula and of the quantum of action. All this would cover the years from 1900

---

([104]) On EINSTEIN and on the interactions among environments, image of knowledge and problem choice see G. HOLTON: *Finding favour with the angels of the lord: notes toward the psychobiographical study of scientific genius*, in Y. ELKANA (ed.): *Chapters in the Interaction between Science and Philosophy* (New York, 1973).

to 1911 when, at the Solvay Congress, PLANCK finally accepted the Einsteinian world view. This line of thought could be determined by comparing Planck's 1908 address « On the unity of the physical world picture » and his return to the same problem in 1938, again at Leyden, when he reviewed the changes, re-emphasized his realistic world view, and juxtaposed it to the growing instrumentalistic philosophy of science.

For the rest, allow me to hint only at some other chapters that I would include in a course on modern physics and that I would like to balance against the conventional formal treatment.

On the metaphysical debate on atomism, I would combine the various atomic models and bring the contemporary arguments against them. Only after these considerations would the usual treatment of the « old quantum theory » follow.

Next would come the Bohr-Einstein-Schrödinger debate on the epistemological consequences of quantum theory and the emergence from this discussion of the Copenhagen School, or the Copenhagen interpretation of quantum mechanics with its emphasis on instrumentalistic instruction. It should be shown that the interpretation itself is a scientific metaphysics (this, for me, is no derogation), which can be debated rationally and critically.

As a last example I will mention the status of the conservation of energy principle in modern physics. The story should start by showing that the principle was considered sacrosanct until around 1920. Then, in 1924, the famous Bohr-Kramers-Slater paper appeared, and it found a ready audience: the greatest physicists of those times were suddenly willing to give up the principle completely. And then, only five or six years later, when the tide turned again, it was decided on purely metaphysical grounds that the conservation of energy principle was still fundamentally valid and that it was better to accept Pauli's postulations of a never-dreamt-of particle (even if there were no signs of any possibility of ever proving its existence experimentally) than to give up the principle that had been so useful to science for a hundred years. The participants in the debate on the Bohr-Kramers-Slater idea were PAULI and DIRAC, EINSTEIN and SCHRÖDINGER, and of course COMPTON.

Only such a historical dialogue would enable the students to realize what is involved in conservation laws.

**4·4.** *Coda.* – These are the lessons I wish to draw from history. To quote PLUMB ([105]) once more: « As the past dies and its hand grows palsied in its grip on religion, morality, education, there is a danger of social incoherence, of an idealization of analytic understanding rather than creative belief ». For us this means a danger of idealizing unimaginative technique and rigid dogma. Let us encourage creative scientific metaphysics and reintroduce fertile pluralism by supporting critical dialogue.

---

([105]) *Op. cit.* (see footnote ([1])), p. 67.

# Electrons or Subelectrons? Millikan, Ehrenhaft and the Role of Preconceptions.

G. HOLTON

*Department of Physics, Harvard University - Cambridge, Mass.*

For today's lecture in this course on the history of recent physics, I have chosen excerpts from an on-going case study which currently intrigues me. While I am neither the first to be intrigued by it nor have finished the work on it, presenting the case in its current state has perhaps the merit that it may show you a particular approach to the history of science. Some scholars like to work on the development of a concept over a long period—for example, a century or more. Others like to work internalistically on the structure of a particular paper. Though I am much interested in the other styles also, my special preference is to look at an event, a nascent moment during a discovery, and then step back away, as it were, to see the trajectory of the contributions that came together to form this event, both within the life of the scientist and in the community of science.

When you do that, you quickly have to expand beyond the rigorously defined scientific content of the situation, and begin to look at the epistemological content, the philosophical presupposition, and also the social situation (*e.g.* colleagueship and teacher-student relations), the way a scientific community fights about a point that is in disagreement, how it knows that agreement has been achieved and points of this kind which often are thought really to belong to the sociology of science. In my view, some of the most interesting problems in the history of science are those where the vague boundaries with philosophy of science and sociology of science are not made into barriers.

The study I shall talk about today centers on events in the year 1910, leading two scientists into exactly opposite directions—one to « success », the other to « failure ». (Failures are rarely analyzed in histories of science, but can be just as instructive as victories.) The disagreement was one between a practically unknown scientist in Chicago, over 40 years old, named R. A. MILLIKAN, and a young but already well-established physicist in Vienna, named F. EHRENHAFT. Their controversy concerned a very basic question. As MILLIKAN himself put it some years later in his book, *The Electron* [1917 and 1924], « Evidence has been presented which purports to show that electric charges exist which are much smaller than the electron. Since this raises what may properly be called the most fundamental question of modern physics, the evidence needs very careful consideration » (p. 161). EHRENHAFT and his student were finding subelectrons, first about one-half the charge of the electron, then a hundredth, a thousandth and so forth. All here who have done the Millikan « oil-drop » experiment in school may think that such results cannot be taken seriously, and therefore that there should not have been such a debate at all. Nevertheless, the case did reverberate through the scientific community, and people like SCHRÖDINGER, BORN, SOMMERFELD participated in the debate. The number of articles in the physics literature of many countries grew to many hundreds. It was properly thought to be important, and it illuminates the state of science at that time.

## 1. – The protagonists.

First, let us look at the chief antagonists, for the biographical details do count. MILLIKAN was born in 1868 and died in 1953. At the height of his career, in his 50's and 60's, he was perhaps the most renowned physicist in the United States, scientist, administrator, educator, public citizen, policy-maker. As his charming and frank autobiography (*The Autobiography of* R. A. MILLIKAN, New York, N. Y., 1950) shows, his origins were rather humble. As was the case of many scientists of his generation, he was the son of a poor, small-town preacher (minister). His grandfather had been among the earliest farming settlers of the Mississippi River country in Western Illinois; in 1825, as a child, he had walked alongside the covered wagon as it made its way from the Berkshire Hills in the East to the frontier, the « Western Reserve ». MILLIKAN himself, as a boy, led a life recognizable even to a European who has never visited the United States, but who has read Mark TWAIN—the steamboats on the Mississippi, family farmwork in their 1-acre yard, the swimming hole, the barefoot existence, the rural, simple, direct, fundamentally pious kind of background.

MILLIKAN went to Oberlin College in 1886, took only one physics course (which he said was « a total loss »), and discovered his interest in the subject only when asked to help teach in a physics course. For graduate work he went on to Columbia University, studied under M. PUPIN, whose only graduate physics student he was for two years. A. A. MICHELSON, whom he met in 1894, made suggestions that helped him form his experimental thesis work. He obtained his Ph. D. degree from Columbia in 1895—but then found, as many do today, that there was no satisfactory job. Therefore, with a loan, he went to Germany in May 1895 for a year or two of study. It was of course the best moment possible. Within a few months, the work of ROENTGEN came on the scene like a storm, followed by that of BECQUEREL—and the whole field of physics exploded into excitement at that very time.

In 1896 he had a job offer from MICHELSON and joined him in his department at Chicago. He eventually of course achieved a very large research output (he listed 21 fields of research in his account in *American Men of Science*)—first at Chicago and, after 1921, at the California Institute of Technology.

Now a brief look at the other protagonist. EHRENHAFT was 11 years younger. Born in Vienna in 1879 into a professional family—his father was Obermedizinalrat, his mother the niece of B. EGGER, a student of FOUCAULT—he was educated at the University of Vienna in the period when E. MACH was still teaching there. In his earliest work (1900) he was one of the first to produce and study inorganic colloids. In 1905 he became Privatdozent; by 1909 he was instructor in statistical mechanics, obtaining his (associate) professorship in 1912 and the directorship of the Third Physical Institute in 1920. He found photophoresis and named the effect in 1918.

By 1910 he was already known for a remarkable piece of work: his experimental study, from 1907 on, of Brownian motion in gases, following up the theoretical publications by EINSTEIN and SMOLUCHOWSKI. Later, his work became more and more controversial; but his early work was of high quality, and marked him as a young scientist from whom much could be expected.

## 2. – Antiatomism and a faculty vacancy.

Let us return to Millikan's intellectual formation. The European continental tradition of physics entered his training also, even before his visit to Europe. PUPIN, his teacher, had joined the Columbia faculty after getting a degree under HELMHOLTZ in Berlin, and he had brought with him a point of view about physics. MILLIKAN says that the course on optics and electromagnetism under PUPIN was an eye-opener, and he came to admire and respect him greatly. But he was that much more amazed by Pupin's attitude to atomism. PUPIN had evidently become a convert to the school of energetics that counted OSTWALD and, to a large degree, MACH among its adherents. Thus PUPIN told him once he did not believe in the kinetic theory at all, and he taught thermodynamics in that spirit.

Indeed, the question of whether the atomic and kinetic theories were essential or true was still hotly argued in 1904, when it was a chief subject of debates at the world scientific congress, held as part of the St. Louis Exposition, and even to the end of that first decade of the new century (as recounted, for example, by Mary Jo NYE in *Molecular Reality* [London, 1972]). The Solvay Congress in late 1911, to a large degree, was still concerned with the fundamental, persisting impasses in a physics based on the atomic hypothesis—and critics such as DUHEM were scoffing at that hypothesis as late as 1913.

It is significant that, whereas many a student absorbs the epistemology of his honored teacher, in the case of MILLIKAN nothing of the sort happened.

Despite Pupin's example, he completely resisted the philosophy of science exhibited in energetics and antiatomism. Let us remember what it involved, and what it was that he was not accepting.

OSTWALD, MACH, STALLO, HELM, the whole group of like-minded scientists around the turn of the century hoped to erect science on a purely phenomenological base, without such unnecessary hypotheses as atomism, to cite the frequently given example. Indeed, there was relatively little direct evidence for the reality of atoms and molecules, or of discreteness itself, of the kind which is so natural to us now. None of these people had seen particle tracks in a cloud chamber. They did not yet have the Geiger counter, and had not heard the individual clicks triggered by individual atomic events. They were working with average values, not individual atomic entities, as most physicists and chemists still do.

The best short description of the tenets of this school, and particularly of its most feared proponent, E. MACH, was given by the philosopher of science M. SCHLICK in an obscure but fine essay published in June 1926 at the unveiling of a monument to MACH in Vienna. There he said (*):

« MACH was a physicist, a physiologist, also a psychologist, and his philosophy... arose from the wish to find a principal point of view to which he could hew in any research, one which he would not have to change when going from the field of physics to that of physiology or psychology. Such a firm point of view he reached by going back to what is given before all scientific research, namely, the world of sensations.... Since all our testimony concerning the so-called external world relies only on sensations, MACH held that we can and must take these sensations and complexes of sensations to be the sole content of those testimonies, and, therefore, that there is no need to assume in addition an unknown reality hidden behind the sensations. With that, the existence *der Dinge an Sich* is removed as an unjustified and unnecessary assumption. A body, a physical object, is nothing else than a complex, a more or less firm pattern of sensations, *i.e.* of colors, sounds, sensations of heat, of pressure, etc.».

Atomism—both of chemical and physical matter, as well as of electricity—therefore was considered to be not only unnecessary, but even a rather dangerous, metaphysical hypothesis. The liberation of science from all metaphysical bonds was Mach's life-long ambition, and reached his largest audience through his book *The Science of Mechanics* (1883). He took that mission with utmost seriousness. Therefore he must be seen not only as an ingenious physicist and philosopher, but also as a powerful figure in the politics of academic life (**).

---

(*) M. SCHLICK: *Ernst Mach, der Philosoph*, in *Neue Freie Press* (*Supplement*), (12 June 1926).

(**) A good indication can be found in J. T. BLACKMORE: *Ernst Mach, His Life, Work, and Influence* (San Francisco, 1972), *e.g.*, Chap. 13, 19.

Some of his loyal students or followers saw to it that his message would have strong proponents at universities where they could also participate in the choice of new appointees.

One of these « agents » was A. LAMPA, who will get into our story soon. LAMPA had been a student of MACH and was a physicist himself as well as an idealistic fighter for the reform of education. As P. FRANK puts it in the book on EINSTEIN (p. 135, German edition), « LAMPA saw it as his main aim, his life's chief goal, to propagate Mach's ideas and to find adherents for them ». Together with another Machist on the faculty, PICK, he was operating at the German University in Prague, where MACH himself had been professor for twenty-eight years.

As it happened, in 1910, the year we are particularly concerned with, a faculty vacancy in physics became available, and LAMPA and PICK began to look for a Machist, a man who could be relied upon to carry on physics in that tradition. One of their two chief candidates was JAUMANN of Brno. To obtain Mach's approval, LAMPA wrote to MACH on 9 February 1910: « I need not reassure you that Jaumann's high talent seems to me beyond doubt and that his whole cast of thought is sympathetic. I consider the ideal of theoretical physics to be the purely phenomenological presentation [Darstellung], as lies at hand for example in thermodynamics. JAUMANN proceeds from the wish to build up such a phenomenological presentation for electricity and all that can be connected with it. He therefore rejects the theory of atoms and of electrons ». LAMPA ends by sharing frankly also some worries about JAUMANN, and by announcing his visit to MACH in Vienna in a few weeks.

Evidently, MACH sent his approval speedily, for by a letter of 18 February 1910 LAMPA thanks him for it; all qualms are now laid to rest, and LAMPA will be able to intervene warmly on behalf of JAUMANN (*).

But the selection process went on for many months more, and JAUMANN was not the only candidate. Another was A. EINSTEIN of Zürich, then still regarded by the Machists to be of their persuasion (**), and himself just then in correspondence with MACH, to whom he signs one of his letter, « Ihr Sie verehrender Schüler » (***). As it turned out, it was EINSTEIN who, after many hitches in the negotiations, assumed the chair in Prague in March 1911 (JAUMANN was asked first but declined it)—and when EINSTEIN left the next year, his

(*) Other aspects of the indirect effect of MACH via LAMPA and PICK on the selection of the candidate are discussed in J. T. BLACKMORE: *op. cit.*, Chap. 17.
The Lampa-Mach correspondence from which excerpts are quoted in translation in this paper is in the E. MACH Archive in Freiburg.
(**) Thus A. LAMPA writes to MACH on 1 May 1910: « I believe that relativity theory is the opening of a phenomenological epoch of physics ». See also, G. HOLTON: *Mach, Einstein and the search for reality*, in *Daedalus* (Spring 1968), p. 636-673.
(***) *Ibid.*, p. 644.

successor was P. FRANK, the physicist and philosopher, who was also fully supported by MACH, LAMPA and PICK.

### 3. – Millikan's commitment.

Let us now leave all these Central-Europeans to their philosophies and academic intrigues, and turn again to MILLIKAN, who of course knows nothing about all these events and currents. We find here a very different atmosphere. MILLIKAN has a simple, unsophisticated, straightforward epistemology of his own—he is very clear about it—namely « concrete visualization » and direct « explanation ». The words « concrete visualization » recur in his writings, and are meant in the most practical, unadorned way imaginable. When MILLIKAN writes about the electron in those early years, he does not think of a particle that has magnetic moment, or angular momentum, or wavelength, or intrinsic self-energy, or any of the properties that we now think of as being associated with and defining the electron. He thinks of the electron as a discrete quantity of electricity which he can *see*. « *He who has seen that experiment* », he writes in his *Autobiography* (p. 80) about the oil drop experiment, using italics for the sentence, « *and hundred of investigators have observed it, have in effect SEEN the electron* ». The word « seen » is indeed printed in capital letters. And two pages later: « *But the electron itself, which man has measured..., is neither an uncertainty nor a hypothesis. IT IS A NEW EXPERIMENTAL FACT that this generation in which we live has for the first time seen, but which anyone who wills may henceforth see* » (p. 82. Italics in original).

This direct and quite workable idea about the electron is supported in his letters, in his notebook, in the anthropomorphic way be perceives the experimental results. He writes that when the small oil droplet was « moving upward [in the electric field, against the gravitational pull] with the smallest speed that it could take on, I could be certain that just one isolated electron was sitting on its back. The whole apparatus then represented a device for catching and essentially seeing an individual electron riding on a drop of oil » (p. 83). Sometimes, while watching a charged oil droplet held in the electric field, it is observed to change suddenly from slow to faster motion, owing to the fact that the droplet has encountered a charged molecule (ion) in the air. The *discontinuity* was the great new fact, and the image that helped interpret it was directly at hand: « ...one single electron jumped upon the drop. Indeed, we could actually see the exact instant at which it jumped on or off » (p. 83) (*).

(*) The power of visualization distinguishes a number of scientists. For example, RUTHERFORD made no bones about this skill and its usefulness. I have analyzed the role of visualization in the « *Gedanken* »-experiments of Einstein, in *On Trying to Understand Scientific Genius*, *American Scholar*, Vol. **41** (Winter, 1971-72), p. 95.

At about the same time, J. PERRIN in France was battling on behalf of the atomicity of matter with all the same strength of preconception and consequent focussing of vision which characterized Millikan's determination to demonstrate the atomicity of electric charge. As NYE writes, « Perrin's primary goal from the very beginning of his scientific career was to prove the reality of the invisible atom, to eliminate as " puerile anthropomorphism " those structures which seemed logically necessary to many others.... One student wrote of Perrin... " He ' sees ' atoms—there is no doubt at all—as Saint Thomas saw seraphim " » (*op. cit.*, p. 65).

The very way in which MILLIKAN got into his research on the charge of the electron illustrates three related factors: his capability to look with fresh, clear eyes at what is going on; his powers of visualization as an aid in drawing the conclusions; and behind all these, almost unconfessed and certainly unanalyzed, a preconceived theory about electricity which gave him eyes with which to look and interpret. It is an important part of our story to understand this interrelationship, hence we shall analyze it within the limits of the brief space here available.

## 4. – On the road to the electronic charge.

It was a series of accidents which set him on his way, and he describes them frankly. Soon after MILLIKAN joined MICHELSON in Chicago in 1896, he was asked by MICHELSON to be responsible for the weekly seminars in physics. At one of these he presented a review of J. J. Thomson's great paper of 1897 « which put together, in matchless manner, the evidence for the view that the "cathode rays" consist not of ether waves, as LENARD and the Germans were maintaining, but rather material particles carrying electric charges, each particle possessing a mass of about a thousandth of that of the lightest known atom... [This paper] impressed me greatly and started me on the researches which have been my life work » (*Autobiography*, p. 59).

As it turned out, for the next 10 years Millikan's work, largely concerned with « the laws of detachment of electrons from metals (1) by light and (2) by electric fields » (p. 64), was not going well. He uses such words as « this apparently fruitless work » (p. 63) and « my own research failures » (p. 69) in describing it. Much of his time and energy was going into writing textbooks and teaching. He realized he still had to establish himself as a research physicist. By 1908 he was already 40 years old. I take it he was in some depression about his chances as a research scientist; but in an act of will, he « kissed textbook writing good-bye... and [while aware of the risk of further failures] started intensively into the new problem that is the subject of this chapter » (p. 69)— « My Oildrop Venture (e) ».

There were three obvious merits in chosing this particular subject. One was that « everyone was interested in the magnitude of the charge of the electron » (p. 69), then known only to very low accuracy. Another was that the experimental method to use seemed quite obvious and rather simple—though, as we shall see later, this turned out not to be the case. And lastly, the theoretical underpinnings or epistemological assumptions seemed also quite clear to MILLIKAN. Thus MILLIKAN recalled later: « ...being quite certain that the problem of the value of the electric charge (Franklin's fundamental atom of electricity—apparently invariant and indivisible—the assumed unit building block of the electrical universe) was of fundamental importance, I started into it » (p. 72).

Not for him all the turmoil and bitter debate raging in Europe concerning the « reality » of molecules, atoms and electrons! The electronic charge plainly existed; it was of « fundamental importance » to find its value; and—as the remarkably off-hand parenthesis in the last-quoted passages indicates—its theoretician and conceptual architect was none other than that great American folk hero, statesman and scientist, the sensible B. FRANKLIN. In Millikan's writings, FRANKLIN is always referred to as the originator of the conception of a granular structure and material reality of the « electrical particle or atom » (*The Electron* (Chicago, Ill., 1917, 1924), p. 15). He is the father of the subject itself: « There are no electrical theories of any kind which go back of Benjamin FRANKLIN (1750) » (*ibid.*, p. 11), and the result of all modern research has merely been « to bring us back very close to where FRANKLIN was in 1750, with the single difference that our modern electron theory rests upon a mass of very direct and convincing evidence » (*ibid.*, p. 24). In 1948, looking back on the 50-year celebration of what he called J. J. Thomson's « unambiguous establishment of the electron theory of matter », MILLIKAN went so far as to urge that since FRANKLIN had started his experiments in 1747, one should have been also celebrating « Franklin's Discovery of the Electron » (*).

In the *Autobiography*, MILLIKAN does not even bother to summarize, as he had in *The Electron*, the great rival theories of electricity, namely those based on the thematic concept of the continuum rather than on the thematic concept of atomism. Maxwell's theory of electricity, for example, strongly suggested the notion that electricity was a continuous displacement, a motion within the electromagnetic ether. To be sure, MAXWELL noted in 1873 in his *Electricity and Magnetism* (Vol. I, 375ff) that electrolysis seems to invite conceptualizing a definite value for an electric charge: « For convenience in description we may call this constant molecular charge (revealed by Faraday's experiments) one molecule of electricity ». But this phrase, « gross as it is and out of harmony with the rest of this treatise », should not mislead us to ascribe reality to granules of electricity. « ...The theory of molecular charges may serve as a method

(*) In an article using this phrase as the title, in *Amer. Journ. Phys.*, **16**, 319 (1948).

by which we may remember a good many facts about electrolysis [with economy of thought]. It is extremely improbable, however, that when we come to understand the true nature of electrolysis we shall retain in any form the theory of molecular charges, for then we shall have obtained a secure basis on which to form a true theory of electric current and so become independent on these provisional hypotheses. »

This view that the atomicity of electricity was at most a heuristic speculation was fairly widespread in England and on the continent, so that A. SCHUSTER could write of the period around 1880-1890: « The separate existence of a detached atom of electricity never occurred to me as possible, and if it had, and I had openly expressed such heterodox opinions, I should hardly have been considered a serious physicist, for the limits to allowable heterodoxy in science are soon reached » (*). In 1897, Lord KELVIN thought careful consideration should be given to the idea that « electricity is a continuous homogeneous liquid » (**); and as late as 1900, as he later confessed, M. PLANCK did not fully believe in the electron hypothesis.

In retrospect, it is all too easy to see evidence that should have clinched the argument in favor of the particle theory of electric charge: J. J. Thomson's measurement of the constant charge-to-mass ratio of cathode rays; Rutherford's measurement of the charge on $\alpha$-particles; the charge on droplets of various liquids by J. J. THOMSON and his student TOWNSEND; and H. A. Wilson's measurement of the electric charge on cloud droplets. But where the error margins were not admittedly enormous, the methods all shared with Faraday's determination of the unit charge exchanged in electrolysis one fatal flaw: they represented the *average* of a determination involving very many charges at the same time. At best they would be indirect measurements; at worst they would be the statistical mean values of a distribution of unknown shape. Nobody before MILLIKAN had measured the charge of an individual object and found it to be equal to one or two or any small multiple number of units of electricity.

## 5. – Some happy accidents.

Nor indeed did MILLIKAN himself have the slightest hope to do this when he set out to measure what he was certain to find: *the* value of *the* electronic charge. When MILLIKAN began this work with his student L. BEGEMAN, they used H. A. Wilson's method. Clouds of droplets were produced in an expansion

---

(*) A. SCHUSTER: *The Progress of Physics during 33 Years 1875-1908* (Cambridge, 1911), p. 59. I thank Mr. B. COLLIER for having drawn my attention to this quotation.
(**) *Nature*, **56**, 84 (1897).

cloud chamber between the parallel plates of a charged condenser. The slowly falling top layers of the clouds, containing the smallest droplets, were observed—one set falling with, another set falling without the aid of an electric field set up across the condenser. Assuming Stokes' law to hold for the droplets, assuming each of the droplets to have formed on a singly charged ion and not to shrink due to evaporation, and assuming further that the different clouds were all similarly formed, one could quickly obtain an equation for the charge of a unit of electricity in terms of the observables (speeds of fall, field strength, densities, viscosity of the gas).

It was a method full of unsatisfactory features, both theoretically and practically. But MILLIKAN thought first only of making a rather small procedural change. Using an exceptionally large (10 000 V) battery he had left over from other experiments, and letting it set up its stronger electric field in *opposition* to the effect of gravity, he expected to hold the top surface of the cloud in balance and so would study its rate of evaporation—thereby improving the calculations by allowing for that source of error.

At the moment when MILLIKAN threw on the electric field by turning a switch, something happened to this man that oriented and gathered his immense energy, his optimism, his talent as an observer and researcher, his ability to use students, his instinct for smelling out the important and basic problems, his great eye for the accident that opens an unsuspected door. He chanced on an event—or rather a sequence of accidents, described by MILLIKAN in his *Autobiography* in the plainest way—which « *made it possible for the first time to make all the measurements on one and the same individual droplet, and ... made it possible to examine the attracting or repelling properties of an individual isolated electron...* » (p. 72; italics in original). When he turned the switch, the cloud dissipated instantaneously and completely. The strong field, acting on the variously (not equally) charged droplets, cleared them out, and thus there was no top surface of the cloud left on which to make one's measurements.

It « seemed at first to spoil my experiment. But when I repeated the test, I saw at once that I had something before me of much more importance than the top surface.... For repeated tests showed that whenever a cloud was thus dispersed by my powerful field, *a few individual droplets would remain in view* » (p. 73)—those which happened to have just the right charge relative to their mass to allow balancing their weight in the electric field.

It was the first time someone had looked at the individual, charged droplet. Indeed, MILLIKAN had discovered a very sensitive balance for holding an object of the order of $10^{-13}$ to $10^{-15}$ grams in view. There was the keyhole for opening up a new experimental field, the more so as serendipity struck a second time: « I chanced to observe... on several occasions on which I had failed to screen off the rays from the radium [for ionizing the air before producing the cloud] that now and then one of [the balanced drops] would suddenly change its charge and begin to move up or down in the field.... This opened the possibility of measuring, with certainty, not merely the charges on individual droplets as I had been doing, but the charge carried by a single atmospheric ion » (*The Electron*, p. 65). All the rest followed fairly obviously, from the replacement of water by a liquid with less vapor pressure, to the long and exhausting work to remove or narrow the sources of uncertainty (*e.g.*, by the modification of Stokes' law for small droplets). But even in the first months, in the Summer of 1909, he saw that the « *charges actually always came out, easily within the limits of error of my stop-watch measurements*, 1, 2, 3 and 4, or some other exact multiple of the smallest charge on a [water] droplet that I ever obtained. *Here, then, was the first definite, sharp, unambiguous proof that electricity was definitely unitary in structure...* » (*Autobiography*, p. 74, italics in original).

His first results, reported at the B.A.A.S. in September 1909, aroused considerable interest. The first major paper (*Phil. Mag.*, **19**, February 1910, pp. 209ff) is still worth reading. He knows what he is doing. Knowledge of « the ultimate, or elementary electric charge... makes possible a determination of the absolute values of all atomic and molecular weights, the absolute number of molecules in a given weight of any substance », etc. It is one of the most important constants, and the atomicity of electric charge gives the best evidence for the atomicity of matter. Conversely, doubts about the former would support doubts about the latter. No such doubts are of course evident in Millikan's case. On the contrary, his belief in a corpuscular theory of electricity (*à la* FRANKLIN) is so firm in his papers, and other writings, that one can see its function as a thematic hypothesis, underlying and informing his research work.

He frankly distinguishes between his data, labeling those for the « best » drops with 3 stars, those which are « very good » with 2 stars, others with 1 star, and noting in one particular case in his confident and straightforward way: « I have discarded one uncertain and unduplicated observation, apparently upon a singly-charged drop, which gave a value of the charge of the drop some 30 % lower than the final value of *e* » (*).

Through the kindness of Dr. J. GOODSTEIN, Institute Archivist at California Institute of Technology, I have been able to see and analyze some pages of Millikan's laboratory notebooks, starting with that for October 1911. (His measurements continued for some years.) It is fascinating to see how he handles his data in a way which in retrospect seems uncanny, and for which one has to have the kind of talent that, unhappily, is hardly trainable. One also sees

(*) 1910 paper, *op. cit.*, p. 220.
Professor P. A. M. DIRAC (private communication after the lecture, and letter of 11 October 1972; see also the record of Prof. Dirac's remarks in this volume) mentioned to me that he noted with interest the fact that Millikan's measurement seems in this case to have been made on the smallest drop. As we shall note later, EHRENHAFT also obtained his smallest charge values on the smallest particles. See also Professor Dirac's discussion of Millikan and Ehrenhaft's work in the third of his lectures at Varenna (this volume).

his exuberance in his comments on specific runs with a given droplet, such as « Beauty. Publish this surely. Beautiful », or alternatively « Error high, will not use » or « Might omit... This is one of the low ones ».

## 6. – Finding the keyhole.

Another kind of omission, one that may possibly have had a good deal to do with the passion infusing Ehrenhaft's response to Millikan's work, occurred in Millikan's February 1910 paper. There, MILLIKAN summarized the results of others and accepted the values for $e$ obtained by workers like PLANCK, RUTHERFORD and GEIGER. But he also specifically rejected the values obtained, by various methods, of four others—and, among them, EHRENHAFT (*Phys. Zeits.*, 1 May 1909).

As MILLIKAN pointed out, the limitation in Ehrenhaft's method was that it did not allow all the measurements to be made on the same charged droplet or fragment. EHRENHAFT in fact had at that point come to about the same stage of work, except that he was still measuring the motion of one particle in the gravitational field, and another particle for motion in the presence of the electric field. Moreover, EHRENHAFT did not make a correction to Stokes' law in his calculation, despite the very small size of the particles used.

One ironic fact is that the value for $e$ obtained by EHRENHAFT in 1909, which MILLIKAN rejected in 1910, was close to Millikan's value of $4.891 \cdot 10^{-10}$ e.s.u. Another is that, at first glance, Millikan's equipment and procedure must have seemed in many ways more primitive than Ehrenhaft's. Millikan's simple apparatus was put together in a rather home-spun way—thus, the atomizer was originally a perfume sprayer bought at a drugstore, and the telescope was an ordinary short-focus tube set up two feet from the 1.6 cm gap in the horizontal, big (22 cm diameter) air condenser. Ehrenhaft's equipment seemed much more refined and sophisticated, built around the newly perfected ultra-microscope (SIEDENTOPF and ZIGSMONDY had caused a sensation in 1903 with the description of the device, which permitted observing objects down to a limit about 500 times below the resolving power of an ordinary microscope, and EHRENHAFT himself, from 1905 on, had perfected its use in the observation of Brownian movement). Ehrenhaft's condenser was about an order of magnitude smaller than Millikan's in each dimension, and the range of sizes of charged objects he used was many times more extensive. Compared with EHRENHAFT, MILLIKAN might have appeared to be looking at the world of charged particles through a curiously chosen, primitive keyhole.

But that was part of Millikan's strength. As he stressed later in his Nobel Prize acceptance speech, the particular dimensions of the apparatus and the voltage of the battery were « the element which turned possible failure into success. Indeed, nature here was very kind. She left only a narrow range of field strengths within which such experiments as these are at all possible. They demand that the droplets be large enough so that the minute dancing movements, the 'Brownian movements', are nearly negligible, that the droplets be round and homogeneous, light and nonevaporable, that the distance of fall be long enough to make the timing accurate, and that the field be strong enough to more than balance gravity by its upward pull on a drop carrying but one or two electrons. Scarcely any other combination of dimensions, field strengths and materials could have yielded the results obtained. Had the electronic charge been one-tenth its actual size, or the sparking potential in air a tenth of what it actually is, no such experimental facts as are here presented would ever have been seen » (*).

To find the right keyhole is of course the essence of genius—Nature seldom offers a convenient, wide-open door or window. Thus GALILEO fastens on the falling or rolling ball as a key to dynamics, and FERMI used the slow neutron. But, needless to say, such a keyhole is only rarely found, and often the approach has to be quite the opposite. EHRENHAFT came to express an additional, epistemological disagreement. To him, the fact that Millikan's measurements were restricted to only a certain region, with droplets which are relatively large, rather than with arbitrarily small droplets, was a detrimental feature. He seemed to prefer to base science on a platform where sensation would be equally validly obtained all along the range, where one did not have to zero in on a relatively small sanctuary, protected from the rest of the world where other effects and laws take over so strongly as to make experimental observation too complex. As to the need to apply a correction term to Stokes' law for smaller droplets or other objects, EHRENHAFT resisted it strongly on the grounds that this approach tends to build one's assumptions concerning the desired outcome into the method of obtaining the correction terms. Nature should be allowed to communicate directly with us through our unmediated sensations.

But precisely in this experiment, recognition and removal of relatively small sources of « error » are needed to uncover a grand simplicity, if indeed one seeks that; conversely, disinclination to treat some observable effects as small but significant errors will effectively cost finding the existence of such a grand simplicity. The point is this: As R. BÄR (**) summarized it in 1922, assume a single droplet has successively the electric charges $n_1 e$, $n_2 e$, $n_3 e$, where $n$ is an integral multiple of the fundamental unit charge $e$. The applied potential difference needed to suspend it against the pull of gravity, given by $mg = neV/d$, will then have successively the values $V_1$, $V_2$, $V_3$. But

$$n_1 : n_2 : n_3 = \frac{1}{V_1} : \frac{1}{V_2} : \frac{1}{V_3} .$$

(*) *Autobiography, op. cit.*, p. 78-79.
(**) *Naturwissenschaften*, 14 April 1922.

For example, if by experiment

$$V_1 = 47.5\ \text{V and}\ V_2 = 71.1\ \text{V}, \qquad n_1 : n_2 = 1/47.5 : 1/71.1,$$

or 3:2 (to 3 parts in 1500). Here, the hypothesis that droplets are charged in whole multiples of one basic charge is strongly supported. But if errors from one or many causes produce only a difference of 1% in the relative values of $V_1$ and $V_2$—for example if $V_2$ remained 71.1 volts but $V_1$ was 47.0 V—then

$$n_1 : n_2 = 1/47.0 : 1/71.1 = 711 : 470$$

(to about 1 part in 3000)—not a convincing proof of the quantization of charges. In short, the experimental proof of the existence of the electron by this method, used by both MILLIKAN and EHRENHAFT, is quite sensitive to the selection and treatment of data (*).

## 7. – Turning from the electron.

It is now important to pinpoint the time when EHRENHAFT turned from his original, expressed belief in atomism and electrons. As was said before, by 1907 he was well known for his work on Brownian movement, and his publications at that time show he was convinced of the « natural truth » of the kinetic theory, with all the atomism it was built upon. His first serious publication (an expansion of a paper read at a meeting of 4 March 1909) on the question of the existence of a smallest, quantized electric charge is in *Phys. Zeits.*, **10** (1909), 308-310, and is still entirely along the lines of current work on the charge of the electron; he there refers to work by THOMSON, RUTHERFORD, WILSON, DE BROGLIE—and to the preliminary first report by MILLIKAN. He derives the value of $4.6 \cdot 10^{-10}$ e.s.u. for $e$ which he calls the « elektrische Elementarquantum ». The paper in which he introduced the use of the ultramicroscope, permitting the viewing of very small particles, has in fact the title « Eine Methode zur Bestimmung des elektrischen Elementarquantums ».

But the switch has evidently been thrown to the opposite pole by 23 May 1910, the date given for the receipt of Ehrenhaft's next major paper (**). It has the remarkable title (in translation) « Concerning a new method for measuring the amount of electricity on single particles [*e.g.* vapor of metals] whose

(*) There is no space here to treat other, similar points of dispute, *e.g.* the method for obtaining the constant $A$ in the correction term for Stokes' law of droplets of small radius.

(**) As published in *Phys. Zeits.*, **11**, 619 (1910). A very preliminary report on his « improved method » of observation was summarized in the *Wiener akademischer Anzeiger*, No. 10, 21 April 1910.

charges are considerably below the charge of the electron and also appear to deviate from multiples of the latter » (*). He acknowledges that he himself had recently worked to find the charge on the electron (Elektronenladung). But now he draws attention to the wide spread or variation between the values obtained for it by different methods—1 to $6 \cdot 10^{-10}$ e.s.u—as well as by different observers using the same method. He notes that if one wants to avoid a style of science which piles up hypotheses and corrections—« Wenn mann sich da von weiteren Hypothesen und Korrekturen fernhalten will » (p. 619)—one is led to the recognition that these apparent variations of the elementary quantum or charge of the electron are possibly *grounded in Nature itself*. The interpretation of the experiments then has to be modified correspondingly.

Using a variety of ultra-microscopic vapor particles and colloidal fragments of metals (platinum, silver, gold) and fog prepared from phosphorus, he obtains the result indicated in the title of the paper. Moreover, he recognizes that if he used the corrections to Stokes' law for small objects (*e.g.*, by E. CUNNINGHAM: *Proc. Roy. Soc.*, **83**, 593 (1910)), the smallest charges he had determined would become smaller still. If charges are multiples of a smallest quantum, it would have to be smaller than $1 \cdot 10^{-10}$ e.s.u.; « an indivisible atom of electricity of the order of $1 \cdot 10^{-10}$ e.s.u. seems not to have to be assumed as existing in Nature » (p. 630).

There are two striking omissions. MILLIKAN, whose paper of February 1910 had recently dismissed Ehrenhaft's first published value for $e$, is not mentioned at all; nor is the consequence of Ehrenhaft's new thinking for the atomic hypotheses of matter itself. The first of these omissions is amply made up for in Ehrenhaft's next publication in the same volume of the *Physikalische Zeitschrift* (**). His conclusion concerning the divisibility of $e$ is the same as before (with extra attention drawn to Millikan's own report of one measurement that was 30% less than $e$). EHRENHAFT announces that the existence of charges with values less than $e$, for which later the word *subelectrons* was introduced, has herewith been ascertained as certain beyond doubt—« zweifellos sichergestellt » (p. 946).

For one only need look at what Nature itself makes directly accessible to the assiduous experimenter. Thus his student PRZIBAM has made 1000 measurements of charges on fog particles obtained by blowing moist air over

---

(*) *Über eine neue Methode zur Messung von Elektrizitätsmengen an Einzelteilchen, deren Ladungen die Ladung des Elektrons erheblich unterschreiten und auch von dessen Vielfachen abzuweichen scheinen.*

(**) Vol. **11** (1910), p. 940-952. The paper has almost the same title as the previously cited one, as well as much of the same content, being evidently a revised version of it and of a talk with the same title listed for May 1910 in *Wiener Berichte*, **119**, 815 (1910). The title of the new paper is *Über eine neue Methode zur Messung von Elektrizitätsmengen, die kleiner zu sein scheinen als die Ladung des einwertigen Wasserstoffions oder Elektrons und von dessen Vielfachen abweichen.*

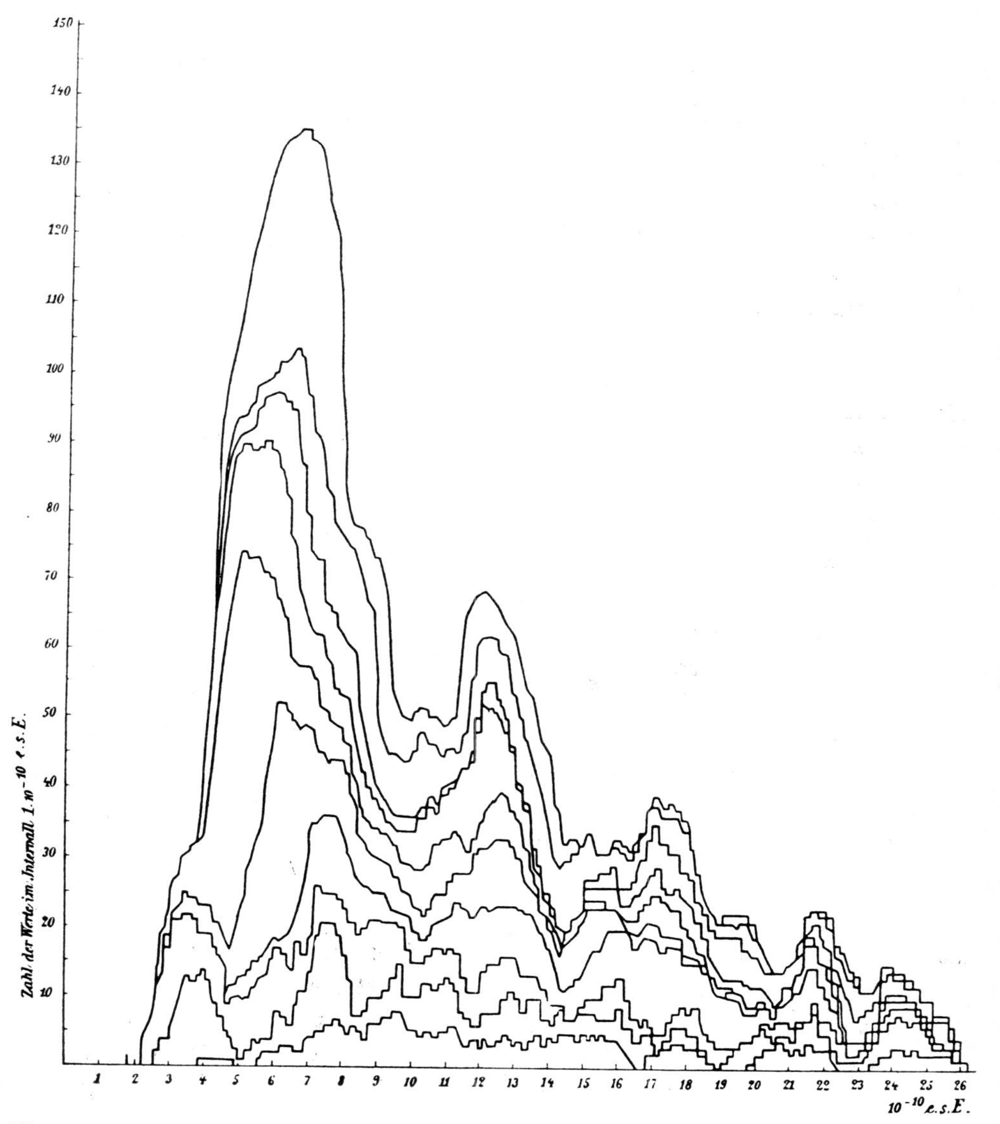


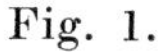
Fig. 1.

white phosphorus (Fig. 1) (*). The abscissa shows observed charges in units of $10^{-10}$ e.s.u., the ordinate the number of values that appeared within a $1 \cdot 10^{-10}$ e.s.u. interval. The 10 superposed curves are the *cumulative* results emerging from the first 100, 200, ..., 1000 observations. Thus the first, lowest curve still may allow one to see a grouping around a definite lowest value of about $4.0 \cdot 10^{-10}$ e.s.u. But the fifth curve already shows a « strong » peak

(*) Reproduced from *ibid.*, p. 947, Fig. 2.

around $7 \cdot 10^{-10}$ e.s.u., a lesser one at $12.1 \cdot 10^{-10}$ e.s.u., a still smaller one at $17.0 \cdot 10^{-10}$ e.s.u. There are—« so far inexplicable »—variations in the placement of the maxima as more data accumulate. But one thing should be obvious: there are values which fall much below the hitherto accepted elementary quantum. « The author », he concludes, « is unable to explain these experiments by adhering to the fundamental hypothesis of the electron theory » (p. 949).

The record of the Academy's discussion period (*ibid.*, pp. 949-952) shows that then—as for many years to come—Ehrenhaft's audience was distinguished, puzzled and unable to produce disproof or even modification of the presentation. PLANCK, BORN, SOMMERFELD, KAUFMANN, BESTELMEYER, SIEDENTOPF and others tried to put their finger on a great variety of possible sources of error in the work, from the assumed shape and density of the metal and fog particles to his replacement of Stokes' law by « empirical formulas ».

## 8. – Challenge to atomism.

In subsequent years many details were added, but the main outlines of the controversy remained as they first appeared in 1910. For example, in Ehrenhaft's paper of 1914, entitled *Über die Quanten der Elektrizität* (*Ann. der Phys.*, **44**, 657 (1914)) we begin to see more explicitly the development of an idea that previously had been only implicit: the use of these experiments as an attack on the credibility or necessity of atomism. Referring to BOLTZMANN, Mach's great antagonist, EHRENHAFT writes at the very beginning of the paper:

« In recent years, more than ever before, the atomistic theories of matter, electricity and radiation have gained ground in physics. A few years ago, Ludwig BOLTZMANN considered it necessary to stress particularly the 'indispensability of atomism' in natural science. Since then these views rapidly changed in this direction. All who are active in physics as an exact natural science are convinced of the great heuristic value of these theories; the greater the successes of such theories the more easily it is understandable that the hypotheses based on this point of view gain such vivid reality among their creators and adherents that often what is only an anthropomorphic concept appears absolutely as a natural truth.

« Now if a theory is to be more than mere speculation, if it tries to yield conclusions about the real existence of conceptions arising from the theory, as often happens these days, the foundations on which it is built must be able to stand up under the critical trial of experiment in every respect. The following study takes on the task—starting from *direct facts*—of submitting to examination—up to the farthest bounds now attainable—one part of the above hypothesis, namely the atomicity of electricity... ».

The author then previews the conclusions that the question is in principle open whether the electron should be thought of as free or as a « quantitative condition bound up with matter ». The last phrase may hint at some theory of indefinitely divisible electric charge, with the absolute amount present on an object being a function of the size, somewhat like the case of a condenser (*). If there *are* quanta, he now believed they « should be sought at most on the order of $10^{-11}$ e.s.u. ». With this, one can turn the tables on MILLIKAN, for now the puzzle that needs explanation is why in Millikan's and similar experiments a specific value of $e$ *is* found (p. 699):

« The higher values of this quantum which are often recurringly found on larger particles under certain conditions and with the same method of production will then perhaps have to be considered as stable spatial distributions of equilibrium of these ' subelectrons ' which arise in certain circumstances. »

As this controversy continued, its longevity became astonishing, as did other features. Thus almost all the arguments in favor of subelectrons came only from EHRENHAFT and his students and colleagues; and while Millikan's droplets were usually larger than $2.5 \cdot 10^{-5}$ cm in diameter, EHRENHAFT and students used smaller and smaller particles (to $2 \cdot 10^{-6}$ cm) and obtained smaller and smaller values of any unit charge that might exist (to $2 \cdot 10^{-3}$ of Millikan's). They deflected all the objections of MILLIKAN and others: that for such very small particles Stokes' law correction becomes indefinite and Brownian movement interferes with the accuracy of measurement; that dust, liquid containing dust, or occluded gases in metal particles may have falsified some measurements; and that particles of metal vapors may have a density very different from that of the solid metal, not to speak of the possibility of leakage of charge at the sharp corners during the measurement, etc. Ehrenhaft's school skillfully, doggedly and at length countered each attempt with yet another modification of their technique, giving similar results.

It never came to a direct disproof. In the 1916 edition of *The Theory of the Electron*, H. A. LORENTZ still had to confess: « The question cannot be said to be wholly elucidated ». In 1922, the review of the case by R. BÄR (*op. cit.*, p. 327), concluded in a way that reminds the historian of science of the tail-end of a number of cases, including D. C. Miller's repeated claim to have found experimental evidence for an ether wind: « Most physicists did not believe in the reality of the subelectrons, but the experiments left at the very least an uncomfortable feeling... ». EHRENHAFT himself, undaunted, continued to publish on subelectrons into the 1940's.

(*) See also *ibid.*, p. 644: « Smallest quantities of electricity are to be expected in all likelihood on bodies of smallest capacity ». And p. 699: « [The studies show] a decrease of the quantum with radius [of particle] ».

## 9. – The world of neglected dimensions.

In the final few minutes here I can only touch on one of the reasons for the extent and considerable interest in the debate at the time. From the beginning of the 20th century physicists and chemists found themselves more and more interested in the study of small, colloidal particles. Thus F. EXNER of Vienna, apparently one of Ehrenhaft's early mentors and thesis supervisors, published observations on the size and motion of colloidal particles in 1900. After the introduction of the ultra-microscope and the theories of Brownian movement, the « colloidal state »—the dispersed state of matter where particles have diameters between $10^{-4}$ and $10^{-7}$ cm—was seen as an exciting new frontier for pure and applied science. In 1908 W. OSTWALD added a chapter on colloid chemistry to the new edition of his influential textbook, *Allgemeine Chemie.* His son Wofgang, editor of the new *Zeitschrift für Chemie und Industrie der Colloide* from 1907, published a text on colloid chemistry in 1909, and another introduction to colloid chemistry in 1915, « with special consideration for its application », with the revealing though theatrical title *The World of Neglected Dimensions.* Under the initiative of P. EHRLICH, an Institute for Colloid Research was founded in Frankfurt am Main in 1919, specifically « to provide a link between pure science and praxis. Research on colloids is a branch of physical chemistry and provides a bridge between the world of organisms and inorganic matter ». It promises much for medical-biological research on one side, and on the other it should help « the colloid industry, which draws its raw material from organic matter such as rubber, fibers, etc. » (*).

Then, too, Jean Perrin's beautiful work, from 1904 to 1911, drew much attention to the « colloidal » world. And EINSTEIN himself, we must remember, thought it worth trying to reach readers outside « pure » physics, by rendering his work on the Brownian movement of small particles in two articles (1907 and 1908) in the *Zeitschrift für Elektrochemie.*

All these trends, while not directly intersecting with the problem of the existence of subelectrons, did help to keep prominent attention focussed on the controversy and on the techniques of observation developed in it. But by 1923 the last doubts about the correctness of Millikan's work were banished virtually everywhere; for in that year, MILLIKAN was awarded the Nobel Prize, specifically « for his researches upon the elementary charge of electricity » as well as for his researches on the photoelectric effect.

---

(*) Announcement in *Phys. Zeits.*, **20**, 120 (1919). Cf. also an interesting article by R. MCCORMMACH on this movement in *Historical Studies in Physical Science* (in press).

## 10. – A battle of two worlds.

In his Nobel Prize acceptance speech, MILLIKAN put an end to his side of the debate with a careful review of his work. It was published in 1925—MILLIKAN had delayed the long journey to Stockholm to deliver the speech before the royal assemblage because, as he wrote, he was busy with his new research and administrative work. A year later EHRENHAFT also gave an address that signalled effectively the end of his side of the controversy. As it happened, that address was part of a ceremony in a public park, on a Saturday in Vienna.

It was at the occasion we have mentioned before—the unveiling of a bust in honor of E. MACH, to commemorate the 10th anniversary of Mach's death. M. Schlick's eulogy, noted above, was delivered there. There was also a message, sent by EINSTEIN, who had admired MACH and had once specially sought him out during a visit to Vienna in 1911. In fact, the visit had been arranged by EHRENHAFT, whose friendship EINSTEIN enjoyed in those early days, and who was a genial host with whom EINSTEIN would stay on passing through Vienna (*). By 1926, of course, EINSTEIN had somewhat different feelings about Mach's epistemology (though always and still respectful of Mach's great intellectual honesty and incorruptible skepticism), and the message was brief and distant.

Felix Ehrenhaft's presentation (**) was also brief but more revealing. Perhaps for the first time, some of the pieces of Ehrenhaft's motivation in his long fight against the atom of electricity came out into the open, and could be fitted together. Like himself, EHRENHAFT saw MACH as a lonely fighter. Even the bust of MACH, which the authorities did not want in the arcade of the university building, stood there « alone and isolated ». In his time, MACH (said EHRENHAFT—quite wrongly) « remained not understood, and had so few followers, and those not among physicists... ».

« ... I only want to draw attention to this: the great difference between MACH and most physicists arises from the fact that through the further development of physics each of the two opposing views shows itself to be ever more fundamental, ever more contrary and unbridgeable, like two professions of faith. MACH [appears] as an advocate of the much more modest, phenomenological point of view which finds satisfaction merely with the description of the phenomena and despairs of other possibilities—and the others are advocates of views which, through statistical methods and speculative discussions concerning the constitution of matter, are reflected in atomism, and who believe themselves able to get down to the true Being of things. »

(*) Cf. P. FRANK: *Einstein* (New York, N. Y., 1947), p. 175.

(**) *Ernst Mach's Stellung im wissenschaftlichen Leben*, in *Neue Freie Presse* (Vienna), Supplement, 12 June 1926, p. 12.

After citing an approving quotation about MACH from P. DUHEM, EHRENHAFT ended with a Wagnerian crescendo:

« MACH had the courage to set himself with mighty arguments against the current of the atomistic Weltanschauung that was sweeping along almost all others—the very same atomistics which, in the smallest, supposedly indivisible constituents of matter and, recently, also of electricity, believes to have attained the magic keys for opening at last all doors of natural knowledge.

« But the world follows a remarkable development. On the one hand, daring researchers storm further into the realm of atomistics, undaunted by such powerful thinkers as Mach; on the other hand one must admit that the great man whom we celebrate today may be victorious in the end.

« Who dares to render judgment in this battle of two Worlds? » (« *Wer wagt es, in diesem Kampfe zweier Welten das Urteil zu fällen?* »)

EHRENHAFT had indeed put his finger on the right point. Whatever else the controversy had been about, it was also about two sets of thematically antithetical positions: atomism as the basic explanatory tool in electrical phenomena *vs.* the continuum, and an un–self-conscious methodological pragmatism *vs.* a conscious, ideological phenomenology.

## 11. – Postscript.

Evidently, at some point after his early success in a physics based on atomism, EHRENHAFT had been converted to Mach's antiatomism. We saw the first indication in the paper EHRENHAFT had sent to the *Physikalische Zeitschrift* on 23 May 1910. But to switch from one theme to its opposite is rare indeed in the history of science. One naturally wonders what, if anything, helped push EHRENHAFT to this point. Perhaps the majestic MACH himself (who, though largely bedridden since 1900, remained a great presence in Vienna until 1913) exerted his influence personally. Perhaps the rebuff by MILLIKAN (published in February 1910) made EHRENHAFT more susceptible to enticement from the other camp.

We do not know, and perhaps will never know. But there is another curious letter in the Mach-Lampa correspondence that falls into the critical period and that may possibly contain a clue. It is from A. LAMPA in Prague to MACH, dated 1 May 1910.

We will recall that the search for a candidate for the Prague professorship has not yet been concluded, though two candidates had already been « cleared » with MACH. However, here LAMPA does not speak directly about that matter.

He first has to tell MACH about an attack on the philosophy they both share, from the pen of M. PLANCK—the last major physicist who still dared to attack MACH openly, though MACH and his circle saw themselves to the end as a little, beleaguered group. PLANCK, LAMPA warns, has published a book, *Acht Vorlesungen über theoretische Physik* (Leipzig, 1910), « in which he maintains *in extenso* the views of his which you have been fighting against ». But PLANCK has embroiled himself hopelessly in contradictions; hence « reading it will give you much pleasure ». While interesting physically, the book is epistemologically « childish ».

Then LAMPA turns to the results of his recent trip to Vienna. Perhaps this was the occasion announced in his earlier letter (of 9 February 1910), where he had written to MACH: « I look forward with pleasure to be able to greet you personally in a few weeks, and to report to you then on the further developments of the case [the pending physics appointment] ». At any rate, he enters on this track suddenly, without preliminaries, as if it was familiar territory to both—and the subject is Ehrenhaft's results that were not sent off for publication until more than three weeks later. « If the provisional measurements should be verified which EHRENHAFT carried out when I was just now in Vienna as part of his continuing research on the charges on colloidal particles, then the electron would be divisible. Even then EHRENHAFT had found particles with half electrons—in the meantime, LANG [V. v. LANG, whose assistant EHRENHAFT was in 1903] has told me, he appears to have observed some with $\frac{1}{3}$, $\frac{1}{5}$ electron. It would be just too beautiful (*Es wäre doch zu schön*) if the electron were to undergo the same fate as the atom did as a result of cathode rays.... »

Indeed, from the viewpoint of the Machist, that would have been just too beautiful—a long-awaited new hope for a battered cause. All other recent events had been discouraging to them, *e.g.* Perrin's successful crusade on behalf of molecular reality. The defection of the great ally, W. OSTWALD, must have been a particularly severe blow. For in his edition of the text *Allgemeine Chemie*, he had recanted his antiatomism. « I am now convinced », he wrote in the Preface, dated November 1908, « that we have recently become possessed of experimental evidence of the discrete or grained nature of matter, which the atomic hypothesis sought in vain for hundreds and thousands of years. [Experiments such as those of J. J. THOMSON and J. PERRIN] justify the most cautious scientist in now speaking of the experimental proof of the atomic nature of matter. The atomic hypothesis is thus raised to the position of a scientifically well-founded theory.... »

In this dark situation, EHRENHAFT must have appeared as a bright new star. He too can hardly have been oblivious to the joyful impact his provisional new finding had upon the Machists, though in his publications he waited for some years before making explicit the antiatomist edge of his subelectron argument. Possibly EHRENHAFT had actually been visited during Lampa's trip to Vienna just in that provisional period; if so, they would have had much to talk about (*).

In this case of the premature subelectron, we have had to select only one of its many aspects—the opposition between contrary thematic choices, and the place of these choices in the interplay between preconception, research design, data evaluation and the institutions for obtaining scientific consensus. These acted upon the observations made on oil droplets and colloidal metal fragments as surely as did the gravitational and electric fields that filled the viewing chambers. Hence one is prepared to find that, in time and on some other level, the nature of electric charge might be viewed in other ways than seemed conclusively settled with the recognition of Millikan's work.

MILLIKAN himself knew this well. In his Nobel Prize address he did not mention his chief antagonist; but at the end of the first part of his lecture, devoted to the determination of the charge on the electron, he wrote with a perception which we now, five decades later, can agree to have been rather prophetic:

« Shall we ever find that either positive or negative electrons are divisible? Again, no one knows; but we can draw some inferences from the history of the chemical atom.... If the electron is ever subdivided, it will probably be because man, with new agencies as unlike X-rays and radioactivity as these are unlike chemical forces, opens up still another field where electrons may be split up without losing any of the unitary properties which they have now been found to possess in the relationships in which we have thus far studied them. »

EHRENHAFT, in the depth of failure, had warned that no final judgment is ever possible between rival scientific world views such as those that were underlying the subelectron debate. MILLIKAN, at the height of his success, was also saying that the case was not quite finished—that he had left us something still to do.

* * *

It is a pleasure to acknowledge help in the search for documents received from Dr. J. R. GOODSTEIN and Prof. D. J. KEVLES of the California Institute of Technology, Dr. C. WEINER and Ms. J. WARNOW of the Center for History of Physics at the American Institute of Physics in New York, Prof. E. HIEBERT

---

(*) LAMPA went on to describe Ehrenhaft's work in the semi-monthly *Das Wissen für Alle*, Vol. **11** (1911), p. 45-47, a journal among whose collaborators was MACH. Cf. P. SPEZIALI: *Albert Einstein, Michele Besso, Correspondence, 1903-1955* (Paris, 1972), p. 26.

of Harvard University, and Prof. M. J. HIGATSBERGER of the University of Vienna. I have discussed this case in my History of Science Seminar and thank the students, above all Mr. B. COLLIER, for useful comments and leads; and the National Science Foundation Program for History and Philosophy of Science for research support. Versions of this paper have been presented at the Physics Colloquium at M.I.T., May 1972, and at the History of Science Society Meeting during the AAAS annual meeting, Washington, D.C., December 1972.

# Ehrenhaft, the Subelectron and the Quark.

P. A. M. DIRAC

*Florida State University - Tallahassee, Fla.*

Dr. HOLTON told us about Ehrenhaft's work, in which he claimed to have discovered the subelectron. He told us about the controversy between EHRENHAFT and MILLIKAN, how MILLIKAN was much the better experimenter, did careful work, and MILLIKAN proved to be right.

Some time later, I think around the mid-1930's, EHRENHAFT got onto another line of work, and he claimed to have discovered single magnetic poles. The magnets which we are familiar with always have a pole with one sign at one end, and a pole of the other sign at the other end. But I had put forward a theory in which there are particles with just a single magnetic pole. EHRENHAFT claimed to have discovered particles with single magnetic poles, and he wrote to me continually about it. He thought that I ought to support his work because of my theory, but the particles which he claimed to have discovered had poles much weaker than those required by my theory, and so I was not able to support him.

I met EHRENHAFT several times on later occasions, at meetings of the American Physical Society. EHRENHAFT was not allowed by the secretaries to speak at these meetings. His reputation had sunk so low, everyone believed him to be just a crank. All he could do was to buttonhole people in the corridors and pour out his woes. He often talked to me like that in the corridors. I formed the opinion that he was in any case sincere and honest, but he must have given the wrong interpretation to his experiments. He kept saying that he had these experimental results and nobody would listen to him. I do not doubt that he really did have the experimental results, but he put the wrong interpretation on them.

I do not want to talk to you now about Ehrenhaft's work on the monopoles. I never looked into this closely and I doubt if it would be worth-while to do so. But I would like to talk some more on the subject of the subelectron. And the reason for this is that the climate of opinion among physicists with regard to this question has changed during the last few years. Physicists have discovered lots of new particles, and they are able to produce them with high-energy machines. They have been studying the interactions of these particles, and the theoretical people have been busy trying to account for the experimental results.

The theoretical people have found that it helps their theories very much if they postulate a new kind of particle which they label a quark, and the most prominent kind of quark should have two-thirds of the charge of the electron. Well, the quark is essentially just a subelectron, but people have no idea what its mass should be, so that the name subelectron is not altogether appropriate. It might very well be very much greater than the mass of the electron. Many theorists nowadays are inclined to believe that these quarks do exist as constituents of the elementary particles of physics, and that it might be possible to detach one quark and have it quite separate.

It would need a great deal of energy to produce a quark, but still quarks might be produced naturally by cosmic rays. Cosmic rays have enormous energy, and if a single quark occurs anywhere, it is a perfectly stable particle. It cannot get rid of its $\frac{2}{3}e$ because electric charge is conserved. The quark can disappear only by combining with other quarks.

The experimenters have done a lot of work trying to discover individual quarks, but so far there is no certain evidence that they exist. However, with this changed opinion about the possibility of subelectrons, it becomes worthwhile to look again at Ehrenhaft's claims, and examine them from the point of view in which one is not prejudiced against the subelectron.

Now, Ehrenhaft's early results were completely wild. Dr. HOLTON told us a good deal about that. They must have been extremely inaccurate. After a while no reputable physics journal would accept his work for publication, but he did get a paper published in *Philosophy of Science* in 1941 (Vol. 8, 403-457). I think that this was his last paper on the question of the subelectron. There is a diagram (Fig. 1) which is contained in that paper (p. 433). It represents the

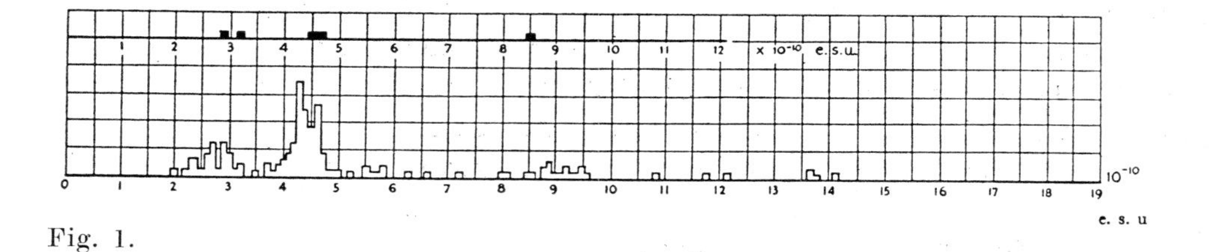


Fig. 1.

results of a large number of experiments. There ought to be a sharp peak at the point corresponding to the charge 4.77. Now, there is a peak there. It corresponds to a charge somewhat less than the true value and it is also a fairly wide peak. The wideness of the peak shows the inaccuracy in Ehrenhaft's work, and the fact that the charge is somewhat less than it ought to be means that there must have been some systematic error which EHRENHAFT had not noticed. These results were all obtained by observing small spheres of red selenium, which moved up and down under the influence of gravity and electric fields. The experiments were similar to Millikan's experiments, with red selenium replacing Millikan's oil drops.

The interesting thing about this diagram is that there is a second peak, also a very wide peak, and that it occurs corresponding to a charge of about two-thirds of the charge of the main peak. There is also a tiny peak corresponding to $2e$.

Now of course, when these experiments were done, it was decades before anyone had any idea of quarks, and there was no reason to prefer the fraction two-thirds to any other fraction. But still this fraction two-thirds is very definitely shown by the diagram, and the question arises, how did it occur? Was it just due to some inaccuracies in Ehrenhaft's work, or did he really have some evidence for quarks?

There is also in the diagram a set of black squares. These black squares refer to the results of some later experiments. EHRENHAFT gives details about these later experiments, and in particular he gives the sizes of the spheres for each of those black squares. And if you look at his figures, you see that the smaller particles have the smaller charge. EHRENHAFT did notice this feature of his experiments, that the smaller particles did have the smaller charge, but he did not seem to realize that the natural interpretation for this would be some systematic error in his experiments which reduced the apparent value of the charge for the smaller particles. He was certainly not a good physicist, or he would have realized that there was very strong evidence for systematic errors in his work.

The question of the reason for these two peaks, I am afraid, cannot be answered. When Hitler's armies marched into Vienna, EHRENHAFT had to leave suddenly because he was Jewish, and he had to leave most of his work behind. All the details of the experiments leading to that curve were lost. So we cannot trace out just why there should be those two peaks.

I was talking about this to Dr. HOLTON and he mentioned a very interesting fact which he had found. In a paper MILLIKAN published in 1910 there occurs this passage. MILLIKAN says: « I have discarded one uncertain and unduplicated observation, apparently upon a singly charged drop, which gave a value for the charge of the drop some 30 % lower than the final value of $e$ ».

Now MILLIKAN was a very careful observer, and MILLIKAN also had great scientific honesty, as you can see from this passage in his work. Many experimenters, when they find a result which is discordant with the main results of their experiments, would assume that there was something wrong with the apparatus on that occasion and that it was just not worth referring to at all in the published records. However, MILLIKAN was not like that, and had this extreme honesty. We can infer also that there was only one drop which did give a serious discrepancy and this one discrepancy showed a charge 30 %

less than the charge of the electron. It sets one to wonder whether MILLIKAN did not have a quark stuck to this drop.

However, a more detailed study of Millikan's paper shows that it is unlikely that he had a quark. He was working at that time with water drops, not oil drops, and he says there may have been errors caused by evaporation. But the disturbing feature is that his anomalous drop was smaller than all the others. It was singly charged, while all the others were doubly or triply charged.

There is thus a similarity between Millikan's results and Ehrenhaft's. They both found anomalous charges for their smaller particles. This does not constitute evidence for quarks. It merely shows there was some experimental error, perhaps the same for both of them, affecting their smaller particles. It must be counted as an unexplained coincidence that their anomalous charges tended to be close to two-thirds of the electronic charge.

That is the situation at the present time. The hunt for the quark is still being carried on intensively by physicists at the present day. It will be ended only when indisputable evidence for the existence of a quark is discovered, or alternatively by the theory being developed sufficiently for one to become convinced that quarks are impossible.

# Personal Notes on Neutron Work in Rome in the 30s and Post-war European Collaboration in High-Energy Physics.

E. AMALDI
*Istituto di Fisica dell'Università - Roma*

## 1. – Recollections of research on artificial radioactivity.

As an introduction to an article on the production and slowing-down of neutrons [1] that I wrote about 15 years ago, I tried to give a rather detailed and accurate account of the succession of discoveries and contributions that opened the field of artificial radioactivity and neutron physics.

In the two lectures that I give on the same subjects today and tomorrow I will be much more biased since they are in great part based on my recollections of what happened in the thirties at the University of Rome, when I was working in the group led by FERMI.

Much information about the work done by FERMI and his collaborators in that period can be found in various publications: among these I should recall Fermi's *Collected Papers*, published a few years ago, jointly by the Accademia Nazionale dei Lincei and the University of Chicago Press. The first volume [2] contains the papers of the «Italian period» (1923-1938); the second the papers of the «American period» (1939-1954) and a number of unpublished reports declassified by the AEC on the occasion.

Each paper or group of papers on a specific subject is preceded by an introduction written by one of the members of the Editorial board (or exceptionally by some other scientist) who was to remember or reconstruct the circumstances, sometimes even of political nature, under which the work was done by FERMI.

These introductions provide a succession of flashes on Fermi's life, which have been in some way co-ordinated in a complete and excellent biography by SEGRÈ in his recent book: *Enrico Fermi: Physicist* [3].

Other information not only on FERMI, but, more in general, on the life at the Istituto di Fisica of Via Panisperna, can be found also in the book by Fermi's wife, Laura [4]. In these two lectures unavoidably I will very often repeat in different words or, sometimes, practically in the same words, what can be found in these various books.

1'1. – The discovery of the artificial radioactivity was announced by I. CURIE and F. JOLIOT in a note presented to *Comptes Rendus* (and *Nature*) in January 1934 [5]. They had observed that boron and aluminium, when bombarded with polonium $\alpha$-particles, gave a positron emission which did not start immediately when the $\alpha$-particle source was placed close to samples of these elements, but increased in intensity from that moment onwards reaching, some time later, a limiting value.

When the Po source was taken away, the positron emission did not cease immediately, but started to decrease exponentially with time as does the activity of a radioactive substance. The lifetime was 14 min for B, 15 s for Al and 2.5 min for Mg, which was found, slightly later [6], to show a similar behaviour with the exception that the emitted electrons were of both signs.

Some detail about how this discovery was made can be found in the speech that F. PERRIN pronounced in July 1964 on occasion of the 30th anniversary of the discovery of artificial radioactivity [7]. This discovery, made and interpreted correctly in a few days, was the result of about two years of experimental work of remarkable quality. Using all possible techniques for the investigation of the atomic radiations known at that time, the CURIE-JOLIOT had studied, about two years before, the penetrating radiations which had been discovered by BOTHE and BECKER to be emitted by beryllium bombarded with $\alpha$-particles. Their experiments, as I will say in more detail tomorrow, had paved the way to the discovery of the neutron by CHADWICK.

Furthermore, about six months before the discovery of artificial radioactivity, they had found that aluminium bombarded by $\alpha$-particles emits not only protons and neutrons [8] but also positrons [9]. Later it became clear that these positrons were due to artificial radioactivity induced in aluminium, but for six months their origin appeared rather mysterious and therefore was a matter of discussion everywhere. In particular this effect was the subject of long—and rather inconclusive—debates at the « Conference on Nuclear Physics » held at Leningrad, in September 1933, and at the 7th « Conseil de Physique Solvay » that took place in Bruxelles about one month later.

The discovery of artificial radioactivity was due to an accidental observation made by JOLIOT. In order to study the emission of positrons by aluminium bombarded with $\alpha$-particles, he used a cloud chamber in a magnetic field with a window in the side wall closed by a thin aluminium foil. The source of $\alpha$-particles (emitted by polonium with 5.3 MeV energy) was usually placed very close to the window, outside the chamber.

One day, in January 1934, JOLIOT noticed that the emission of positrons persisted when the Po source was taken away. As he later told PERRIN, he immediately understood the importance of his observation and the necessity of trying to study the new phenomenon from different points of view. Thus, he went to look for his wife Irène, who was working in a nearby laboratory, with the idea of associating her in all physical and chemical tests that he could foresee would be necessary in order to provide decisive proofs about the qualitative nature and the quantitative aspects of the new phenomenon.

According to F. PERRIN, the first observation was made on Friday morning and the note announcing the discovery of nuclei showing a new type of radioactivity, the radioactivity by positrons, was presented to the Academie des Sciences by J. PERRIN, the successive Monday. It contained the correct interpretation of the phenomenon and the results of a few physical and chemical tests which became classical examples that were followed by all other physicists and chemists working later in this same field. One year later, Irène and Frédéric JOLIOT-CURIE received the Nobel Prize for Chemistry for this discovery.

In Stockholm, on receiving the Nobel Prize, they gave two lectures by dividing the subject as one would not have expected. Irène, who was a pupil of her mother, was mainly a chemist: she treated the physical aspects, in particular the radioactivity by positrons. Frédéric had studied at the École de Physique et de Chimie de la Ville de Paris and was mainly a physicist and engineer; he discussed the chemical aspects, underlining the extraordinary consequences that were opened by the possibility of producing artificially a number of radioisotopes.

The last point should be emphasized also today. With the exception of elastic and inelastic scattering, all processes produced by the absorption of an incident $\alpha$-particle give rise to a nuclide of atomic number different from that of the target element. Thus it becomes possible to separate it by applying the same classical procedures that had been used years before by Marie and Pierre CURIE for the discovery of radium.

In fact, by detecting the decay electrons, one can easily test the various chemical fractions and recognize in some of them the presence of exceedingly small amounts of transmutation products [8].

Although its basic principle had been used before in the case of natural radioactive substances, radiochemistry started to become an important branch of modern science with extraordinary applications in many fields of chemistry, biochemistry, biology, technology, etc., only with the discovery by the JOLIOT-CURIES of artificial radioactivity.

1·2. – This discovery gave to FERMI and, in general, to the group working at the University of Rome the occasion to initiate really important new experimental work. For some years there had been talk of the advisability of gradually abandoning atomic physics—the field in which every one had worked for some years—and concentrating the main research effort on nuclear physics. These ideas were beginning to take practical shape towards the end of 1931. In October, on my return from Leipzig, where I had spent about ten months working on X-ray diffraction by liquids under DEBYE, I was pleased to accept the task of systematically presenting, in a series of seminars, the content of the classical treatise by RUTHERFORD, CHADWICK and ELLIS [10], which I had begun to study a few months earlier. FERMI, RASETTI, SEGRÈ, MAJORANA and a few others attended the seminars. My lecture was often interrupted by observations of the most varied kind from members of the group, which gave rise to long discussions and gave FERMI the opportunity of developing *extempore* the theory of some of the phenomena mentioned. Exceptionally, MAJORANA made observations almost always very penetrating. Many of the ideas and approaches that emerged in these discussions were presented a few years later in the book on the nucleus by RASETTI that was published in Italy as well as the United States [11] and that, in the late thirties, became one of the most popular texts of the new nuclear physics.

The attendance at the seminar fell off in one or two months: in November SEGRÈ went to work in Hamburg at the Institute directed by STERN to learn the technique of molecular beams, and RASETTI went to the Kaiser Wilhelm Institut für Physik in Berlin-Dahlem, directed by Lise MEITNER, to work on the penetrating radiation emitted by beryllium bombarded with $\alpha$-particles, already mentioned above. At the same time FERMI and I in Rome began to construct a cloud chamber of the Blackett type and average dimensions in order to become familiar with one of the most important techniques then in use for studying radioactivity and transmutations.

On the return of RASETTI from Berlin, in the Fall of 1932, FERMI and RASETTI organized a program of research in nuclear physics. A rather large cloud chamber, essentially designed after those in use in Berlin-Dahlem, was constructed and worked excellently as soon as it was assembled. A gamma-ray crystal spectrometer was built by FERMI and RASETTI, who also developed the technique of growing bismuth monocrystals of large dimensions [12].

Various types of counters were also put into operation and RASETTI separated a strong source of RaD from a radium solution. He further separated polonium and, mixing the latter with beryllium powder, prepared a neutron source comparable to the most powerful ones then in use elsewhere. These developments were made possible by a grant from the Consiglio Nazionale delle Ricerche, which had raised the research budget of the department to an amount of the order of $ 2000 to $ 3000 per year, corresponding to about ten times the average budget of the physics departments in Italian universities in those years.

During the Summer and Fall of 1933 SEGRÈ and I did not participate in these developments. We were occupied in finishing some spectroscopic work on a new phenomenon that we had recently observed and which had been correctly interpreted by FERMI only in November of that year. I will say a few words on this work in tomorrow's lecture since its theoretical interpretation turned out to be very useful in the description of the behaviour of slow neutrons.

The switching from atomic to nuclear physics was taking place gradually also in the theoretical activity of FERMI and others. In 1930 FERMI started to work on the hyperfine structure of spectral lines, a subject he further developed in collaboration with SEGRÈ in 1932 [13].

From this work as well as from his interventions in the discussions taking place at various conferences and seminars in Italy and abroad, Fermi's competence on the properties of nuclei had started to be recognized, so that he was asked to report on the status of the physics of the nucleus at a nuclear conference held in Paris in 1932 as part of a large international conference on electricity. In his report he mentioned Pauli's hypothesis on the existence of the neutrino in order to explain the apparent nonconservation of energy and momentum in beta-decay.

MAJORANA, after the discovery of the neutron, had proceeded to develop a nuclear model based on neutrons and protons [14] without electrons and FERMI towards the end of 1933 wrote his paper on the beta-decay, where he introduced a new type of force, the « weak interaction » described by a proper Hamiltonian [15]. I will not try to enter into a discussion of this fundamental paper because it would bring me too far away from my main subject. What I have said is enough to give an idea of the experimental and theoretical background that already existed at the University of Rome when the Joliot-Curies announced the discovery of artificial radioactivity produced by $\alpha$-particle bombardment.

1·3. – Shortly after the first papers of these authors were read in Rome, FERMI, in March 1934, suggested to RASETTI that they try to observe similar effects with neutrons by using the Po+Be source prepared by RASETTI.

In a few weeks several elements were irradiated and tested for activity by means of a thin-walled Geiger-Müller counter, with a totally negative result, obviously due to lack of intensity. RASETTI left for a vacation in Morocco, while FERMI continued the experiments. The idea then occurred to FERMI that in order to observe a neutron-induced activity it was not necessary to use a Po+Be source. A much stronger Rn $\alpha$+Be source could be employed, since its beta and gamma radiations (absent in the Po+Be sources) were no objection to the observation of a delayed effect. Radon sources were familiar to FERMI since they had been supplied previously by Professor TRABACCHI (head of the Laboratorio Fisico dell'Istituto di Sanità Pubblica) for use with the gamma-ray spectrometer mentioned above [12].

All one had to do was to prepare a similar source, consisting of a glass bulb filled with beryllium powder and radon (Fig. 1 and 2). When FERMI had his stronger neutron source (about 30 mCu) he systematically bombarded the elements in order of increasing atomic number, starting from hydrogen and following with lithium, beryllium, boron, carbon, nitrogen and oxygen, all with negative results. Finally, he was successful in obtaining a few counts on his Geiger-Müller counter when he bombarded fluorine and aluminium. These results and their interpretation in terms of (n, $\alpha$) reactions were announced in

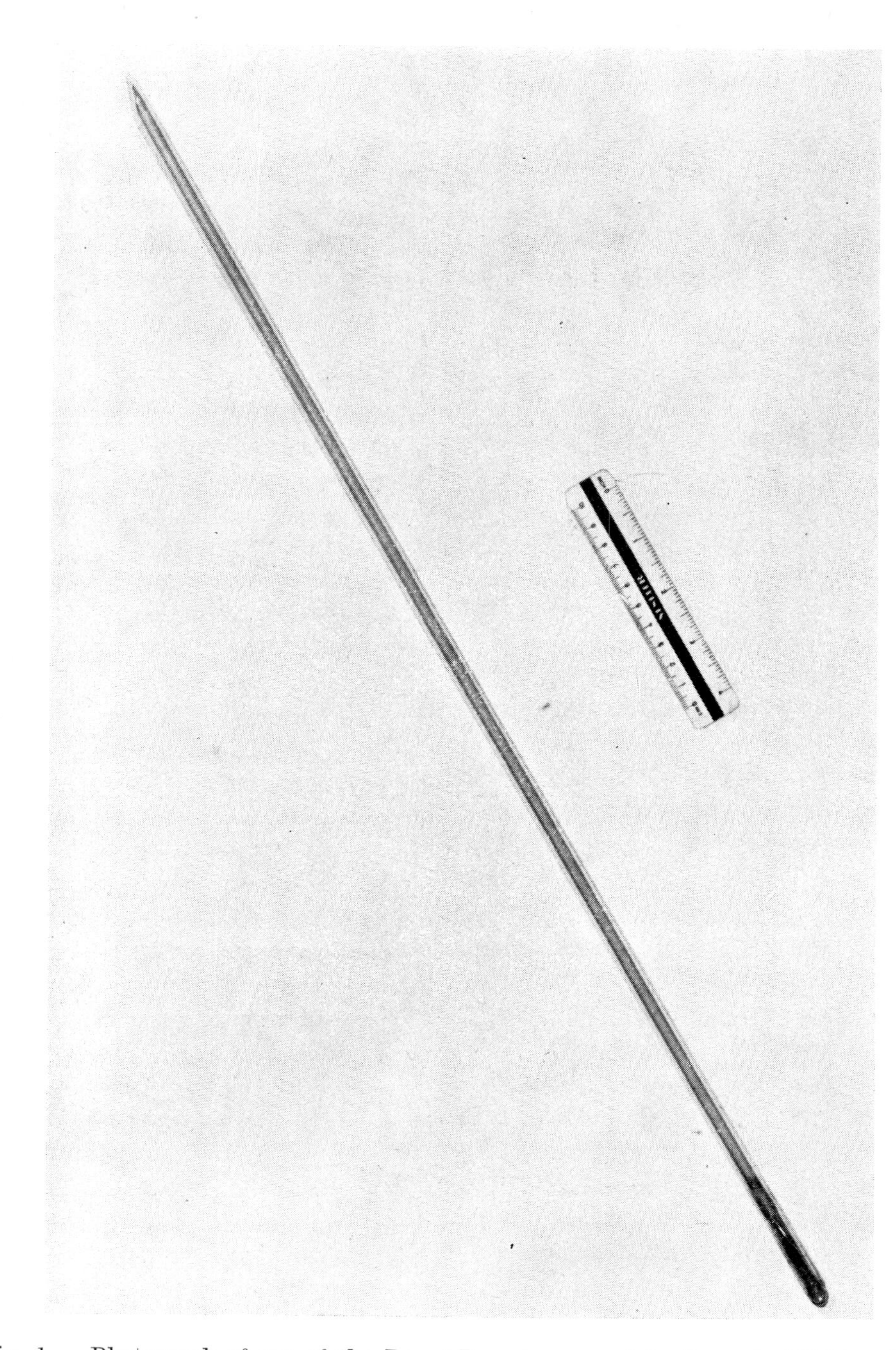

Fig. 1. – Photograph of one of the Rn $\alpha$+Be sources used in Rome in 1934-36. The long glass tube is used only for handling the source without having the hand of the operator exposed too heavily to the gamma-radiation emitted by the decay products of Rn.

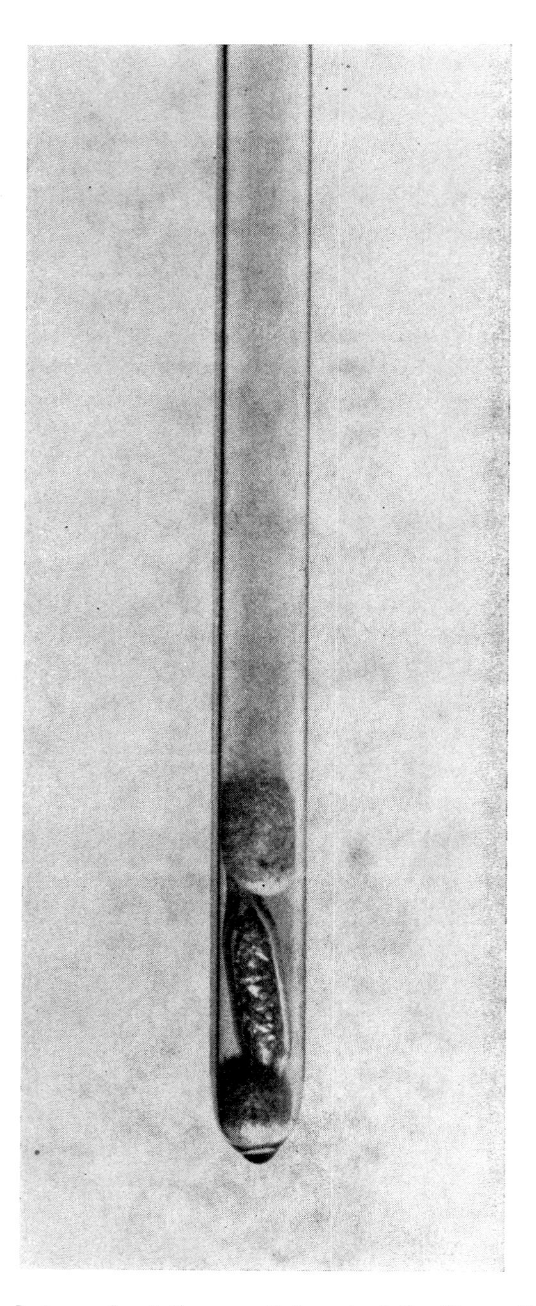

Fig. 2. – Enlarged photograph of the essential part of the Rn $\alpha$+Be shown in Fig. 1. It consists of a glass bulb $(1.0 \div 1.5)$ cm long filled with beryllium powder and radon, kept in a fixed position at the end of the handling tube.

a letter to *Ricerca Scientifica* on 25 March 1934 [16]. The title « Radioattività indotta da bombardamento di neutroni - I » indicated his intention to start a systematic study of the phenomenon which would have led to the publication of a long series of similar papers.

FERMI wanted to proceed with the work as quickly as possible and therefore asked SEGRÈ and me to help him with the experiments. A cable was sent to RASETTI asking him to come back from his vacation. The work immediately was organized in a very efficient way: FERMI did a good part of the measurements and calculations; SEGRÈ secured the substances to be irradiated, the sources and the necessary equipment, and later became involved in most of the chemical work. I took care of the construction of the Geiger-Müller counters (Fig. 3 and 4)

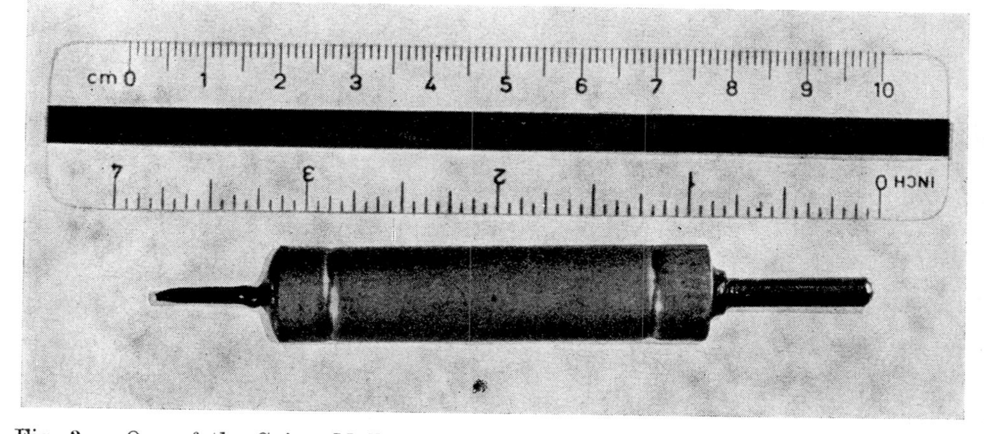


Fig. 3. – One of the Geiger-Müller counters used in 1934 by Fermi's group. The wall was of aluminium between 1 and 2 tenths of a millimetre thick; the small cylinder was obtained by cutting the bottom of a box of medicinal tablets.

and of what we now call electronics. This division of the activities, however, was not rigid at all and each of us participated in all phases of the work. We immediately realized that we needed the help of a professional chemist. Fortunately, we succeeded almost immediately in convincing D'AGOSTINO to work with us. He had been a chemist in the laboratory of Professor TRABACCHI and, at the time I am talking about, he held a fellowship in Paris at the laboratory of Madame CURIE where he was learning radiochemistry. He had come back to Rome for a few days during the Easter vacations but we showed him our work and, on request of FERMI, he remained with us and never went back to Paris.

During the succeeding months our group published in quick succession a long series of experimental results: about sixty elements were irradiated with neutrons and in about forty of them at least one new radioactive product was discovered and often identified.

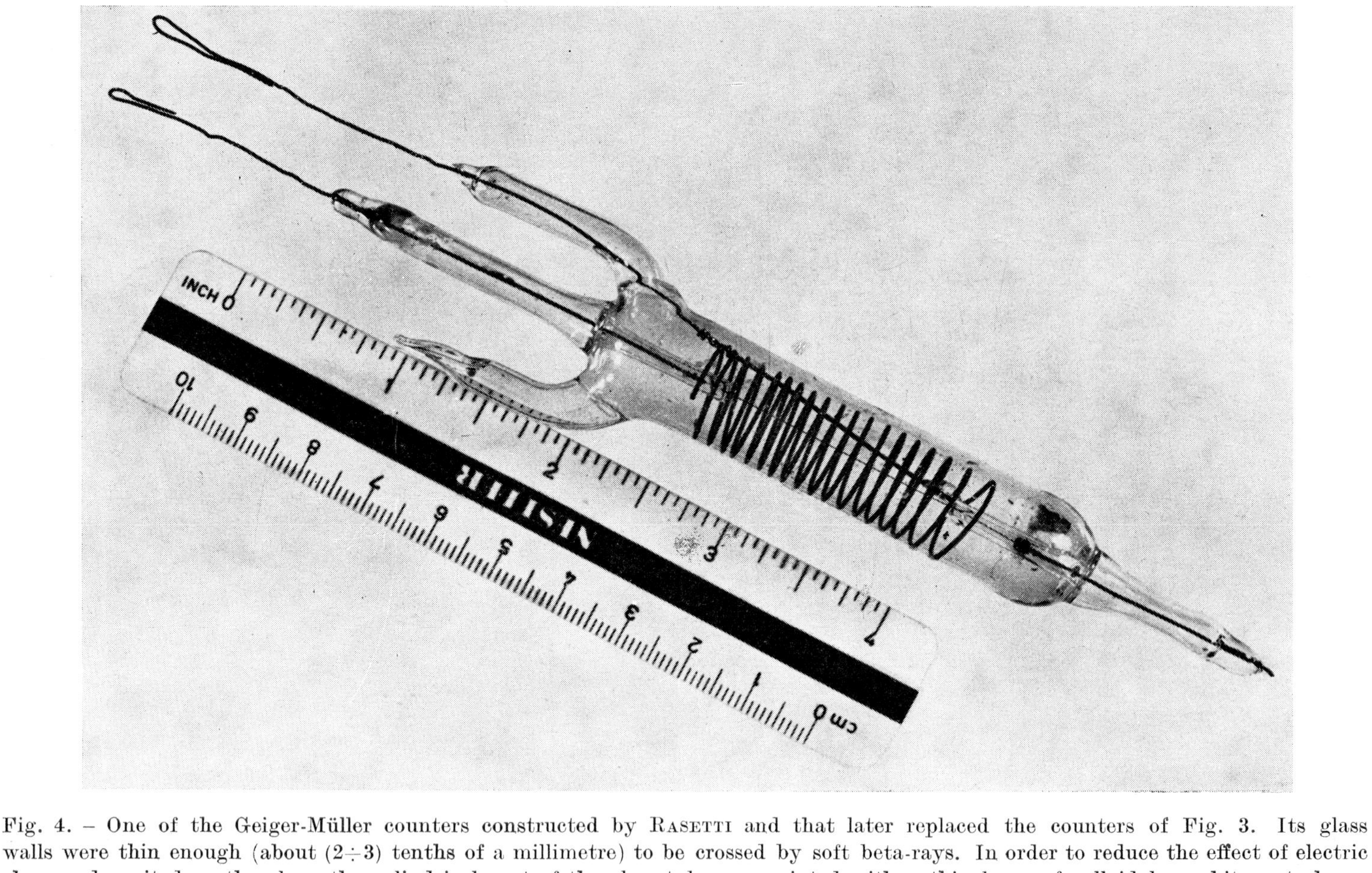


Fig. 4. – One of the Geiger-Müller counters constructed by RASETTI and that later replaced the counters of Fig. 3. Its glass walls were thin enough (about $(2 \div 3)$ tenths of a millimetre) to be crossed by soft beta-rays. In order to reduce the effect of electric charges deposited on the glass, the cylindrical part of the glass tube was painted with a thin layer of colloidal graphite not shown in the photograph.

These results were sufficiently abundant to allow the beginning of a systematic classification of nuclear reactions produced by neutrons. We had found that all elements, whatever their atomic weight, could be activated by neutrons. The nuclide product was sometimes an isotope of the target nucleus, on other occasions it had an atomic number lower by one or two units. From this point of view a marked difference was found in the behaviour of the light and heavy elements. For light elements the active products had in general an atomic number smaller than that of the target nucleus, while for heavy elements the active product was always an isotope of the bombarded nucleus. The results obtained with the light elements can in general be explained as due to (n, p) and (n, $\alpha$) reactions, in which the nuclear charge of the nucleus decreases by one or two units, respectively. In these processes it is the outgoing particle which has to cross the electrostatic potential barrier; the higher this is, the heavier the residual nucleus is. The energies of the neutrons emitted from Be bombarded with the $\alpha$-particles from Po or Rn in equilibrium with its product (and all neutron sources used at that time were of this type) are a few MeV so that in the case of heavy elements the penetrability of the electrostatic barrier turns out to be so small that the corresponding cross-sections are negligible.

The same argument applied to the incident particle explained why the artificial radioactivity produced by $\alpha$-particles could be observed by the JOLIOT-CURIES, as well as by others, only in the case of light elements.

The interpretation of the nuclear reactions in which $Z$ does not change met, on the contrary, some difficulties, the complete solution of which took some time.

The importance of the work on artificial radioactivity produced by neutrons was obvious to us as well as to all nuclear physicists. It was only one month after the beginning of this work that Lord RUTHERFORD wrote the following letter to FERMI:

23rd April 1934

Dear FERMI,

I have to thank you for your kindness in sending me an account of your recent experiments in causing temporary radioactivity in a number of elements by means of neutrons. Your results are of great interest, and no doubt later we shall be able to obtain more information as to the actual mechanism of such transformations. It is by no means clear that in all cases the process is as simple as appears to be the case in the observations of the Joliots.

I congratulate you on your successful escape from the sphere of theoretical physics! You seem to have struck a good line to start with. You may be interested to hear that Professor DIRAC is also doing some experiments. This seems to be a good augury for the future of theoretical physics!

Congratulations and best wishes,

Yours sincerely,
(RUTHERFORD)

Send me along your publications on these questions.

Two remarks may be in order at this point: the first is that our group was probably the first large physicists' team working successfully for about two years in a very well organized way. The second that we were perhaps the first to introduce the use of preprints. In order to communicate rapidly our results to our colleagues we wrote almost weekly short letters in Italian to the *Ricerca Scientifica*, the journal of the Consiglio Nazionale delle Ricerche, and obtained what we would now call preprints of these letters that where mailed to a list of about forty of the most prominent and active nuclear physicists all over the world, and the letters appeared a couple of weeks later in the journal. This procedure was facilitated by the fact that my wife, Ginestra, was working at that time at the *Ricerca Scientifica*.

**1'4.** – Proceeding according to increasing atomic number, before Summer 1934 we irradiated finally thorium and uranium. We observed a number of new activities which were not easily interpreted. We thought that the irradiation of uranium should produce transuranic elements for which we expected properties similar to those of rhenium, osmium, iridium and platinum. This erroneous expectation was then common and, since we had proved that a few of the activities produced were not due to isotopes of elements with atomic number from 86 to 92, we concluded that these activities seemed to be due to elements with atomic number higher than 92. More precisely we thought we had succeeded in separating an ekaRe ($Z = 93$) and an ekaOs ($Z = 94$). The possibility of fission suggested by a German chemist, Ida NODDACK [17], was not considered seriously by us although later we were not able to understand the reason.

One should say that our results were confirmed during 1935 and 1936 by HAHN and MEITNER [18] who even extended them: they thought at that time they had identified two beta-radioactive families, originating by neutron capture in two different uranium isotopes and in each of which appeared, besides an ekaRe ($Z = 93$) and an ekaOs ($Z = 94$), also an ekaIr ($Z = 95$) and in one of the two an ekaAu (given as uncertain). Only after the discovery of fission by HAHN and STRASSMANN in 1939 was it shown by workers in various countries, independently of each other, that the transuranic elements would not behave like Re, Os, Ir and Pt, but would form a second family of rare earths indicated later as «actinides». This conclusion was reached in particular with semi-empirical arguments by ABELSON and MCMILLAN [19] in 1940 and was proved in 1941 by GÖPPERT-MAYER [20] who, at Fermi's instigation, calculated the energy of the $5f$ atomic orbits by the Fermi-Thomas statistical method, as FERMI had done for the $4f$ orbits of the rare earths in 1928 [21].

The work accomplished by Summer 1934 was summarized in a paper that was brought by SEGRÈ and me to Lord RUTHERFORD in Cambridge at the beginning of July 1934. When FERMI wrote to RUTHERFORD if we could spend the summer at the Cavendish Laboratory, he answered with the letter shown in

Fig. 5 [22]. The manuscript given by us personally to RUTHERFORD was presented by him to the Royal Society and was published very quickly in the *Proceedings of the Royal Society* [23].

## 2. – Recollection of early research on the properties of the neutron.

**2·1.** – When, at the beginning of July 1934, SEGRÈ and I arrived in Cambridge, the Cavendish Laboratory appeared to us as the very world capital for nuclear physics. RUTHERFORD was working with OLIPHANT on the D+D reactions but with his strong personality dominated the whole laboratory. CHADWICK was working with M. GOLDHABER; they had discovered shortly before the photodisintegration of the deuteron [24]. COCKCROFT, DEE, ELLIS and FEATHER were there and ASTON was going on improving the accuracy of his measurements of atomic masses. In his laboratory there was a young American, BAINBRIDGE, who had recently made an important step forward in this fundamental technique. From time to time, going around in the Cavendish, one could meet J. J. THOMSON who had retired only a few years before.

The only people working on artificial radioactivity produced by neutrons were T. BJERGE and C. H. WESTCOTT, the first from Copenhagen, the second from Canada. We visited systematically all research groups active at that time in the Cavendish, as well as KAPITZA who was in charge of the Mond Laboratory and had—not long before—constructed his famous rotating generator for producing extremely intensive magnetic fields.

The Cavendish was the birth place of the neutron and this was not purely accidental. The process of generation of this particle actually started in Charlottenburg, where in 1930 BOTHE and BECKER [25] had discovered the emission of a very penetrating radiation from beryllium under α-particle bombardment. During 1932 this radiation had been studied also by RASETTI while he was working in Lise Meitner's laboratory in Berlin-Dahlem [26].

I remember very clearly when, towards the end of January 1932, we began to receive in Rome the issues of the *Comptes Rendus* containing the notes by the JOLIOT-CURIES on the properties of this penetrating radiation.

In the first of these notes it was shown that the penetrating radiation emitted by Be under bombardment with polonium α-particles could transfer kinetic energies of about 5 million electronvolt to the protons present in small layers of various hydrogenated materials (such as water or cellophane). In order to interpret these observations the JOLIOT-CURIES at first [27] put forward the hypothesis that the phenomenon was similar to the Compton effect, namely that the incident photon undergoes an elastic collision with a proton. They had calculated, by applying the laws of energy and momentum conservation, that the incident photons must have an energy of at least 50 million electronvolt. Later, however, they realized that the Klein-Nishina formula applied

June 20.1934

Dear Professor Fermi

Your letter has been
forwarded to me in the country where
I am taking a holiday. I saw your
account in 'Nature' of the effects on
uranium. I congratulate you & your
colleagues on a splendid piece of
work. I have had two of my men
Westcott & Bjerge repeating some
of your experiments with our Em + Be
tube and promised them when I
returned in a week or two to try
out for them the effect of the
2 million volt neutrons from
the D + D reaction on a few
elements. We & myself have been
naturally interested in the energy
of the neutron required for these

Fig. 5. – Letter of RUTHERFORD to FERMI [22].

transformation. We do not, however, propose at the moment to make systematic experiments in this direction as our installation is required for other purposes. I cannot at the moment give you a definite statement as to the output of neutrons from our tube but it should be of the same order as from an Em + Be tube containing 100 millicuries & may be pushed much higher.

If your assistants come to Cambridge say in the 2nd week of July, I shall be delighted to give them the benefit of our experience and to see the mode of operation of our installation for transmutation in general. I hope one or both speak English as the knowledge of Italian in the Laboratory is very limited!. The two men to see are Dr Oliphant & Dr Cockcroft.

Excuse this hastily written letter. I do not take a secretary with me on holiday!

Yours ever

Rutherford

to the protons gave a value for the cross-section too small by many orders of magnitude [28]. Consequently, they suggested that the observed effect was due to a new type of interaction between gamma-rays and protons, different from that responsible for the Compton effect [29].

I like to recall that when MAJORANA [30] read these notes he said, shaking his head, « They haven't understood a thing. They are probably observing recoil protons produced by a heavy neutral particle ». A few days later we got, in Rome, the issue of *Nature* containing the letter to the Editor from CHADWICK [31], dated 17 February 1932, entitled « Possible existence of a neutron », in which he demonstrated the existence of the neutron on the basis of a classical series of experiments, in which recoil nuclei of some light elements (such as nitrogen) were observed in addition to recoil protons.

In order to understand how MAJORANA could guess this discovery, which was suggested but certainly not demonstrated by the Joliot-Curies' results, it should be remembered that he was familiar—through a paper published a few years earlier by GENTILE jr.—with the nuclear model proposed in 1927 by Lord RUTHERFORD [32]. It attempted to make the value of the radius of the uranium nucleus, deduced from the deviations observed in the elastic scattering of particles with respect to the predictions based on Coulomb repulsion alone ($3.2\cdot10^{-12}$ cm), agree with the value deduced from the energy of $\alpha$-particles emitted by radioactive nuclei ($\simeq 6\cdot10^{-12}$ cm). According to this model, the nucleus consisted of a central part with a positive charge $Ze$, around which turned neutral satellites kept in their circular orbits by an attractive force originating from the electric polarization to which the satellites were subjected under the effect of the central electric field.

RUTHERFORD had actually suggested already in 1920 that, if the nucleus consisted of protons and electrons, inside it there might be neutral particles consisting of a proton and an electron closely bound together. This idea had remained so much alive in Cambridge that some of Rutherford's students and colleagues, and particularly CHADWICK, had tried several times between 1920 and 1932 to show experimentally that there existed neutral particles with a mass of the same order as, or greater than, that of the proton. The story of these attempts has been given by CHADWICK himself in «Some Personal Notes on the Search for the Neutron » published in the proceedings of the 10th International Conference on the History of Science [33].

In 1928 GENTILE had shown [34] the inconsistency of the model suggested by RUTHERFORD the year before. However, the idea that there might exist in Nature neutral particles of subatomic dimensions had also remained in the air in Rome.

Soon after Chadwick's discovery, various authors understood that the neutron must be one of the components of the nucleus [35] and began to propose various models which included particles, protons, electrons and neutrons.

The first to publish the idea that the nucleus consists solely of protons and neutrons was probably IWANENKO [36]. Neither I nor his other friends remember whether MAJORANA came to this conclusion independently. What is certain is that before Easter of that same year MAJORANA had tried to work out a theory of light nuclei assuming that they consisted solely of protons and neutrons and that the former interacted with the latter through exchange forces. He also reached the conclusion that these exchange forces must act only on the space co-ordinates (and not on the spin) if one wanted the α-particle, and not the deuteron, to be the system saturated with respect to binding energies. He was, however, dissatisfied with his results and refused to publish them in spite of the insistence of FERMI and all his friends [37]. He published his paper on the nuclear forces, later called Majorana forces, only in late 1933 from Leipzig [38]. He was persuaded to do this by the sheer weight of the authority of HEISENBERG who had published his famous three papers on the nuclear forces already in 1932 [39] without knowing anything of Majorana's work.

Going back to the Summer of 1934, SEGRÈ and I brought with us the manuscript of the long paper in which our work on artificial radioactivity produced by neutrons was summarized. We handed it over to Lord RUTHERFORD who received us, commenting favourably on our work and making jokes one after the other while smoking his pipe. We could not understand what he was saying but anyhow he took care that the paper was published as soon as possible [23].

After a visit to most laboratories I devoted some time to learn from Wynn WILLIAMS how to construct linear amplifiers for measuring with good accuracy the energy lost by a single α-particle or proton in the gas of a small ionization chamber. SEGRÈ and I, however, spent most of the time discussing our common problems with BJERGE and WESTCOTT.

**2·2.** – One of the questions that was left unsolved in our paper in the *Proceedings of the Royal Society* [23] was whether the reactions that produced an isotope of the target were (n, γ), *i.e.* radiative capture, or (n, 2n). They could also have been processes of inelastic scattering of the incident neutron with formation of an isomer of the target nucleus. The last process however was not considered in Summer 1934 and in any case it would have been excluded by the same experiments and considerations reported below. The objections against the (n, 2n) reactions were based on the clearly endoenergetic character of this process. Therefore, the (n, γ) reaction would have remained as the most natural interpretation if it had not been for the following fact based on the « one-particle model » universally accepted at that time. The probability of emission of one or more photons of a few MeV energy during the short time that a neutron takes to cross a nucleus is so small that the calculated cross-section for a (n, γ) process turns out to be negligibly small.

The most important argument, and the decisive one in favour of radiative capture, consisted in testing whether the same unstable nuclide of atomic number $Z$ could be produced by bombarding elements of the same atomic number $Z$, as well as elements of atomic number $Z+1$ and $Z+2$. For example if the *same* $_{11}$Na is produced in the three reactions

$$^{27}_{13}\mathrm{Al}(\mathrm{n}, \alpha)^{24}_{11}\mathrm{Na}\,, \qquad _{12}\mathrm{Mg}(\mathrm{n}, \mathrm{p})_{11}\mathrm{Na}\,, \qquad ^{23}_{11}\mathrm{Na}(\mathrm{n}, ?)_{11}\mathrm{Na}\,,$$

necessarily the last one should be a radiative capture.

The first step to prove that this was just what happened was made by BJERGE and WESTCOTT [40]. While we were in Cambridge they found that, after neutron bombardment, sodium shows a weak activity, the period of which (about 10 h) is, within the error of measurement, the same as that of the long periods produced in magnesium and aluminium, which were known [23] to to be due to an isotope of Na [41].

Back in Rome we completed the proof by showing, with the help of D'AGOSTINO, that this long activity produced in Na was actually due to an isotope of the same elements.

I have insisted on this particular problem because I believe that the proof that neutrons can undergo radiative capture with an appreciable cross-section was one of the first experimental evidences that the « one-particle model » was inadequate for describing many important properties of the nuclei.

We also found a second case of « proven » radiative capture, which was based on the discovery of a new radioisotope of Al with a lifetime of almost 3 minutes. But, when, a few days after our results were made known, I tried to repeat the measurements, I did not find this new activity anymore.

Unfortunately, we had communicated our result to FERMI who, on his way back from South America (where he had spent the Summer), was attending the International Conference on Physics in London. He had even mentioned our experiment at one of the meetings. The fact that we were not able to confirm our result was hurriedly communicated to FERMI who was angry and embarrassed at having presented an erroneous result at the Conference.

We were unhappy and confused because we were not able to understand the origin of our fault. This was the first hint of unexpected complications which were fully clarified in about one and a half months.

**2·3.** – In the paper published in the *Proceedings of the Royal Society* [23] the activity of the various artificial radioactive bodies had been classified only qualitatively as strong, medium and weak. This classification was clearly unsatisfactory and, therefore, at the beginning of the academic year 1934-35, we decided to try to establish a quantitative scale of activities which for the moment could be in arbitrary units. This work was assigned to me and PONTECORVO, one of our best students who had taken the degree (laurea) in July 1934 and after the summer vacations had joined the group. We started by studying the conditions of irradiation most convenient for obtaining well-reproducible results. For this type of work we used the activity of 2.3 min lifetime of silver [23].

We immediately found, however, some difficulty because it became apparent that the intensity of activation depended on the conditions of irradiation. In particular there were certain wood tables near a spectroscope in a dark

Riassunto – Em+Be 15. 30 mc

| P.t. 39^5 | NE. | A1 | A2 | B1 | B2 | C1 | C2 | D |
|---|---|---|---|---|---|---|---|---|
| 110 | 7,5 | | | | | | | |
| | | 153 | 162 | 38 | 22 | 42 | 25 | 34 |
| | | 131 | 145 | 32 | 23 | 39 | 31 | 27 |
| 116 | 9,9 | | | | | | | |
| | | 149 | 139 | 54 | 25 | 25 | 27 | 30 |
| | | 132 | 114 | 38 | 25 | 21 | 23 | 24 |
| 109 | 9,1 | | | | | | | |
| 115 | 8,6 | | | | | | | |
| | | 131 | 110 | 41 | 28 | 28 | 26 | 31 |
| 107 | 8,2 | | | | | | | |
| 114 | 11 | | | | | | | |
| | | 150 | 143 | 37 | 19 | 31 | 21 | 33 |
| 112 | 8,0 | | | | | | | |
| | | 846 | 813 | 220 | 142 | 186 | 153 | 179 |
| | | 738 | 705 | 112 | 34 | 78 | 45 | 71 |
| | | | ↑ | | ↑ | | ↑ | |

Fig. 6. – Photograph of page 3 of the notebook *B*1. It shows the summary of measurements taken under various conditions sketched in the lower part of the page.

room which had miraculous properties, since silver irradiated on those tables gained much more activity than when it was irradiated on a marble table in the same room.

In order to clarify the situation I started a systematic investigation. According to the notebook $B1$ where the measurements of that period are recorded [42] these measurements were started on 18th October 1934. Figure 6 reproduces page 3 containing the summary of a typical series of measurements made inside and outside a lead housing (castelletto), the walls of which were 5 cm thick. The Rn $\alpha$+Be source was always in position $A_1$ (or $A_2$). The irradiation was made by placing the Ag cylinder inside the lead housing in the four positions $A_1$, $B_1$, $C_1$ and $D$. The distance from $D$ to $A_1$ was the same (11 cm) as from $B_1$ to $A_1$. Outside the lead housing the measurements were made in the three positions, $A_2$, $B_2$ and $C_2$, which were at the same relative distances as $A_1$, $B_1$ and $C_1$. The results of the measurements, given in the lower part of the page, clearly show that the activation decreases with the distance from the source more slowly inside than outside the « castelletto ». In particular $B_1$ is about 4 times larger than $B_2$ and $D$ is about equal to $C_1$ and twice $C_2$. The fact that $D$ was equal to $C_1$ was an indication that in $D$ the absorption of lead was roughly compensated by the « scattering in » of the neutrons moving from the source in directions different from that of the detector. It was then decided to test this conclusion in a better geometry, *i.e.* to compare the activation of the same Ag cylinder under the two conditions shown in Fig. 7. The lead wedge

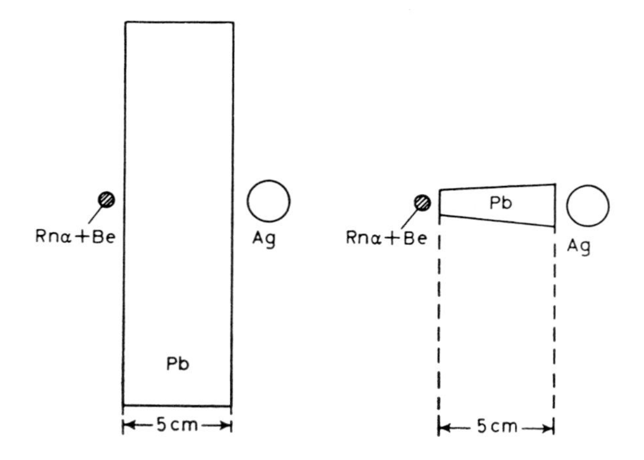


Fig. 7. – Arrangements used by FERMI and collaborators for studying the transmission of lead under conditions similar but « cleaner » than those shown in Fig. 1.

necessary for the arrangement shown on the right-hand side of Fig. 7 was ready the morning of October 20, but it was actually used only a few days later.

In the morning of October 22 most of us were busy doing examinations and FERMI decided to proceed in making the measurements. ROSSI from the University of Padua and PERSICO from the University of Turin were around in the Institute of Via Panisperna and PERSICO was, I believe, the only eye-witness of what happened. At the moment of using the lead wedge FERMI decided suddenly to try with a wedge of some light element, and paraffin was used first [43]. The results of the measurements are recorded on pages 8 and 9 of the same notebook *B*1 (Fig. 8). They are written at the beginning by FERMI and towards the end by PERSICO. Towards noon we were all summoned to watch the extraordinary effect of the filtration by paraffin: the activity was increased by an appreciable factor.

The work was, as usual, interrupted shortly before one o'clock and when we came back, as usual, at 3 p.m., FERMI had found the explanation of the strange behaviour of the filtered neutrons. The neutrons are slowed down by a large number of elastic collisions against the protons present in the paraffin and in this way become more effective. This last point, *i.e.* the increase of the reaction cross-secticn by reducing the velocity of the neutrons, was at that time still contrary to our expectation. The same afternoon the experiment was repeated in the pool of the fountain in the garden of the Institute and we also had succeeded in clarifying the reasons for the discrepancy of the two sets of measurements on the activation of Al that I mentioned above [44].

FERMI went on to hypothesize that the neutrons could be thermalized and that same day an experiment was devised to test the validity of this assumption.

The evening of the 22nd October all the group came to my house and a letter announcing our results was written to the *Ricerca Scientifica* [45].

**2'4.** – The discovery of the hydrogen effect obviously opened a number of problems and we had to modify our previous program. The first step was to measure what we called the « coefficient of aquaticity », *i.e.* how much immersion in water would increase the activity when a thin cylinder of the target material was placed around the neutron source. These measurements gave us confirmation that the (n, $\gamma$) reactions were the only ones sensitive to hydrogeneous substances and by early November we were convinced that slowing-down was the correct explanation of the phenomenon observed.

Then we concentrated our study in trying to understand the behaviour of slow neutrons rather than the nuclide they produced. Among other things we made an experiment aimed at establishing whether the neutrons were thermalized by comparing the activation of a piece of Rh surrounded by hydrocarbons at two different temperatures: room temperature and 200 °C [45]. We did not observe any effect; shortly afterwards MOON and TILLMAN, working in London, realized that such an effect can be observed only in the vicinity of the boundary between two media at different temperatures, and succeeded in obtaining a positive result [47].

FERMI, PONTECORVO and RASETTI [48] also found that some substances like B, Cl, Co, Y, Rh, Ir, Ag, Cd, etc. had very large capture cross-sections $\sigma_c$:

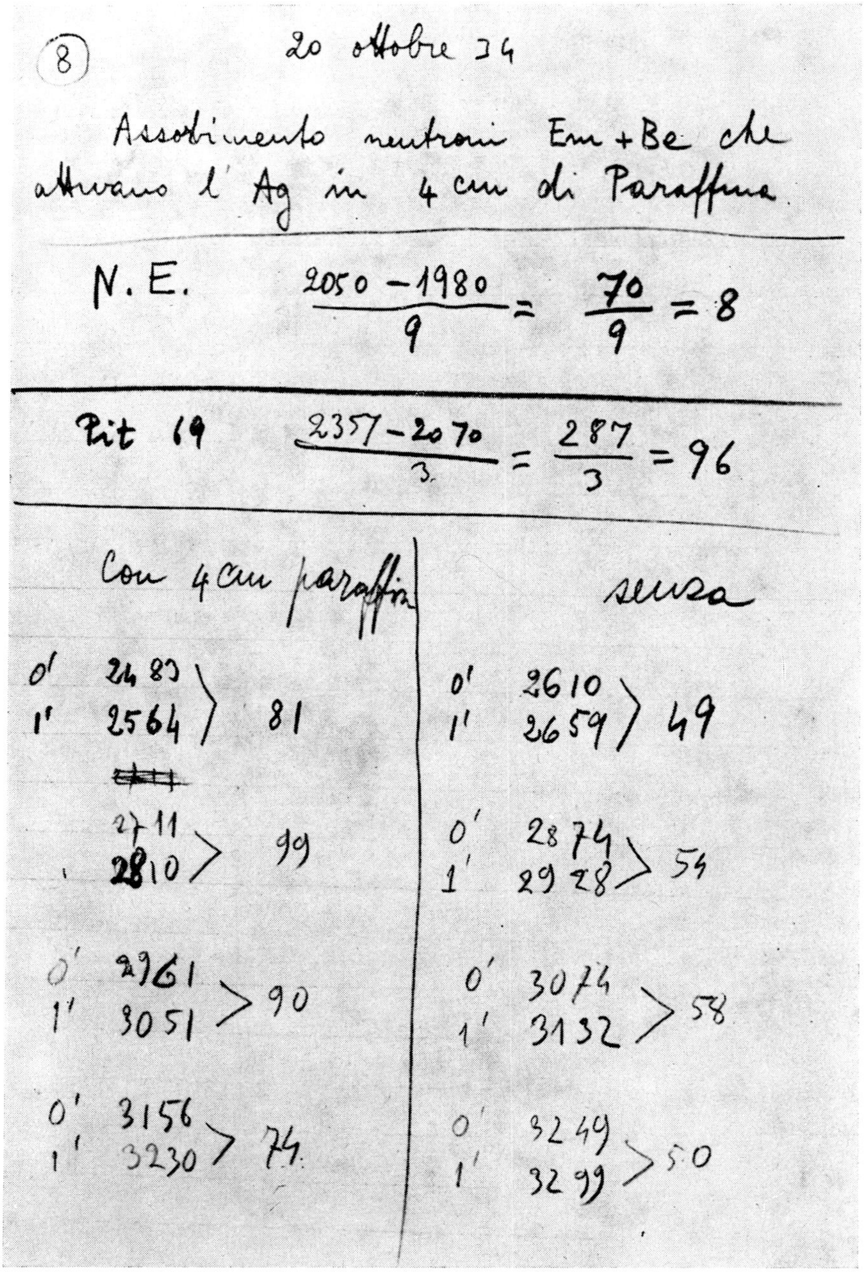

(8) 20 ottobre 34

Assorbimento neutroni Em + Be che attivano l'Ag in 4 cm di Paraffina.

N.E. $\frac{2050 - 1980}{9} = \frac{70}{9} = 8$

Pit 19 $\frac{2357 - 2070}{3} = \frac{287}{3} = 96$

| Con 4 cm paraffina | | | senza | | |
|---|---|---|---|---|---|
| 0' | 2483 | > 81 | 0' | 2610 | > 49 |
| 1' | 2564 | | 1' | 2659 | |
| | 2711 | > 99 | 0' | 2874 | > 54 |
| | 2810 | | 1' | 2928 | |
| 0' | 2961 | > 90 | 0' | 3074 | > 58 |
| 1' | 3051 | | 1' | 3132 | |
| 0' | 3156 | > 74 | 0' | 3249 | > 50 |
| 1' | 3230 | | 1' | 3299 | |

Fig. 8. – Photograph of pages 8 and 9 of the notebook *B*1 of FERMI *et al.* showing the record of the first observation of the effect of hydrogeneous substances on the radioactivity induced by neutrons.

| Con 4 cm. paraff. | | | senza | | |
|---|---|---|---|---|---|
| 0' | 3334 | > 62 | 0 | 3409 | > 41 |
| 1' | 3396 | | 1' | 3450 | |
| 0' | 3477 | > 72 | 0 | 3520 | > 99 |
| 1' | 3549 | | 1 | 3619 | |
| | | 468 | | | 302 |

468
302
166

468
302
170

(80)

0' 35

Si

4770

the measurements were still very rough but sufficient to establish that in some cases $\sigma_c$ was by $10^3$ or even $10^4$ greater than the geometric cross-section of the nuclei.

The explanation of these anomalous capture cross-sections clearly required quantum mechanics: for particles of such a small velocity that their wavelength $\lambda$ is much larger than the radius $R$ of the target obstacles, the upper limit of the cross-section is not $\pi R^2$ but $\lambda^2/4\pi$ multiplied by a numerical factor which can be not much smaller than 1. The same type of reasoning brought FERMI as well as others to foresee a general law for the dependence of the capture cross-section $\sigma_c$ on the velocity $v$ of the neutrons. The probability $P$ of capture of a neutron by a nucleus is given by $P=\sigma_c \cdot v$ and if one describes the process in a simple approach, as that provided by the « one-particle model », one easily finds that

$$\lim_{v \to 0} \sigma_c \cdot v = \text{constant}\,.$$

This means the so-called $1/v$-law, *i.e.* that for very low velocities $\sigma_c$ is proportional to $v^{-1}$.

We also studied the gamma-rays emitted by various nuclei in the radiative capture of neutrons and tried to slow down neutrons by collisions with substances other than hydrogen. Thus we found some effect of inelastic collisions which explained our original observations of the effect of surrounding the source and the detector with lead. This was accomplished by December 1934, *i.e.* within about six weeks of the discovery of slow neutrons.

The large capture cross-sections for slow neutrons observed in boron and lithium did not correspond to any radioactivity nor to a prompt emission of gamma-rays (as we had observed, for example, in cadmium). The behaviour of boron (and less extensively of lithium) was then studied by means of an ionization chamber connected to an electrometer as well as by means of a small ionization chamber connected to a linear amplifier built following Wynn Williams' general guidelines. Thus we arrived at the correct conclusion that slow neutrons produce in $^{10}$B an $(\alpha, \text{n})$ reaction.

The results of all these experiments and of many others that I have no time to mention were published in a second paper in the *Proceedings of the Royal Society* [49] and in a series of more detailed articles in the *Nuovo Cimento* [50].

**2·5.** – I have not yet mentioned Prof. CORBINO, the head of the Department of Physics of the University of Rome and the person responsible more than any other for the creation, in his Institute, of a group of young physicists around FERMI. In 1926 he had succeeded in creating a chair of « Theoretical Physics » for FERMI and, a couple of years later, a chair of « Spectroscopy » for RASETTI. SEGRÈ and I were his assistants but he pushed us to work with FERMI. He protected all of us from the criticisms of the traditional university environment, which in many cases was not too favourable to us. He was an extremely intelligent person. He had been a very distinguished physicist, but from a certain moment on he had devoted most of his activity to politics and industry. He had become a member of the Italian Senate, and had been Minister of Public Education and later of Economics and Industry.

Shortly after the discovery of the effect of hydrogeneous substances he suggested that slow neutrons might have important practical applications and that it could be advisable to take out a patent on this work. This was done and resulted in Italian patent n. 324458 of October 26, 1935, which was later extended to other countries.

Coming back to our work, I like to recall that in January-February 1935 we made, however, an attempt in a different direction, which, although unsuccessful, is of some interest. In order to explain the great number of new activities induced in thorium and uranium which had been isolated by various groups working in Rome, Paris (I. CURIE and co-workers) and Berlin (HAHN and MEITNER) one had invoked a number of transuranic elements and their possible decay products. It was then rather natural to consider that besides beta activities there must be new radioactive alpha-emitters. Thus the decision was taken that I had to look for activities of this type by means of a small ionization chamber connected to the linear amplifier. Helped from time to time according to necessity by one or the other of my colleagues, I began to irradiate some foil of uranium (or thorium) in the form of oxides and put them immediately after irradiation in front of the thin-window ionization chamber. Since we could not observe any activity, we thought that this might be due to the fact that the corresponding lifetimes were too short, perhaps fractions of a second. Therefore, the uranium foil was placed in front of the ionization chamber and irradiated there with a neutron source surrounded by paraffin. A piece of lead was placed between the neutron source and the chamber in order to reduce the background due to gamma-rays.

We also thought that, if alpha-emitters existed, and they had a short lifetime, they had to emit (according to the Geiger-Nuttal law) alpha-particles of a range considerably longer than that of the particles emitted spontaneously by uranium and thorium. Thus all the experiments were carried out with uranium and thorium covered with an aluminium foil equivalent to 5 or 6 cm of air.

The experiments gave negative results, but, if we had occasionally forgotten the aluminium foil, we should have observed the recoiling nuclei due to fission already in January or February 1935. Some years later, talking with HAHN after he had discovered in collaboration with STRASSMANN the fission in 1939, I learned that a very similar experiment, also with negative results, was made at about the same time in Berlin-Dahlem by VON DROSTE. Looking backward it is difficult to say what would have been our reaction if we had observed fission fragments at the beginning of 1935.

**2·6.** – A very serious difficulty was clearly emerging from our experiments [49] as well as from those of other groups, in particular the group working at Columbia University: DUNNING, PEGRAM and MITCHELL [51]. There was no correlation between the scattering and the capture cross-section in cases where the capture cross-section was very large as, for instance, for cadmium. FERMI tried to explain the phenomenon but without success. As I will say later, it was explained by BOHR in January of the following year, 1936.

For interpreting the elastic cross-sections in the second paper in the *Proceedings of the Royal Society* [49] the concept of scattering length was used as a very convenient artifice for describing what happens in the limit of infinite wavelength of the neutron; here infinite means, of course, very large with respect to the dimensions of the nucleus. The concepts of scattering length and pseudopotential had been, however, developed by FERMI in 1933 [52] in order to explain some spectroscopic phenomena (pressure shift of the spectral lines) that had been found experimentally by SEGRÈ and myself [53].

Before summer 1936 we succeeded in extending our previous results and we made a mechanical experiment by which one could compare the velocity reached by the neutrons with the velocity of the edge of a fast rotating wheel [54, 55].

An equivalent, although different, experiment was made at about the same time in Copenhagen by FRISCH and SØRENSEN [56]. During the summer the Columbia University group with the addition of SEGRÈ, who was visiting there, constructed and operated the first mechanical velocity selector for slow neutrons [57].

After the summer vacation of 1935, FERMI and I found ourselves alone in Rome. Most of the group had dispersed by now. The general atmosphere in Italy was chiefly to blame for this as the country prepared for the Ethiopian war. RASETTI had gone to the U.S.A. and planned to stay at Columbia University for at least one year. SEGRÈ, too, had left for a Summer in the United States, and in the meantime had been appointed professor at the University of Palermo. Upon his return to Italy, he also left Rome to go to Sicily. D'AGOSTINO no longer worked with us; he had taken a position at the Istituto di Chimica del Consiglio Nazionale delle Ricerche. PONTECORVO had returned to Rome shortly after us and for a few months worked with WICK on the back-scattering of slow neutrons by various elements [58]. Later he won a Ministero dell'Educazione Nazionale scholarship for study abroad and left Rome in the spring of 1936 to work with the JOLIOTS at the Curie Laboratory in Paris. Thereafter, his visits to Rome were very brief and infrequent.

Upon resuming work, FERMI and I turned our attention to some results of BJERGE and WESTCOTT [59] and of MOON and TILLMAN [60], who had observed that the absorption of slow neutrons by various elements differed slightly depending on the element used as detector.

This fact was not explained by the current theory of the absorption of neutrons by nuclei. As I said before, this theory predicted for all nuclei a capture cross-section inversely proportional to the velocity of the neutrons. This energy dependence was supposed to be valid for such a large energy interval as to certainly cover the energy range of slow neutrons.

We went to work with even greater energy than in the past, as if by our own more intensive efforts we intended to compensate for the loss of manpower in our group. We had prepared a systematic plan of attack which we jokingly summarized by saying that we would measure the absorption coefficient of all 92 elements combined in all possible ways with the 92 elements used as detectors. In jest we added that after combining all the elements two by two, we would also combine them three by three. By this we meant that we would also study the absorption properties of the neutron radiation filtered in several ways.

Actually, after having measured the absorption coefficient of eleven different elements in all possible combinations with 7 detectors, we were convinced that the observations of the groups quoted above were correct, and that in general the rule was valid that the absorption coefficient of a given element was greater when the same element was used as detector. We began to study the particular cases of silver, rhodium and cadmium in great detail. The absorption properties of cadmium were investigated more thoroughly. We performed on the neutrons filtered by a cadmium layer absorption measurements of different elements with various detectors, as we had already done in studying the unfiltered neutrons. Thus, early in November 1935, we established that, if the neutrons were previously filtered with cadmium, the self-absorption effect mentioned above was considerably greater [61].

SZILARD [62] independently had the same idea of studying selective absorption of the neutrons filtered through cadmium and showed that In and I had a behaviour similar to that found by us for Ag and Rh. Indeed the idea of the existence of selective absorption was already contained in the papers of the two British groups [59, 60], but the experiments carried out by us and by SZILARD made this interpretation compelling.

I think it is interesting to note that on this occasion I, as many others, tended to make a simple picture of the phenomenon. I tried to interpret the different groups of neutrons as different bands of energy. FERMI, however, did not want to accept this description. He, too, was convinced that this was obviously the simplest hypothesis, but maintained that it was not strictly necessary, at least for the moment, and was therefore harmful if introduced into our mental picture. He insisted that one must proceed by reasoning with the observed experimental facts only. The correct interpretation of the nature of the neutron groups would finally emerge as a necessary consequence of the data. He was afraid that a preconceived interpretation, however plausible it sounded, would sidetrack us from an objective appraisal of the phenomenon that confronted us.

Therefore, we began a systematic study of the absorption and diffusion properties of the various neutron groups, labelling them with letters, both for brevity and in order to avoid any trace of interpretation. The expression « group *C* » was used for neutrons strongly absorbed by cadmium; « group *D* » for neutrons strongly absorbed by rhodium, but not by cadmium; « group *A* and *B* » for the two components which we believed we had characterized in the radiation strongly absorbed by silver, but not absorbed by cadmium.

In a second letter to the *Ricerca Scientifica*, dated December 12, 1935 [63], the groups of slow neutrons are clearly defined and their absorption and other properties are studied. In this paper experiments are reported which were performed to establish the number of neutrons belonging to each group (numerosity) among those that emerge from the surface of the moderator (paraffin or water), the reflection coefficient of the various group (albedo) [64] and their diffusion distance.

Thus we showed evidence that group *C*—the neutrons strongly adsorbed by cadmium—had properties very different from those of the neutrons that passed through cadmium. The albedo of the neutrons of group *C* was very high (0.83), while that of groups *D*, *A*, etc. was negligible. The diffusion length in paraffin was about 3 cm for group *C*, while it was about six times less for the other groups.

These latter results were further clarified by a more accurate experiment performed in January 1936, in which the diffusion length was determined from the escape probability of a neutron originally found at a depth $x$ within a medium filling a half-space [65]. The expression of such a probability is derived from the diffusion equation which had been adopted for the description of the properties of group *C*. In this same paper, the interpretation of the neutron groups, as due to energy difference, is discussed as the most likely interpretation, without, however, precluding others. This work concludes with reference to an experiment well under way, but not yet completed, whose purpose was to clarify this point: if various groups differ only in energy, the neutrons which at a certain moment belong to a group, as a result of further slowing-down, must transform into neutrons of another group. At that time we had already learned from several experiments by other workers [66] that group *C* included thermal neutrons. Therefore, if the interpretation of the groups in terms of different bands of energy was correct, all the other groups ought to be transformable by slowing-down into group *C*. The definitive results of these experiments were given in a final long paper [67]. In the meantime an experiment of the same type had been published by HALBAN and PREISWERK [68].

In order to establish precisely the diffusion properties of group-*C* neutrons, it was necessary to supplement the measurements of the diffusion length mentioned above with a measurement of the mean free path. The result of a first measure of this quantity showed a clear difference between the values of the group-*C* neutrons mean free path and the mean free path of all other neutron groups [69, 67].

The explanation of the neutron groups in terms of differences in energy had in the meantime been imposed by various experiments in particular by those of the type referred to above: the transformability of various groups into group *C*.

**2·7.** – At the same time an important step towards solving the difficulties mentioned above (*i.e.* the existence of resonances and the fact that the very large capture cross-sections were not associated with appreciable scattering) was made in two independent papers by BREIT and WIGNER [70] and by BOHR [71].

The first authors, guided by the analogy with certain molecular resonances, assumed that the neutron, once it had penetrated the nucleus, could give part of its energy to one of the nuclear components and thus create an excited meta-stable state whose lifetime was long enough to produce the desired small width of the level. They derived an expression for the radiative capture cross-section which is valid when the neutron is captured in a single resonance level and which is usually indicated as the « one-level formula of BREIT and WIGNER ».

The paper by BOHR, presented at the Danish Academy on 27th January 1936, is also based on the idea of an intermediate level. But instead of postulating its existence and of deriving the expression of the corresponding cross-sections, BOHR developed a new conception of the mechanism of nuclear processes which justifies the existence of many excited levels in nuclei of intermediate and high atomic mass number, whose lifetime is sufficiently long to match the experimental results. The extraordinary stability of the intermediate level is explained by BOHR by noticing that as soon as the incident neutron has entered the nucleus it starts to collide with the constituent neutrons and protons. As a consequence its energy is rapidly shared among many particles no one of which acquires an energy large enough to leave the nucleus. This situation can last a long time, until one of the particles, through a momentary fluctuation which concentrates in it a sufficient energy, flies out of the nucleus.

According to this point of view a nuclear process can be described as taking place in two independent steps; the first is the formation on an « excited compound nucleus » as a consequence of the capture of the incident neutron, the second is the decay of the compound nucleus either by emission of a particle, or—more frequently—by irradiating a photon.

However, the difference in mean free path reported in ref. [69] was interpreted by FERMI as due to the chemical bond, since he could not imagine any nuclear phenomenon that could possibly give rise to a similar effect. The theory of this phenomenon was given by FERMI in a fundamental extensive paper containing the theory of a number of phenomena observed with slow neutrons [72].

At this point, perhaps, it is useful to remember that, while this work progressed, measurement techniques were becoming considerably more refined, During the first period of investigation by the group at the University of Rome.

the activity measurements were taken exclusively by means of Geiger-Müller counters. However, after the discovery of the effect of hydrogenous substances, the activity had become so high that it was frequently possible to use an ionization chamber connected to an electrometer (Fig. 9). This technique was then

Fig. 9. – Ionization chamber and electrometer used by FERMI and collaborators to measure the activity induced by slow neutrons. Note on the left the scale on which the wire of the Edelman electrometer was projected and the nomogram for deduction of the activation of Rh to standard measuring conditions (Fig. 10). The white shield on the background was introduced, for a better illumination of the instrument, only when this photograph was taken (in 1962 on occasion of the celebration of the 20th anniversary of the operation of the first nuclear pile by FERMI *et al.* in Chicago).

developed and perfected by experimenting with new types of ionization chambers and new ways of using the electrometers. These had been calibrated with great care in order to know well their characteristics and utilize them to their maximum potentialities. The preparation of nomograms (Fig. 10) and graphs allowed a rapid computation, from the readings made on the electrometer scale, of the activity of the radioactive body being measured.

Once the interpretation of the phenomena observed on the basis of the « compound nucleus » resonance levels, according to Bohr's hypothesis, was

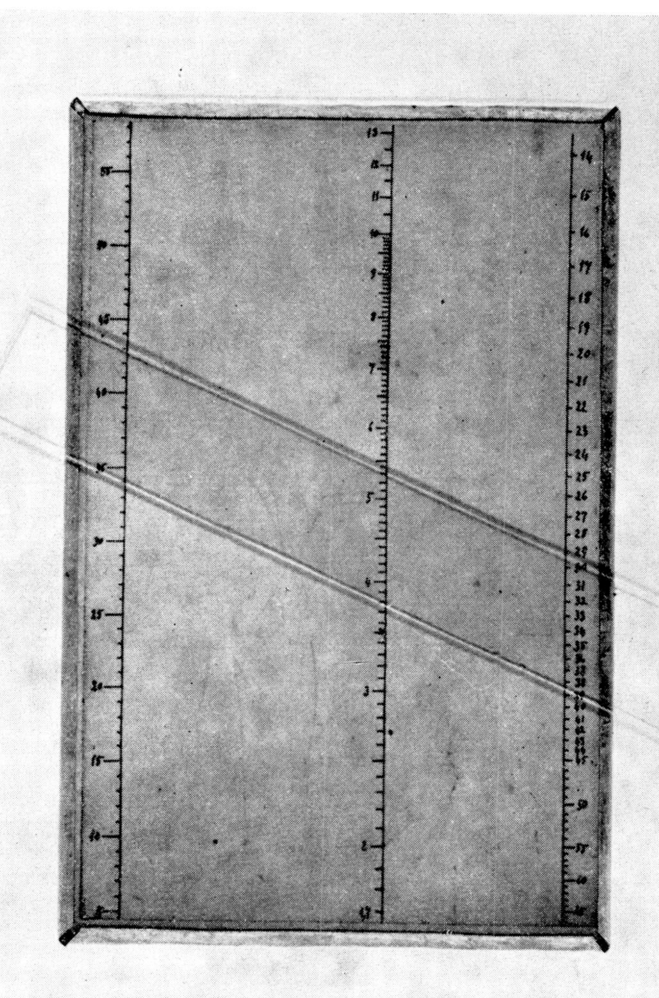

Fig. 10. – Photograph of the nomogram calculated and drawn by G. C. WICK to help the work of FERMI and AMALDI. It regards the activity of $T_{\frac{1}{2}} = 44$ s of Rh. On the central scale the total activity (*i.e.* the activity measured from the end of the irradiation, $t = 0$, to $t = \infty$) is read on the straight line determined by the time passed between the end of the irradiation and the beginning of the measurements (left-hand scale) and the duration of the measurement (right-hand scale).

accepted, the problem of determining the width and energy of these resonance lines naturally arose. This was done by FERMI and me in a series of experiments [67, 73] the theory of which was given by FERMI in the extensive theoretical paper quoted above [72]. This work demonstrates how the mean value of the square of the distance traveled by the neutrons, before reaching the resonance energy of the detector, increases as the resonance energy of the detector decreases. In this way a quantitative relation between spatial distribution of resonance neutrons and their energy was established.

2'8. – The academic year, which had slipped by in an atmosphere of frenzied work and isolation, was by now drawing to a close. RASETTI wrote to us every now and then about what was happening at Columbia University; reprints by HALBAN and PREISWERK kept us informed of the work in Paris and a correspondence with PLACZEK kept us in contact with Copenhagen. Through this latter correspondence we learned of Bohr's work, as well as of that of FRISCH and PLACZEK, concerning the $1/v$ absorption law in boron [74]. Through this latter correspondence the joke that, « as the captain's age can be determined from the length of the ship's mast, so the energy of a neutron group can be determined by the distance it travels as it slows down », spread from Rome. The expression « age », used later by FERMI to represent the quantity $\langle r \rangle^2/6$, might date back to this period. At first the expression « the age of the captain » was used to refer to experiments concerning the transformation of one group into another of lower energy (Fig. 11).

Besides FERMI and me, WICK and PONTECORVO were also working at the Institute, as I said before. There was also MAJORANA, and occasionally SEGRÈ came to visit us from Palermo.

We worked with incredible stubbornness. We would begin at eight o'clock in the morning and take measurements almost without a break until six or seven in the evening, and often later. The measurements were taken with a chronometric schedule as we had studied the minimum time necessary for all the operations. They were repeated every three or four minutes, according to need, for hours and hours and for as many successive days as were necessary to reach a conclusion on a particular point.

Having solved one problem we immediately attacked another without a break or feeling of uncertainty: « Physics as soma » [75] was the phrase we used to refer to our work performed while the general situation in Italy grew more and more bleak, first as a result of the Ethiopian campaign, and then as Italy took part in the Spanish Civil War.

This was more or less the end of the golden period for the investigation of the neutron properties at the University of Rome. The work went on for a few years and some interesting results were still obtained. But the leading role that the group had maintained for about three academic years had finished. This was due certainly in part to the political situation becoming worse and worse in Europe in general and in Italy in particular. It was due, perhaps, also to the fact that it was becoming always more difficult to compete with other groups that, in the meantime, had equipped their laboratories with accelerators of various types which provided neutron sources much better than those at our disposal.

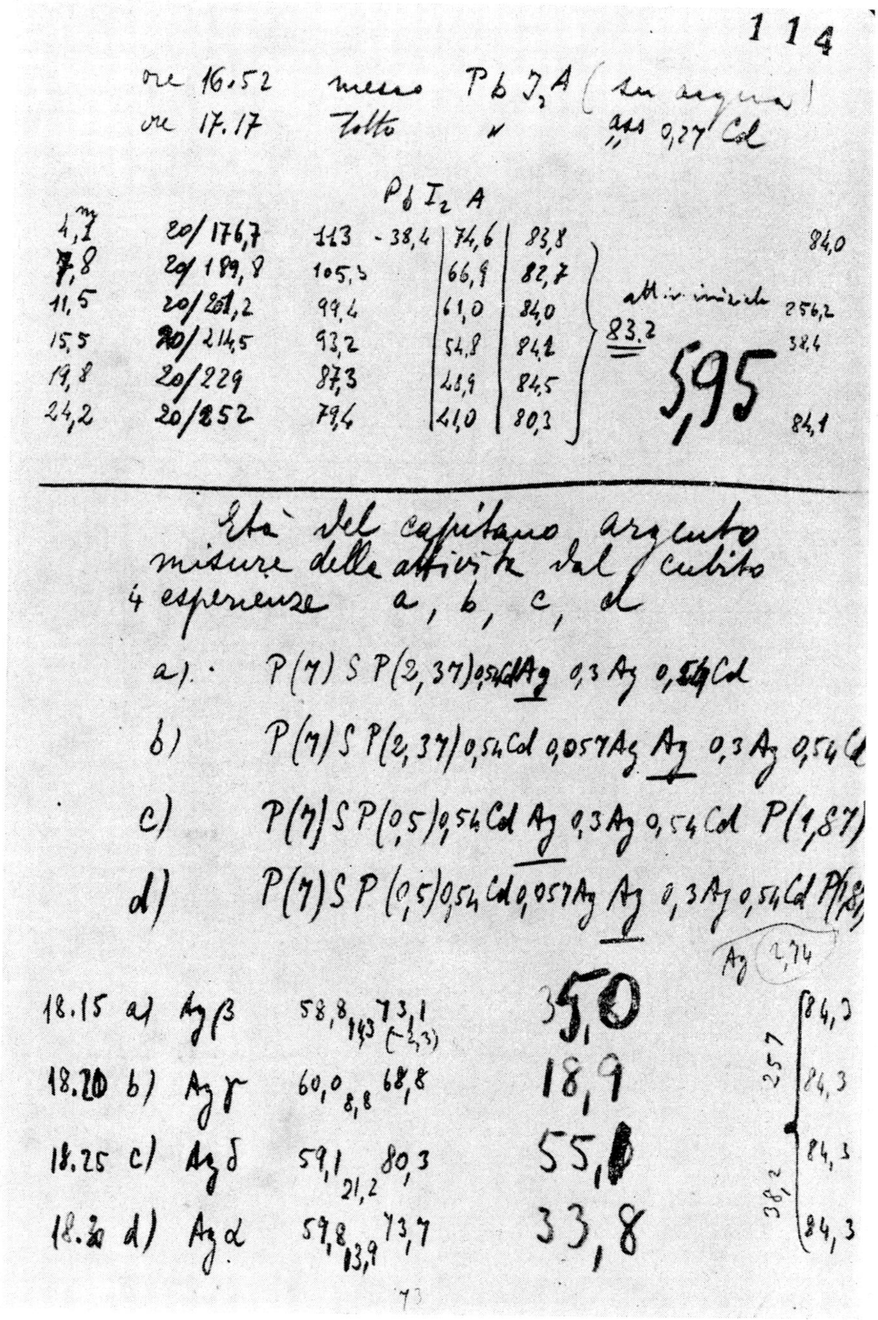

114

ore 16.52 messo Pb $I_2$ A (su sughero)
ore 17.17 tolto " 0,27 Cd

Pb $I_2$ A

| | | | | | |
|---|---|---|---|---|---|
| 4,1 | 20/176,7 | 113 - 38,4 | 74,6 | 83,8 | |
| 7,8 | 20/189,8 | 105,3 | 66,9 | 82,7 | |
| 11,5 | 20/201,2 | 99,4 | 61,0 | 84,0 | |
| 15,5 | 20/214,5 | 93,2 | 54,8 | 84,2 | |
| 19,8 | 20/229 | 87,3 | 48,9 | 84,5 | |
| 24,2 | 20/252 | 79,4 | 41,0 | 80,3 | |

attiv. iniziale 83,2 — 5,95

84,0 — 256,2 — 38,4 — 84,1

Età del capitano argento
misure delle attività dal cubito
4 esperienze a, b, c, d

a) P(7) S P(2,37) 0,54 Cd Ag 0,3 Ag 0,54 Cd

b) P(7) S P(2,37) 0,54 Cd 0,057 Ag Ag 0,3 Ag 0,54 Cd

c) P(7) S P(0,5) 0,54 Cd Ag 0,3 Ag 0,54 Cd P(1,87)

d) P(7) S P(0,5) 0,54 Cd 0,057 Ag Ag 0,3 Ag 0,54 Cd P(1,87)

Ag (2,74)

| | | | | | |
|---|---|---|---|---|---|
| 18.15 | a) | Ag β | 58,8 14,3 73,1 (-2,3) | 35,0 | 84,3 |
| 18.20 | b) | Ag γ | 60,0 8,8 68,8 | 18,9 | 84,3 |
| 18.25 | c) | Ag δ | 59,1 21,2 80,3 | 55,0 | 84,3 |
| 18.30 | d) | Ag α | 59,8 13,9 73,7 | 33,8 | 84,3 |

25,7 — 38,2

73

Fig. 11. – Photograph of p. 114 of the notebook $B3$ handwritten by FERMI.

## 3. – First international collaborations between Western European countries after World War II in the field of high-energy physics.

**3'1.** – In this Section, I will try to outline the stage of development of the study of subnuclear particles in Europe immediately after the Second Wordl War and to summarize the most important steps that brought the creation of CERN. The two subjects are closely related and the case of CERN is of particular interest for a number of reasons. Among the various international research bodies created in Western Europe after the Second World War, CERN was the first one in order of time. It entered into operation at a very early date and acted from the beginning with the full satisfaction of the governments of all participating countries as well as of the involved scientific circles.

It would be very interesting but it is not easy to find out and disentangle all the elements that made such an early and—at the same time—lasting success possible. Among these, certainly one should keep in mind the general idea—held in many political circles of Western European countries—of the necessity of moving towards some form of political unification of at least a considerable part of the old continent.

A second favourable element was the scientific, technical and administrative experience in various countries, during and immediately after the war, of wide and complex organizations operating in the field of the nuclear sciences and their applications. I will not say more on this point since it will be treated by KOWARSKI. I should mention, however, that this experience had brought about the creation in the U.S.A. of a few big research laboratories such as the Argonne National Laboratory and the Brookhaven National Laboratory. The latter had been created and run—very successfully—by the Associated Universities Inc. Furthermore the dimensions of the geographic region involved and of the laboratory were very similar to those of a possible future European research establishment. In the latter case a program of research had to be chosen completely free from any limitation or restriction originating from military, political or even industrial secrecy.

To these one should add a further element of paramount importance: immediately after the Second World War cosmic-ray research had a very high level in Western Europe and was in part carried out through successful international collaborations.

Today we are used to considering cosmic rays and high-energy physics as two separated chapters of physics, but originally it was not so. For many years the only available source of high-energy particles was the cosmic radiation and high-energy physics was the part of cosmic-ray research regarding the study of the behaviour of the component particles. Even today, for the investigation of particles of energies greater than $10^{13}$ eV, the only thing that one can do is to use cosmic rays.

3·2. – It would be very interesting to review in detail the development of cosmic rays from the point of view of high-energy physics, starting from the discovery of the positron and the mu-meson in the thirties. It would be a very exciting story but too long, at least in the frame of my lecture. In the first part of which I will try to underline—mainly from my personal direct knowledge and experience—the transition undergone in Europe by high-energy physics from being a part of cosmic-ray research to becoming an independent and strong activity going on in many European universities and national laboratories, but mainly in the Meyrin Laboratories of CERN.

The whole story can be seen in the right perspective only if one realizes that, in spite of the enormous destruction suffered by most European countries, research in the field of cosmic rays was still at a very high level at the end of the Second World War.

The opportunity for a first encounter between physicists was provided by the conference organized as a joint effort by the British Physical Society and the Cavendish Laboratory. The Conference took place in Cambridge on 22-27 July 1946 and its subject was «Fundamental Particles and Low Temperature». Contact between physicists in different parts of the world had been impossible for years and this conference provided a welcome opportunity to renew old friendships and to hear what others had been doing [76]. Theoretical papers were presented by BOHR on problems of elementary-particle physics, PAULI on difficulties of field theories and field quantization, DIRAC on difficulties in quantum electrodynamics, BORN on relativistic quantum mechanics and the principle of reciprocity and by BHABHA on the relativistic wave equations for elementary particles. The experimental contributions referred to classical problems. For example, JANOSSI presented results of the investigation into the production of mesons, LEPRINCE RINGUET and WILSON discussed measurements of the mass of mesons, CLAY and WATAGHIN cosmic-rays showers, bursts and showers of penetrating particles, BERNARDINI the meson decay and its secondary electrons. There was also a number of papers on nuclear physics and slow-neutron physics. Among these one should recall a few papers by FERMI and others on the diffraction and reflection of slow neutrons.

Shortly after the conference, however, a number of papers started to appear which opened completely new perspectives. A few examples can be recalled here. For instance, towards the end of 1946 CONVERSI, PANCINI and PICCIONI [77], in the wake of the cosmic-ray work started in Rome by BERNARDINI, found the unexpected result [78] that negative mesons reduced to rest in carbon are captured by the nuclei at a very low rate so that they undergo spontaneous decay. The natural interpretation of this result was that the mesons constituting the hard component of cosmic rays had an interaction with nuclei much weaker than expected at that time. There was, however, a possible different interpretation based on the suspicion that for some reason the time required by a meson to be reduced to rest in carbon was longer than its lifetime. This possibility was pretty soon excluded by FERMI, TELLER and WEISSKOPF [79].

Thus it was definitely established that the « mesons » constituting the hard component of cosmic rays were not the particles hypothesized by YUKAWA, as quanta of the field responsible for nuclear forces. The experiment of CONVERSI *et al.* has been frequently quoted by many authors, like BETHE, SCHWEBER and DE HOFFMANN [80] and ALVAREZ [81], as the one marking the origin of modern high-energy physics.

At the beginning of October 1947, LATTES, OCCHIALINI and POWELL—working in Bristol with the newly developed nuclear research emulsions—observed a new particle that they called « $\pi$-meson » [82]. It decayed into a lighter particle correctly identified by the same authors as the weak interacting particle [77, 79] constituting the hard component of cosmic rays, and that they called « $\mu$-meson ».

Already at the beginning of the same years, PERKINS [83] and OCCHIALINI and POWELL [84] had observed in nuclear emulsions a few tracks of slow mesons ending in a « star », *i.e.* a group of divergent tracks due to protons and other charged nuclear fragments, and had interpreted them as due to mesons.

One week after the appearance of the discovery of the $\pi \rightarrow \mu$ decay, the Bristol group published a second paper [85] in which they pointed out that the number of mesons giving rise to the $\pi \rightarrow \mu$ decay was of the same order of magnitude as the number of mesons giving rise to nuclear disintegrations at the end of their track. They identified the star-generating mesons as negative $\pi$-mesons, and from this as well as from other considerations they arrived at the conclusion that the $\pi$-meson had all the main properties of the Yukawa particle.

Almost at the same time ROCHESTER and BUTLER [86] at the Manchester University observed in a cloud chamber, triggered by a number of counters, two « V events » identified later as decays of a $\theta^0(\equiv K^0)$-meson and a $\Lambda^0$-hyperon.

These are only a few outstanding examples. Many more papers could be quoted, the quality of which was a clear proof that the European tradition in the investigation of the structure of matter and its tiniest components was still alive in many places, in spite of the incredible material and moral destruction that had taken place throughout a great part of Europe during the recent war. An important international encounter was the « Symposium on Cosmic Rays » organized in Krakow on October 1947 by the International Union of Pure and Applied Physics [87]. Besides many problems already discussed during conferences in previous years, the evidence for the existence of the $\pi$-meson was presented.

A few groups operating cloud chambers at mountain altitudes made many remarkable contributions to the study of the new particles. Among these, three should be mentioned because of the extent and importance of their work. Two groups were led by BLACKETT from Manchester University and later from

Imperial College, when he moved to London in 1953. The first group started to operate in 1950 the same cloud chamber used in 1947 by ROCHESTER and BUTLER [86] at the Laboratoire du Pic du Midi in the French Pyrénées (2867 m) [88]. The other group started to operate in 1951 a much larger cloud chamber [89] at the Hochalpine Forschungs-Station at the Jungfraujoch (3460 m) in Switzerland [90]. The third group was led by LEPRINCE-RINGUET from the École Polytechnique. It started to operate at the Laboratoire du Pic de Midi two large cloud chambers [91] placed one on top of the other in an arrangement that allowed determination of the momentum as well as of the range of the tracks with good accuracy [92, 93].

At the Laboratorio della Testa Grigia (3500 m) [94] near the Teodule Pass and, later, at the Osservatorio della Marmolada (2030 m) [95] in the Italian Alps other investigations with nuclear emulsions, counters, ionization chambers and cloud chambers were carried out [96, 97].

Some of the problems tackled in those years were discussed at the 8th Solvay Conference on Elementary Particles that took place in Bruxelles in September 1948 and in a rather large International Conference on Cosmic Rays that was held in Como about one year later [98].

All the work with cloud chambers and counters at mountain altitudes was rather expensive and required the physicists involved to remain for rather long periods far away from their institutions. From both points of view, the nuclear-emulsion technique—developed mainly at the Bristol University and later also at the Université Libre de Bruxelles—was the essential vehicle providing physicists in many European Universities the possibility of participating in the most important developments of high-energy physics in spite of the poverty of the budgets of the corresponding laboratories and without imposing long absences from their teaching duties. We should be grateful to OCCHIALINI and POWELL, who played a fundamental role in promoting the development of this technique, which during the early fifties allowed an extensive study of the particles produced by collisions of cosmic-ray primaries in the upper atmosphere and, shortly afterwards, the participation of many European groups in interesting research on the new particles produced by high-energy accelerators built in the U.S.A. [99].

This participation was due mainly to the generosity of a few U.S.A. laboratories, among which particularly the Lawrence Radiation Laboratory in Berkeley, and the Brookhaven National Laboratory, L.I., should be mentioned. But without the emulsion technique such participation would not have been possible at all.

The mountain laboratories mentioned above as well as the nuclear-emulsion laboratories of the Universities of Bristol, Bruxelles (where OCCHIALINI was working since 1948) and a few other Western European cultural centres had become in those years points of encounter of young physicists originating from many different countries. The life in common in mountain huts and the co-ordination of the experiments planned by different groups were elements that paved the way to the idea of wider and more ambitious collaborations which—as I will say in a few minutes—were popping out in various places before and around 1950.

**3·3.** – The exploration of high altitudes by means of balloons to study cosmic radiation had been tried extensively with success in the U.K., particularly by the Bristol Group, around 1950. This group thus arrived at the discovery of the $\tau$- [100] and the charged K-meson [101].

In 1951 it was suggested that a study of high-energy events produced by cosmic radiation could be more conveniently carried out at lower latitudes. At a latitude of 40°, for example, only that part of the primary radiation which has an energy above 7 GeV per nucleon is allowed by the magnetic field of the Earth to enter the atmosphere. Thus, at these latitudes the magnetic field of the Earth removes a large number of primaries which are not very efficient in producing the new secondary particles because of their relatively low energy. The nearer one goes to the equator the stronger this effect is.

Thus a first international expedition was planned in 1952 [102] under the sponsorship of CERN that recently had entered its « Planning Stage » as I will explain below. Naples and Cagliari were chosen as the most convenient of the available bases in the Mediterranean. Thirteen universities took part in the expedition which met with some success. In particular it was the first successful attempt to recover balloons at sea. Furthermore, the expedition made a survey of the winds at high altitudes which was of great importance for the expedition of the subsequent year. The results of this first expedition were briefly reported to the Third Session of the Council of CERN held on 4-7 October 1952 in Amsterdam [103].

A second expedition took place, also to Sardinia, in June-July 1953. This was much on the same lines but on a larger scale than the first one [104]. Eighteen laboratories from European countries in addition to one from Australia took part in it. 25 balloons were launched and over 1000 emulsions were exposed 7 hours at altitudes between 25 and 30 km (Fig. 12). This corresponds to 9.27 litres of nuclear emulsions weighing about 37 kg. They were successfully recovered at sea on account of the employment of a seaplane and a corvette (Pomona) generously placed at disposal of the expedition by the Italian Airforce and the Italian Navy (Fig. 13).

While in the 1952 expedition only glass-backed emulsions had been used, in the 1953 expedition « stripped emulsions » were introduced. Thus, the tracks were followed almost always from one emulsion to the adjacent one, a point of great importance when high-energy events are studied. The development required special care and in particular the construction of special developing systems at the Universities of Bristol, Padua and Rome, where the emulsions of the whole expedition were processed.

In October 1953, a meeting was held at the Department of Physics of the University of Bern (Switzerland) for distributing the packages of exposed and developed emulsion among the participating universities and, in April 1954, an international conference was organized in Padua [105] to discuss the first results and plan jointly the most efficient methods for the investigation.

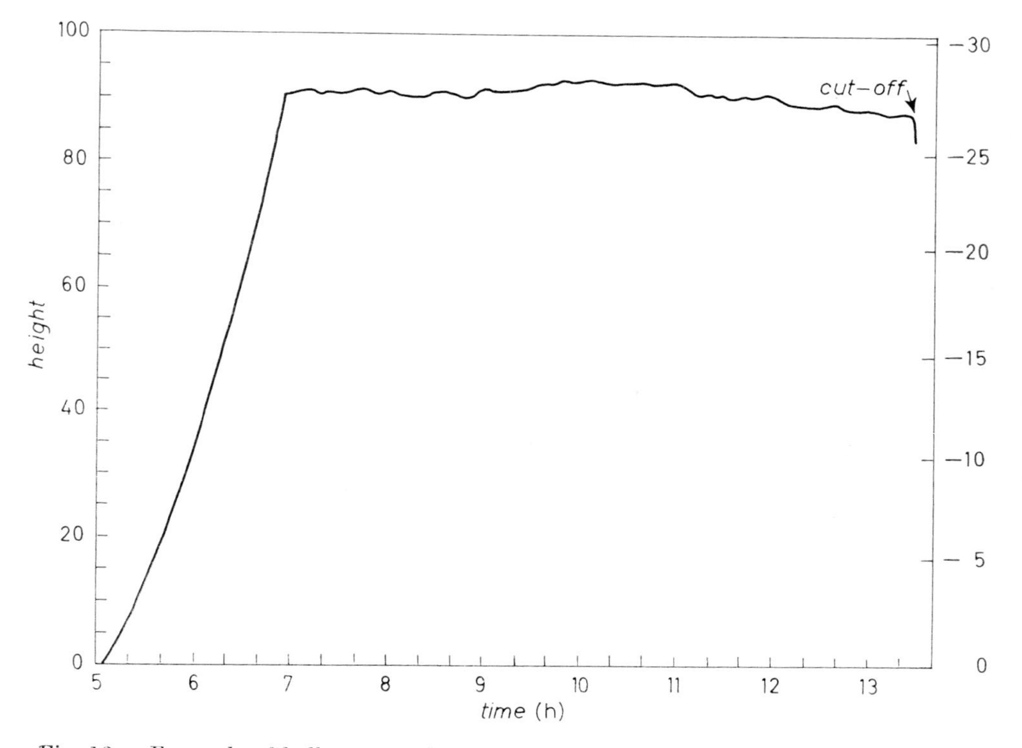


Fig. 12. – Example of balloon record of the altitude as a function of time: flight No. 15.

Further results appeared in the normal scientific literature and in lectures given by various participants at the second Course of the Varenna International Summer School that took place during the Summer 1954 [106]. At the same course, FERMI gave a series of « Lectures on pions and nucleons » which were his last contribution to the teaching of physics before his death in Chicago on November 29 of the same year.

Among the main results of the expedition one can recall a number of determinations of the values of the masses of heavy mesons giving rise to different decays which brought suspicion that all these processes could be alternative decays of the same particle. Also the identification of the various hyperons and the determination of their masses and decay modes were considerably improved [107].

In July of the same year, 1953, a very important conference had been held

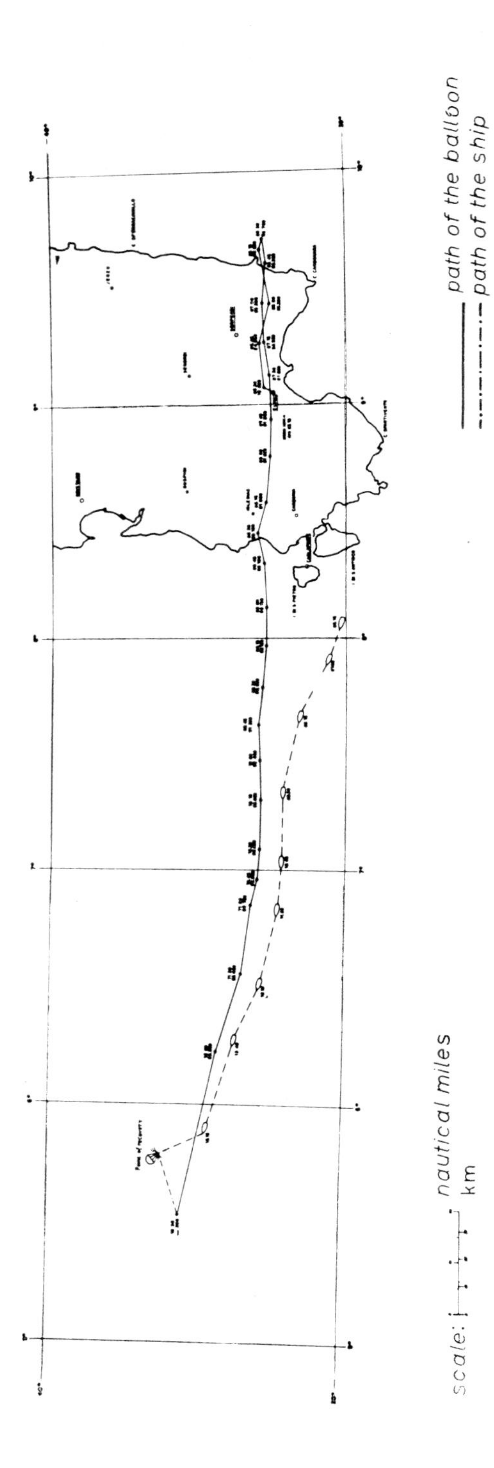


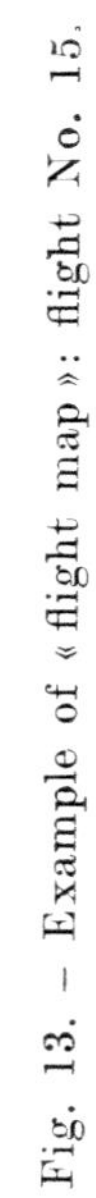


Fig. 13. – Example of « flight map »: flight No. 15.

at Bagnères de Bigorre [108] in the French Pyrénées where the new heavy particles were discussed at length. At this meeting, for example, the Dalitz-Fabri plot for the $\tau$-meson was discussed by DALITZ, MICHEL presented the selection rules for decay processes and the Rome Group introduced the logarithm of $\mathrm{tg}\,\theta$ as a very appropriate random variable for the description of high-energy events in which many secondary particles are emitted.

A third expedition was organized by the Universities of Bristol, Milan and Padua in October 1954 from Northern Italy (Novi Ligure). It consisted in the launching of a single stack of 15 litres of nuclear emulsions (corresponding to a weight of 63 kg) which is indicated in the literature as « *G*-stack » [109].

In the landing, which took place in the Apennines, the stack was partly damaged. The delicate operation of recovery of the damaged emulsions was made by the Milan group, while its processing, involving an exercise in small-scale chemical engineering, was made at the Universities of Bristol and Padua.

The great advantage in employing a very large stack is evident for studying the different modes of decay of the heavy mesons produced in the collision of a high-energy particle with a nucleus of the emulsions. A substantial fraction of the secondaries come to rest inside the stack so that their modes of decay can be observed and the energy determined with high accuracy from their range.

A great part of the results [110] was presented at the International Conference on Elementary Particles held in Pisa in June 1955 to celebrate the Centenary of the *Nuovo Cimento* [111].

One of the most important results of the *G*-stack study was the final recognition that the values of the masses of heavy mesons giving rise to different decay processes were identical within rather small experimental errors. Thus the interpretation of all these processes as alternative decays of the same particle was strengthened, contributing in an essential way to the general recognition of the so-called $\theta$-$\tau$ puzzle. This found its explanation only in 1956 with the discovery by LEE and YANG [112] of the nonconservation of parity by weak interactions.

A striking fact that emerged in Pisa was that the time for important contributions to subnuclear particle physics from the study of cosmic rays was very close to an end. A few papers presented by physicists from the U.S.A. showed clearly the advantage for the study of these particles presented by the Cosmotron of the Brookhaven National Laboratory (3 GeV) but even more by the Bevatron of the Lawrence Laboratory in Berkeley (6.3 GeV).

In order to complete this outline of the status of elementary-particle research in Europe the creation of International Summer Schools should also be mentioned. The idea was of MRS. C. MORETTE, who married Dr. B. S. DE WITT shortly before the beginning of the first course of the International Summer School she founded in 1951 at Les Houches [113]. This example was followed by the International Summer School founded in Varenna in 1953 by POLVANI, President of the Società Italiana di Fisica. As mentioned by C. DE WITT in the ad-

Fig. 14. – The preparation of the launching of the $G$-stack (by courtesy of FRANZINETTI).

dress she gave at the ceremony of opening of the first course at Varenna [93], the program of the Italian School was in some way complementary to that of Les Houches since it regarded mainly the experimental aspects of modern physics. The subjects of the first two courses (1953 and 1954) both regarded elementary particles.

At this point it appears in order to add the following remarks. The nuclear-emulsion technique was used in those years for the study of subnuclear particles in cosmic rays more extensively in Europe than in the U.S.A. where the group of SCHEIN, working at the Chicago University, was, however, one of the most efficient and outstanding.

The cloud chamber technique was employed by an even larger number of groups in the U.S.A. Among these one should recall ANDERSON *et al.* at Pasadena, BRODE, FRETTER *et al.* at Berkeley, THOMPSON at the Indiana University and ROSSI, BRIDGE *et al.* at M.I.T.

Many very important results were obtained by all these groups which are not mentioned at length here only because the lecture refers mainly to the work done in Europe. Furthermore, the collaboration was always very active between the groups working in the two continents. In particular, the collaboration was very tight between the groups of M.I.T. and the École Polytechnique.

3'4. – This very superficial review of the state of development of high-energy physics in Europe in the period 1945-1950, with its extension to later times, is sufficient, I believe, to explain how in different places—in a more or less independent way—the idea of a wide European collaboration started to be considered with the aim of tackling and solving in common those problems that no single European country would have had the technical, organizational and financial strength to face and solve successfully. But how did one arrive at the creation of an organization of the type of CERN and what were the procedures followed and the problems faced in order to reach such a goal?

If one looks at the chronicle of the time one finds that the prehistory of CERN can be divided in three periods [114-116]. The first one includes the first initiatives and extends from about 1948 to 15th February 1952. On that date the representatives of eleven European governments [117] signed in Geneva the so-called *Agreement* establishing a provisional organization with the aim of planning an International Laboratory and organizing other forms of co-operation in nuclear research.

The second period, the so-called *planning stage* extends from February 1952 to 1st July 1953 when the *Convention* establishing the final organization was signed by the representatives of twelve European governments.

Finally, the third period, usually called *interim stage*, runs from July 1953 to 29th September 1954, when a prescribed point of the ratification procedure by the Parliaments of the Member States was reached and the Convention entered into force.

It is impossible to establish the beginning of the first stage since around 1948 the idea of starting an international collaboration among European countries in the field of nuclear physics and elementary particles was making its first appearance, in more or less nebulous forms, in many places. Many scientists were aware of the continually increasing gap between the means available in Europe for research in general and in particular for research in the field of nuclear physics and elementary particles, and the means available in the United States where a few high-energy accelerators had started to produce results, while others had already reached advanced stages of construction or design [99]. It was becoming more and more evident that such a situation would be changed only by a considerable effort made in common by many European nations.

Most of the historical remarks that I will give in the following are based mainly on my diary of that period and on my personal recollections, with the result that the presentation may have some advantage of liveliness but the obvious disadvantage of being onesided.

3'5. – I remember that in the years 1948-1950 the various aspects of the problem including energy and cost of machines were examined in Rome in frequent discussion between FERRETTI and myself and in letters exchanged with BERNARDINI, who in those years was at Columbia University and thus had the opportunity of contributing to stimulating the interest of RABI in this subject.

I remember that I became aware that similar problems were discussed in other European countries, in particular in France, when I heard of the « European Cultural Conference » held in Lausanne in December 1949. At the meeting a message from DE BROGLIE was read by DAUTRY, Administrator of the French Commissariat à l'Energie Atomique. In the message the proposal was made to create in Europe an international research institution without mentioning, however, nuclear physics or fundamental particles [115].

At that time the opinions were still rather divided about the type of research to be tackled by the new organization and the nature of the collaboration to be established.

In June 1950 the General Assembly of UNESCO was held in Florence and RABI, who was a member of the delegation from the U.S.A., made a very important speech about the urgency of creating regional centres and laboratories in order to increase and make more fruitful the international collaboration of scientists in fields where the effort of any one country in the region was insufficient for the task. In the official statement approved unanimously by the General Assembly along the same lines, neither Europe nor high-energy physics were mentioned. But this specific case was clearly intended by many people taking part in the Assembly, in particular by AUGER, who was Director of Natural Sciences of UNESCO, and by RABI, with whom I had discussed the subject at length a few days before.

A further endorsement came from IUPAP which was at that time under the presidency of KRAMERS. I was one of the vice-presidents and I had asked KRAMERS, at the beginning of the Summer 1950, to include the discussion of Rabi's proposal, with specific reference to Europe and high-energy physics, in the agenda of the meeting of the Executive Committee of IUPAP that was to take place at the beginning of September of the same year in Cambridge, Mass. Although KRAMERS could not preside at the meeting because of bad health, the problem was discussed at length under the chairmanship of Sir DARWIN [118] assisted by the Secretary General FLEURY. As a conclusion I was asked to get in contact with RABI and with physicists from various European countries in order to clarify the aims and structure of the new organization and to help in the co-ordination of the different efforts. My first step was to write to AUGER [119] who in the meantime had presented the problem to the conference on nuclear physics held in Oxford during the month of September, where he had found enthusiastic support from many parts, in particular from the young physicists.

AUGER had now the authority to act but there was no money appropriated on the scale required for a detailed expert study of such a project.

In December 1950, however, the European Cultural Centre (which was founded at the already mentioned Lausanne meeting of 1949) called at Geneva a commission for scientific co-operation. AUGER, KRAMERS, FERRETTI, PREISWERK and RANDERS were present. As a result of the meeting funds were made available, immediately by the President of the Italian Council of Research Prof. COLONNETTI, and soon afterwards by the French and Belgian Governments. The total sum collected by AUGER was very modest, about $ 10 000; it was, however, sufficient to initiate the first steps for arriving at the planning and construction of a large particle accelerator.

At the beginning of 1951 AUGER established a small office at UNESCO and invited me to Paris at the end of April to discuss the constitution of a working group of European physicists interested in the problem.

The first meeting of this «Board of Consultants» was held at UNESCO, in Paris, at the end of May 1951. Two goals were immediately established: a long-range, very ambitious, project of an accelerator second to none in the world, and in addition the construction of a less powerful and more standard machine which could allow at an earlier date experimentation in high-energy physics by European teams.

Many other features of the new organization were also examined, as preparatory work for a *conference of delegates of governments* that was convened by UNESCO in Paris, in December 1951. This conference, which took place under the chairmanship of DE ROSE, a French diplomat who a few years later was elected President of CERN, led to the signing of the Agreement, which took place, as I said before, in Geneva in February 1952.

The first problem tackled by the Council of the provisional organization created by the agreement was the nomination of the officers responsible for the appointment of the remainder of the staff and for planning the laboratory.

BAKKER from the Netherlands was nominated Director of the Synchro-Cyclotron Group, DAHL from Norway (with GOWARD from the U.K. as deputy) Director of the Proton-Synchrotron Group, KOWARSKI from France (with PREISWERK from Switzerland as deputy) Director of the Laboratory Group, which had to take care of site, buildings, workshops, administrative forms, financial rules, etc., BOHR Director of the Theoretical Group and finally myself as Secretary General with the task of maintaining cohesion between the four groups of which the provisional organization was composed.

Almost all these people had worked on the Board of Experts, nominated by UNESCO, during the first stage and contributed later to the creation of the new organization and to the development of its activities.

In July of the same year, 1952, an international nuclear-physics conference was held in Copenhagen; on that occasion the type of accelerator to be built as the main goal of the new European organization was amply discussed. A report of the conclusions reached by the participants was presented by HEISENBERG to the Council which held its Second Meeting in Copenhagen immediately after the Conference. Thus the decision was taken that the Proton-Synchrotron Group should explore the possibility of constructing a 10 GeV proton-synchrotron which, at that time, represented the biggest machine in the world.

During the month of August DAHL and GOWARD went to Brookhaven in order to study in detail the *Cosmotron* (with a maximum energy of about 3 GeV) that was very close to completion [99]. During their two-week visit, and in some way in connection with the discussions going on in relation to the European project, COURANT, LIVINGSTONE and SNYDER came out with the « strong-focusing principle » [120].

This important discovery came soon enough to allow a change of the plans of the provisional organization: with the approval of the Council, the PS Group embarked on the study of a strong-focusing PS of $(20 \div 30)$ GeV instead of the weak-focusing 10 GeV machine considered until then.

During the summer of 1952 four sites were offered for the construction of the new laboratory: one near Copenhagen, one near Paris, one in Arnhem in the Netherlands and one in Geneva.

After long and lively discussions the site in Meyrin near Geneva was unanimously selected [103].

Another point I would like to touch about those times concerns the participation of the European nations.

All the European members of UNESCO had been invited to the Conference opened in Paris in December 1951 and closed in Geneva on 15th February 1952, but no response came from the countries of Eastern Europe. Furthermore, you have probably noticed that while the agreement was signed at that date by the representatives of eleven countries, the Convention establishing the permanent organization was signed by twelve countries.

The difference was due to the U.K. which, at the beginning, was rather cautious in committing itself to take part in the new organization. The U.K. government preferred to remain in the formal position of an observer during the first two stages, while, by signing the Convention, it became a full-right member of the permanent organization.

I remember that in the Autumn of 1952 it was decided that BAKKER, DAHL and myself should go to Brookhaven to take part in the dedication of the cosmotron that was foreseen for December 15th.

Our trip was already arranged when I was called on the telephone from London by Sir John COCKCROFT who had been from the start very much in favour of the participation of the U.K. in the new venture.

As a consequence of our conversation on the telephone I decided to leave earlier and to pass through London on my way to Brookhaven.

In London I went to the D.S.I.R. where I met its chairman Sir Ben LOCKSPEISER, who a few years later was elected president of the permanent CERN.

After rather long discussions about various organizational and financial aspects of the project as a whole, Sir John brought me to Lord CHERWELL [121] who, at that time, was a member of the Churchill government.

Lord CHERWELL appeared to be very clearly against the participation of the U.K. in the new organization. As soon as I was introduced in his office he said that the European laboratory was to be one more of the many international bodies consuming money and producing a lot of papers of no practical use. I was annoyed and answered rather sharply that it was a great pity that the U.K. was not ready to join such a venture which, without any doubt, was destined to full success, and I went on by explaining the reasons for my convictions. Lord CHERWELL concluded the meeting by saying that the problem had to be reconsidered by His Majesty's Government.

When we left the Ministry of Defence, where the meeting had taken place, I was rather unhappy about my lack of self-control, but Sir John and Sir Ben were rather satisfied and tried to cheer me up.

A few weeks later Sir Ben LOCKSPEISER wrote an official letter asking that the status of observer be given to the U.K. in the provisional organization, and the D.S.I.R. started to regularly pay «gifts», as they were called, corresponding exactly to the U.K. share calculated according to the scale adopted by the other eleven countries.

In spite of its very particular legal position, the U.K. gave in practice fundamental support to the provisional organization and was the first among the European countries to ratify the Convention.

3·6. – The four groups at the beginning had started to work at the institutions of the corresponding directors. In October 1953 the PS staff was assembled in Geneva, partly at the Institute of Physics of the University and partly in temporary huts built in its vicinity. At that time the transition took place from mainly theoretical work to experimentation and technical designing. In the same month of October an International Conference on protonsynchrotrons was organized at the Institute of Physics by the PS staff. In March 1954 GOWARD died after a short and tragic illness. He was succeeded by ADAMS who became Director of the PS Division in 1955 when DAHL went back to Norway to direct the design and construction of the first Norwegian nuclear research reactor.

During October 1953 an administrative nucleus began to function in a temporary Geneva office and, from January 1954 on, in the Villa Cointrin at the Geneva airport.

First instrumentation workshops, then a library and a few laboratories were gradually set up also at the airport in the Summer 1954. On August 13 of that same year the bulldozers started to break the ground at Meyrin and the construction of the European laboratory began.

The SC group remained centered in Amsterdam and the theoretical study group in Copenhagen, where the training of young theoreticians recruited as CERN Fellows went on. The training of experimentalists went on at Uppsala [122], Liverpool [99] and the Jungfraujoch. Through an agreement between CERN and Imperial College the cloud chamber at this high-altitude station was operated jointly by the two institutions starting from 1955 [123].

At the end of the interim period (September 1954) the total staff of CERN, not counting the holders of fellowships and the consultants on a part-time fraction, amounted to 120, of which about one half belonged to the PS group. The total floor area at disposal of the staff was some 2500 square metres.

On September 29, 1954, when a prescribed point in the ratification procedure was reached all the assets of provisional CERN became suddenly masterless. For eight days I had the honour—as Secretary General—of sole responsibility of ownership on behalf of a new-born permanent organization. Then the first meeting of the permanent Council assembled on the 7th October in Geneva, the Secretary General presented his final report, BLOCH was nominated Director General and CERN entered its final permanent form. The SC was operated for the first time in 1958 and the proton beam circulated in the PS on November 29, 1959, *i.e.* almost one year before the time schedule.

## REFERENCES


[1] E. AMALDI: *The production and slowing-down of neutrons*, in *Handbuch der Physik*, Vol. **38**/2, edited by S. FLÜGGE (Berlin, 1959), p. 1.

[2] E. FERMI: *Collected Papers*, Vol. **1**, Italy 1921-1938, Accademia Nazionale dei Lincei and the University of Chicago Press (1962), edited by E. AMALDI, H. L. ANDERSON, E. PERSICO, C. S. SMITH, A. WATTENBERG and E. SEGRÈ.

[3] E. SEGRÈ: *Enrico Fermi, Physicist* (Chicago, 1970).
[4] L. FERMI: *Atoms in the Family* (Chicago, 1954).
[5] I. CURIE and F. JOLIOT: *Compt. Rend.*, **198**, 254 (1934); *Nature*, **133**, 201 (1934).
[6] I. CURIE and F. JOLIOT: *Compt. Rend.*, **198**, 559 (1934); *Journ. Phys. Rad.*, **5**, 153 (1934).
[7] Allocution de M. F. PERRIN, presenté à la *Séance Commémorative du XXX Anniversaire de la Découverte de la Radioactivité Artificielle, Grand Amphitéâtre de la Sorbonne, Vendredi 3 Juillet 1964*, published at page 21 of Vol. 1 of the proceedings of the *Congrès International de Physique Nucléaire, Paris, 2-8 Juillet 1964.*
[8] The JOLIOT-CURIES proved, for example, that the radioactive nuclide produced by bombarding aluminium with α-particles has the chemical properties of phosphorous. Therefore the reaction produced necessarily involves the absorption of the incident α without emission of charged particles. Furthermore aluminium consists of a single isotope, ${}^{27}_{13}Al$; therefore one can conclude that the observed transformation is probably due to an (α, n) reaction:

*a*) $${}^{27}_{13}\mathrm{Al}(\alpha, \mathrm{n})^{30}_{15}\mathrm{P}\ .$$

Since the nuclide ${}^{30}_{15}P$ does not exist in Nature, it is very reasonable to consider that it is unstable and, therefore, decays with the emission of positrons

*b*) $${}^{27}_{13}\mathrm{P} \rightarrow {}^{30}_{14}\mathrm{Si} + \mathrm{e}^{+}\ .$$

${}^{30}Si$ is stable and was already known to be the final product of the reaction

*c*) $${}^{27}_{13}\mathrm{Al}(\alpha, \mathrm{p})^{30}_{41}\mathrm{Si}\ .$$

[9] I. CURIE and F. JOLIOT: *Journ. Phys. Rad.*, **4**, 494 (1933).
[10] Sir E. RUTHERFORD, J. CHADWICK and C. D. ELLIS: *Radiations from Radioactive Substances* (Cambridge, 1930).
[11] F. RASETTI: *Il nucleo atomico* (Bologna, 1936); *Elements of Nuclear Physics* (New York, 1936).
[12] E. FERMI and F. RASETTI: *Ric. Scient.*, **4** (2), 299 (1933).
[13] E. FERMI: *Nature*, **125**, 16 (1930), and p. 328 of ref. [2]; *Zeits. Phys.*, **60**, 320 (1930) and p. 336 of ref. [2]; E. FERMI and E. SEGRÈ: *Zeits. Phys.*, **82**, 11 (1932) and p. 514 of ref. [2].
[14] For more detail on this point see the second Section.
[15] E. FERMI: *Nuovo Cimento*, **11**, 1 (1934); *Zeits. Phys.*, **88**, 161 (1934).
[16] E. FERMI: *Ric. Scient.*, **5** (1), 283 (1934) and p. 645 of ref. [2].
[17] I. NODDACK: *Ang. Chem.*, **47**, 653 (1934).
[18] O. HAHN and L. MEITNER: *Naturwiss.*, **24**, 158 (1936); *Ber. Dtsch. Chem. Ges.*, **69**, 905 (1936); L. MEITNER: *Kernphysik*, edited by E. BRETSCHER (Berlin, 1936), p. 24.
[19] E. MCMILLAN and P. H. ABELSON: *Phys. Rev.*, **57**, 1185 (1940).
[20] M. MAYER: *Phys. Rev.*, **60**, 84 (1941). The same problem has been discussed by Y. SUGUIRA and H. UREY (*Kgl. Danske Vid. Selsk. Mat.-fys. Medd.*, **7**, 3 (1926)) on the basis of the old quantum theory.
[21] E. FERMI: *Zeits. Phys.*, **48**, 73 (1928).
[22] The letter of RUTHERFORD shown in Fig. 5 reads as follows:

June 20th, 1934

Dear Professor FERMI,

Your letter has been forwarded to me in the country where I am taking a holiday. I saw your account in « Nature » of the effects on uranium. I congratulate you and your colleagues on a splendid piece of work. I have had two of my men WESTCOTT and BJERGE repeating some of your experiments with an Em+Be tube and promised them when I returned in a week or two to try out for them the effect of the 2 million volt neutrons from the D+D reaction on a few elements. I myself have been naturally interested in the energy of the neutron required to start transmutations. We do not, however, propose at the moment to make systematic experiments in this direction as our installation is required for other purposes. I cannot at the moment give you a definite statement as to the output of neutrons from our tube but it should be of the same order as from an Em+Be tube containing 100 millicurie and may be pushed much higher.

If your assistants come to Cambridge say in the first week of July, I shall be delighted to give them the benefit of our experience and to see the mode of operation of our installations for transmutations in general.

I hope one or both speak English as the knowledge of Italian in the laboratory is very modest. The two men to see are Dr. OLIPHANT and Dr. COCKCROFT. Excuse the hand written letter, I do not take a secretary with me in holiday!

Yours sincerely,
RUTHERFORD

[23] E. FERMI, E. AMALDI, O. D'AGOSTINO, F. RASETTI and E. SEGRÈ: *Proc. Roy. Soc.*, A **146**, 486 (1934).

[24] J. CHADWICK and M. GOLDHABER: *Nature*, **134**, 237 (1934).

[25] W. BOTHE and H. BECKER: *Zeits. Phys.*, **66**, 289 (1930). See also the report by BOTHE on page 153 of the *Proceedings of the International Conference*, held in Rome in 1931 as well as W. BOTHE and H. BECKER: *Naturwiss.*, **20**, 349 (1932); *Zeits. Phys.*, **76**, 421 (1932).

[26] F. RASETTI: *Naturwiss.*, **20**, 252 (1932).

[27] I. CURIE and F. JOLIOT: *Compt. Rend.*, **193**, 1412 (1932); F. JOLIOT: *Compt. Rend.*, **193**, 1415 (1932).

[28] According to this formula the cross-section is inversely proportional to the square of the mass of the target particle.

[29] I. CURIE and F. JOLIOT: *Compt. Rend.*, **194**, 708 (1932).

[30] E. AMALDI: *La vita e l'opera di Ettore Majorana*, Accademia Nazionale dei Lincei (1966): translated into English as: *Ettore Majorana, man and scientist*, in *Strong and Weak Interactions. Present Problems*, edited by A. ZICHICHI (New York, 1966).

[31] J. CHADWICK: *Nature*, **129**, 322 (1932); see also by the same author: *Proc. Roy. Soc.*, A **136**, 692 (1932).

[32] Sir E. RUTHERFORD: *Phil. Mag.*, **4**, 580 (1927).

[33] J. CHADWICK: *Actes du X Congrès International d'Histoire des Sciences* (Ithaca, N. Y., 1962), (Paris, 1964), Vol. **1**, p. 159.

[34] G. GENTILE: *Rend. Acad. Lincei*, **7**, 346 (1928).

[35] F. PERRIN: *Compt. Rend.*, **195**, 236 (1932); see also W. HEISENBERG: *Rapports et Discussions du VII Congrès de l'Institut International de Physique Solvay* (1934), p. 289.

[36] D. IWANENKO: *Nature*, **129**, 798 (1932).

[37] More detail about the refusal by MAJORANA to publish his results can be found in ref. [30, 2, 3].

[38] E. MAJORANA: *Zeits. Phys.*, **82**, 137 (1933); *Ric. Scient.*, **4** (1), 559 (1933).

[39] W. HEISENBERG: *Zeits. Phys.*, **77**, 1 (1932); **80**, 156, 587 (1932).

[40] T. BJERGE and C. H. WESTCOTT: *Nature*, **134**, 286 (1943).

[41] In this case the proof is very strong, because aluminium is formed by the single

isotope $^{27}$Al and $^{22}_{11}$Na was already known to emit positrons instead of electrons, as would be expected for nuclides having an excess of neutrons with respect to the corresponding stable isotopes. In the case of the (n, 2n) processes one would expect, at least in general, an emission of positrons.

[42] This as well as most notebooks of that period have been collected by Mr. Lodovico ZANCHI and me and later have been deposited at the Domus Galilaeana in Pisa: E. AMALDI: *The Fermi Manuscripts at the Domus Galilaeana*: *Physics*, **1** (2), 69 (1959); T. DERENZINI: *Analisi dei manoscritti di Enrico Fermi alla Domus Galilaeana*, *Physics*, **6** (1), 75 (1954).

[43] In the introductory remarks to Fermi's paper on *The Magnetic Fields in Spiral Arms* (*Fermi's Collected Papers*, Vol. **2**, p. 927) CHANDRASEKHAR wrote that Fermi told him once more or less what follows:

« I will tell you how I came to make the discovery which I suppose is the most important one I have made. We were working very hard on the neutron-induced radioactivity and the results we were obtaining made no sense. One day, as I came to the laboratory, it occurred to me that I should examine the effect of placing a piece of lead before the incident neutrons. Instead of my usual custom, I took great pains to have the piece of lead precisely machined. I was clearly dissatisfied with something: I tried every excuse to postpone putting the piece of lead in its place. I said to myself: "No, I do not want this piece of lead here; what I want is a piece of paraffin". It was just like that with no advance warning, no conscious prior reasoning. I immediately took some odd piece of paraffin and placed it where the piece of lead was to have been ».

In a short note dedicated to Norman FEATHER—to mark his twenty-five-year tenure of the chair of Natural Philosophy at the University of Edimburgh—Maurice GOLDHABER has circulated at the end of 1971 a few *Remarks on the Prehistory of the Discovery of Slow Neutrons*. Having read in Segrè's book (ref. [3]) the conversation of FERMI with CHANDRASEKHAR reported above, GOLDHABER apparently had the impression that the idea of using paraffin instead of lead could suddenly have come in the mind of FERMI as a consequence of what had been written by CHADWICK and GOLDHABER at the end of the paper on the photodisintegration of the deuteron (ref. [24]). What actually CHADWICK and GOLDHABER state is that the value obtained for the cross-section of the inverse process—*i.e.* the radiative capture of fast neutrons by protons to form deuterons—calculated by applying the detailed balance to their experimental results was too small to provide an explanation for the gamma-ray emission by hydrogeneous substances observed by D. E. LEA. The idea that neutrons could lose energy through a number of collisions is not mentioned although it is possible that CHADWICK and GOLDHABER had thought about it.

In order to logically arrive at recognizing the high efficiency of slow neutrons starting from the photodisintegration of the deuteron, a rather long chain of assumptions of unknown facts was necessary in 1934:

*a*) The extrapolation of the cross-section of photoeffect to very low energy gives a negligible value, as stated by GOLDHABER in his *Remarks*. Only the photomagnetic effect is important at very low energy and this was introduced only in 1936 by FERMI who proved that the corresponding cross-section follows the $1/v$ law and reaches the value of $\sim 0.3$ barns at energies around $kT \simeq 0.025$ eV.

*b*) The slowing down is determined by the *elastic cross-section* $\sigma_e$ which is not related in a simple way to that of the previous process. At energies around

1 eV $\sigma_e$ is about 20 barn, as it was shown by us in Rome—as well as by other authors—in the course of 1935-1936.

*c*) From 1 eV to thermal energy the elastic cross-section increases by another large factor due to chemical bond of hydrogen in molecules. This effect was found experimentally by FERMI and myself only in 1936 and explained correctly by FERMI the same year.

In conclusion it appears inconceivable that FERMI arrived at the idea of the slowing-down through this complicated reasoning. On the contrary it seems to me that the fact that in the mind of FERMI came suddenly the idea of using paraffin instead of lead is remarkable but not so extraordinary if one considers the many experiments made by our group on irradiation under various conditions. These were discussed every day with all the group: at the beginning RASETTI tried to explain our strange results as due to some mistake made by PONTECORVO and me. We protested and proved we were not wrong and this gave rise to long and vivacious discussions in which FERMI took always an active part.

[44] E. AMALDI, O. D'AGOSTINO and E. SEGRÈ: *Ric. Scient.*, **5** (2), 381 (1934).

[45] E. FERMI, E. AMALDI, B. PONTECORVO, F. RASETTI and E. SEGRÈ: *Ric. Scient.*, **5** (2), 282 (1934) and p. 757 of ref. [2].

[46] O. D'AGOSTINO, E. AMALDI, F. RASETTI and E. SEGRÈ: *Ric. Scient.*, **5** (2), 467 (1934) and p. 655 of ref. [2].

[47] P. B. MOON and J. R. TILLMAN: *Nature*, **135**, 904 (1935); *Proc. Roy. Soc.*, A **153**, 476 (1936). The indication of a small effect was found independently by J. R. DUNNING, G. B. PEGRAM, D. P. MITCHELL and G. A. FINK: *Phys. Rev.*, **47**, 888 (1935).

[48] E. FERMI, B. PONTECORVO and F. RASETTI: *Ric. Scient.*, **5** (2), 380 (1934) and p. 759 of ref. [2].

[49] E. AMALDI, O. D'AGOSTINO, E. FERMI, B. PONTECORVO, F. RASETTI and E. SEGRÈ: *Proc. Roy. Soc.*, A **149**, 522 (1935) and p. 765 of ref. [2].

[50] E. FERMI and F. RASETTI: *Nuovo Cimento*, **12**, 201 (1935) and p. 795 of ref. [2]; B. PONTECORVO: *Nuovo Cimento*, **12**, 211 (1935); E. AMALDI: *Nuovo Cimento*, **12**, 233 (1935); E. SEGRÈ: *Nuovo Cimento*, **12**, 232 (1935).

[51] J. R. DUNNING, G. B. PEGRAM, G. A. FINK and D. P. MITCHELL: *Phys. Rev.*, **48**, 265 (1935).

[52] E. FERMI: *Nuovo Cimento*, **2**, 157 (1934) and p. 706 of ref. [2].

[53] E. AMALDI and E. SEGRÈ: *Nature*, **133**, 141 (1934); *Nuovo Cimento*, **11**, 145 (1934).

[54] E. AMALDI, O. D'AGOSTINO, E. FERMI, B. PONTECORVO and E. SEGRÈ: *Ric. Scient.*, **6** (1), 581 (1935) and p. 669 of ref. [2].

[55] E. AMALDI: *Phys. Zeits.*, **38**, 692 (1937).

[56] O. FRISCH and E. T. SØRENSEN: *Nature*, **136**, 258 (1935).

[57] J. DUNNING, G. B. PEGRAM, G. A. FINK, D. P. MITCHELL and E. SEGRÈ: *Phys. Rev.*, **48**, 704 (1935).

[58] B. PONTECORVO and G. C. WICK: *Ric. Scient.*, **7** (1), 134, 220 (1936).

[59] T. BJERGE and C. H. WESTCOTT: *Proc. Roy. Soc.*, A **150**, 709 (1935).

[60] P. B. MOON and J. R. TILLMANN: *Nature*, **135**, 904 (1935); **136**, 106 (1935). See also L. ARTSIMOVICH, I. KOURTCHATOV, G. LATYSCHEV and W. CROMOW: *Zeits. Sovjet.*, **8**, 472 (1935); B. PONTECORVO: *Ric. Scient.*, **6** (2), 145 (1935); D. P. MITCHELL, J. R. DUNNING, E. SEGRÈ and G. B. PEGRAM: *Phys. Rev.*, **48**, 175 (1935).

[61] E. FERMI and E. AMALDI: *Ric. Scient.*, **6** (2), 344 (1935) and p. 808 of ref. [2].

[62] L. SZILARD: *Nature*, **136**, 951 (1935). Some time later a similar behaviour was

found also for Au by O. R. Frish, G. Hevesy and H. A. C. Mc Kay: *Nature*, **137**, 149 (1936).

[63] E. Amaldi and E. Fermi: *Ric. Scient.*, **6** (2), 443 (1935) and p. 816 of ref. [2].

[64] The expression « albedo » is used by astronomers for the fraction of incident light diffusely reflected from the surface of a planet or satellite. The most common use refers to the Moon. It is, however, also used for other surfaces, for instance, for snow.

[65] E. Amaldi and E. Fermi: *Ric. Scient.*, **7** (1), 56 (1936).

[66] The first indication was provided by the fact that a velocity distribution roughly Maxwellian was observed (ref. [55]) using a velocity selector in which the « selection » was made by means of Cd absorbers. This argument was confirmed and strengthened by the observation of P. Preiswerk and H. von Halban (*Nature*, **136**, 951 (1936)) in the case of Ag and of F. Rasetti and G. A. Fink (*Phys. Rev.*, **49**, 642 (1936)) in the case of Ag, Rh, In and I, that while all these elements show a thermal effect of the type of that investigated by Moon and Tillman (ref. [47]), no effect at all is observed if the same elements are screened with Cd. This means that the absorption lines of these elements are above the thermal region, which, on the contrary, is practically all included in the region absorbed by Cd.

[67] E. Amaldi and E. Fermi: *Ric. Scient.*, **7** (1), 454 (1936) and p. 841 of ref. [2]; *Ric. Scient.*, *Phys. Rev.*, **50**, 899 (1936) and p. 892 of ref. [2].

[68] H. von Halban and P. Preiswerk: *Nature*, **136**, 951, 1027 (1935); *Compt. Rend.*, **202**, 840 (1936).

[69] E. Amaldi and E. Fermi: *Ric. Scient.*, **7** (1), 223 (1936) and p. 828 of ref. [2]; **7** (1) 393 (1936) and p. 837 of ref. [2].

[70] G. Breit and E. Wigner: *Phys. Rev.*, **49**, 519 (1936).

[71] N. Bohr: *Nature*, **137**, 344 (1936); *Naturwiss.*, **24**, 241 (1936).

[72] E. Fermi: *Ric. Scient.*, **7** (2), 13 (1936) and p. 943 of ref. [2]. The English translation by G. N. Temmer of this fundamental paper is given at p. 990 of ref. [2].

[73] E. Amaldi and E. Fermi: *Ric. Scient.*, **7** (1), 310 (1936).

[74] O. R. Frish and G. Placzek: *Nature*, **137**, 357 (1936). The same arguments were developed independently by D. F. Weeks, M. S. Livingston and H. A. Bethe: *Phys. Rev.*, **49**, 471 (1936).

[75] This expression comes from Aldous Huxley's novel « Brave New World » and refers to a pill with sexual hormones base used by men in the year 2000 to combat spleen.

[76] Report on an *International Conference on Fundamental Particles and Low Temperature*, held at the Cavendish Laboratory, Cambridge on 22-27 July 1946, Vol. **1**, *Fundamental Particles* (London, 1947).

[77] M. Conversi, E. Pancini and O. Piccioni: *Nuovo Cimento*, **3**, 372 (1945); *Phys. Rev.*, **71**, 209 (1947).

[78] S. Tomonaga and G. Araki: *Phys. Rev.*, **58**, 90 (1940). These authors had foreseen that only positive mesons at rest should undergo spontaneous decay, whereas the negative ones should be captured in a time far shorter than the meson lifetime.

[79] E. Fermi, E. Teller and V. Weisskopf: *Phys. Rev.*, **71**, 314 (1947); E. Fermi and E. Teller: *Phys. Rev.*, **72**, 399 (1947).

[80] S. S. Schweber, H. A. Bethe and F. De Hoffmann: *Mesons and Fields* (Evanston, Ill., 1956).

[81] L. Alvarez: *Recent Developments in Particle Physics* of « Les Prix Nobel of 1968 » (Stockholm, 1969), p. 125.

[82] C. M. C. Lattes, G. P. Occhialini and C. F. Powell: *Nature*, **160**, 453 (1947).

[83] D. H. Perkins: *Nature*, **159**, 126 (1947).

[84] G. P. Occhialini and C. F. Powell: *Nature*, **159**, 186 (1947).

[85] C. M. C. Lattes, G. P. Occhialini and C. F. Powell: *Nature*, **160**, 486 (1947).

[86] C. D. Rochester and C. C. Butler: *Nature*, **160**, 855 (1947). The chamber, of rather modest dimensions (28 cm in diameter and about 7 cm deep), was placed in a rather intensive magnetic field (7500 G). It was crossed by a lead plate about five cm thick. Already in 1944 L. Leprince Ringuet and M. Lhéritier (*Compt. Rend.*, **219**, 618 (1944)) had otained at the École Polytechnique in Paris a mass value very close to that established a few years later for the K-meson, from the application of the principles of conservation of energy and momentum to a cloud chamber picture of a collision of a positive particle with an electron. The result was, however, affected by such a large experimental error that it could not be considered as a proof of the existence of new particles of mass about three times greater than that of the $\pi$-meson.

[87] *Symposium on Cosmic Rays*, Krakow, October 1947, IUPAP Document RC 48-1.

[88] R. Armenteros, K. H. Barker, C. C. Butler, A. Cachon and A. H. Chapman (*Decay of V-particles*, *Nature*, **167**, 501 (1951)) identified protons in the decay of the $V^0$; R. Armenteros, K. H. Barker, C. C. Butler and A. Cachon (*The properties of neutral V-particles*, *Phil. Mag.*, **42**, 1113 (1951)) separated the $V_0^1(\equiv \Lambda^0)$ from the $V_0^2(\equiv K^0 \rightarrow \pi^+ + \pi^-)$; R. Armenteros, K. H. Barker C. C. Butler, A. Cachon and C. M. York (*The properties of charged V-particles*, *Phil. Mag.*, **43**, 597 (1952)) observed the first $\Xi$.

[89] The dimensions of the chamber were $(55 \times 55)\ \text{cm}^2 \times 16$ cm. It was placed in a 5000 G magnetic field.

[90] J. P. Astbury, J. S. Buchanan, P. Chippendale, D. D. Millar, J. A. Newth, D. I. Page, A. Ritz and A. B. Sahiar: *The mean life of charged V-particles*, *Phil. Mag.*, **44**, 242 (1953).

[91] The dimensions of the two chambers were $(68 \times 64)\ \text{cm}^2 \times 30$ cm. The upper chamber had a 3600 G magnetic field, the lower one contained 15 copper plates 1 cm thick.

[92] B. Gregory, A. Lagarrigue, L. Leprince Ringuet, F. Muller and Ch. Peyrou: *Nuovo Cimento*, **11**, 292 (1954). This paper contains the suggestion of the $K_{\mu 2}$ decay mode and mass measurements by momentum and range. R. Armenteros, B. Gregory, A. Hendel, A. Lagarrigue, L. Leprince Ringuet, F. Muller and Ch. Peyrou: *Nuovo Cimento*, **1**, 915 (1955). It contains the final proof of the $K_{\mu 2}$ decay mode and the recognition that the $K_\mu$-particle is essentially positive.

[93] A review of the cloud chambers operated in the various laboratories with a discussion of the possibilities and limits of this technique is given in P. M. S. Blackett: *Lectures* (p. 264) *to the First International Summer School in Varenna*: *Rendiconti del Corso tenuto nella Villa Monastero a Varenna 19 Agosto-12 Settembre 1953*, Vol XI (Serie IX), *Suppl. Nuovo Cimento*, **1**, 141 (1953).

[94] G. Bernardini, C. Longo and E. Pancini: *Ric. Scient.*, **18**, 91 (1948); G. Bernardini and E. Pancini: *Ric. Scient.*, **20**, 966 (1950).

[95] A. Rostagni: *L'Energia Elettrica*, **28**, 211 (1951). A better laboratory was rebuild in 1953. For more detail on all these laboratories see: *The World's High Altitude Research Stations*, Research Division, College of Engineering, New York University.

[96] The following researches were made at the Laboratorio della Testa Grigia-Nuclear disintegrations produced by cosmic rays were studied by G. Bernar:

Dini *et al.* with nuclear emulsions (*Phys. Rev.*, **74**, 845, 1878 (1948); **76**, 1792 (1949); **79**, 952 (1950); *Nature*, **163**, 981 (1949)), by I. F. Quercia *et al.* (*Nuovo Cimento*, **7**, 457 (1950)) and E. Amaldi *et al.* (*Nuovo Cimento*, **7**, 697 (1950)) with ionization chambers. Extensive showers by E. Amaldi *et al.* (*Nuovo Cimento*, **7**, 401, 816 (1950)) and C. Ballario *et al.* (*Phys. Rev.*, **83**, 666 (1951)); the electromagnetic component produced in nuclear explosions by G. Salvini *et al.* with a cloud chamber (*Nuovo Cimento*, **6**, 207 (1949); **7**, 36, 943 (1950); *Phys. Rev.*, **77**, 284 (1950)); penetrating showers from hydrogen and other elements by M. Conversi, G. Fidecaro *et al.* (*Nuovo Cimento*, **1**, 330 (1955)); the determination of the mean lifetime of the new strange particles by means of a cloud chamber by P. Astburry, C. Ballario, R. Bizzarri, A. Michelini, E. Zavattini and A. Zichichi *et al.* (*Nuovo Cimento*, **2**, 365 (1955); **6**, 994 (1957)).

[97] The following researches were made at the Laboratorio della Marmolada: P. Bassi *et al.* studied with counter techniques the positive excess of mesons (*Nuovo Cimento*, **8**, 469 (1957)), with ionization chambers the zenith distribution of the nucleonic component (*Nuovo Cimento*, **9**, 722 (1952)) and of the particles of extensive showers (*Nuovo Cimento*, **10**, 779 (1953)). The last problem was investigated also by M. Cresti *et al.* (*Nuovo Cimento*, **10**, 779 (1953)). M. Deutschmann, M. Cresti *et al.* investigated by means of a cloud chamber the decay of $V^0$ (*Nuovo Cimento*, **3**, 180, 566 (1956); **4**, 747 (1956)). H. Massey, E. H. S. Burhop *et al.* the production of secondary particles in hydrogen by means of a high-pressure ($\sim$ 100 atmospheres) cloud chamber (*Nature*, **175**, 445 (1955); *Suppl. Nuovo Cimento*, **4**, 272 (1956)).

[98] *Congresso Internazionale di Fisica sui Raggi Cosmici, Como 11-16 Settembre 1949*; *Suppl. Nuovo Cimento*, **6**, 309 (1949).

[99] The accelerators capable of producing at least $\pi$-mesons that entered into operation before 1956 are the following:

| Institution | Type of machine | Energy (MeV) | Date first operated |
|---|---|---|---|
| In the U.S.A. | | | |
| Univ. of Calif. Berkeley | synchrocyclotron | 350 | 1946 |
| Univ. Rochester | synchrocyclotron | 240 | 1948 |
| Columbia Univ. | synchrocyclotron | 400 | 1950 |
| Univ. of Chicago | synchrocyclotron | 460 | 1951 |
| Carnegie Inst. Tech. | synchrocyclotron | 450 | 1952 |
| Brookhaven Nat. Laboratory (Cosmotron) | proton synchrotron | 3200 | 1952 |
| Univ. of Calif. Berkeley (Bevatron) | proton synchrotron | 5700<br>6200 | 1954<br>1955 |
| Univ. of Calif. Berkeley | electronsynchrotron | 320 | 1949 |
| Cornell Univ. | electronsynchrotron | 300 | 1951 |
| M.I.T. | electronsynchrotron | 300 | 1952 |
| Univ. of Mich. | electronsynchrotron | 300 | 1953 |
| Purdue Univ. | electronsynchrotron | 300 | 1954 |

| Institution | Type of machine | Energy (MeV) | Date first operated |
|---|---|---|---|
| Univ. of Ill. (Urbana) | betatron | 300 | 1950 |
| In Europe | | | |
| Univ. of Liverpool | synchrocyclotron | 400 | 1954 |
| Univ. of Glasgow | electronsynchrotron | 350 | 1954 |

For more information, see, for example, M. S. LIVINGSTONE and J. P. BLEWETT: *Particle Accelerators* (New York, 1962).

[100] R. M. BROWN, U. CAMERINI, P. H. FOWLER, H. MUIRHEAD, C. F. POWELL and D. M. RITSON: *Nature*, **163**, 47 (1948).

[101] G. O'CEALLAIGH: *Phil. Mag.*, **42**, 1032 (1951); M. G. K. MENON and C. O'CEALLAIGH: *Proc. Roy. Soc.*, A **221**, 2941 (1954).

[102] *Report on the Expedition to the Central Mediterranean for the Study of Cosmic Radiation*: *Napoli 18-28 May, Cagliari 30 May-13 July*, CERN/16, Rome, 30 September 1952. As much as I know a single copy of this document exists in the files of CERN. It contains also a number of photographs (in original) of the two ships (A. S. Altair and cannoniera Bracco) and of the airplane that the Italian Navy and Airforce had put at disposal of the expedition in Naples and Cagliari. The Universities that took part in the expedition were Bristol, Bruxelles, Cagliari, Genova, Glasgow, Göttingen (Max Planck Institut), London (Imperial College), Lund, Milano, Padova, Paris (École Polytechnique), Roma, Torino. The expedition started in Naples and shifted to Cagliari because of the strength and direction of the winds at high altitude. A total of thirteen balloons were launched. The document includes a report on the technical details by J. H. DAVIS from Bristol and C. FRANZINETTI from Rome.

[103] *Minutes of the Third Session, Amsterdam 4-7 October 1952*, CERN/GEN/4, Rome, 15 February 1953. At this meeting, that took place under the chairmanship of P. SCHERRER, the Council decided to propose Geneva as a site for the European Laboratory.

[104] J. DAVIS and C. FRANZINETTI: *Suppl. Nuovo Cimento*, **2**, 480 (1954). The same report, in a slightly abbreviated form, appeared also as a CERN document: CERN/GEN/11, *Report on the Expedition to the Central Mediterranean for the Study of Cosmic Radiation*. The Physics Laboratories of the following Universities participated in this expedition: Bern, Bristol, Bruxelles (Université Libre), Catania, Copenhagen, Dublin, Genova, Göttingen (Max Planck Institut), London (Imperial College), Lund, Milano, Oslo, Padova, Paris (École Polytechnique), Roma, Sidney, Torino, Trondheim, Uppsala. The balloons were constructed at Bristol and Padova under the direction of Dr. H. HEITLER and the gondolas and the radio equipment were built in Milano and Roma. Altogether the corvette Pomona covered over 5000 nautical miles for the recovery of the equipment. Until it was recovered it was kept underwater—at a depth convenient for assuring a low and constant temperature—by a buoy, whose position was tracked by a radio-wind transmitter that was heard by the receiver on the ship about 5 nautical miles away. Furthermore a radio cut-off apparatus was developed which allowed the cut-off of the «gondola» from the balloon to be operated at any time from the ship. Among the technicians who made an essential contribution to the success of the expedition A. PELLIZZONI from Roma and M. ROBERTS from Bristol should be particularly mentioned.

[105] *Rendiconti del Congresso Internazionale sulle Particelle Elementari Instabili Pesanti e sugli Eventi di Alta Energia nei Raggi Cosmici, Padova 12-15 Aprile 1954, Suppl. Nuovo Cimento,* **2**, 163 (1954).

[106] *Rendiconti del Corso che fu tenuto, nella Villa Monastero, a Varenna dal 18 Luglio al 7 Agosto 1954 a cura della Scuola Internazionale di Fisica, Suppl. Nuovo Cimento,* **2**, 1 (1955).

[107] It may be interesting to recall that also two stars connected by a heavy track were observed and tentatively interpreted as due to the production and annihilation of an antiproton (E. AMALDI, C. CASTAGNOLI, G. CORTINI, C. FRANZINETTI and A. MANFREDINI: *Nuovo Cimento,* **1**, 492 (1955)). When, about one year later, the properties of artificial antiprotons became known, doubts were raised about such an interpretation mainly because of a large-angle scattering undergone by the presumed antiproton (see for example at p. 419 of C. POWELL, P. H. FOWLER and D. H. PERKINS: *The Study of Elementary Particles by the Photograph Method* (London, 1959)). The observation, however, was at the time of some interest and is still today in the frame of the history of antiparticle hunting.

[108] *Congrès Internationale sur le Rayonnement Cosmique,* organisé par l'Université de Toulouse, sous le patronage de l'UIPPA, avec l'appui de l'UNESCO, *Bagnères de Bigorre, 6-11 Juillet 1953.*

[109] A short information on the expedition can be found in the « Introduction » by C. F. POWELL to the extensive paper of ref. [110]. This Introduction has been included at p. 318 in *Selected Papers of C. F. Powell,* edited by E. H. S. BURHOP, W. O. LOCK and M. G. K. MENON (Amsterdam, 1972). As stated by C. F. POWELL if anybody has played a distinctive and leading part in this joint effort, it was Dr. A. MERLIN from Padua.

[110] *Observations on Heavy Meson Secondaries* (*G*-stack collaboration), p. 398 of ref. [110]. Some emulsions were distributed also to the Institute for Advanced Studies and the University College in Dublin, and to the University of Genova.

[111] *Conferenza Internazionale sulle Particelle Elementari, Pisa, 12-18 Giugno 1955, Suppl. Nuovo Cimento,* **2**, 135 (1956).

[112] T. D. LEE and C. N. YANG: *Phys. Rev.,* **104**, 254 (1956).

[113] École d'Eté de Physique Theorique de l'Université de Grenoble.

[114] E. AMALDI: *Suppl. Nuovo Cimento,* **2**, 339 (1955).

[115] L. KOWARSKI: *An account of the origin and beginning of CERN*; CERN 61-10, 10 April 1961.

[116] E. AMALDI: *CERN, past and future,* p. 415 of *Topical Conference on High-Energy Collisions of Hadrons, CERN, Geneva 15-18 January 1968.*

[117] Belgium, Denmark, France, Greece, Italy, Netherlands, Norway, Sweden, Switzerland, West Germany, Yugoslavia.

[118] C. G. DARWIN, professor of Theoretical Physics at the University of Cambridge, is well known for many important contributions among which one should recall the treatment of the anomalous Zeeman effect by means of the Dirac equation and in general the study of solutions of the Dirac equation in the case of low velocities, so that the four-component Dirac spinors can be replaced by two-component spinors, of which one is large, the other small.

[119] Letter of E. AMALDI to P. AUGER: Rome, le 3 Octobre 1950

Prof. P. AUGER
12 Rue Emile Faguet
Paris XIV
Cher Professeur AUGER,

À l'occasion de la réunion du Comité Exécutif de l'Union Internationale de Physique Pure et Appliquée qui a eu lieu à Cambridge Massachusetts (U.S.A.) les jours 7-8-9 septembre, on a parlé de la proposition présentée par le Prof. I. I. RABI à l'Assemblée Génerale de l'U.N.E.S.C.O. à Florence, au sujet de la construction d'un Laboratoire Européen de Physique Nucléaire. Après quelques discussions, l'Exécutif a décidé de faire préparer deux rapports sur cet argument: l'un devrait être rédigé par RABI, qui devrait préciser aussi bien que possible sa pensée, l'autre par moi, qu'on a chargé de prendre contact avec les physiciens européens, de manière à rejoindre, si possible, un accord sur plusieurs points fondamentaux.
Je savais déjà que Vous vous interessés de l'argument pour l'U.N.E.S.C.O.; à présent je viens d'apprendre de FERRETTI qu'à l'occasion du Congrès d'Oxford Vous avez provoqué et dirigé une discussion intéressante à ce sujet. Jusqu'ici j'ai écrit seulment à RABI. Avant de m'adresser à d'autres physiciens européens, j'ai pensé vous écrire afin d'éviter que mon action puisse paraître en désaccord avec celle que vous êtes en train de développer.
Je Vous serais trés reconnaissant si Vous vouliez me communiquer aussitôt que possible, quels sont les moyens que Vous suggérez pour obtenir que l'action de l'U.N.E.S.C.O. et de l'U.I.P.P.A. résultent s'additioner efficacement.
J'ai appris de FERRETTI les lignes générales que Vous avez données à la discussion à Oxford et je suis complètement d'accord avec Vous. Surtout je suis convaincu de l'importance spécifique et générale de ce qu'un tel projét puisse se réaliser.
Je penserais d'écrire aus physiciens suivants:

| | | |
|---|---|---|
| Angleterre | — | J. D. COCKCROFT |
| France | — | ? |
| Suisse | — | P. SCHERRER |
| Allemagne | — | W. HEISENBERG |
| Belgique | — | M. COSYNS |
| Hollande | — | C. J. BAKKER |
| Danemark | — | N. BOHR |
| Suède | — | K. SIEGBAHN |
| Norvège | — | ? |
| Espagne | — | ? |
| Autriche | — | ? |

en présentant la chose et en posant quelques problèmes, par exemple:

1) lieu
2) direction
3) financement.

Quant au premier point, il me semble essentiel que le lieu soit beau et assez central, et il me sembre qu'on pourrait choisir une localité située en Suisse. Pour ce qui se rapporte au deuxième point je n'ais aucune idée particulière. Pour le troisième, je désirerais connaître votre avis et celui de RABI, pour pouvoir prèsenter des proposition concrètes. En effet je pense que, si on écrit d'une façon générique, sans proposer des solutions raisonnables, les sceptiques vont faire mourir la chose avant qu'elle naisse.

Votre avis à ce sujet me sera très précieux afin que je puisse accomplir la tâche qu'on m'a donnée aussi bien que possible, surtout pour l'effective réalisation de ce projet.

Avec mes meilleures amitiés,

(Prof. EDOARDO AMALDI)

[120] E. D. COURANT, M. S. LIVINGSTONE and H. S. SNYDER: *Phys. Rev.*, **88**, 1190 (1952).

[121] Before becoming an eminent personality in public life, Lord CHERWELL was well known as a physicist under the name of F. A. LINDEMANN. Among other contributions, his work on the theory of melting is now classical.

[122] A 200 MeV cyclotron was in operation at the Werner Institute since 1953.

[123] W. A. COOPER, H. FILTHUTH, J. A. NEWTH, G. PETRUCCI, R. A. SALMERON and A. ZICHICHI: *Nuovo Cimento*, **4**, 1433 (1956); **5**, 1388 (1957).

# Niels Bohr and the Atomic Bomb: The Scientific Ideal and International Politics, 1943-1944.

M. J. SHERWIN (*)

*Department of History, Princeton University - Princeton, N. J.*

**1.** – When the Second World War ended Manhattan Project (**) scientists organized to gain a measure of control over the atomic-energy policies of the United States government [1]. Their primary foreign-policy objective was the international control of atomic energy, and, in so far as they contributed to pressuring the government into presenting a proposal for international control to the United Nations in June, 1946, they were successful. Like other schemes for international co-operation, however, the internationalization of the atom was a casualty of the emerging cold war. Rather than serving as a new area of international co-operation, atomic-energy research became the basis of national military programs, and for an armaments race fraught with unprecedented dangers. Although scientists in America were disappointed that their government did not support international control more vigorously, by 1949 blame for the break-down of negotiations in the United Nations was generally attributed to Russia. Most observers in America assumed that their government had promoted international control from the time it was first proposed.

In 1950, however, the eminent Danish scientist BOHR published *An Open Letter to the United Nations* which raised questions about President Roosevelt's commitment to internationalizing the atom [3]. After escaping late in 1943 from Nazi-occupied Denmark, BOHR reported, he tried to convince ROOSEVELT and CHURCHILL to open negotiation with STALIN during the war for the international control of atomic energy. Bohr's proposal was not accepted, and for

(*) I am grateful to Cornell University's Program on Science, Technology and Society and its Peace Studies Program for supporting my research. This paper is a chapter from my book, *A World Destroyed: The Atomic Bomb and the Grand Alliance* (New York, 1975).

(**) The Manhattan Project is a popularized version of the official code name of the wartime atomic energy program, the Manhattan Engineer District.

over two decades scientists who regretted that decision believed that ROOSEVELT and CHURCHILL misunderstood what BOHR had in mind. As OPPENHEIMER, the wartime director of the atomic-bomb laboratory at Los Alamos, New Mexico, has written: It was « easy, as history has shown, for even wise men not to know what BOHR was talking about » [4]. Bohr's efforts to avert a post-war atomic armaments race between the great powers were explained as a tragedy; his obscure prose, and his hardly audible speech were the tragic flaws that prevented sympathetic statesmen from heeding his advice [5].

But tragic interpretations of history can also result from incomplete evidence, and Bohr's story is a case in point. During the past several years the curtain of secrecy that has hidden so many critical documents related to important decisions taken during World War Two has begun to open, allowing historians to fill out and, in many cases, causing them to revise the diplomatic history of the Grand Alliance [6]. The account of Bohr's activities during World War Two can still be viewed as a tragedy, but one caused by a lack of vision on the part of ROOSEVELT and CHURCHILL. Bohr's wartime proposals, it now appears, were rejected because ROOSEVELT and CHURCHILL opposed them.

Prior to Bohr's arrival in America late in 1943, he spent several months in England discussing the postwar implications of the Manhattan Project with Sir John ANDERSON, the Minister responsible for atomic-energy affairs. Andersons' high regard for the competence of Soviet physicists led him to consider seriously the advantage of Bohr's proposal for laying the foundation of international control during the war. He therefore encouraged BOHR to find some means of bringing his views to the attention of President ROOSEVELT. « That is why I went to America », BOHR commented about his work as a consultant for the Manhattan Project, « they didn't need my help in making the atom bomb » [7]. As OPPENHEIMER later said, « Officially and secretly he came to help the technical enterprise, (but) most secretly of all ... he came to advance his case and his cause » [8]. In the broadest sense, Bohr's cause was to ensure that atomic energy « is used to the benefit of all humanity and does not become a menace to civilization » [9]. Specifically, he warned that « quite apart from the question of how soon the weapon will be ready for use and what role it may play in the present war ... » some agreement had to be reached with the Soviet Union about the future control of atomic energy [10].

Bohr's ideas on the international control of atomic energy are significant beyond any actual effect they might have had on policy. He was universally admired by his colleagues both for his accomplishments as a scientist and his qualities as a human being, and his influence among them, even on political and social issues, stemmed both from his humanity and from his professional achievements. He was responsible for several important advances in atomic physics—the quantum theory of atomic systems—that had led to the discovery of nuclear fission. As a leader in the field of nuclear physics, Nobel Laureate in 1922, founder and director of the internationally famed research institute in his native Denmark, as well as the social conscience of the international community of physicists during the interwar years, Bohr's judgments on political matters always received a respectful hearing and a respectable following. They were based on his experiences as a scientist and on a set of idealistic scientific values that he had internalized. Whether or not other scientists ultimately agreed with his political proposals, they shared the basic view about science from which his political ideas grew. His proposals therefore reveal more than the insights and oversights of an individual scientist; they represent the transfer of the scientific ideal into the realm of international politics.

BOHR believed that the intellectual traditions of science preserved the fundamental values of western civilization: individual freedom, rationality and the brotherhood of man. Science and progress, science and rationality, science and peace, all went hand in hand. « Knowledge », he once wrote with reference to the discovery of nuclear fission and its consequences, « is itself the basis of civilization », but he was quick to add, « any widening of the borders of our knowledge imposes an increased responsibility on individuals and nations through the possibilities it gives for shaping the conditions of human life » [12]. His faith in science as a constructive force in the world was reinforced by a personal philosophy of social action, a concern for the human condition which rested upon his belief that each individual was a responsible member of society. Every person was obliged to confront the historical process, not merely as an observer, but as an active participant. In 1933 he remarked, « Every valuable human being must be a radical and a rebel for what he must aim at is to make things better than they are » [13].

Throughout Bohr's life his political interests and activities exemplified his commitment to these values. « The image of the scientist cut off from the world's problems was never true about father », remarked his son AAGE, himself a physicist. « He always had a great interest in social problems and problems concerning the relationships between nations » [14]. This image was shared by all who knew him. WEISSKOPF echoed the science community's sentiments when he said that « BOHR was not only a great scientist, he was also a man of unusual sensitiveness of feeling for the world in which he lived. The relation of science with the world of men was for him an important question » [15]. His concern for the influence of science on society was not suddenly unlocked by the discovery of nuclear fission. Long before that discovery he had acted upon his belief that science should form a bridge rather than a barrier between nations. At the end of the First World War he worked against the war guilt hysteria that led to the exclusion of German scientists from international scientific societies. During the 1930's, Bohr's institute in Copenhagen became the European haven for refugee scientists from Fascism. As the head of the Danish Comittee for the Support of Fugitive Intellectuals and Scientists, which he had helped to organize in 1933, BOHR became the head dispatcher of an underground railroad that delivered many of Europe's most brilliant scientists to England and America [16]. Throughout the interwar period, he helped to encourage peace through international science.

In 1944, Bohr's concerns and his understanding of the new discoveries that he had done so much to introduce into the world led him to reject traditional approaches to international relations for the post-war years. To his mind, economic, ideological, territorial, and military questions had to be reconsidered within new limitations imposed by the threat of a post-war nuclear arms race. In formulating his proposals he did not ignore the role of traditional considerations in international relations; he sought only to re-assess them in light of an unprecedented situation. His proposals were based on his estimation of the cause and the consequences of an atomic arms race, a race which he considered inevitable if attempts to obtain international control were unsuccessful. This assumption, coupled with a perceptive understanding of the uncertain security of a nuclear balance of terror, led him to reject the past as a diplomatic guide for the future.

Simply put, BOHR believed that the development of the atomic bomb necessitated a new international order. At the heart of his proposal was a scientist's natural distrust of secrecy, and, like WILSON, a desire to reverse radically the historic traditions by which nation states conducted their foreign policies. International control of atomic energy was only possible in an « open world », a world in which each nation could be confident that no potential enemy was engaged in stock-piling atomic weapons. He would urge ROOSEVELT to consider « any arrangement which can offer safety against secret preparations » [17]. International inspectors must be granted full access to all military and industrial complexes and full information about scientific discoveries [18]. In essence, his argument was based on the proposition that the values of science had to control international relations after the war, if the accomplishments of scientists were not to destroy the world.

Behind Bohr's faith in the possibility of arranging for international control was his estimate of the potential influence statesmen had over the course of international relations. He assumed that relations between states were guided by calculated decisions. The judgments of statesmen were often in error, and their understanding of their own and the world's best interests were often dangerously myopic. However, governments composed of men, not immutable laws of history or of international politics, controlled the cycle of war and peace. He believed, therefore, that if statesmen could be made to understand the political and military implications of atomic energy, they could respond to a new international situation just as scientists responded to new discoveries. BOHR believed that under the threat of a nuclear arms race, creative statesmanship—diplomacy based on the possibility of a new and more hopeful future rather than on lessons from the past—could bring the great powers into harmony. There were no historical precedents to encourage him, but the lack of precedent seemed irrelevant. The threat of atomic warfare was also unprecedented.

What was necessary, he once remarked to OPPENHEIMER in jest, referring to the quantum theory, was « another experimental arrangement » [19]. But in a deeper sense he was not jesting, for his proposals called for a political quantum leap into the era of the atomic bomb. In 1944, such a leap did not seem impossible to BOHR; on the contrary, if the world was to survive, he considered it a necessity. Attempting to apply the ideals of science rather too directly to the conduct of international affairs, he took as his model an idealized version of the international scientific community with scientists responsible for educating statesmen to the necessity for a new diplomatic morality and a new international order.

**2.** – A close examination of Bohr's proposals for the international control of atomic energy, as he expressed them in his numerous wartime memoranda and letters, is necessary to clearly understand exactly what he proposed to CHURCHILL and ROOSEVELT during the Spring and Summer of 1944.

The problem that BOHR saw emerging with the development of the atomic bomb was an atomic arms race after the war. He never suggested that the military use of the bomb in war might influence post-war relations with the Soviet Union. « What role it (the atomic bomb) may play in the present war, » he wrote to ROOSEVELT, was a question « quite apart » from the post-war issue; it was a military matter rather than a political matter [20]. Looking beyond the war, however, BOHR argued that an agreement for international control could be accomplished only by promptly inviting Soviet participation in atomic-energy planning, before the bomb was a certainty and before the war was over. No other problem surrounding the atomic bomb —certainly not its use—posed so profound a dilemma for ROOSEVELT, and no proposal for the resolution of that dilemma was as thoughtful as the one BOHR offered.

The point of Bohr's proposal concerned *what ought to be done during the war to encourage the possibility of the post-war control of atomic energy* [21]. He began with two assumptions: 1) the bomb was a creation out of proportion to anything else in human experience, and 2) it could not be monopolized. He therefore concluded that its development would endanger rather than enhance the future security of the United States and Great Britain if it were not effectively neutralized. A world in which rival nations could employ atomic bombs would stand in perpetual danger of total destruction. In such a world traditional concepts of security through military protection were clearly impractical. BOHR believed that the atomic bomb limited the alternatives of statesmanship. A world in which each great power would feel confident that no other nation was producing nuclear weapons was one choice; a world dominated by the constant spectre of total destruction was the other. There could be no middle ground; the new weapon was too effective for that. If national security was not guaranteed to the great powers by some form of international control of atomic energy, he concluded, they would inevitably follow policies that planted the seeds of their own destruction.

Bohr's second point—the heart of his proposal—followed logically from his analysis. Since the atomic bomb, in his judgment, would be the critical factor in setting the post-war international political climate, it was necessary that STALIN be informed about the existence of the Manhattan Project before the war ended. It had to be made clear to Soviet leaders that an Anglo-American alliance supported by an atomic monopoly was not being formed against them. Timing was critical. Discussions had to be initiated before developments proceeded so far as to make an approach to the Russians appear more coercive than friendly.

BOHR understood that the initiative he urged did not guarantee the post-war co-operation of the Soviet Union, but he also believed that co-operation was impossible unless his proposal was adopted. He did not ignore the profound risks involved in executing his plan, though he was overly optimistic about the possibility of overcoming those risks. The stakes at issue appeared to him slight in comparison. He suggested that the Soviets be informed *simply of the existence of the Manhattan Project*, but not of the details of the bomb's construction. Should their response to this limited disclosure be favorable, then the way was open for further planning and increased co-operation. « In preliminary consultations (with the Russians) », he wrote to ROOSEVELT, « no information as regards important technical developments should, of course, be exchanged; on the contrary, the occasion should be used frankly to explain that all such information must be withheld until common safety against the unprecedented dangers has been guaranteed » [22]. BOHR wanted ROOSEVELT to offer STALIN an atomic-age *modus vivendi*: international control of atomic energy and thereby security, in exchange for surrendering traditional national secretiveness which could offer little security in a nuclear armed world. A modified version of this concept dominated Secretary of War STIMSON's views during the last year of the war. In June, 1946, it emerged again in the United States government's plan for the international control of atomic energy— the Baruch Plan [23]. BOHR, however, was insisting in the Summer of 1944 that such an offer had to be made *before* the weapon was a certainty and *prior* to the end of the war. He believed that this was the best time for initiating an effort to resolve this immense problem.

Bohr's third and final point revealed an exaggerated faith in the political influence of the international community of scientists, and it reflected how completely he had transferred the values of science to his judgments about diplomacy. He suggested to ROOSEVELT that « helpful support may perhaps be afforded by the world-wide scientific collaboration which for years had embodied such bright promises for common human striving. On this background personal connections between scientists of different nations might even offer means of establishing preliminary and noncommittal contact » [24].

He was certain that among the eminent scientists in Russia « one can reckon to find ardent supporters of universal co-operation » [25]. That their influence with the Soviet government on political matters might be nil was a possibility he did not raise. The scientific perspective which led him to predict the course of international affairs led him to underestimate the very difficult political obstacles that had to be overcome. These obstacles were not limited to Stalin's suspicions of his Allies; they were first raised by CHURCHILL and ROOSEVELT, who intended to use the atomic bomb as a diplomatic counter for bargaining with the Soviet Union after the war.

**3.** – Once in America, BOHR accomplished the difficult task of bringing his ideas to Roosevelt's attention with the assistance of FRANKFURTER, a Roosevelt appointeee to the Supreme Court and an unofficial adviser to the President. BOHR had met FRANKFURTER in 1933 at Oxford University. They had met again in 1939 when BOHR came to the United States to attend meetings of the National Academy of Sciences and the Americal Physical Society. They had discussed international cultural relations at length at Frankfurter's home on this occasion. In July of the same year they had met again in London at a meeting of the Society for the Protection of Science and Learning, an organization established to support German scientists and scholars persecuted by the Nazi regime. At these meetings BOHR had been impressed with the breadth of Frankfurter's interests, and perhaps overly impressed with his influence on ROOSEVELT. In 1944 they were brought together in Washington, once again, at a tea at the home of the Danish Minister to the United States [26].

Some time before BOHR arrived in America, FRANKFURTER had learned about the Manhattan Project. He had been approached by a number of scientists and asked « to advise them on a matter (the atomic bomb) that seemed to them a matter of the greatest importance to our national interest ». Thus, even before he and BOHR renewed their friendship in 1944, FRANKFURTER had « become aware of X », as he referred to the atomic bomb in his wartime correspondence. BOHR and FRANKFURTER lunched together in the Justice's chamber within a few days after their Washington meeting. In the course of their conversation FRANKFURTER « made a very oblique reference to X »; BOHR replied in a similar vein, « but it soon became clear to both of us », FRANKFURTER recalled later, « that two such persons ... could talk about the implications of X without either making any disclosure to the other ». In this manner, the relationship was formed that was to carry the earliest known analysis of the implications of the atomic bomb's development to Roosevelt's attention. Frankfurter's recollection of the conversation is of interest:

The conversation proceeded on this basis and BOHR then expressed to me his conviction that X might be one of the greatest boons to mankind or might become the greatest disaster. This has been the core of his talk with me in the many conversations that I have since had with him .... Throughout he has been concerned entirely with the political problem of so controlling X as to make it a beneficial instead of a disastrous contribution by science .... He was a man weighed down with a conscience and with an almost overwhelming solicitude for the danger to our people [27].

Aage BOHR, who had come to the United States with his father, recalled the elder physicist's pleasure after he returned from his meeting with FRANKFURTER: « It was an exciting day. The result of the meeting corresponded to his best expectations » [28].

Affected by Bohr's dedication and perceptions, FRANKFURTER arranged an appointment with ROOSEVELT, probably late in February, 1944. He told the President the full details of Bohr's concerns, emphasizing his central worry: « that it might be disastrous if Russia should learn on her own about X rather than that the existence of X should be utilized by this country and Great Britain as a means of exploring the possibility of an effective international arrangement with Russia for dealing with the problems raised by X ». FRANKFURTER also pointed out that BOHR believed that it would not be too difficult for Russia to gain the information necessary for building her own atomic weapons.

FRANKFURTER left the White House after speaking with ROOSEVELT for an hour and a half, feeling that the President was « plainly impressed by my account of the matter ». When FRANKFURTER had suggested that the solution to the problem of the atomic bomb might be more important than all the plans for a world organization, ROOSEVELT had agreed. Moreover, he had authorized FRANKFURTER to tell BOHR that upon his scheduled return to England he might inform « our friends in London that the President was most eager to explore the proper safeguards in relation to X ». FRANKFURTER also vividly recalled that ROOSEVELT had told him that the problem of the atomic bomb « worried him to death », and that he was very eager for all the help he could have in dealing with it [29].

The alternatives placed before ROOSEVELT posed a difficult dilemma. On the one hand, he could continue to exclude the Soviet government from any official information about the development of the bomb, a policy that would probably strengthen America's post-war military-diplomatic position. But such a policy would only encourage Soviet mistrust of Anglo-American intentions and was bound to make post-war co-operation more difficult. On the other hand, ROOSEVELT could use the atomic-bomb project as an instrument of co-operation by informing STALIN of the American government's intention of co-operating in the development of a plan for the international control of atomic weapons, an objective that might never be achieved.

Either choice involved serious risks. ROOSEVELT had to balance the diplomatic advantages of being well ahead of the Soviet Union in atomic energy production after the war against the advantages of initiating wartime negotiations for postwar co-operation. The issue here, it must be emphasized, is not whether the initiative BOHR suggested would have led to successful international control, but, rather, whether ROOSEVELT demonstrated any serious interest in laying the groundwork for such a policy.

Several considerations suggest rather forcefully that ROOSEVELT already was committed to a course of action that precluded Bohr's international approach. First, FRANKFURTER appears to have been misled. Though Roosevelt's response had been characteristically agreeable, he told V. BUSH, the director of the Office of Scientific Research and Development, that he was very disturbed that FRANKFURTER had learned about the project at all. Moreover, ROOSEVELT knew at this time that the Soviets were finding out on their own about the development of the atomic bomb. Security personnel had reported an active Communist cell in the Radiation Laboratory at the University of California. Their reports indicated that at least one scientist at Berkeley was selling information to Russian agents [30]. « They (Soviet agents) are already getting information about vital secrets and sending them to Russia », STIMSON told the President on September 9, 1943. If ROOSEVELT was indeed worried to death about the effect the atomic bomb could have on Soviet-American post-war relations, he took no action to remove the potential danger, nor did he make any effort to explore the possibility of encouraging Soviet post-war co-operation on this problem. The available evidence indicates that he never discussed the international control of atomic energy with his advisers after this first or after any subsequent meeting with FRANKFURTER [31].

How is the President's policy to be explained if not by an intention to use the bomb as an instrument of Anglo-American post-war diplomacy? Perhaps his concern for maintaining the tightest possible secrecy against German espionage led him to oppose any discussion about the project. Or he may have concluded, after considering Bohr's analysis, that Soviet suspicion and mistrust would be further aroused if STALIN were informed of the existence of the project without detailed information about the bomb's construction. The possibility also exists that ROOSEVELT believed that neither Congress nor the American public would approve of a policy giving the Soviet Union any measure of control over the new weapon. Finally, ROOSEVELT might have thought that the Spring of 1944 was not the proper moment for such an initiative.

Though it would be unreasonable to state categorically that these considerations did not contribute to his decision, they appear to have been secondary. ROOSEVELT was clearly, and properly, concerned about secrecy, but the most important secret with respect to Soviet-American relations was that the United States was developing an atomic bomb. And that secret, he was aware, already had been passed on to Moscow. Soviet mistrust of Anglo-American post-war intentions could only be exacerbated by continuing the existing policy. Moreover, an attempt to initiate planning for international control of atomic energy would not have required the revelation of technical secrets. Nor is it sufficient to cite Roosevelt's well-known sensitivity to domestic politics as an explanation for his atomic-energy policies. He *was* willing to take enormous political risks, as he did at Yalta, to support his diplomatic objectives [32].

Had ROOSEVELT avoided all post-war atomic-energy commitments, his failure to act for international control could be interpreted as an attempt to reserve his opinion on the best course to follow. But he had made commitments to CHURCHILL in 1943 supporting an Anglo-American atomic monopoly, and he continued to make others in 1944 [33]. On June 13, for example, ROOSEVELT and CHURCHILL signed an « Agreement and Declaration of Trust » specifying that it is « the intention of the Two Governments to control to the fullest extent practicable the supplies of uranium and thorium ores within the boundaries of such areas as come under their respective jurisdictions ... (and) in certain areas outside the control of the Two Governments and of the Governements of the Dominions of India and of Burma ». The Declaration noted quite specifically that the arrangement would extend beyond the end of the war: « The signatories of the Agreement and Declaration of Trust will, as soon as practicable after the conclusion of hostilities, recommend to their respective Governments the extension and revision of this wartime emergency agreement to cover post-war conditions and its formalization by treaty or other proper method. This Agreement and Declaration of Trust shall continue in full force and effect until such extension or revision » [34].

Thus while BOHR was in England seeking an interview with CHURCHILL under the illusion that he had Roosevelt's endorsement to initiate discussions on the international control of atomic energy, ROOSEVELT was preparing to sign an agreement that could hardly have been more incompatible with any wartime plan for post-war co-operation with the Soviet Union. ROOSEVELT chose not to initiate discussions with STALIN about post-war plans for the international control of atomic energy despite the likelihood that such a course necessarily prejudiced his objective of peaceful post-war collaboration with the Russians—a sacrifice that in other respects he was never willing to make. Roosevelt's position on this issue cannot be explained by suggesting that he was unaware of the implications of his inaction; he already had demonstrated through numerous comments and policy decisions that he considered the atomic bomb a matter of extraordinary importance. The so-called « Agreement and Declaration of Trust » signed with CHURCHILL on June 13, 1944, reinforces this view and the interpretation that in the Spring of 1944 ROOSEVELT favored a post-war Anglo-American atomic-energy monopoly rather than a plan for the international control of atomic energy [35].

**4.** – « It was with the greatest expectations that we came to London », Aage BOHR has recalled. « It was a fantastic matter for a scientist to try to affect the world's politics in this way, but it was hoped that CHURCHILL, who had such an outstanding imagination, and who often had shown great foresight, would be able to be enthused about the new implications » [36]. Their overly optimistic judgment was based partly on their hopes, which were raised unrealistically by the excitement of their mission. Beyond this, however, was a very deep personal stake and involvement on Bohr's part which he revealed many years later to a colleague and friend. « He (BOHR) had come (to the United States) with a plan, in the execution of which *he himself* would play an essential part, » NIELSEN remarked [37]. If scientists were to be the agents for ushering in a new, peaceful world order, BOHR, scientist-diplomat of the new age, was to lead the way. Had his mission been a less personal crusade, BOHR might better have been able to deal with CHURCHILL. As it was, he was completely unprepared for the Prime Minister's reaction.

Upon Bohr's return to England in April, 1944, he carried with him a letter written by FRANKFURTER containing a « formula » for a discussion with CHURCHILL— a summary of what FRANKFURTER presumed were Roosevelt's true interests in the post-war international control of atomic energy [38]. Before BOHR arrived, ANDERSON had already sought unsuccessfully to bring CHURCHILL around to accepting Bohr's position [39]. Lord CHERWELL, too, and Sir Henry DALE, President of the Royal Society, had tried their hands also at convincing CHURCHILL of this view. Others of influence, including the distinguished South African statesman SMUTS, were persuaded to add their names to a campaign waged in favor of Bohr's proposal. But the Prime Minister, busy and difficult to persuade, was unmoved by their arguments. CHURCHILL plainly preferred an Anglo-American monopoly of the atomic bomb to the post-war international control of atomic energy [40].

BOHR had to wait many weeks before he could see CHURCHILL. While he waited, his most profound fear was confirmed. It seemed that the Soviets were aware of the joint American-British-Canadian effort to build an atomic bomb. His suspicions were initially aroused by a letter he received from KAPITZA, a Soviet physicist and long-time friend, inviting him to settle and work in Russia [41]. They were reinforced shortly afterward in a conversation with the Soviet Counselor ZINCHENKO at the Russian Embassy, where BOHR delivered his carefully worded reply to KAPITZA, drafted with the assistance of British security agents [42]. Though « no reference was made to any special subject », in Kaptiza's letter, BOHR later wrote to ROOSEVELT, the letter itself and the pre-war work of Russian physicists made it « natural to assume that nuclear problems will be in the center of (Russian) interest ». Moreover, Bohr's conversation with ZINCHENKO left BOHR with the impression that the Russians were « very interested in the effort in America about the success of which some rumors may have reached the Soviet Union » [43]. It was now more important than ever, BOHR reasoned, to demonstrate British and American good faith. The Soviets must not be panicked into choosing the path of an atomic arms race which, his knowledge of the capabilities of Soviet science assured him, they were capable of entering [44].

By the middle of May, BOHR was able to obtain an interview with CHURCHILL. It is no exaggeration, however, to describe their meeting as a disaster. The Prime Minister showed little patience with a scientist intruding into his jealously guarded arena of international affairs, especially one so apparently naive as to urge him to inform the Russians about the development of the atomic bomb. Bohr's scheduled thirty minute interview was not long underway before CHURCHILL lost interest and became embroiled in an argument with Lord CHERWELL, the only other person present. BOHR was left out of the discussion, frustrated and depressed, unable to bring the conversation back to what he considered the most important diplomatic problem of the war. The allotted time elapsed, and in a last attempt to communicate his anxieties and ideas to the Prime Minister, BOHR asked if he might forward a memorandum to him on the subject. A letter from BOHR, CHURCHILL bitingly replied, was always welcome, but he hoped it would deal with a subject other than politics. BOHR has succinctly summed up their meeting: « We did not even speak the same language » [45].

CHURCHILL was unmoved by Bohr's argument because he rejected the assumption upon which it was founded: that the bomb could change the very nature of international relations. To accept it the Prime Minister would have had to execute a political and intellectual *volte-face* of which he was incapable. Reasoning from his estimates of the potential menace of the atomic bomb, BOHR wanted CHURCHILL to reject his understanding of international politics; to reject, in effect, the very axioms that had guided him to the wartime leadership of his country. The argument that a new weapon invalidated traditional power considerations ignored every lesson CHURCHILL read into history. Like BOHR, CHURCHILL believed that the atomic bomb would be a major factor in post-war diplomacy, but he did not believe that it would alter the basic nature of international relations, nor apparently did he believe that international control was practicable regardless of how straightforward and cooperative the United States and Britain might be. « You can be quite sure », he wrote in a memorandum less than a year later, « that any power that gets hold of the secret will try to make the article (atomic bomb) and this touches the existence of human society. This matter is out of all relation to anything else that exists in the world, and I could not think of participating in any disclosure to third (France) or fourth (Russia) parties at the present time » [46]. Nothing could persuade the Prime Minister that his nation would be better served by exchanging familiar calculations of political behavior and military power for a scientist's utopian predictions about the effects on international relations of an enigmatic new weapon. As CHURCHILL strained to see the future, his perceptions were influenced by his considerable knowledge of the past. The monopoly of the atomic bomb that England and America would enjoy after the war would be a significant diplomatic advantage in settling post-war geo-political rivalries between Britain and the Soviet Union. CHUR-

CHILL did not believe that anything could be gained by surrendering that advantage.

Less than a week after the Allied Armies landed on the beaches of Normandy, BOHR was back in the United States—less optimistic, but still resolute. Once again, he contacted FRANKFURTER, who reported their conversation to the President. An interview between BOHR and the President was arranged for August 26, in preparation for which BOHR spent many long, hot June evenings in his Washington apartment composing a summary of his ideas for presentation to the President [47].

To Bohr's delight, though ultimately to his disillusionment, the meeting with ROOSEVELT contrasted markedly to his confrontation with CHURCHILL. They talked for over an hour about the atomic bomb, Denmark and world politics in general. The President was as usual very agreeable. On the post-war importance of atomic energy, BOHR told his son that ROOSEVELT agreed that contact with the Soviet Union had to be tried along the lines he suggested. The President said he was optimistic that such an initiative would have a « good result ». In his opinion STALIN was enough of a realist to understand the revolutionary importance of this development and its consequences. The President was also confident that the Prime Minister would come to share his views on the atomic bomb. They had disagreed before, he said, but in the end they had always succeeded in resolving their differences. Another meeting, ROOSEVELT suggested, might be useful after he had spoken with CHURCHILL about the matter at the Second Quebec Conference scheduled for the following month. In the meantime, if BOHR had any further suggestions, the President would welcome a letter [48].

Roosevelt's enthusiasm for Bohr's proposals was more apparent than real. The President did not mention them to anyone until he discussed atomic-energy matters with CHURCHILL on September 18 at Hyde Park, following the second conference at Quebec. Then he endorsed Churchill's point of view, agreeing too that BOHR ought not to be trusted.

The decisions agreed upon at Hyde Park on atomic-energy policy were summarized and documented in an *aide-memoire* signed by ROOSEVELT and CHURCHILL on September 19, 1944 [49]. It bears all the markings of Churchill's attitude toward the atomic bomb and his opinion of BOHR. It stated, among other points, « that the suggestion that the world should be informed regarding tube alloys (atomic bomb), with a view to an international agreement regarding its control and use, is not acceptable. The matter should continue to be regarded as of the utmost secrecy ». But BOHR had never suggested that « the world » be informed about the atomic bomb. He had argued that peace was not possible unless the Soviet government was officially notified about the project before what he considered the inevitable moment when any discussion of the matter would appear more coercive than friendly [50].

The *aide-mémoire* covered three additional matters. One was the continued co-operation on atomic-energy affairs between the British and American governments in the post-war years. Another offered some insight into how ROOSEVELT intended to use the weapon in the war: « When a bomb is finally available it might perhaps, after mature considerations, be used against the Japanese who should be warned that this bombardment will be repeated until they surrender » (*). But it was the last paragraph which revealed how successful CHURCHILL had been in bringing his point of view to dominate: « Enquiries should be made », it noted, « regarding the activities of Professor BOHR and steps taken to ensure that he is responsible for no leakage of information particularly to the Russians ».

Whatever caused ROOSEVELT to suspect BOHR—the physicist's conversations with FRANKFURTER, his correspondence with KAPITZA, his interest in informing the Soviets about the Manhattan Project—there can be no doubt that they were encouraged by CHURCHILL. It is clear that the Prime Minister considered the Danish scientist a security risk—or worse. After the Hyde Park meeting, he wrote CHERWELL asking how BOHR became involved with the Manhattan Project, and stating that both he and ROOSEVELT were very concerned about Bohr's communications with FRANKFURTER and KAPITZA. « It seems to me », he concluded, that « BOHR ought to be confined or at any rate made to see that he is very near the edge of mortal crimes » [51].

Churchill's opinions about Bohr's proposals were as definite as his views of the scientist himself; he never wavered in his convictions about them: In March, 1944, he rejected—with a curt « I do not agree »—a memorandum from

(*) The reasons for the shift of atomic bomb targets from Germany to Japan is unclear. Though racism has been charged, no direct evidence has been found in support of this thesis. The following reasons, probably in descending order of importance, may explain the shift which was settled late in the spring of 1944: 1) the war in Europe was expected to end first and since the training of the 509th Composite Squadron was an extremely arduous matter, it was safer to pick Japan as a target; 2) shipping an atomic bomb to a Pacific island to bomb Japan was safer than shipping it to England; 3) the war in the Far East was the « American war » and the bomb was an American product. General Groves (and perhaps others as well) wished to use an American aircraft to deliver the weapon. Using the bomb in the Far East guaranteed the use of the B-29.
There is an interesting but ambiguous reference to the shift of the atomic bomb from Germany to Japan in a memorandum by BUSH dated June 24, 1943. Summarizing a discussion with the President, BUSH noted: « We then spoke briefly of the possible use against Japan, or the Japanese fleet, and I brought out, or I tried to, because at this point I do not think I was really successful in getting the idea across, that our point of view or our emphasis on the program would shift if we had in mind use against Japan as compared with use against Germany », (*Memorandum of Conference with the President*, AEC Doc., No. 133). See also HEWLETT and ANDERSON: *The New World*, p. 252.

CHERWELL and ANDERSON urging him to inform the War Cabinet about the matter and to take up the problem of the atomic bomb in post-war international relations. A month later, in April, he refused even to listen to Bohr's opinions. He also rejected a number of other suggestions, urged upon him by members of his inner circle of advisers, to face up to the fact that « plans for world security which do not take account of tube alloys (atomic bomb) must be quite unreal » [52].

It seems probable that Churchill's hostility to Bohr's proposal, and to BOHR himself, were both related to the Prime Minister's determination to secure for Britain a position of equality with the United States in a post-war atomic energy partnership. In Churchill's view, Britain's future position as a world economic and military power depended on that relationship. As early as 1943 CHURCHILL had stated his belief that atomic energy was « necessary for Britain's independence in the future ... », a view confirmed by Lord CHERWELL, his science adviser, who told HOPKINS that Britain was considering « the whole (atomic energy) affair on an after-the-war military basis » [53]. Bohr's ideas challenged the validity of any such special relationship.

Bohr's proposals, moreover, were vulnerable to the internal logic of Churchill's position and his concerns for his nation's security. BOHR had not dealt directly with the possibility that *any* discussion of atomic-energy matters with the Soviet government was bound to have a coercive element about it given the Anglo-American lead. Granted that from the standpoint of seeking cooperation the earlier the Soviets were told, the better. But ROOSEVELT and CHURCHILL had to choose between the apparent advantages of having the weapon first and alone against the advantages of international control. It is likely that they would have opted for an Anglo-American monopoly regardless of what BOHR argued in his various memoranda, letters and discussions, and it must be noted that his proposal failed to face the most difficult question: Was international control achievable? *He assumed* it was possible, and he suggested the first step in that direction— informing STALIN of the Manhattan Project's existence and of Anglo-American post-war intentions. But ROOSEVELT and CHURCHILL did not share his assumption that the advantages of informing STALIN during the war outweighed the advantages of keeping the secret.

Though personal suspicion of BOHR was removed within the next six months, by the determined efforts of his friends and colleagues in England and America, his opportunity to have any immediate influence on the atomic-energy foreign policies of the Allied governments was lost. His ideas and his formulations, however, remained for Bohr's colleagues to espouse after the war. But the opportunity Bohr's idea offered to gauge and perhaps influence the Soviet Union's response during the war to the international control of atomic energy could not be regained. Instead of promoting their most laudable post-war aims, ROOSEVELT and CHURCHILL pursued an atomic-energy policy that made those aims more difficult to achieve.

## REFERENCES

[1] The most complete study of the post-war political activities of the atomic scientists is A. K. SMITH: *A Peril and a Hope: The Scientists' Movement in America, 1945-1947* (Chicago, 1965). See also D. A. STRICKLAND: *Scientists in Politics: The Atomic Scientists Movement 1945-46* (Lafayette, Ind., 1968).

[2] R. GILPIN: *American Scientists and Nuclear Weapons Policy*, chap. 3 (Princeton, 1962); J. I. LIEBERMAN: *The Scorpion and the Tarantula: The Struggle to Control Atomic Weapons 1945-1949*, chap. 20 (Boston, 1970).

[3] N. BOHR: *Open Letter to the United Nations* (Copenhagen, 1950).

[4] J. R. OPPENHEIMER: *Niels Bohr and Atomic Weapons*, in *The New York Review of Books*, Vol. **3** (December 17, 1966), p. 6.

[5] R. MOORE: *Niels Bohr* (New York, 1966), p. 344.

[6] Recently opened manuscript collections of particular importance for this study include: The Manhattan Engineer District files (hereafter MED), National Archives, Washington, D.C.; Atomic Energy Commission Archives (hereafter AEC), Washington, D.C.; Atomic Energy papers in President's Safe File (hereafter PSF), Franklin D. Roosevelt Library (hereafter FDRL), Hyde Park, New York; Harry Hopkins atomic-energy papers (hereafter HHP), FDRL; Henry L. Stimson Diary, Yale University Library, New Haven, Conn.; J. Robert Oppenheimer Papers (hereafter JROP), Library of Congress, Washington, D.C.

[7] As quoted in R. NIELSEN: *Phys. Today*, **16**, 29 (1963).

[8] J. R. OPPENHEIMER: *Niels Bohr and Atomic Weapons*, p. 7.

[9] N. BOHR: Memorandum dated May 8, 1945, JROP, Box 34.

[10] BOHR to ROOSEVELT, July 3, 1944, JROP, Box 34. An edited version of this memorandum to ROOSEVELT and another one dated March 24, 1945 are printed in BOHR: *Open Letter*. See also *Bull. Atomic Sci.*, **6**, 213 (1950).

[11] J. A. WHEELER: *Phys. Today*, **16**, 37 (1963).

[12] N. BOHR: *Open Letter*.

[13] R. NIELSEN: *Niels Bohr*, p. 27; BOHR made this remark in April, 1933, at a Meeting in Chicago.

[14] A. BOHR: *The War Years and the Prospects Raised By Atomic Weapons*, in *Niels Bohr*, edited by S. ROZENTAL (New York, 1967), p. 190.

[15] V. WEISSKOPF: *Phys. Today*, **16**, 58 (1963).

[16] S. ROZENTAL: *The Forties and Fifties*, in *Niels Bohr* (1967), p. 153.

[17] BOHR to ROOSEVELT, March 24, 1945, JROP, Box 34.

[18] BOHR to ROOSEVELT, July 3, 1944, and *Ibid.*, JROP, Box 34.

[19] J. R. OPPENHEIMER: *Niels Bohr and Atomic Weapons*, p. 8.

[20] BOHR to ROOSEVELT, July 3, 1944, JROP, Box 34.

[21] Bohr's memoranda to ROOSEVELT, especially his memorandum of March 24, 1945, suggest many of the control techniques later incorporated into the United States Government's proposal to the United Nations Atomic Energy Commission (June, 1946) for the international control of atomic energy. These include: 1) technological inspection, 2) advanced scientific research and development by the inspection agency and 3) the distinction between safe and dangerous activities.

[22] N. BOHR: Memorandum of May 8, 1945, summarizing his communications with ROOSEVELT, JROP, Box 34. For additional information on Bohr's proposals

see the Felix Frankfurter Papers, Library of Congress. FRANKFURTER had the complete file of Bohr's wartime correspondence with ROOSEVELT, but gave it to OPPENHEIMER to use for The Pegram Lectures: *Three Lectures on Niels Bohr and his Times* (Brookhaven, 1963).

[23] R. G. HEWLETT and O. E. ANDERSON: *The New World, 1939-1946: A History of the United States Atomic Energy Commission*, Vol. **1**, Chap. 15 (University Park, Pa., 1962); see also A. K. SMITH: *A Peril and a Hope*, Chap. 15 and 16.

[24] BOHR to ROOSEVELT, July 3, 1944, JROP, Box 34.

[25] N. BOHR: *Notes Concerning Scientific Co-operation with the USSR. Written in Connection with Memorandum of July 3, 1944-September 30, 1944*, JROP, Box 34.

[26] The following account of Bohr's activities is drawn form Felix Frankfurter's correspondence, especially, FRANKFURTER to Lord HALIFAX, April 18, 1945, JROP, Box 34.

[27] *Ibid.*

[28] A. BOHR: *The War Years*, p. 194.

[29] FRANKFURTHR to HALIFAX, April 18, 1945, JROP, Box 34.

[30] V. BUSH: *Memorandum of Conference*, September 22, 1944, AEC doc. No. 185; United States Atomic Energy Commission, *In the Matter of J. Robert Oppenheimer: Transcript of Hearing before Personnel Security Board* (Washington, 1954), testimony of General Leslie R. GROVES, 171-74; see also 163-80 and testimony of J. LANSDALE jr., 258-81.

[31] H. L. STIMSON: Diary, Sept. 9, 1943. FRANKFURTER met several times with the President and discussed Bohr's proposal with him in detail. See *Roosevelt and Frankfurter: Their Correspondence, 1928-1946*, edited by M. FREEDMAN (Boston, 1967), p. 725.

[32] D. SHAVER CLEMENS: *Yalta* (New York, 1970).

[33] On August 19, 1943, at the Anglo-American summit conference in Quebec, ROOSEVELT and CHURCHILL signed a secret agreement that guaranteed the British a close relationship with the United States in atomic-energy matters. The Quebec Agreement, officially entitled *The Articles of Agreement Governing Collaboration between the Authorities of the U.S.A. and the U.K. in the Matter of Tube Alloys (atomic energy)*, is reprinted in M. GOWING: *Britain and Atomic Energy* (London, 1964), Appendix 4. For a more complete treatment of the early ROOSEVELT-CHURCHILL negotiations on atomic energy matters during the war see M. J. SHERWIN: *The Atomic Bomb and the Origins of the Cold War*, in *American Historical Review*, Vol. **78** (October, 1973).

[34] *Agreement and Declaration of Trust*, June 13, 1944, in General L. R. GROVES: *Diplomatic History of the Manhattan Project* (hereafter DHMP), annex No. 22 MED files.

[35] L. R. GROVES: DHMP, annex No. 17 and 18, MED files.

[36] A. BOHR: *The War Years*, p. 196.

[37] R. NIELSEN: *Niels Bohr*, p. 30.

[38] FRANKFURTER to HALIFAX, April 18, 1945, JROP, Box 34.

[39] M. GOWING: *Britain and Atomic Energy*, p. 346.

[40] *Ibid.*, p. 346.

[41] Memorandum discussing Bohr's relationship to KAPITZA, June 28, 1944, unsigned but it is likely that FRANKFURTER wrote it as it is on stationary with a U.S. Government watermark and appears with Frankfurter's other papers (as well as Bohr's) in JROP, Box 34.

[42] « Report, written immediately after the conversation took place between BOHR

and Counsellor ZINCHENKO at Soviet Embassy in London on April 20, 1944 at 5 p.m. », JROP, Box 34.

[43] BOHR to ROOSEVELT, July 3, 1944, JROP, Box 34.

[44] By February 1943, the Soviet Union had already begun to work on the development of a « uranium bomb ». I. N. GOLOVIN: *I. V. Kurchatov* (Moscow, 1967), in Russian; see review by A. KRAMISH: *Science*, **157**, 912 (1967).

[45] M. GOWING: *Britain and Atomic Energy*, p. 355. See also BOHR to CHURCHILL, May 22, 1944, JROP, Box 34.

[46] As quoted by M. GOWING: *Britain and Atomic Energy*, p. 360.

[47] BOHR wrote the memorandum and forwarded it to FRANKFURTER for editorial advice. It included a last paragraph that underlines Bohr's view of the relationship between science and the future of the world: « Should such endeavors be successful, the project will surely have brought a turning point in history and this wonderful adventure will stand as a symbol of the benefit to mankind which science can offer when handled in a truly human spirit ». BOHR had doubts about the wisdom of such a sentimental passage; FRANKFURTER had none—he edited out the entire paragraph. See letters attached to Bohr's memorandum of July 3, 1944, JROP, Box 34.

[48] A. BOHR: *The War Years*, p. 197. BOHR sent a letter to ROOSEVELT on September 7, 1944, Roosevelt Papers, Official File 2240, FDRL.

[49] The complete text of the *aide-mémoire* is reprinted in M. GOWING: *Britain and Atomic Energy*, p. 447.

[50] For a different view of this episode see R. G. HEWLETT and O. E. ANDERSON: *The New World*, p. 236.

[51] As quoted in M. GOWING: *Britain and Atomic Energy*, p. 358.

[52] *Ibid.*, p. 352.

[53] CHURCHILL quotation in H. BUNDY: *Memorandum of Meeting at* 10 *Downing Street on July 23, 1943*, AEC doc. 312. CHERWELL quotation in BUSH: *Memorandum of Conference with Mr. Harry Hopkins and Lord Cherwell at the White House, May 25, 1943*, in HHP, A-bomb folder. See also BUSH: *Memorandum of Conference with the President, June 24, 1943*, AEC doc. 133.

# New Forms of Organization in Physical Research after 1945.

L. KOWARSKI
*CERN - Geneva*

## 1. – Big science.

I think it was KLEIN who remarked that it seems that physicists lecture standing and humanists lecture seated. So I seem to be an exception. I am a physicist, but perhaps this will remind us that physicists are human. This on the whole will be the main conclusion, I hope, of my three talks.

We have heard here many fine lectures on history of physical ideas, more or less taken in a self-contained world consisting of such ideas. That was one kind. We have also heard lectures about the connection between the world of physical ideas and the outside environment of society: ideologies, philosophies, technologies and so on. My subject is in a way also a consideration of this environmental aspect of physics, but on a somewhat lower plane. It deals with those *institutions* which make the physicist's work possible.

After all, a physicist has to be taught and given equipment, and journals and meetings and laboratories and enough income to feed him (even physicists have to eat), and all this has to be done institutionally. And all this influences the way physics is done.

So I propose to talk, as WEINER just said, in a historical perspective of the physics institutions. An idea of how important they have been can be derived from some beautiful examples we have heard in this school. Take for instance the difference between the work of MILLIKAN and EHRENHAFT, as explained by HOLTON: you see quite clearly how their respective institutional environments acted on their results. Or take the contrast between EINSTEIN, whose institutional background was, so to speak, mostly negative, and DIRAC. He never saw, it seems, a theoretical physicist other than himself, until several years after his epoch-making papers. On the other hand, DIRAC had all the benefit of the wonderful Cambridge environment, with its Kapitza Club and so on. All this was, indeed, reflected in their work.

Now, I do not propose to discuss here the problem of institutional (or organizational) framework of physics work in all its generality. I shall choose the period after 1945, and I shall explain at the next point why this date is significant.

There is another reason for me to start with 1945: I propose to relate some definite facts, and to state my opinions about them. A big proportion among these facts will be those I have witnessed or taken part in, and about which I could form an opinion, for what it is worth. What I cannot do, because I am ignorant of sociology and philosophy, is to draw conclusions from these facts and even from those first-hand opinions which could be formed immediately without going into further abstractions. This is a limitation which, perhaps, prevents my lectures from directly contributing to the kind of question which has been raised so persistently among the audience in this school. However, I hope that the facts and the opinions I will present to you may prove to be a useful raw material for discussing such topics on a more ideological level and if anyone of you finds that this raw material can indeed be so used, I shall be glad to answer questions and go into details as required.

Let us go back now to the intrinsic importance of the date 1945. GOLDSTEIN remarked yesterday that science and technology did not end in 1945, and I heartily agree with him, as usual. They certainly did not end: they started anew. Science and technology as we know it today can be in a rough approximation considered as having been born in 1945. 1945 therefore appears as one of the crucial dates in history like, shall we say, the date of the French Revolution or the Russian Revolution or discovery of America, or the liberal wind which swept Europe so suddenly in 1848.

1945 has an importance both in the factual world of science and technology, and in the public opinions and emotions about science and technology, that is in the public myth, as opposed to facts. But since public opinion, after all, through its positive or negative attitude toward science and technology is the ultimate source of scientific and technological activities, the myth is not less important than facts.

1945 was the date when public opinion had an occasion (I shall explain what occasion) to catch up with the changing situation in the relations between science and technology, which was gathering speed during the first half of the 20th century. Both science and technology in their Western form, as we have heard from other speakers, were already in existence for quite a few centuries, but for most of this period they grew up in a kind of almost peaceful co-existence, but not more. WATT, BESSEMER, FORD, even EDISON were certainly no scientists, and in the overwhelming majority of their work they made no direct use of science. There might have been certain subtle influences which have been discussed by other speakers, but there was no direct impact.

The situation began to change only toward the end of the 19th century. It changed in Germany; there science began to get closer to the process of technological innovation. This interaction gradually spread to other countries, and erupted in full force during World War II. The famous time lag between scientific discovery and technological application, under the tremendous urgencies of the World War, suddenly was reduced to practically nil, and advanced science finally did become the immediate origin of new technology.

The governments realized that perfectly well, and scientists were mobilized in the belligerent countries. But the public opinion was still clinging to its opinion of scientists as unworldly dreamers and absent-minded professors, as people who knew nothing about practical affairs, knew nothing about mundane goods, and therefore luckily could be allotted a fairly small portion of these.

This misconception obviously could not go on forever. The time lag between the public myth and the completely changed facts was ready to be cancelled by some crystallizing event. This event was of course Hiroshima. Public opinion realized quite suddenly, at one stroke, that impractical scientists have become very powerful magicians. Nothing is better calculated to inspire awesome respect than proven evidence that a certain group of people are able to cause 100 000 deaths by two bolts out of the blue.

Of course, this respect went at first with an eager wish to believe that the new force was not exclusively malevolent. Today it feels strange to remember how, on the whole, public opinion expected, hoped, prayed for the new scientific magic to prove itself beneficial, and that is how the myth of powerful magicians who were certainly to be feared, but hopefully might become dispensers of good things, came to be firmly established in 1945. Science, in fact, was quite ready to assume the task of dispensing not only horrors but also benefactions. The World War once out of the way, the new technologies were available for civilian use and for enrichment of material (possibly even not only material) life. All those advances which the scientists had translated into technology, in the space of a couple of years during the World War—antibiotics, new kinds of electronics (which immediately brought about the spread of television), computers, jet propulsion, new fibers—new sources of wealth everywhere, and all of them ready to be tapped.

This new public expectation produced an obvious readiness of the public to support the scientists and, in fact, to mobilize them in public interest. It would be quite wrong to believe that it was only the ruling class, whatever it is, who was responsible for this mobilization; the expectation and the will to support were rooted in all social levels. In this all-pervading atmosphere, immediately after 1945, science began to expand; science itself, and not only its derivative technology.

It expanded in various ways: in human numbers, and in its readiness to spend money, and in the size of units in which scientific research was done after the war.

Let us consider these three factors separately. The expansion of the numbers of scientists was perhaps the least significant of the three. After all, scientists have to be trained, and new generations simply have to be born—not much room there for sudden acceleration. In a first approximation, very rough approximation, it can even be said that the numbers of scientists, after the war, did not grow substantially faster than they had already done for several previous generations. We probably all have heard the observation that science nowadays has expanded so rapidly that the overwhelming majority of all scientists who ever lived are living today. It was the merit, I think, of DE SOLLA PRICE, to have pointed out for the first time that this was already true for quite a few generations before, because the numbers of scientists were already growing exponentially. Exponential growth does mean that its last doubling period produces as many as those which have ever existed before. If a working scientist lives through three doubling periods of science, which approximately is true if you consider that Price's rough estimate of the doubling period of the scientific population is about 15 years, and normal working life of a scientist is three times that—then it can be said that, at the end of his working life, he sees around him eight times as many scientists as there had been in the beginning. This estimate of the now-living was valid also before the War.

The only difference is that growth from four to eight is less noticeable than growth from 16 to 32, which occurs a few doubling periods later. And so, although the doubling time does not change from before, the level reached becomes far more conspicuous.

The other factor of growth was the financial support, and there the year 1945 was definitely a breaking point. On the whole it can be said that, since then, the affluence of scientists, at least in the form of their means of work, increased enormously and far more quickly than ever before. The expenditure per scientist, which was rising rather slowly before World War II, began to grow with a doubling time of about ten years. And if we combine the doubling of the head count in 15 years, with the doubling of the expenditures per head in 10 years, we see that the total expenditure for the whole population doubles every six years.

To sum up this simple arithmetic, it can be said, again in rough orders of magnitude, that between 1945 till about 1971, when the game was definitely up, a little over four six-year periods had elapsed, and the total expenditure in the United States (and elsewhere) grew therefore by a factor of about 20.

Now, this twentyfold growth witnessed by a single generation is quite a lot. « Upon what meat doth this our Caesar feed that he is grown so great? »—in money terms, this impression was irresistible.

There was also a third feature: the size of institutional units, in which scientific research was done. A great sociologist of science, the late BERNAL, in his memorable book published in 1939, remarked that the typical research institute counts about five to fifty qualified, that is scientifically trained, personnel. After the War, this became peanuts. Big Science laboratories began to employ hundreds in a single self-contained unit and some of them went into thousands. Total personnel, that is all kinds and not only those trained in science, had of course to be far more numerous. It is easy to quote examples of establishments which reached 5000 soon after the War; at the same time, their yearly budget grew from millions of dollars to tens of millions.

Two more remarks about the first years of the spread of Big Science: one is that under the influence of the wartime habits, big institutes were first aimed at definite applications, rather than intended to foster some self-motivated discipline of scientific research. To use a term which was created later on, they were « mission-oriented ». An interesting consequence was that, from the start, they became more easily multidisciplinary than it was customary in the universities, where this style is still searching its way, or in the more modern, no longer exclusively mission-oriented, Big Science institutes.

The second remark is that the first modern-type mission-oriented establishments were set up during the World War for immediate emergencies and therefore were located in the main belligerent countries: United Kingdom, United States, Soviet Union, Germany. (In Germany, by the way, they achieved successes in rocketry and radar, but not in nuclear energy. This was already mentioned in one of yesterday's lectures.)

But after the War, they were soon imitated in other countries where wartime habits were much less determinant. Even there they were an element of preparation for future wars. In addition, practically every advanced country entered the race for new kinds of industrial technologies for peaceful use.

On the whole, the setting-up of Big Science units in all these countries was strongly fostered by the nation states. GOLDSTEIN explained this yesterday; I might slightly differ from him by insisting on the preponderant role of the prestige component among the motives by which all nation states are moved to action. Prestige is not just vanity: a nation which holds prestige can or hopes to hold its own by sheer bluffing in its foreign policy. Moreover, a government will risk a serious loss of credibility if it leaves the essential military strength in the sole possession of the United States and the essential civilian innovations also in the same hands, plus a few other countries such as the United Kingdom.

Prestige of the nation, and also the prestige of the state: and this was why state-fostered Big Science establishments began to spring up in a number of advanced countries, well before 1950.

Among the scientists themselves, there were two strains, two tendencies. Those who had taken part in wartime work, had witnessed its efficiency, and were eager to apply the same methods to more specifically scientific aims. On the other hand, those less experienced in wartime work, or not at all, wanted only to return to their pre-war habits. A curious opposition could at first be felt between these two strains. But then—I am glad to say or sorry to say, I do not know—the opponents noticed that, after all, there was something to be said for big establishments—above all, greater financial freedom. They were then won over pretty quickly.

Towards the end of the 1940s, the mood became subtly optimistic in the scientific world. It is interesting to note that in those years, there was no established ideology or methodology of Big Science. The big units were created without any prior blueprint of how to do it. The same BERNAL whom I already mentioned made a remarkable statement as early as 1946, when reviewing wartime science in the British yearbook « Progress in Physics ». He said that the achievement in particular of nuclear energy, was made possible only because a new form of organization had been invented, and this quite possibly was a bigger invention than atomic energy itself. GOLDSTEIN—I think it was he—yesterday quoted POGO, to the effect that we were looking for the enemy and we found it was us. When, later on and in a much changed mood, I was looking for those enemies who over-sold Big Science units, I had the same experience. I was unable to find anything earlier than a paper which appeared in 1949 in the *Bulletin of Atomic Scientists* of Chicago, and whose author happened to be myself.

In this paper which was called « Psychology and structure of large-scale physical research », I depicted the reasons for setting up these big units and how they are built, and already then I mentioned difficulties arising from certain contradictory trends.

For example, there was the fact that the goals of most of these establishments at that time were strictly applied, yet since the time lag between pure science and its application had dwindled to practically nothing, « pure » scientists had to be employed. They had to be recruited and made to work in conditions certainly different from all their previous habits. This created, at first, many difficulties.

Another opposition was inherent in the necessary kind of hierarchy. All these mobs of scientists and all the money which was so eagerly thrust on them by the public could no longer be handled according to the same happy-go-lucky principles which had been prevailing before the War. Now there has to be a strong directing hand to ensure efficiency. And yet no strong hand will succeed unless there exists a dedication to the common goal on all levels, a feeling that « It depends on me ». Therefore, one has to somehow combine autocracy with democracy.

The third conflict was that between the possible ways of structuring this multitude of scientists. The traditional way is to group them around the disciplines: physicists here, chemists there, metallurgists in another corner. However, a typical immediate goal is to build a reactor, an accelerator, a computer, and, for each of these goals, close coexistence of different disciplines has to be ensured. How to subdivide the big establishment into smaller manageable units? By separate disciplines, or separate projects?

It is interesting to note that, on the whole, this problem remains largely unsolved. It began to be realized that the hierarchy, instead of the usual military system one-way downwards, tended to become two-dimensional, and this was duly pointed out in my paper.

An interesting consequence was that WEINBERG, who at that time was not yet director of the Oak Ridge National Laboratory (I think he was then something like director of research or chief scientist), noticed that these ideas were relevant to the task of directing the Oak Ridge staff towards their new goals. In particular, the concept of two-dimensional hierarchy was quite explicitly understood in that laboratory.

Another consequence was that AUGER (of whom there will be a considerable mention in my next lecture), in his capacity as the then director of Science Division in UNESCO, took note of the same 1949 paper and saw that it offered a sort of rough codification of large-scale physical research. We shall see later how he became acutely interested in precisely this kind of research. And that is how this paper played a certain role in getting me involved in the early foundation of CERN.

1949 was still a very optimistic time for Big Science. A smaller paper, a sort of appendix to the previous one was written by me for « Financial Times » (England) when, in 1956, they published an overall study of atomic energy. There I more or less repeated the description and the formulation of the contradictory trends inherent in Big Science units. However, many things had happened in the intervening seven years and, as a result, the tone was markedly less buoyant. Two new contradictions were put forward. One was the growing necessity for a new form of co-operation between scientists to which they were not used at all—in one word, committees. Scientists noticed that they had to spend more and more time sitting in committees and trying to arrive at decisions by the committee way, and you know what this means: when a committee designs a horse it comes out with a camel.

By 1956, committees had grown noticeably more irksome than in 1949. The other snag was that, at the same time, the Big Science idea began to spread more and more toward the purer brands of science. The proportion of unruly longhairs (at that time they were still called longhairs, not eggheads; today, of course, long hair has no intellectual connotation), shall we say creative people, grew and began to cause new problems for the autocratic directors of establishments. Some of the establishments were still run by strong scientific personalities like COCKCROFT or WEINBERG (who by that time *had* become director of Oak Ridge) but others were not, and their hold on the unruly became problematic.

So much for my 1956 paper. The next descriptive literature to appear was some writings by WEINBERG himself. He started that batch in 1961 with an article dealing with the « Impact of large-scale science on the United States. » He still used the words « Large-Scale Science », but, in fact, in this paper he also mentioned the term « Big Science » which, I think, was his creation. This same term was made popular by PRICE in his Columbia lectures on « Little Science, Big Science, » which were published in 1963 as a book.

There may be something slightly sarcastic in this expression « Big Science » when it is officially used. Large-scale science, which is what I called it twelve years before, sounds more noble. But WEINBERG, for the first time, began to describe that impact, about which he wrote his paper, as something no longer wholly positive. In his 1962 lectures which often dealt with « Problems of Big Science », he squarely posed the problem: is Big Science now invading Little Science? He began to discuss the fundamentally negative characteristics of Big Science. In this connection he used three somewhat fancy words; what they really meant was his complaint of three separate excesses. First, « too many journals »—too many publications and reports, and also too many journalists interested in what these Big Scientists were doing. The necessity to write papers and journals and reports, and also the urge to satisfy the journalists' curiosity, began to distort somewhat the traditionally serene scientific atmosphere.

The second excess was « too much money »: scientists began to feel that any problem can be attacked by spending money rather than by hard thinking. And the third was, far too much administration, too many cramping plans and projects. We shall return to this in my last lecture. The fact that much of scientific research was still run in the style of Little Science was presented by WEINBERG as a hopeful factor, but he made two warnings at the end of his lectures. One was that Big Science had seriously begun to invade the universities; he considered this as a wholly harmful phenomenon. Secondly, he stated a common historical trend: when civilized communities begin to ascribe exalted importance to spectacular monuments of cultural or other achievements, the end is in sight. As he put it, « The French revolution was the bitter fruit of Versailles and the Rome Colosseum helped not at all in staving off the barbarians ». In the same way, the overgrown marvels of Big Science, the accelerators, the reactors, the rockets and all the rest, might turn out to be the first signs of something ominous approaching.

This warning note was sounded as early as 1961, and, as many of you know, WEINBERG continued to treat these question in a series of papers and concepts, « Criteria of Choice », « Technical Fixes » and so on, and I hope more to come.

At the end of 1964, another practitioner's description of Big Science appeared, first as a lecture given at MIT, and then as an article in *Science* Magazine (1965) under the revealing title « Megaloscience ». The author was ADAMS, a prominent figure in Big Science since the early 1950s and still prominent today, an important thinker on science policies (he took a significant part in the recent report on « Science, Growth and Society, » produced by an OECD committee under the chairmanship of Harvey BROOKS of Harvard). ADAMS used the word « megaloscience » as a further downward notch on the concept of Big Science, as a hint of a mental disorder akin to megalomania. He frankly painted megaloscience as a somewhat sickly phenomenon.

I will not repeat his description of new types of scientists, such as the scientist-manager or scientist-pilgrim who goes from conference to conference and, yes, from one summer school to another, but I insist on one point. He noticed that scientific expenditure, at that time, still was doubling every five years (which is approximately the same figure as the six years I have mentioned earlier). Then he said: we have now reached a point when society cannot afford any more doubling. There will possibly still be one more, but certainly not two. And the convenient atmosphere of infinite expansion in which science had lived for the time span of a generation now is quickly coming to an end.

He was remarkably accurate in this prediction. Taking his doubling period, five years, and the time of his lecture, end of 1964, we arrive at the end of 1969. Well, by the end of 1969 it became quite obvious that the exponential expansion had come to an end.

His reasoning went on: we can meet this threat in a manner befitting us as scientists: we shall analyse it, plan for what to do about it, and manage to smooth the transition. Or we can neglect it, and fall victims to those violent fluctuations which, as we know from many electronic and biological phenomena are apt to mark the sudden end of an exponential phase of growth.

His warning was not heeded, and the consequences now are upon us. In the moods of the scientific world today, and in particular even in the mood of this audience before me, I see one reflection of these violent oscillations ADAMS warned us about, nearly eight years ago.

At about the same time, in the later sixties, it became clear that many of the mission-oriented establishments were facing a crisis of a kind which arrives quite naturally when such an establishment comes to the end of its function. This happened, for example, in atomic energy. The task of starting up the civilian use of atomic energy for power stations has largely been completed, at least in the most advanced countries. (In other countries, it was never properly started, so its termination did not matter.)

A mission may also have to be ended if its goal is finally found to be impossible, or far harder to achieve than it appeared originally. I suspect, for instance, that we shall soon realize that some of the lines taken by atomic energy will be found to be far too arduous. One obvious example is the thermonuclear fusion. This effort may yet lead to practical results, but certainly on a time scale enormously longer than what the creators of such establishments as Culham in England had in mind at the time of setting-up. ADAMS, by the way, was Culham's founder and first Director.

The third possibility is the loss of public interest in the goal. Examples would take too much time, but many of us can think of some.

When governments, and even Big Science directors themselves, began to think of what to do with their exhausted missions, a great variety of options became available. First and most obviously, one could simply disband. Somehow this course of action is not taken very often. Once a big unit is there, it is surprisingly difficult to terminate its existence. Culham, for example, was not disbanded outright but its budget was halved, which perhaps made the prospects of a useful return on the remaining half even more remote.

Those who refused to be disbanded tended, quite naturally, to deepen their original goal, so as to justify more and more work on it, regardless of whether the public opinion was approving it or not, and hoping that the government would still support it for a while. Obvious examples could easily be quoted.

The fourth possibility is that of a mission-oriented establishment suddenly going disciplinary. Why not, like a university, become simply a collection of laboratories, physics here, biology there, and so on. This actually happened in some cases.

The most lucky institutes are those who have found new missions of a kind similar to the original one. The Apollo project hopefully may yet lead to the exploration of Mars, although Washington at present seems to think otherwise. And finally, there is the possibility of finding new missions of a different kind. Then it is usually found that the establishment is obliged to switch over not to one mission of a new kind, but to a whole combination of them; therefore, it begins to « diversify ».

For example, the present-day trend is to transform atomic-energy laboratories into multidisciplinary establishments dealing with environment questions. Whether it will succeed or not, we do not know yet.

Meanwhile, similar factors were at work, in more traditional strongholds of the scientific world, introducting new attitudes and, to some extent, driving the minds away from the ideal of bigger and bigger research units. With the increasing role of data processing as a sequel to data acquisition, a growing part of scientific work could be done away from the places where data are actually created and recorded. Geographical dispersal of Big Science establishments became possible. I was one of the heralds of this movement, and as happens to many heralds, I was heard to a far smaller extent than I had hoped. Still, the idea is marching on.

There is also the growing realization that researchers should stop running away from teaching, that combination between research and teaching is a necessary factor in the life of a researcher, a fact which tended to be forgotten in the 1950s and 1960s. Here I must say that some of the attitudes shown here by the younger part of the audience, obnoxious as they are, are definitely useful in this respect—they help to remind the scientists, the professors, that after all, they do have something to give you, that it is not enough to sit in their big laboratories and write research papers.

The already-mentioned tendency to go interdisciplinary was, on the whole, better suited to universities (where a great variety of disciplines already coexist) than to the new Big Science institutes. Yet the latter, in their limited interdisciplinarity, often showed themselves more flexible than the older, more academic, establishments.

Currently it is found that Big Science perhaps is more significant when it is pursued on an international plane rather than nationally. This leads to a more pronounced translation of certain branches of Big Science onto an international plane. This obviously is happening in high-energy physics, where more and more money is given to international institutions such as CERN, and at the same time, national accelerators are being closed down.

This brings us at last to the international aspect, which we shall consider in the next lecture.

## 2. – International enterprises in science.

Yesterday we examined the importance of new forms of organization and that of the breaking point of 1945. We saw how the Big Science organizations began to spring up here and there and practically everywhere in the advanced countries, in response to various forces, among which I count national prestige, in its widest implications, as the most determining factor.

Today we shall review another of these main post-war forms: the international enterprise in science, which began to be noticeable a few years later than the national Big Science form. The two forms have somewhat different motivations, somewhat different ways of getting a start, and therefore it is suitable to consider them separately. Yet there are, of course, close links between the two forms.

One is that to set up an international enterprise (I will define this term a little more closely in a moment), to create a new international bureaucracy, to overcome all the divergent nationalist obstacles, is such a bothersome job that it is worth-while only on a certain scale of size. And, therefore, practically every international enterprise that was set up after the War belonged to the Big Science class of undertakings. The converse link became obvious only in the later years. By that time the Big Science ideology and practice had begun to show some decline, and the concept of a Big Science establishment became a little more diffuse. I mentioned these recent developments at the end of the first lecture, and the gist is that the push toward creation of Big Science establishments is getting to be counterbalanced by forces going in the opposite direction.

However, international arguments, which we shall examine in a moment, add an extra push in favour of creating big establishments. The somewhat unexpected consequence, which probably will be more in evidence in the next few years, is that not only international science in its enterprise form always tends to be big, but also Big Science tends to be more international.

Before we comment any further on this collusion, I have to insert a little flashback. Science of course has traditionally tended to transcend national borders ever since it took its more or less modern form, practically since the 17th century. This activity always was very international in character, which was shown in various ways. There were communications, journals, short and long stays of visiting scientists and scholars in one country or another, international conferences, typical wanderings like the beginning of Weisskopf's career we have heard about—all this existed centuries before our time. But the actual fitting together of tangible assets, assembling a lot of buildings and equipment on one spot under international supervision, that supremely concrete form of collaboration practically did not exist before the Second World War. There have been a few earlier examples on a modest scale. Among the first truly international scientific enterprises, we may quote the Paris (or near Paris) Bureau of Weights and Measures, set up in 1875, and an even earlier interesting initiative which produced the Biological Marine Station in Naples (1870).

Another prominent example, closer to our time, was the high-mountain station on Jungfraujoch in Switzerland. Incidentally, AMALDI today talked about an international launching of balloons, but balloons are only one way of getting high up. The other is to climb mountains. A laboratory of this kind was started in a true international spirit, although under close Swiss supervision, well before World War II.

After the War, the immediate reaction was to set up Big Institutes on strictly national lines, for reasons I already discussed yesterday. As long as one of the main reasons was to provide a symbol or mascot of national greatness, or aspiration to greatness in science, the enterprise had to be financed at home by national taxpayers, and therefore the immediate post-war movement was away from international ventures—not from international spirit in general but from its concrete embodiment in common enterprises. In some branches even the spirit itself fell under the influence of secrecy—first of all in precisely that field of physics which carried most glamour, nuclear physics. Other branches were less affected, but what went for the most renowned, most pace-setting science, which at that time was nuclear physics, to some extent determined the mood for other sciences.

Two additional forces began to work in favour of international enterprises in science very soon after the end of the War. One was that, in opposition to the growing nationalist importance of show-case science, there was also a growing push, especially in Europe, toward multinational groupings. We must not forget that the first sizable manifestation of European technological unity was set up in 1950: that was the Coal and Steel Community. And of course its preparation was already well under way in the late 1940s. The new across-the-borders push needed not only such humble tangibles as coal and steel, but also its glamourous mascots. This rival market for scientific glamour began to develop almost at once.

The other incentive was at least as pressing. Much as the nationalists were eager to foster home-grown manifestations of Big Science as mascots of national capability, they had to keep in mind that in post-war Europe to set up a truly worth-while big institute was a hard task for any single nation. One had to weigh the advantages of being nationalistic on a limited scale, as against taking pride in being a member of a multinational group which *could* attempt something really important.

This brings me almost directly to the history of CERN. AMALDI gave here the most important dates, and he situates the start of CERN somewhere around 1948. I would put it one year earlier. A significant event occurred, in the immediate post-war years; I am pointing it out because I was there. AMALDI has less reason to dwell on it, because he was not there. The event was the negotiation about international control of atomic energy which took place in the United Nations, immediately after that organization had started functioning in 1946.

So vital an issue had, of course, to be discussed by professional diplomats and (as WEISSKOPF explained a moment ago) for scientists it was a working assignment and not just a subject of talk in their leisure hours. A lot of such talk did however take place on this occasion between many of the advanced countries, not only Western but also the Soviet Union and Poland. It was an amazing possibility for their physicists to meet openly, internationally and with full permission of their vigilant authorities; to be able to touch (sometimes delicately) on the domain of nuclear physics and even on the dangers of nuclear weapons!

There was an everyday interpenetration between the physicists and the diplomats. They worked together for the long weeks and months of these negotiations, and this created an unprecedented meeting ground and an opportunity to appreciate each other's craft.

Such was the background of the encounters which took place between OPPENHEIMER and a then junior French diplomat who had nothing to do with physics before, and who was entrusted by the French government with the diplomatic side of the negotiations. His name was DE ROSE.

They became friends and together they began, in 1947, to approach the problem of bringing to life a big international nuclear enterprise in Europe. They managed to blend their professional points of view and this, in my opinion, became one of the most important factors of the subsequent situation.

Going back to Amaldi's time-table, we can trace the sequels of these conversations, on their Oppenheimer end, through RABI, who was in close contact and yet somewhat less controversial in governmental eyes as compared to what OPPENHEIMER was beginning to be already then. From RABI to BERNARDINI, from BERNARDINI to AMALDI, and so on. DE ROSE, on his end, was working towards the French science administrators like DAUTRY. Through DAUTRY he almost immediately enrolled the active interest of the glamourous French theoretical physicist DE BROGLIE, much in the same way as SZILARD, in 1939, had enrolled the help of EINSTEIN to convince ROOSEVELT. In my opinion (but since I was fortunate enough to be entrusted with the writing of the first history of the origins of CERN, this opinion has acquired a more or less official status) that is the real root and beginning of CERN.

The role of AUGER, which was pointed out very clearly by AMALDI, carries the same kind of significance because in Auger's case a very high level of competence in physics and a position of bureaucratic international power happened to be actually blended in one person. It was AUGER who proclaimed, against everybody's insistent advice (and none was more insistent than mine, at that time), that Europe should immediately aim at building the most powerful accelerator in the world.

I was not the only one trying to persuade him against this aim. As you have heard, BLACKETT took the same stand. But fortunately both BLACKETT and myself were convinced by AUGER pretty soon.

All I have now to add to early CERN history as described by AMALDI is to insist a bit more on AMALDI than AMALDI did. His, again, was a unique role of being practically the only, at that time, really active scientist in cosmic-ray research, the forerunner of high-energy physics, who found himself deeply involved in these negotiations, together with AUGER. He helped AUGER decisively, in these early stage at the very centre of things. And that is why AMALDI became the top official of the provisional CERN from 1952 to 1954, and played a very effective and beneficial role in this position.

It is also interesting, since we are here in a school of history, to point out the contribution made by BOHR. He was quite active in some of the aspects of CERN's creation, but above all, his mere presence, as a manifestation of his enthusiasm for this idea, showed the world that this was an enterprise which had the blessing of pure science on its highest level, and that proved to be enormously important.

The outcome of these formative years was that CERN developed into what may only be called a smashing success. I am considering here not only the scientific side, but also and mainly, in connection with my subject here, the organizational side. The success was so big that it became almost embarrassing, because people thought always of CERN and CERN again and other international enterprises were overshadowed.

Several features of this organizational success are worth a mention in detail. Could impractical scientists, especially, when hampered by international forms of bureaucracy, be relied upon to work out such a practical result? CERN did prove to be able to build its huge machines, in close co-operation with its industrial suppliers. Many lessons were learned here, especially those drawn from American and British wartime experience, and in fact the cost-benefit analysis (I believe that is the expression) of early CERN shows a high level of achievement. CERN also became very important politically: first of all it shows that these things *can* be done under a truly international direction, and yet remain completely practical, on par with any nationally directed big enterprise (perhaps even better)? This demonstration reflects positively on the whole idea of international undertakings, not only in science but also in other domains. This conclusion has a definite political meaning.

Another political matter of great importance was that CERN pioneered in its active sponsorship of East-West collaboration. This happened as early as 1956, probably somewhat earlier than the deStalinization speeches of KHRUSHCHEV. The question was asked this morning, how did it happen that East European countries were not involved in the foundation of CERN. Well, in the beginning, Eastern Europe, led by the Soviet Union, understood quite correctly that CERN was a mascot for European unity, which at that time meant Western European, and the Soviet Union was not very happy about this political grouping going to get a powerful mascot. The idea of Europe from the Atlantic to the Urals was not very popular then. Hence, in those years, a strong Communist opposition to the CERN idea.

And then, rather suddenly, perhaps on the principle « if you can't lick them, join them », this opposition was changed almost overnight. The pioneer of this change was JOLIOT, who in 1951-1953 came out strongly against CERN, but after that began to support it. The change of position was so sudden that I ever remember one tense moment when Joliot's wife was still taking part in the anti-CERN propaganda, yet JOLIOT was already beginning to react favourably. But these awkward moments are characteristic of sudden political changes and just show how difficult it is for private life to adapt itself to grand historical events. A theme which, after all, underlies the whole message of the novel « Dr. Zhivago ».

Soon afterwards, the Soviet Union itself began to look at CERN benevolently. A favourable occasion was provided by the first « Atoms for Peace » conference, held in 1955 at the United Nations, Geneva, when the bulk of atomic secrecy came to an end. By 1956, the Soviet Union was ready to pay CERN the great compliment of setting a parallel multinational institution of its own in Dubna, near Moscow.

The co-operation between CERN scientists and Dubna scientists was very good from the start, and in this way CERN was able to play, probably quite unexpectedly, what I might describe as a fairly major role in the first signs of the ending of the Cold War.

Next on our list of political successes is the fact that CERN, although it was anchored in Europe and was a European international undertaking, became noticeable in countries outside of Europe. American scientists began to come to CERN regularly. After all, CERN did, for a while, have the most powerful accelerator in the world, Auger's ambition did come true. American participation was financed by various outlandish sources such as the Ford Foundation. Other advanced countries, like Israel and Japan, which are not exactly European, somehow also managed to get close to the orbit of CERN's activities. The typically British ingenuity in bringing into play the British Commonwealth concept made it possible for people from South Africa or Australia to come under the CERN umbrella; all of them British subjects, were they not? And this again reinforced the world role of CERN.

Among the first scientific conferences held under the CERN flag, that of 1952 in Copenhagen has been mentioned by AMALDI. A conference on accelerator techniques took place in Geneva, some of it already in CERN's provisional buildings, in October 1953, when it became completely obvious that Europe was discussing these techniques on a footing of full equality with America. Already in 1953, and the ground was not yet broken on CERN's present site!

So much for CERN as a success story. To whom goes the credit? This point is worth a brief mention.

CERN started as definitely a Big Science institute, with its concentration of equipment, buildings, scientists, and so on. To some of the European scientists, this looked very unfamiliar. It is often said that CERN was created by a push from the European physicists, who were eager to create an institution capable of giving them the necessary equipment. This is only partly true. It is certainly true of those European physicists who had some first experience, either wartime work or post-war work mainly in America, and knew something about these new forms of organization in science, why they were necessary and how they could be made fruitful. It was less true of those who had remained in their countries in strong academic positions, inherited from pre-war times, and who had no similar opportunity to change their habits. WEISSKOPF, in his lecture here, mentioned PAULI, and PAULI at least was mildly sympathetic. Others were not; I will not mention names. Some of them can be found in that vast crowd of eminent physicists who, today, like to reminisce about how they founded CERN.

On the whole, the image of a concerted and organized push from European physicists can hardly be accepted. A better image would be a sort of complicity between the diplomats who wanted their mascot and those of the physicists who happened to think in the same direction and were willing to cooperate.

Next point: Why high-energy physics? Why did this first undertaking originate in that particular field? Here we find a remarkable illustration of the duality between international aspirations and the home push in the opposite direction. It was obvious in the late 1940s that the international mascot had to be nuclear. It was the age of nuclear magic which I mentioned yesterday. Nuclear physicists were the most glamourous magicians; only they could provide a truly valid mascot and therefore only they could command enough public support in terms of finance. Yet this same glamour, so desirable internationally, tended to put nuclear physics out of the reach of any international action. After all, secrecy had to be kept nationally, because the glamour was inseparable from the thought of applications, and those, military or commercial, were too close to strictly national interests.

And so the dilemma arose, how to have it both ways. The contradiction looked almost insolvable, and yet there was a way out. There was a certain duality in nuclear physics itself between those of its branches which were close to applications, and those more remote. Brookhaven, which was founded a few years earlier than CERN, pursued both kinds of interest, reactors as well as accelerators.

Ideas in the corridors of power about what constituted the power of nuclear physics were, fortunately for the physicsts, very vague. High-energy physicists were always insisting that the problem which interested them by then was not the nuclear structure, but the nature of the particles themselves, a kind of knowledge for which no possible applications were in sight, and that therefore they were quite harmless, that there was no reason for secrecy. Then the politicians would say, « Yes, no possible applications now, but you can't say there will be none later on. Look at your colleagues, the nuclear physicists of, then, fifteen years ago, look how the desirable applications have grown. Very desirable, some of them ».

Public opinion followed the politicians, and the public vaguely expected that particle physicists would produce some beneficial magic, possibly, hopefully, not military, but at least some useful gadgets, some new technology. Well, of course, if the physicists want to keep their purity and say that they are far from all applications, all right, we understand. The general attitude was full of this sort of understanding.

Some of the physicists were shrewd enough to guess what this attitude meant, and then somehow they chose not to disillusion these vague expectations too bluntly. All this, one might say, amounted to a remarkable case of double connivance. To describe it as, actually, a sort of mild deception would be, perhaps, a little too strong. Let us say, a slight dose of hoax enabled the politicians and the physicists to have it both ways so decisively, that the support for CERN was wholehearted and unquestioning. It still is, after some 20 years of operation which, so far, has not led to any visible applications. This interesting sociological phenomenon is worth recording.

By 1955, most of the « atomic » secrecy had disappeared. It happened because it was found that in spite of all precautions taken by the « Sacred Trust » concluded between the United States, Canada and the United Kingdom, the Soviet Union did quite nicely all by itself. It developed its first atomic bomb only about four years after Alamogordo. There was an attempt to put this at the door of the atomic spies; they must have transmitted every secret. But then the hydrogen bomb arrived, and there, to everybody's confusion, the Russian were rather ahead of the Americans. The spy argument seriously did not hold water.

By the time of the first « Atoms for Peace » conference held by the United Nations in Geneva (1955), all sorts of practical people, politicians and public opinion in general began reluctantly to understand that secrecy on nuclear activities could be very substantially relaxed.

This meant that the example of the second half of Brookhaven, the one which stood closer to applications, could at last be followed in Europe on a multinational scale. Nobody thought of this task as a possible part of CERN; rather as a kind of counterpart. Besides, in 1955 the mascot began to be most sorely needed not for Western Europe in general, but for the Six of the Common Market.

From 1956 onward, an enterprise devoted to the study of nuclear energy and the promotion of its applications began to be fostered in Europe under the name of Euratom. It turned out to be one of the saddest stories in the history of International Science, or of Big Science or, perhaps, of applied science in general. There the mascot aspect was pushed far more bluntly than in the case of CERN. There were some other reasons, which I will mention later on. The thing never managed to take off the ground, although Euratom did recruit many highly competent scientists and engineers. It also built up a Big Science laboratory in Ispra, not completely anew like CERN, but by developing an already existing Italian centre of nuclear research.

It turned out, however, that on the European nuclear-energy scene nationalistic forces got the upper hand. Being a French citizen, I am at liberty to say that the strongest push away from the international approach to nuclear energy came from France. At the same time, another international enterprise in atomic energy was started in a far more modest way, using what I described two days ago as the « brave soldier Schweik » tactics, by OECD, which at that time was called OEEC (Organization for European Economic Co-operation). It had a nuclear branch, headed by a French lawyer, HUET, and it managed, in a curiously nonbureaucratic way, to set up several enterprises—genuine international enterprises in atomic energy, which flourished in the late 1950s and early 1960s. The most conspicuous among them was the development of a new type of nuclear-power reactor which became well known under the name of Dragon.

Today, it appears that this type has a promising future in industrial-commercial applications. This detail is useful in order to counter the argument that CERN was so successful because it was so remote from any possible applications. The Dragon story does show that branches of scientific technology which are very close to profitable applications *can* be successfully developed in an international way. Another (unfortunate!) example is the Concorde supersonic airliner. Whatever its commercial prospects (and may they be few) it *is* a technical success, and yet it is genuinely international. Two nations only, but even two nations when they start working together pose all the problems which arise in co-operation between 10 or 12.

I can mention only briefly a few nonnuclear international enterprises. Obviously as soon as space became important, there had to be a European mascot in space, and there both AUGER and AMALDI, with their CERN experience were able to play again a considerable role.

As AMALDI said in his lecture here, this venture soon took a turn which was somewhat unexpected to its early promoters. A large sector of the European space effort had to carry, like Euratom, the burden of an original sin, this time of British origin. ELDO was eagerly supported by the British government in an endeavour to make use of a military rocket, the Blue Streak. It has been developed in England at a considerable cost, but had to be scrapped because the British had very sensibly decided that it was better for them to buy this kind of equipment in America, which by that time had become far more knowledgeable.

And so the Blue Streak was foisted on the European space research, with the result that so far the organization for launching European satellites has not succeeded in putting a single one in orbit.

The purely scientific companion organization (ESRO), of which AUGER was the first director, was more successful, and although its achievements fell somewhat short of expectations, they were by no means negligible. ESRO *can* be considered as, shall we say, a mitigated success.

There have been other European initiatives, such as that in molecular biology, usually strongly influenced by the idea that an international organization *has* to deal with something big. Somehow, the biologists were not successful in proceeding from their small objects of research to big equipment, at least not as successful as the particle physicists had been. The European Molecular Biology Organization (EMBO), according to some last-minute news, *will* have a laboratory, somewhere in Germany, but it certainly will not be on a scale similar to CERN's.

Another interesting example, and probably a somewhat more heartening one, is the European Southern Observatory. Since European astronomers have to observe certain celestial phenomena which, unfortunately for them, can be seen only from the Southern hemisphere, there is a need for a common European observatory located in that hemisphere. It is now being built in Chile, with European-made equipment. An interesting feature is that this equipment is being built with substantial help from CERN, to such an extent that ESO is in some ways, at present, practically a division of CERN. Nobody ever mentions this, because it goes against the CERN Convention, but everybody is looking at this phenomenon with warm benevolence.

There have also been attempts to set up world-wide organizations, but they have not been so successful as the best among the European attempts. When the International Atomic Energy Agency was founded in Vienna, there was some idea of having a laboratory attached to it, but it never properly got off the ground because this organization was mainly involved in more political subjects such as the safeguards against the proliferation of nuclear arms.

However, the same organization together with UNESCO was successful in setting up an International Centre for Theoretical Physics in Trieste, which is a rather modest institution. It spends less than a million dollars a year. For a comparison we only have to look at CERN, which, not counting the new accelerator at all, is now spending something of the order of 100 million dollars a year. Euratom in its heyday spent about the same. And the nuclear agency of OECD was spending somewhere in the neighborhood of 5 to 10 million dollars a year, counting all its enterprises.

And yet, in spite of its small budget, the Trieste Institute is very important as probably the only space in the world where theoretical physicists from all countries come and feel that they are at home. It sets a valuable political pattern, showing the true way for developed countries to help the developing countries.

I shall now end with a brief mention of what I consider as the main reasons for success of some organizations and unsuccess of others.

Some of the arguments usually invoked (such as, you are successful because you are remote from applications) I do not consider as valid at all. The remoteness helps, but also, when a scientific enterprise is close to applications, this fact also helps in a different way.

What then are the reasons? Well, it has been accepted in the official CERN history, that, if you want to promote international co-operation in the form of a scientific enterprise, the objective must be sharply defined, and ambitious at the same time. Auger's successful ambition to build the biggest accelerating machine in the world was typical of that principle. If Europe decides to build something really big, for example a space shuttle, which is probably still within Euopean means, it can do it; of course Europe would have to go about it a little differently from the ways it had adopted for ELDO. The objective must also be sharply defined. There lies one of the reasons why molecular biology had some difficulties and why Euratom failed. The aim also must be realistic, and here, of course, opinions can be different. Auger's ambition proved to be just at the limit of the possible. The Euratom ambition (to build by 1967 an imposing number of atomic-power stations) proved to be totally unrealistic. So much for the first condition, the sharply defined and ambitious objective.

The second calls for mixing scientific competence with administrative competence. Opinions to the contrary are sometimes heard, but experience does seem to show that this *was* one reason for CERN's success, and also of the somewhat more modest success of OECD's nuclear enterprises; conversely, the lack of this mixing was largely responsible for the unsuccess of some other organizations. This second condition is probably as important, if not more important, than the first one.

In the case of nuclear physics or subnuclear physics, the successful blending of competences stems directly from the post-war United Nations negotiations on the control of atomic energy. There scientists and diplomats could meet, learn from each other and they actually started to work together. This tradition of mixing was still alive at the time of CERN's foundation, a few years later. DE ROSE, whom I mentioned before, played a very considerable role at that time, and also in later stages of CERN, adding up to some twelve years of its existence. His example was followed by many diplomats and administrators from other countries.

There are other conditions. My usual full list contains ten of them, but I will mention here only two more. One is the participation of respected and strong personalities. At CERN we had, closely involved in our councils, five Nobel Prize winners in physics: at first, RABI; then in Britain, BLACKETT and COCKCROFT, then HEISENBERG, who in 1952 was in charge of the first launching of a scientific program. Finally, BOHR, whose role I described a moment ago.

These five people were of the highest calibre—and, in addition to their effective contribution, which was great, the simple fact that they *were* people of high calibre, extended to CERN the benefit of the enormous respect in which they were, quite rightly, held.

The fourth and last condition that I will mention here is that attention should be paid to what we might call political winds. I do not mean that action has to be unprincipled and ready to bow to every political pressure. But just as we have to be technologically realistic, the claims of political realism should not be wholly dismissed. CERN was founded at the time of a great push towards European unity. Euratom, unfortunately, had to start when this wave was already receding. The new CERN machine, the 300 GeV project, has noticeably suffered from the European dissensions in early 1960's and then profited from the obviously impending enlargement of the Common Market, which itself was only one manifestation of that « negotiations era » that began to be visible in Europe and around Europe in the late 1960s and early 1970s.

A vivid example, on which we may suitably end today, is that of a little-known successful enterprise which, at its most decisive period, involved only two nations, but, like CERN, will probably extend to many others. The high-flux neutron reactor, which recently has been completed in Grenoble (France), was started as one of the OECD nuclear enterprises, and was promoted originally by Britain, at the moment when Britain was expecting to join Euratom and the Common Market (in 1962). Obviously, the political wind was then favourable. In 1963 it became clear that Britain would not join the communities and immediately the chances of this enterprise went down, to the point that Britain simply refused to support this project. But by that time, DE GAULLE wanted to produce a mascot for his Franco-German friendship policy, and so from late 1963 on, the project was taken out of OECD's all-European domain and became a bilateral Franco-German affair. It took some time to get off the ground, for political rather than technical reasons, but eventually it has been completed. Now that it promises to become a valuable research tool, and that, at the same time, the political wind turns towards greater co-operation, there are many visible signs of this reactor becoming again multinational and not only binational (*).

(*) *Note by the author*: At the end of 1972 Britain officially joined the Grenoble project as its third full-scale participant.

## 3. – Impact of the new forms on « pure » physics.

In the previous two lectures we have examined Big Science and its other form, International Science, as new organizational phenomena. Like every new phenomenon, they have their own new ways of being both useful and obnoxious. I will not dwell today on the positive side of their impact because many of its aspects are self-evident, and for the remainder, those responsible for the launching and direction of the new organizational forms in pure physics are seldom short of eloquence.

Therefore, I am going to concentrate, as you will see, on the negative sides of the impact, and now, please, please do not consider what I am about to say as an indictment of today's pure science. The negative features which I propose to describe are the negatives not of pure science, but of big organizational forms, and a warning is necessary in order to know how to improve these forms. Big Science in its pure-research aspect certainly can stand some improvement.

You may have noticed that WEISSKOPF yesterday, although on the whole he is probably in a more mellow mood than I am about to be, did manage to put in a few firm warnings concerning the state of pure physics toward the end of his career, as compared to what it was at the beginning. All these negative features seem to be due to the Big Science organizational form.

Let us now turn to this analysis, and I will try to conclude on a hopeful note. I hope I shall then sound convincing.

The new era of big organizations, as it has already been explained at some length, was born under the pressure of wartime urgencies. It was, therefore, essentially directed towards applications: Big Science started definitely as applied science. Then the war and its urgencies ended, and for some time the very advanced scientists, who had become so mixed up with large-scale practical questions, almost for the first time in the history of science, were still under the spell of this push towards wider issues.

Thus, in the immediate post-war years, to quote a few examples, we see for instance, that WIGNER, a very pure scientist indeed, is at Oak Ridge, directs research there, and is decisively instrumental in promoting the concept of the water-cooled reactor. Hardly a pure-science activity. OPPENHEIMER immediately joins the stronghold of pure science at Princeton, the Institute of Advanced Studies; yet, at the same timepoint (late 1940's) still is very active in the corridors of power, especially military power, since he is *the* father of the atomic bomb. JOLIOT (in France) is the high commissioner for atomic energy, which is definitely an applied science function, and so on.

This went on for several years, and therefore, the spread of the new forms to the domain of pure science was somewhat delayed.

And then the immediate effects of Hiroshima and its psychological sequels began to wear off slightly. Advanced science moved to fields first less connected and then not connected at all with foreseeable applications. We have seen the role this phenomenon played in the foundation of CERN; we may take note of this episode as one of the signs of an impending divorce between advanced science and large-scale public affairs.

At the same time, after the shock of the sudden eruption of advanced scientists into the corridors of power, the Establishment, the customary dwellers of the corridors of power, began to ask questions, « What are these strange people doing here? ».

I would not describe this Establishment as a dominant class in the Marxist sense of class. It was a group of people far more loosely defined, but still when it came to real wielding of power, it was obvious enough that *they* were wielding it, and not these professors who so recently still were considered as absent-minded and impractical.

Slogans began to be heard which were unthinkable after the immediate shock of Hiroshima. An American politician declared: « Scientists must be on tap, not on top ». Some other politicians began to say: « Thank God that atomic energy was developed by practical military men like General GROVES and not by a bunch of crypto-Communist professors ».

And the Oppenheimer affair, in which many people saw and still see a personal victimization of a prominent man by reactionary forces, to my mind is something far more far-reaching. OPPENHEIMER was chosen again as a mascot, but this time as a scapegoat, to notify that henceforth people of his kind were no longer welcome on high-power levels. When one reads the proceedings of the Oppenheimer affair, one feels how strong was the wish to remove him and how flimsy were the arguments used to remove him. In this respect, the Oppenheimer affair is well worth some further study.

That was the mood in which advanced science, mostly physics (but we *are* concerned here with physics) began to go back to the Ivory Tower, to the aloofness from big public issues. Yet the scientists had learned a lot during the years of their contact with big means and big hold on public opinion. When they went back to the Ivory Tower and back to their pre-war intellectual preoccupations, they were by no means ready to abandon the new fruits of their out-of-Tower experience.

A significant feature of this new phase was a contradictory attitude prevalent in the public opinion: although, on the whole, it was turning away from considering scientists as big public figures close to the power source, they still saw them as, potentially, an important source of practically valuable magic, and, therefore, were still ready to give them money, even if the motives were slightly changed.

Again, the example of CERN illustrates another and similar paradox: public opinion not wanting that really important practical research be pursued internationally, and at the same time, quite ready to give scientists big money.

This contradiction went on for something like another 15 years, which is the period we are now concerned with.

As to the scientists themselves, back in the Ivory Tower—the new forms of running scientific research appeared to them both as a necessity and as a tempting way to solve some of the problems with which pure science was plagued before the war.

First, let us say a word about the necessity. I think it was again WEISSKOPF who said in his lecture here, very briefly, that since the War, science continued to move towards new fields, and the availability of new means accelerated this movement. Scientific inquiry tended to get further and further away from the human scale of events, and events on an inhuman scale demand big means to be produced and studied or approached. Science began to move into sizes of micro-micro-centimetres, time intervals of micro-micro-seconds, very high temperatures, very high pressures, very high vacuum, new states of matter such as plasmas which were never heard of before on the human scale of events. Science went into the depths of the oceans, heights of the atmosphere, and soon into cosmic space.

All this required money, and if pure science was to pursue its legitimate aim of extending human knowledge ever further away from human scale, great means were necessary and therefore the great organizational forms to handle them were adopted in pure science, just as they had to be previously adopted for applied science.

There also was a temptation. What did scientists spend this big money on apart from the new costly equipment? Well, they could now hire assistance in far greater numbers. The proportion of pure scientists, qualified scientists, in the total population around the big equipment began to fall. Now scientists had money to hire, for example, professional administrators to help with purely or almost purely administrative questions such as budgets and personnel handling, which formerly the Herr Professor was handling himself. Workshops where scientists used to go down to produce a new piece of physical apparatus began to be staffed with numerous mechanics, and scientists themselves were seen there more and more seldom. Clerical assistance: before the war a scientist knew that he would be lucky if his manuscript could be typed at all, in the long queue of tasks that was facing the unique typist in a scientific institute. After the war, any first draft would be promptly typed and then a second draft, and then the final typescript would turn out to be not so final after all, and so on. The whole procedure of producing scientific publications was thereby changed—and made irresistibly easy. Laboratory libraries began to be far better staffed and stocked, and so the scientists no longer needed to go out to remote libraries to consult material which became available nearer home.

Procurement: many items of equipment which used to be manufactured by the scientists themselves, for instance like Geiger counters and their amplifiers, began to be easily procurable commercially—at a price, of course—but now the price could be afforded, and somebody could even be sent out to run the errand.

All this meant that individual scientists, in terms of scientific work proper, became more productive, and therefore fewer scientists could be hired and more assistants, per unit of production. And that is why this phenomenon that the proportion—not fewer scientists in absolute numbers, of course, since the production rose so markedly—of trained scientists in the total payroll began to fall down.

All this was a temptation, and temptations often have to be looked at carefully. After all, a scientist who loses the habit of producing his own pieces of equipment or of making his own bibliographical searches and so on, is to some extent a narrowed-down scientist, is he not? We shall have to return to this question.

Temptations also occurred on a far more worldly level. Scientists began to be accustomed to be paid much better than before the War, much closer to people at whom they used to look, not enviously (because even before the War they fondly believed that they had chosen the better part), but with some idea of « it would be nicer if I could have some of these things too ». In my memory, the Radium Institute in Paris shortly before the War, say, in mid-thirties, would boast perhaps two car-owners among its staff. Today, well, look at the parking lots of any big pure-science institute; it seems obvious that the rise of this particular car population has been far more rapid than that of the car population at large.

Life became more pleasant. Scientists—perhaps not all of them, but certainly the more senior ones—began to travel first class on the railways. The conferences and schools which afforded opportunities for quasi-tourist travel began to be held in, shall we say, pleasant places. We have only to look around us here. And in fact, the now familiar term, « dolce vita », which originally was coined in Italy in a quite different context, became a not-too-far-fetched description of today's professional life in the world of science.

One result of this wave of temptations was a change in what we might call the psychological pre-determination of a typical scientific career. WEISSKOPF told us about his own case in which the prudent family advice went unheeded. In a similar mood, JOLIOT once told me: « One had to be crazy to choose our job, » « Il fallait être fou pour choisir notre métier ». It began to be open to people who are less crazy, or to use an expressive French word, « ne sont pas fous ».

And this, of course, was to be expected. As I pointed out before, the continuing doubling of the scientific population had soon to reach the level when the supply of sacrificially-minded people in the community was getting scarce. In order for the scientific population to grow, one had to tap the resources of less dedicated people, and Big Science institutions provided an inducement for them.

A characteristic new phenomenon which made its conspicuous appearance in the Ivory Tower was the increase of teamwork, as opposed to individual work. Before the War, and even more so before the 20th century, it was considered as a matter of course that science and scientific work, creative work, was and must be individual work. We have studied here the rise of 20th-century physics; this school has dealt mostly with theoreticians, but what we have to say now could also apply to experimentalists like RUTHERFORD or THOMSON; our typical examples were those of a master, a recognized master who either worked alone or was surrounded by a few journeymen and apprentices. The structure was strictly medieval. It is interesting to me to hear today, in our off-class discussions here, that many students tend to identify the process of scientific production with the production of goods in general. As I have pointed out in my own writings, up to World War II science was not at all in tune with the surrounding capitalistic form of goods production. It worked in the mode of a medieval craft. Prior to the War, this gap had not even begun to close up.

The outstanding discoveries and advances in pre-war science were made either by individual scientists, or by teams consisting of a master and his pupils—and everybody knew who was the master. One could observe some curious phenomena such as the fashion for husband-and-wife teams, and, of course, it is always difficult to say in any family who is the master. This fashion was partially set by the historical accident of the Pierre and Marie CURIE partnership. On the whole, no outstanding paper was signed by more than three names, and three names began to be more frequent as we got closer to the War years.

There was one remarkable exception, and that was Fermi's Rome team, where there were six qualified scientists, all of them outstanding. FERMI, in this way, was a forerunner, like in so many other ways. There was, I think, a certain division of labour. FERMI himself was originally a purely theoretical physicist. There was one pure chemist: we might ask AMALDI to confirm this point. Others, such as SEGRÈ or RASETTI, were more pure experimentalists. At any rate, there was no question about who was the master, and FERMI, putting himsel in the third place among the six in strictly alphabetical order, was—well, there was no mistake who was the leader and where the main ideas came from.

During the War, the insistence on applications and therefore the necessity of a multidisciplinary approach, led to the formation of bigger teams, which, for the first time, began to rely on co-operation between equals—shall we say, a prominent physicist with a prominent chemist, or builder of electronic apparatus. We have only to think of such an enormous figure as LAWRENCE, who was primarily an innovator in electrical engineering.

This meeting of equals prepared the way to the post-war developments. More numerous groups of people were now needed to perform a new, more complicated kind of experiment. An entirely new form of co-operation appeared: the big team in which there was no longer a recognized master. The multiplication went to lengths which probably some people would defend as necessary, but which I consider ridiculous. In one typical paper—I had the patience to count the signatures—there were 29 of them, all physicists, all presumably claiming the status of creative scientist. I just cannot believe that multitudinous claim.

Then, within these teeming teams, a certain kind of stratification became noticeable. First of all, a big team has to have at its top decision level a person who is able to handle big crowds, and this person is not necessarily the most creative member of the team scientifically. In fact, many teams began to evolve the function of the idea man, which was no longer coincident with that of the boss.

There were also questions of whether, for example, an outstanding equipment designer or inventor could find his place among the 29 signatories. We will come back to that in a moment.

Bosses, because they had access to higher levels of the administration of the institute, and a more immediate contact with money, began to be more and more influential, and the phenomenon of the bosses who were not recognized creative masters became familiar and accepted as a matter of course.

Let us think a moment about all these new bosses, on all levels: group leaders, division leaders, directors—Directors General, as they call themselves today. How are they selected nowadays? Another completely new phenomenon can be observed here. Money is given by people who are not scientists, and there is a good English proverb: « Who pays the piper calls the tune ». In this particular question of selecting the most powerful office holders, tunes were now called by people who had no first-hand scientific knowledge or dedication. Somewhat strange criteria were resorted to. First of all, there was a rush towards visible glamour. For instance, Nobel Prize men were eagerly sought to be put at the head of such and such big institution. CERN lived through a remarkable episode at its beginning. Its Director General was a great theoretical physicist, BLOCH, who was not quite clearly told what kind of job was being offered to him, accepted, came to Geneva, saw what he was supposed to do—and immediately resigned.

(BLOCH actually stayed officially at CERN for something like 11 months, but in fact his resignation became known, for example to me, six weeks after his arrival, and I suspect he signified his intention to quit rather earlier than that.)

In cases where a Nobel Prize man could not be found, the next best would be to look for people with degrees from glamourous universities such as Oxford and Cambridge in England, or Ecole Polytechnique in France, or Ivy League in the United States, and particularly those who had earned top grades. Ability to win top grades sometimes *is* helpful in creative science, perhaps not always.

There were also other criteria. Preference began to be given to safe people, people who were not too insistently renowned for having original ideas. Of course, this is a well-known phenomenon; that is why, for example, England has kept CHURCHILL away from power for so many long years, until people were compelled to call him again under the wartime emergency. No similar emergencies have appeared so far in Big (pure) Science.

Who else? People who decide on such things are sometimes surprisingly prone to yield to the pressures of those forceful self-promoters who simply want to get an important bossing job. In former times this was mitigated by the insistence on the candidate's reputation as a creative scientist. By now this reputation began to be less important; therefore, pushing individuals of any other kind now have a better chance.

WEISSKOPF pointed out to me yesterday, in a private conversation, that things were not so bad after all, that, for instance, GOLDHABER in Brookhaven is definitely a very creative scientist. An even more recent important example is WILSON, at the head of the National Accelerator Laboratory in America. One can, then, say that all these negative trends which I have mentioned are not all-powerful; better tendencies also are at work and sometimes they are victorious.

WEISSKOPF also quoted the example of PANOFSKY, whose main contribution to physics is probably not on the level of individual creative endeavour. He is a remarkable leader who has spent much of his time in directing the teams which have to design and handle big equipment. And here we come to the last category of candidates, which is a positive one. Even if the money-giving Establishment is not able to recognize top creative ability, it is sometimes responsive to top qualities of leadership. In this respect PANOFSKY is an outstanding example of a very successful choice. An even more remarkable case, on a much grander scale, was that of KURCHATOV, in Russia. At an early stage he was selected as the leader of Russian atomic-energy research. The choice fell with a conscious intent on, not one at the very top of creative Russian science, but on an outstandingly able leader. Judging by the subsequent performance, it was indeed a very successful choice.

One more thing we have to retain from this list of criteria: a certain preference for people who play safe. This trend is equivalent to a push, not, of course, a decisive one, in the direction of discouraging great innovation, great originality. And that is how, by the combined processes of teamwork and of boss selection, Big Science began to influence « pure » research in favour of playing safe and putting less insistence on markedly original innovation.

What other consequences can we list? Because greater numbers, recruited as lower-level assistants, began to be employed, because the physicists no longer had to spend a great part of their time on tasks other than thinking about their experiment or theory, whole classes of people who are indispensable to the production of physics began to be no longer recognized as physicists.

Let us have a closer look at these lay members of the parish. For example, even if the Director General still is a creative scientist, the next layer of his assistants may comprise many professional administrators who, on their own, have no great knowledge of scientific values. Yet, among these administrators, great personalities can sometimes be found, creative in their own administrative function, which, nevertheless, remains institutionally divorced from physical creation.

There exist other lay categories, many of them much closer to research work. For example, scientific specialities other than that which constitutes the recognized aim of a given pure-science institute. My guess would be that a pure chemist no longer can be considered as a full-fledged member of a team engaged in physical research, even if he is a very outstanding chemist. The same goes for people skilled in electronics, in such important and very scientific subjects as health physics, and so on.

Then, a bit closer to the actual physics experimentation, there are the inventors, designers and builders of new kinds of techniques. Here is another guess: GLASER was about the last inventor of a detecting device who could get a Nobel Prize; stratification into castes was not yet powerful enough at the time to prevent his getting the prize. ALVAREZ, a great physicist and a great inventor, devoted much of his time in his later years to the development of bubble chamber techniques and data-processing techniques; accordingly, there have been some difficulties in giving him a Nobel Prize. Fortunately, he finally did get it.

There is also the design, building and innovation in great equipment, such as accelerators, reactors, telescopes and so on. Let us turn again to WEISSKOPF: last year, at the CERN accelerator conference, he made a speech in which he passionately pleaded for recognition of the fact that, at CERN itself, at least as much creativity has gone into accelerator design and improvement as into running these machines for experimental physics. This means that WEISSKOPF himself is very much aware of the present situation: the original designers of new equipment, who, traditionally, always were considered as creative scientists, are no longer so considered.

ADAMS, who is a leading initiator of guidelines for scientific research on a large scale, was listed in an important recent dictionary, « Who's Who in World Science » not as a nuclear physicist but as a nuclear engineer, and one knows what this is supposed to mean nowadays.

All maintenance and operating staff of the big equipment was, of course, immediately excluded from the caste of « pure scientists » and this had one curious consequence to which I shall turn in a moment.

Next in the list: bibliographical research, digestion of scientific information, and science writing for the public. All this began to be considered as work not suitable for a true physicist, in spite of all the precedents to the contrary of which there was no lack in the 19th and early 20th centuries.

When the computer appeared, there was a strong tendency to consider computer specialists, not the rank and file programmers but the inventors of program techniques, as remote from physics. This valuation was in fact responsible, to a considerable extent, for the delay in the penetration of computer methods into physics, or in the creation of that new branch of science which, today, is called computational physics. The delay was finally overcome because the younger physicists recognize their need for learning computer techniques. There is still the danger that « pure » physics today will not benefit from the best available computational techniques, because the best computational minds resent being put outside the caste of pure scientists, and therefore will not agree to work in those research institutes where the caste feelings run high.

So the question now arises, in a place like CERN, or Brookhaven, or Batavia, or Dubna, or Serpukhov, what is a physicist? I once described a typical experiment in high-energy physics as an effort exercised in five branches of endeavour, which can be listed as a sequence of initials. It is a little hard, but not impossible, to remember the order in which they occur: *A*, *B*, *C*, *D*, *E*. *A* for accelerators, *B* for beam handling, *C* for chambers and counters, *D* for data processing and *E* for evaluation. All five are absolutely essential, all five are parts of physical research. However, the present-day tendency is to consider that only *C* and *E* are « true » physics—*A*, *B* and *D* are not. (Concerning *D*, as I have just said, some mitigation begins to appear on the horizon.)

Similar phenomena have happened before in other walks of life; wherever it occurs, a too narrow definition of the top-caste creators produces a serious threat to the whole creative profession. I am not a historian of art, but I do have an impression that in the late 19th and early 20th century, the question, what is an architect, became a burning issue. In earlier times architects used to build houses. Now this job was passed to contractors and engineering firms, while the architect spent his time on devising sets of Greek columns to be put in front of otherwise nonarchitecturally conceived buildings. One can observe this phenomenon, for example, in Dubna. You see there perfectly serviceable, prosaic-looking buildings, good buildings in fact, which obviously were not meant as architectural showpieces. And then the architects came and added their classical porticoes to some of these structures.

I am not saying that this sort of thing has already come about in physics, but the trend can be seen, and it is due to the influence of Big Science methods.

WEISSKOPF again: in a private conversation which he probably does not remember, he once told me that some of the features observable in a place like CERN or Brookhaven began to remind him of organic chemists in Germany in the early 20th century. Theirs was an intensive work, employing an enormous number of highly trained scientists, and yet, somehow something had gone out of it. Yet the organic chemists had one excuse which pure physicists do not have today: the chemists' work *was* closely connected with important applications.

Once more, please do not conclude that I am giving you a picture of a whole great science going to the dogs. I am only trying to sound a warning against the negative influences of Big Science on pure research. There is no built-in doom; after all, the role of the architect in the later 20th century was completely and gloriously revived. Organic chemistry itself managed to become a far more exciting and creative science in more recent years when it got preoccupied with such things as the quantum theory of the chemical bond, or the behaviour of free radicals.

These examples show that there are countervailing forces that are at work; we have to keep in mind both kinds of forces in order to be able to direct their interaction towards what we wish.

In today's crisis, it is noticeable that the younger entrants into the scientific profession begin to turn away from those fields which have been most contaminated by Big Science. It is certainly true of several branches of physics, not only high-energy physics. And then, what is to be done?

Well, I am ready to suggest several answers, especially for pure science. First, we have to put to good use those new tendencies which I mentioned at the end of my earlier lecture and which work against the excessive grip of Big Science methods. These tendencies are the geographical dispersal, made possible by the improvement of data transmission and processing; the recognition that research science must not get away from teaching and therefore must return to the universities; the renewed understanding of the fact that big machines are just machines, and not the irresistible magnets which have the power of gathering all the research physicists around them in Big Science clusters. Universities offer as attractions, not only the opportunities for teaching and contacts with the young, but also the interdisciplinary atmosphere which is becoming more and more essential for research as we see it today.

In high-energy physics in particular, there are other ways than those which require the clustering around the accelerator. Some high-energy physicists, using the newly developed equipment (far more sensitive than that of 20 years ago), have gone back to balloons, to high-mountain physics with cosmic rays. I know a few who are eagerly preparing high-energy detectors for cosmic rays on the Moon surface. The ways of complete originality are, thank God, still open.

A physicist who is not content with this prospect can turn to newer branches of physics, less contaminated by the spirit of playing safe in a big, organized crowd. For example, a prominent theoretical physicist, DYSON, a year or so ago in an article in *Physics Today*, pointed out that completely new fields are open for adventurous physicists. He listed a few for example, construction of new types of molecules which might be important in biophysics. It sounds fantastic and science fiction, but, after all, a bit of science fiction would not do any harm, if one looks at it realistically. Also, of course, astrophysics, and various kinds of radio astronomy and so on.

Then there are the social urgencies. We know that technology has committed many excesses since the War, that the environment is hard pressed. Some prophets who may be only partly wrong, say that the whole survival of industrial civilization is seriously threatened, perhaps even of mankind itself. Lines of action have been designed, in particular those considered at the recent

United Nations Conference on Environment (Stockholm, June 1972); there will be, among others, many tasks open to people with physical knowledge and training.

As a last hopeful note, I may put a question without knowing any answer. There is a need for new democratic forms of management in pure science, even if Big Science equipment is involved. What these forms are, I do not know; you should know and think about them. They must be genuinely new, not a return to the sociological theories of the 19th century. They must be inspired by today's methods and today's problems. There is a growing importance of treating the problems in a systemic way instead of linearly, factor by factor, as the habit of science has been until today. These new methods of thinking will be essential for the new democratic structures that need to be evolved.

I will end with one anectode which is suitable to this place; it is here that I listened to FERMI in 1954, almost the last time I saw him alive.

FERMI made a statement, which was taken down by one of his pupils, and I know where I could find the quotation. The exhortation was: « Go out and perform this experiment. If you are lucky, you will find that it contradicts the theory. Then you have made a discovery. If you are unlucky, you will find yourself in total agreement with the theory, and this will mean that you have performed a measurement. »

So it was FERMI, and he should know who made this distinction between experiments and measurements. For the latter we now have, at CERN, Brookhaven and so on, another expression: we are talking of « runs » of experimental equipment.

On the whole, pure science organized on Big Science lines, tends more and more to perform not experiments, but runs which yield measurements. Such results are valuable, but they are not enough. In this connection, we may mention one interesting point of difference between America on the one hand, and Europe and the Soviet Union on the other hand. One of America's most endearing features is that it is rather an anarchic country. There are all sorts of tendencies which clash, and there is no uniform direction, no Gleichschaltung, to use the European word. The results, in my observation, are obvious. In the antithesis, which I have put at Fermi's door, between an experiment which leads to discovery and a measurement, America's vocation seems to be to perform less precise, less reliable measurements, and to make rather more discoveries. CERN and the Russians perform better measurements; I am more familiar with CERN and I have a fairly strong impression that, in this respect, it is superior to American laboratories. The measurements are done more competently, more accurately, and one would almost say, somehow with less expectation to stumble on a discovery, as they do sometimes in Brookhaven or Berkeley.

There may be a lesson in it, and since I derive this last part of the lecture directly from Fermi's exhortation, I better stop here and now.

# Science, Politics and International Affairs.

W. Goldstein

*Graduate School of Public Affairs*
*State University of New York at Albany - Albany, N. Y.*

## Précis.

These three lectures examine the political forces that have helped shape the development of advanced science efforts and of science policy formulation in the industrial nations of the world since 1945. The lectures distinguish between three modes of political influence that have contributed toward the international growth of science and to the strengthening of the national environment for scientific work. The three themes should be viewed as a unified whole though they are pursued in the following pages as separate subjects. This fragmentation could not be avoided if the enquiry was to be pursued with rigor. The basic themes have been summarized in the brief *precis* below. A synthesis is provided in the concluding sections of the last presentation to bring the three themes of analysis toward a set of general findings.

### 1. – International tension.

From the atomic bomb of 1945 to the nuclear arms race and the industrial concerns of the 1970's the stimuli provided by the tensions of the Cold War and international rivalry have been of great significance. The growth of Big Science was encouraged by military funding and by *étatiste* or state-directed programming in the middle or superpowers in the international system. As a result, national loyalties and Establishment initiatives were pursued by the leading scientists of each science community. The speculative issue can therefore be raised: if the stimuli furnished by military threat and by national science institutions had not materialized, what other courses of development might contemporary science policy have taken?

### 2. – The national (or étatiste) political environment.

The dual concentration of elite technology and of political power promoted the emergence of national science establishments in the communist and capitalist states, alike. Scientists came to play a prominent and controversial role in the deliberation of national policy and in the identification of program priorities. Many were co-opted into the key institutions or policy agencies of the nation state. All too frequently this resulted in the erosion of academic autonomy and in an increase in state funding or political authority. Though variations in this development pattern occurred in the United States, the Soviet Union and in the middle powers of Europe, the trend toward state science remained relatively constant. In retrospect, it appears that no alternative (or non-elite) pattern could have emerged, given the political and organizational interests operating in each state.

### 3. – Science and social change.

Various professional and academic groups came to challenge the dominant values and agencies of science policy after twenty years of Cold War and national science funding. The challengers were usually dissidents bent upon strategic nuclear policy, or the allocation of state funds or the co-optation of colleagues into official agencies. The dissenting movements discovered that substantive and policy issues divided the scientific community but that the tactics available for pursuing change were severely limited. A great deal of anxiety has been voiced recently about the distrust of scientists, the repudiation of their policy recommendations and the emergence of a populist or « counter » culture. But there has not been a major disruption of the political or scientific efforts of the technocratic state. Possibly, it might have to be concluded that scientists and scientific institutions are impervious to the rumblings of anti-elitist or anti-war movements in the most advanced and wealthy of societies.

There are no simple formulae to guide the young scientist as he sets off in quest of a career and a work pattern independent of *étatiste* funding and political pressure. Thus these lectures can only describe how science and politics conflict. They cannot instruct the individual scientist how the conflict should be resolved in his own personal career, or in the national-political environment in which he lives. The aim of these three essays is to clarify why *raisons d'état* have conditioned (if not replaced) the organized thrust to acquire and develop scientific knowledge in industrial societies.

Considerable benefit has been gained by the author in developing this analysis. Useful discussions were held with members of the Seminars in Technology and Social Change at Columbia University and at the Institut Universitaire des Hautes Études Internationales at Geneva; and also at the Center for the Study of Democratic Institutions at Santa Barbara, in California, and at the MITRE Corporation in Washington, D.C. Many points were clarified when the lectures were first presented to the Enrico Fermi summer school on the history of physics at the Villa Monastero at Varenna in August, 1972. The author remains responsible, of course, for the findings presented here.

## I. - Science, International Politics and War.

### 1. – The stimulus to contemporary science.

Warfare in the mid-twentieth century has provided a powerful and unprecedented stimulus to the development of science in the leading industrial nations of the world. The reasons for this were obvious. First, war called for an enlargement of the administrative powers and of the executive authority of the State. In turn, wars required the construction of a monolithic bureaucracy, an ever larger budget and a constant increase in the ability of governments to intervene in the economic, social or intellectual life of the community. The degree of bureaucratic intervention varied from one state to another but everywhere the pattern was similar: giant Pentagons of power were established to subsidize and regulate the working life of an industrial nation.

The secondary consequences of war were also of great moment. National regimes, whether capitalist or collectivist in ideology, were driven by a compelling and expansive imperative of international power: the need to promote the security and the industrial capabilities of the state. Government agencies were formed to harness science and technology for the greater power and glory of the «nation at arms». The threat of nuclear war and of the «race» to industrialize required the fullest mobilization of the science community. Astronomical sums of public revenue were invested in weapons research, in secretive R and D laboratories and in the financing of graduate schools. Loyalty oaths and security clearances were demanded from many practising scientists. If they failed to submit to the *étatiste* intervention in science affairs they ran the risk of excluding themselves from the public funding and the mission-oriented research that provided the central thrust of what became known as Big Science. Since most of the challenging work done was accomplished in the facilities or on the grants of Big Science, few scientists could afford to exclude themselves from the new Establishments, especially in high-energy physics, nuclear engineering and applied technology.

The growth in the scale and in the cost of scientific enterprise produced two significant results. High-energy accelerators, expensive radio telescopes, computers and laboratory equipment were largely funded by state authorities. This public aid helped generate an intensified pace of research and of graduate training. It also allowed science to flourish at a level that had never been known before. But the involvement of public authorities in science required that a professional hierarchy or an Establishment organization should be built to provide a peer-group control for the science community. Senior scientists or officers of the National Academies were co-opted into the political leadership of government. In many cases funds and equipment were allocated by scientists who best understood the defense and industrial needs of the state. Program priorities were set, technology assessments were forecast and skilled manpower was allocated to those activities that the government and the science establishment regarded as the most vital to the interests of the nation. The *utilitarian* pay-offs of science came to be judged as the primary criteria of science policy and of scientific research [1]. That science could remain « pure » was no longer feasible.

It has been argued by many scholars of science policy that a Faustian exchange characterized the flourishing of modern science within the nation state. Expensive contributions were made from public revenues to support new installations and to finance applied technology. Graduate education was rapidly increased and science enjoyed a new affluence and prestige. But the price paid by professional scientists for this wealth was a heavy—if not soul destroying—commitment to the purposes of the state [2]. Thousands of talented scientists were employed in military research and weapon procurement programs; and the drive to acquire industrial technology energized university and laboratory staff. The needs for applied technology, in all too many instances, took precedence over the conduct of basic or theoretical work, thus warping the quality of the work that was done. Science peer-groups were co-opted into the highest executive councils or government offices; there, they were expected to endorse the *étatiste* values and priorities of those who held power [3].

In playing the tunes that were paid for, though, science lost much of the spirit of free enquiry and cross-national exchange that it had known in previous years. Programs frequently aimed at the fulfillment of military or industrial

---

(1) The utilitarian concepts of « rationality and efficiency » in the allocation of science policy priorities are usefully examined in the introductory essay of Edward SHILS (ed.): *Criteria for Scientific Development: Public Policy and National Goals* (Cambridge, Mass., 1968). Sir Geoffrey VICKERS has contributed a philosophical enquiry in the utilities of science in two distinguished books that deserve close study: *Value Systems and Social Process* (New York, 1968) and *Science and the Regulation of Society* (New York, 1970). Among the latter papers, also see A. J. TOYNBEE: *Science in Human Affairs: An Historian's View* (1968).

(2) R. A. RETTIG: *Science, technology and public policy: some thematic concerns*, in *World Politics*, Vol. **23** (Jan. 1971), pp. 273-93, presents a lucid analysis and an extensive bibliography on recent science-politics conflicts.

(3) A striking analysis of the political uses and usurpations of science is to be found in C. H. G. OLDHAM: *Science and Social Change: Politics and the Organization of Science*, in K. H. SILBERT (ed.): *The Social Reality of Scientific Myth* (New York, 1969).

crash programs. In the anxieties generated by the Soviet's launching of the first Sputnik, most states determined on the instant to redeploy their skilled technicians and R and D budgets for the purposes of maximizing national defense and industrial productivity. In the best and also in the worst sense of the term, scientists become servants of the vested interests that had furnished them with abundant resources (4).

The pattern of mobilization and deployment varied slightly among the leading industrial nations of the post-war world. Market economies relied upon different techniques and agencies than those found in collectivist societies. The Soviet or the French patterns were characterized by a *dirigisme* that had long guided the formulation of public policy in Moscow and Paris. The *ad hoc* institutions of Washington and London, by contrast, were less rigid in centralizing authority but no less effective in mobilizing scientific talent. Clearly, differences of ideology were of negligible significance. Marxist regimes inverted the doctrinal relationships prescribed between the superstructures of government and the base in society; while capitalist regimes thwarted market forces to subsidize science as a public enterprise (5). As a result, four political considerations of power guided the development of science policy in all leading nations:

*a*) *The traditional values* that had previously governed the selection of economic or social goals (*e.g.*, private entrepreneurship, as in Western Europe, or *dirigiste* planning, as in collectivist societies) were applied to the formation and funding of new science programs. Market and socialist societies pursued similar values to control their scientific and their economic development.

*b*) *The forms of organization* that had customarily prevailed (*e.g.*, in autonomous universities in the U.S. or U.K., or state-run colleges elsewhere) were now given responsibility for recruiting professional staff and for implementing crash programs in high technology. Thus the Soviet or the National Academies or the Écoles Polytechniques took over the leadership, if not the operation, of elite science programs.

*c*) *The industrial and defense choices*, and the science priorities adopted by the policy elites, were fundamentally determined by the development doc-

---

(4) A caustic view of the excess of commercialism and entrepreneurship perpetrated by large industrial labs appears in D. S. GREENBERG: *Bootlegging holds a firm place in the conduct of research*, in *Science*, No. 153 (1966), pp. 848-49.

(5) For a brief but excellent review of the political, organizational and ethical problems faced by science establishments across the world, see H. and S. ROSE: *Science and Society* (London, 1970). On the administration of the American science establishment and its management of R and D, there are useful articles collected in W. R. NELSON (ed.): *The Politics of Science* (New York, 1968).

trines or the Cold War formulae adopted within each nation state. In this regard it can be said that the impact of arms races did more for science than the lobbying of EINSTEIN, FERMI or KAPITSA could ever have achieved.

d) *The industrial and educational goals* selected in the past were also of critical importance. Each advanced economy needed to maximize its industrial productivity, its technological « balance of payments », the upward social mobility of its professional population, and the rewards to be distributed to its new scientific « meritocracy ». The planning of science policy was expected to secure many of these economic and educational objectives simultaneously.

Though the organization of science agencies and the choice of R and D goals differed sharply from one society to another, one principle crossed all ideological and national frontiers. In profit-seeking and in nonmarket economies alike, the institutional needs of Big Science tended to make the career opportunities of scientists and research workers dependent upon the financial support and the political influence of secular institutions. These included *a*) corporate industry, *b*) the professional military, *c*) power-minded peer groups (such as national academies) or *d*) the Executive organs of the nation state. Each of these agencies exerted a major, but not exclusive, influence over the budget allotments and the performance goals of the nation's science communities. Specifically, these agencies sought—either through co-operation or competition with professional bodies—to determine three key questions in science policy:

*a*) *What priorities* in resource allocation (of funds, equipment or skilled manpower) should be adopted in encouraging basic or applied work? In promoting one discipline or field at the expense of another, or in distributing public revenues for crash R and D programs, consideration was given first to the closing of a missile or technology « gap » so that a rival nation's advance in ABM or mass production techniques should not injure the state's defense or industrial posture.

*b*) *What future goals* should be identified, both for society as a whole and for various elite bodies (or pressure groups) of professional staff and policy leaders? In forecasting the science *and* the educational needs of a technological society, priorities were abruptly switched from nuclear engineering to aeronautics or sophisticated computer research simply because dictates of national interest were read into the procedures of long-range forecasting.

*c*) What political control mechanism should be devised to monitor quality control, to accelerate defense or industrial programs, to consolidate (or break up) vested power interests? If the function of skilled technology assessment was to safeguard the « public interest », should this be performed by politicians responsible to a lay electorate or by scientists who squabbled among themselves over the optimum division of public funds?

While a few academic scientists struggled to assert that no scientific work could be usefully planned or centrally directed ([6]), each central government forged ahead to assemble the apparatus necessary for a state-funded and state-directed program in science and technology. On most matters, the organizational and administrative complexities of the science Establishment were too strong to be disregarded. While 137 000 people had been recruited into the atomic bomb project in 1945, the work force recruited by giant industry and by defense or space agencies in the 1960's far surpassed this figure. The sorcerers' apprentices constructed vast and powerful bureaucracies which no scientist could ignore. Worse, politicians campaigned on their record of bending science policy to the purposes of national defense; industry advertised the contributions that their science staffs had made to the nation's prowess and profit; academics stalked the « corridors of power » seeking the R and D funds needed to build labyrinthine laboratories and research empires; and the general public, overwhelmed by the magic and glamor of sophisticated technology, came to view « science » as a mystical, but vital source of national prestige and well-being ([7]). As KOWARSKI at CERN put it: scientists were viewed as magicians or saints when the nuclear era began but they became bishops and ecclesiastics as the organizational momentum of state science developed.

## 2. – Science and war.

The greatest stimulus to science organization and funding appeared as a consequence of the national perception of military threat. In super or middle powers, alike, extensive resources were diverted to missile, nuclear energy and electronic warfare programs ([8]). Basic work was generously promoted in the

---

([6]) The imperative need to *not* plan science policy—in the formal or political sense of the term—is argued in two essays by M. POLANYI in *Minerva*, Vol. **1**, No. 1 (1962) and Vol. **5**, No. 4 (1965). The latter makes the point that scientists must co-ordinate their own affairs through a process of self-authority and mutual consensus if they are to avoid the dreaded planning rigidities of the state. POLANYI believed that the *Republic of Science* should enjoy a greater autonomy than it had so far manifest if its works was to be freed from political controls.

([7]) An exemplary case study of the political strategies and instruments available for the manipulation of scientific research appears in H. P. GREEN and A. ROSENTHAL: *Government of the Atom* (New York, 1963). Looking only at the Joint Committee on Atomic Energy that regulated Congressional and Executive science decisions throughout the Cold War, the authors describe how leading Congressmen became movers and seekers of nuclear physics.

([8]) Only in Japan were physicists criticized or ostracized by their colleagues (in the Japan Physical Society) if they went to work on defense contracts. Their experiences are investigated in P. M. BOFFEY: *Japan on the threshold of Big Science?*, in *Science* No. 167 (1970), pp. 31-35. A popular but unique analysis of the conditioning of scientific

hope that spin-off benefits would quickly be reaped in hardware improvements and defense capabilities. The « space race » sustained the pattern of *dirigiste* funding and policy control once the strategic arms race of the 1950s had been stabilized (and until it was resumed in the late 1960s). A qualitative arms race replaced the quantitative competition to stockpile weapons that had previously prevailed. This turn was highly beneficial to the science Establishments, though it also created a certain amount of dislocation and hardship. As defense or industrial anxieties switched from one field to another, various disciplines (*e.g.* high-energy nuclear physics) found themselves deprived of support just at a time when their graduate students were ready to begin work [9].

TABLE I. – *Allocations of* R *and* D *expenditure, by country and program.*

| Country | % of Gross National Product spent on R and D | $ spent on R and D per capita | % of R and D budget spent on nuclear, space and all defense programs |
|---|---|---|---|
| USA | 3.4 | 110 | 62 |
| UK | 2.3 | 40 | 40 |
| France | 1.6 | 27 | 45 |
| W. Germany | 1.4 | 24 | 17 |
| Japan | 1.4 | 9 | — |
| Soviet Union | 2.5 (estimate) | — | — |

*Note.* – The military boost to science expenditures in the U.S., U.K. and France was dependent upon various cycles of weapon procurement and strategic doctrine. Only Germany and Japan escaped from this cyclical pattern; though they enjoyed a less affluent level of funding as a result of their unambitious armament programs, their rate of economic growth was boosted by deploying fewer industrial resources for military production. Their experience, in promoting industrial rather than military R and D, can serve usefully in qualifying a popular assertion: that the public financing of weapons R and D helped generate valuable spin-off benefits in manufacturing industry. This assertion, entertained both by Marxist critics and by Jean-Jacques SERVAN SCHREIBER (in his poetic work, *Le Défi Americain*), used to be valid. However, the recent increase in state-supported civilian R and D has presented a useful argument to inhibit expensive arms races and military research. It is more profitable to sell IBM computers to the Chinese, it seems, than to build an ABM system against them.

---

thought exercised by nuclear war and the digital computer (for the sake of political contrast) should be sought in A. WILSON: *The Bomb and the Computer* (London, 1968).

(9) The national impetus and the international limitations of science are explored in E. B. SKOLNIKOFF: *Science, Technology, and American Foreign Policy* (Cambridge, Mass., 1967). Though international co-operation in science projects continued at a slow pace, the emphasis upon « utilitarian » or applied scientific work has greatly increased in recent years. In the United States in 1973, for example, $ 7.9 billion was allocated for the Defense Dept., $ 3 billion for the Space Administration and $ 1.4 billion for the A.E.C. out of a public R and D budget of $ 16 billion; 60% of the total was appropriated for development and 40% for research. For a full analysis of these figures see *Science*, No. 179 (1973), pp. 544-552; all nonmilitary projects received $ 6 billion of which the National Science Foundation received $ 0.4 billion.

The relationship between security perception and science spending can be traced across twenty-five years of Cold War. Though the pattern of allocation changes as weapons systems became more sophisticated, or as the level of international tension rose (and then slightly fell), the correlation between military procurement and R and D support tended to remain constant. Statistical data are not as ample as one would wish, especially outside the OECD countries, but the emerging pattern of science/war relationships can be shown in the Tables. The data are derived from OECD reports for the years 1963-66, at which time R and D programs were moving toward a 20-year climax of defense spending.

### 3. – Science and international trade.

The post-war obsessions with the strategic arms race were temporarily dampened either because of the escalating costs of new weapons systems or because the political leadership became appalled by the growing size of the R and D empires supported by the defense, space and nuclear agencies. As a result, politicians and scientists began to attend to the economic benefits and trade pay-offs to be sought from national science programs. Politically-minded scientists and industrial research entrepreneurs came to realize that think-tanks and research institutes could profitably contribute to the training of skilled staff, to the improvement of industrial productivity, to science exports and licensing fees and to the amelioration of the nation's technological « balance of payments » [10]. Four high-technology industries gained impressive support in high political places: electronic data processing, electrical and aerospace engineering and satellite communications. Strikingly, all four had built strong political connections and pressure group syndicates during the preceding decades of the strategic arms race. Moreover, all four had successfully identified themselves in political terms as « critical growth sectors » of the nation's wealth. They were vital industrial components of that « white heat of technological revolution » (as Britain's last Labour Prime Minister put it) that politicians extolled in their appeal to the lay electorate.

The implications for science policy derived from this new fascination with economic and technological growth were of major significance. An advanced economy would be badly deprived if it failed to acquire its own autonomous,

---

[10] The number of new patents registered in each country and the technology « balance of payments » indicators provide an aggregate accounting of science productivity. But they are also subject to fluctuations in patent law, currency devaluations and industrial variances. The figures issued by OECD suggest that in various industrial technologies—such as petro-chemicals, electrical engineering or machinery—the U.S., Germany and Japan earn considerable sums of foreign currency for license fees and scientific royalties. The consequences of those transfer payments are noted in R. VERNON: *Sovereignty at Bay* (New York, 1971).

national capability to build computers, jet aircraft or communications satellites. Subsidized grants, tax waivers and R and D payments were therefore awarded to promote each of these fields of mission-oriented research. In most cases governments provided 50% or more of the support funds—even though the largest part of the work was performed either in industrial or university laboratories. Scientists committed to basic research found themselves losing funds as R and D allocations flowed to applied or engineering programs. Ministries of Science or Education reminded the universities and academies when it came to recruiting and then placing their graduates that careers in technology were at least as important to the national interest as those in « pure » science. Moreover, public money was spent on science schools and research at the expense of many other social needs. The funds that had been allocated to mission research were expected to generate profitable pay-offs for the nation—even if the profits appeared first in the annual accounts of such technology users as Imperial Chemical Industries, Rhone Poulenc or IBM. (These firms employed the largest number of graduates and they drew most heavily upon the nation's research accomplishments.) By contrast, the costly demands for public housing or medical services were deferred for yet another generation so that the privileged sectors in the work force could achieve ever higher levels of productivity ([11]). Ironically, the socialist states were no less biassed in their use of social transfer payments, to favor one group (of skilled professionals) over all others, in the hope of improving the nation's international trade position.

As a consequence of this switch in funding priorities a renewed concern was shown for the allocation of professional manpower. The availability of qualified scientists and engineers (Q.S.E.s) became a significant index of a nation's ability to compete in the trade and prestige stakes of the international economy ([12]). For the middle 1960s OECD reported the following ranking among the industrial nations that were competing against each other to maximize the productivity of their economy and the skill of their work force:

---

([11]) The detrimental consequence of Big Science or military funding projects on American universities and research institutes are enumerated in a popular manner by H. L. Nieburg: *In the Name of Science* (Chicago, 1966). A more sceptical view has been taken by H. Brooks in two recent articles: *Can science survive in the modern age?*, in *Science* (October 1, 1971), p. 21 and *What's happening to the U.S. lead in technology?*, in *Harvard Business Review* (May-June 1972), pp. 110-18.

([12]) The most reliable, though slightly out-of-date, figures are to be found in C. Freeman and A. Young: *The* R *and* D *Effort in Western Europe, N. America and the Soviet Union* (Paris, 1965). The U.S. sum of nearly $ 20 billion spent on R and D (or 3% of GNP) is only 250% and not 400% larger than European budgets as the R and D dollar buys less in a high-cost economy. The comparative figures have changed in recent years as the European nations gave increased subsidies to their science laboratories, as did the Japanese, too. The scientific spill-overs into industry from military research were not as great as some advocates of the « technology gap » at first believed, but their social and academic side effects in the 1950s and 1960s should not be minimized in retrospect.

TABLE II. – *Allocations of* R *and* D *by country and factor.*

| Country | Q.S.E. per 10000 population | Industrial R and D as % of industrial output | % of R and D spent on economic or social programs |
|---|---|---|---|
| U.S.A. | 36 | 6 | 38 |
| U.K. | 29 | 3 | 60 |
| France | 18 | 3 | 55 |
| W. Germany | 18 | 2 | 83 |
| Japan | 19 | 3 | — |

*Note.* – The international « division of labor » held that national economies would specialize their output and skills in order to take maximum advantage of their resource and investment assets. Closing the « technological gap » became a primary imperative for any nation intent upon securing a greater wealth —or power position—in the international market. Public money was devoted to aerospace or computer engineering so that nations could keep up with the SST or data processing race among industrial nations —and thus improve their technological balance of payments.

## 4. – Science and international tension.

Financial and manpower resources can best be mobilized in an advanced society, it appears, when some form of threat is perceived to the nation's security or well being. No major move was made to encourage science programs or to finance new research institutes between World Wars I and II, or prior to the onset of the strategic weapons race; nor until recently has there been so strong a concern to boost industrial research and productivity. In recent years, however, many of the science disciplines and programs financed by universities or corporate industry tended to enlarge their staff or their research facilities once they felt sure that support would be given by national institutes or academies, or nonmilitary agencies and private foundations. At the same time, it should be noted, the universities' efforts in « Little Science » and disinterested research were often conspicuous in quality and innovation even though they lacked the funds that were lavished upon applied or Big Science programs ([13]).

The optimum mobilization of a nation's science efforts became a matter of prime national importance in a world order dominated by considerations of national power and high technology. As the cycle of state spending on science moved from weapons procurements to high-technology support, the reliance

([13]) Alvin WEINBERG, the Director of the Oak Ridge laboratories, wrote in his *Reflections on Big Science* (New York, 1967) of his distaste for the perpetual self-advertising and self-aggrandizement of the largest R and D empires. That they required ever greater support in order to stay alive (and grow) become an irony of organization but a possible defect for good research. It now remains to be seen whether the public funds allocated for cancer, ecological or fuel energy research will generate useful new knowledge or further bureaucratic empires.

of scientists upon state funds continued to grow. The political threat posed by an industrial technology « gap » was as great as the threat that had been perceived in the strategic arms race. Capitalist and collectivist governments continued, therefore, to deflect mission-oriented research for *étatiste* purposes long after the Cold War dampened down.

The need for state funding and for public support of Big Science, both in the USA and elsewhere, stemmed first of all from the security fears and from the perceptions of strategic vulnerability that worried the political and military leadership. As tension levels rose there was a nuclear warhead and then a missile « gap »; then came the multibillion dollar absurdities of the space race, and latterly the fashionable theology of a technology « gap ». Ministries of Science began to take over some of the finances and functions previously handled by defense ministers. New modes of supporting scientific training and research institutes were developed by industrial or state bodies. Key growth industries were favored with tax allowances, subsidies and training funds in order to boost the nation's competitive capabilities; and the science community found itself as deeply committed to political funding and support as it had been in the worst years of the Cold War.

A quarter of a century after the Alamogordo bomb had been set off it was realized that the acquisition of an active science program was almost as vital, if not quite as expensive, as the acquisition of a full-scale defense capacity ([14]). The needs of a vigorously modernizing economy could not be disregarded by science peer groups or by their protectors. Consequently, progress in the various science disciplines (and of their respective applied technologies) were of critical import to central government, to the military, to industry and the university.

Thus the *utilitarian* aspects of science policy continued to play a major role in molding the values of the public, the political leadership and of many scientists, themselves. The need to justify the receipt of public funds and to organize the resource allocations of Big Science exacted a high toll from the science community in each national community. The immediate utility of science subsidies had to be politically justified, whether they were given for high-energy physics, molecular biology or laser research. Unfortunately, in the Faustian exchange made by many scientists—in surrendering their ancient academic privileges and scholarly autonomy for larger budgets and for entry into the R and D establishments of the state—there was a great suspicion that

([14]) In his *Address to the French Nation* in 1964 President De Gaulle urged on his countrymen that «... it is necessary to produce always more ... to push relentlessly our technical and scientific research ... to avoid sinking into a bitter mediocrity and being colonized by the activities, inventions, and capacities of other countries. » Cited in R. Gilpin: *France in the Age of the Scientific State* (Princeton, 1968), p. 39. National prestige and prowess in scientific achievement are also investigated in W. T. R. Fox: *Science, technology and international politics*, in *International Studies Quarterly*, Vol. **12**, No. 1 (1968).

the quality of research and training had been warped by the dependence upon political support. All too often, however, funds could be raised from no sources other than the state or industry and there was no alternative course of action for scientists to pursue. It gradually dawned upon the scientific community that if their work required public support and massive outside funding there was no option available except to follow the dictates of the nation's political-scientific leadership ([15]).

## II. - The National Political Environment.

**1.** – In the first lecture it was noted that no advanced industrial state today can afford to ignore the imperative thrust for power and scientific knowledge. Imperatives dictate that, regardless of political ideology or political system, a vigorous science policy is critical to the modernization of an industrial economy or to the acquisition of a full-scale defense capability. Most industrial states have come to justify the public funding of programs in science and technology for the strictly utilitarian purposes of advancing these national capabilities. The political leaders and industrial managers of capitalist and socialist regimes, alike, have been drawn into science policy involvments; so, too, have thousands of academic scientists or professional research workers. As a result, when eminent scholars or Deans have been co-opted into the leadership elites of the state their professional advice has received more serious attention than they ever gained before.

Logically, the thrust to build a national science establishment was inescapable and irreversible. Massive funds could not be invested in R and D upon a random or haphazard basis. Stability and order in program allocations were vital; therefore the official authority of the state had to be summoned. Nor could the formulation of program goals and priorities be left to the arbitrary decisions of the scientists themselves. As LYSENKO showed in fact, and SNOW in fiction, scientists were at least as incapable of responsible self-government as any other elite group in society ([16]). In most societies the executive or the

---

([15]) The tensions subsisting within the Unites States and its science communities during the 1960s were categorized in W. GOLDSTEIN: *The science establishment and its political control*, in W. R. NELSON (ed.): *American Government and Social Change* (New York, 1969). A more recent and cross-national enquiry appears in C. P. HASKINS: *Science and policy for a new decade*, in *Foreign Affairs*, Vol. **49**, No. 2 (1971); and D. S. GREENBERG: *The Politics of American Science*, (London, 1969).

([16]) Two different models of national science communities can be cited simply to suggest polar opposites. C. P. SNOW: *Science and Government* (Cambridge, Mass., 1962) depicts organized and militant camps of « committee scientists » as they struggle to influence

representative leaders of public power were either asked or forced to involve themselves in the promotion and administration of science efforts. This assured the multiple communities of science a permanent and regulated access to the channels of political power. It also left an indelible stamp of the nation's authority upon the decisions made by top science leaders (17). The co-optation of scientists thus divided the science communities between those who enjoyed the least and the greatest access to power. To no one's surprise but their own, the latter group found themselves as status-bearing members of the Establishment formations of the bureaucratic state. In short, in order to play an official role they first had to organize themselves as a military formation, like the Joint Chiefs of Staff.

The first task of the new politics-science oligarchies in each state was to resolve three troubling issues (noted in the first lecture) in an authoritative manner:

*a*) How were resources to be allocated between fields or programs, between basic and applied work, and between research laboratories and technology units?

*b*) What goals should guide the state's technology assessments and long-range forecasts, both in the promotion of research and of specialized education?

*c*) Which control mechanisms could safeguard *both* the development of the sciences and the residual public interest? *Quis custodiet ipsos custodes?* Should the experts be on top or on tap when the state came to determine its science priorities or the risky applications of sophisticated technology?

Various formulae were proposed in the market or collectivist economies of the industrial world. Some deprived the science leadership of all but technical responsibilities; others accorded them a major role in the determination of policy. It was obvious that in an era of thermonuclear energy, high-speed computer systems and particle accelerators that scientists could not be left to run their own affairs. In 1939 the Physics Department at Columbia Uni-

---

Government policy and win political preferment. By contrast, D. J. DE DOLLA PRICE: *Little Science, Big Science* (New York, 1963) investigates the multiple « invisible colleges », professional associations and « commuter groups » who operate in the power channels of scientific research.

(17) The consonance between the values of science and political institutions is elegantly noted in R. K. MERTON: *Social Theory and Social Structure* (Grenoble, 1957). The unity of values is also examined in such classic works as B. BARBER: *Science and the Social Order* (Grenoble, 1952); J. BEN-DAVID: *The Scientist's Role in Society: A Comparative Study* (Princeton, 1971); M. WEBER: *Science as a vocation*, in H. H. GERTH and C. WRIGHT MILLS (eds.): *From Max Weber: Essays in Sociology* (New York, 1958) and W. HAGSTROM: *The Scientific Community* (New York, 1965).

versity enjoyed a budget of only $ 10 000 a year. Thirty years later, M.I.T. and Berkeley each received over $ 50 millions annually from public sources. The setting of goals and distributive criteria could not be entrusted to government agencies or to market forces. The analogy was drawn with the principles of military control. Generals could not be free to govern armies—whether in France or China—and politicians could not do the job for them. Some form of compromise had to be developed that would accord with the traditional value preferences and the institutional forms prevailing in each society. The basic question that remained unresolved concerned the *dirigiste* energies of central government. Should each state build its own Pentagon of science as a national body; or should it leave the scientists to scramble competitively along whatever corridors of power they managed to penetrate?

## 2. – The power to make decisions.

The sociology of knowledge has placed a great emphasis upon the organization and peer-group control of research and teaching institutions. Studies have also shown that the « information industries » of the modern state, regardless of its professed ideology, would develop similarly complex and monolithic techniques of bureaucratic control ([18]).

How best to control science and technology had to be considered from two perspectives. Could quality controls exercised within each discipline ever be preserved and academic standards be firmly respected? and were the overall determination of goals and the implementation of programs to remain in the strong grasp of the state bureaucracy? An eminent physicist, WEINBERG, has suggested that this clash of perspectives can best be resolved by applying two separate criteria of scientific judgment—one internal, the other external—when any complex policy decisions have to be made ([19]). A careful distinction between these criteria must thus be made.

*a*) *The internal criteria* relate to the level of achievement, of professional training and of research competence manifest in any particular field, program

---

([18]) The issue of whether scientists should build a political base in, rather than an administrative concordat with the state, to resolve their organizational problems, was examined in W. SAYRE: *Scientists and American science policy*, in R. GILPIN and C. WRIGHT (eds.): *Scientists and National Policy Making* (New York,1964).

([19]) Alvin M. Weinberg's distinction between scientific and institutional values appears in *Criteria for scientific choice*, in *Minerva*, Vol. **1**, No. 2 (1963). He notes that only in a decentralized system can the two be sensitively meshed. In the United States, he regrets, this has not occurred. « There is already evidence that our ratio of money to men in science is too high and (greater) ... than the number of really competent men can justify ».

or discipline. If work in high-energy physics, radio astronomy, cancer research or molecular biology happens to be well advanced, then public revenues can be safely invested in supporting further R and D. If the field should not be ready to expand an investment of funds will be wasted; alternatively if it were ready but funds were denied there would be inequity and wastage. Peer-group monitoring within the discipline, in short, should police any development of research and it should warn of any imbalance between talent and funds that could be foreseen.

*b) The external criteria* relate, by contrast, to the judgment and perceived needs of the funding of policy agencies. If they choose to concentrate the distribution of funds in one discipline or in one laboratory to the relative detriment of all others, their decisions must be justified by an informed estimate of the « national interest » or of politically determined priorities. By co-opting the leadership of the scientific communities into this adjudicating process, the official agencies can add a note of technical sophistication to the political legitimacy of their decisions. Thus the eventual choice of science priorities would, in theory, reflect the wisest calibration of scientific merit, research promise, political need, economic pay-off, status reward and institutional support. « To each according to his ability » *and* « according to the state's needs » would provide complementary criteria for a just decision.

This dichotomy, though widely sustained, was responsible for the promotion of two unforeseen and unfortunate consequences. First, a marriage of convenience was effected between the elite representatives of *both* scientific and political power. Influential « committee scientists » were too often co-opted into high political councils ([20]). There they gave secret and influential advice; and they were able to determine which projects or research empires should flourish and which should be cut back. For their part, the political leaders found that they could select the scientific advice that best suited their needs; hostile critics could be frozen out and sympathetic allies could be promoted. As a result, a virtual monopoly of advice (or censorship) could control the technical advice flowing into the political system. In this co-opting manner the marriage of specialized skill and political authority could strengthen the executive competence and the hierarchical channels of the national Establishment.

---

([20]) A fascinating case history of political intrigue in the science hierarachy has been written by P. M. BOFFEY and B. NELSON. They looked at the controversial decision of President NIXON to deny to Franklin LONG, a noted scientist, the Directorship of the U.S. National Science Foundation. (See their articles in *Science* on 18 April, 25 April and 2 May, 1969.) It is worth noting that the National Academy of Sciences went so far as to criticize the President's top science advisors for vetoing Professor Long's appointment.

The second consequence of the attempt to juggle with internal and external criteria was equally unfortunate. In all too many cases the *illusion* of dichotomy was preserved. In all too few cases was it recognized that the latter (the external criteria) had achieved a powerful pre-eminence. As far as one can dare hazard a general proposition across political systems, it appears that the development and utilization of specialized or complex knowledge tends to be determined by social and not technical criteria.

Several examples can be cited to demonstrate the persistence of the trend. In Britain, for example, the work of three prominant scholars— HALDANE, BERNAL and NEEDHAM (admittedly, all Marxists)—has documented how external criteria governed scientific growth in various era. In the Soviet Union, the rise and fall of Academician LYSENKO illustrated the same tendency. In the national security regimes of the 1950s and 1960s the pattern, though never complete, was remarkably consistent. The innovation of new weapons technologies or strategic doctrines called for an increase in basic science or in graduate education funds, first in one field and then in another. Huge sums were squandered because the competence in a new field had not been properly established; or because the graduates in an older field completed their training just after a switch in funding priorities had been made to favor another discipline. The predominance of external over internal criteria was equally manifest in the demands of industry or of the military. Successive cycles of under- and then over-employment appeared in the high-technology, as well as in the defense-related industries. As a result, the misallocation of resources did much harm to the reputation of science and to impair the effectiveness of the science leaders who had been co-opted into the values and offices of power ([21]).

## 3. – The structure of science communities.

The need for Big or state-supported science to create its own political establishement was self-evident. No permanent hierarchy of wealth and power can exist for long without an elite structure—as churches, universities, bureaucracies or hippie communes have all discovered. The science establishment

---

([21]) The professionalizing of science, the closed structures of its « disciplines », and the uneasy relations between its « peer-groups » and « invisible colleges » are admirably portrayed in B. BARBER and W. HIRSCH (eds.): *The Sociology of Science* (New York, 1962). M. POLANYI: *op. cit.*, has noted that as the volume and complexity of scientific knowledge increases, there is a greater need to stress its « coherence » to political authority rather than its « correspondence » to internal tests of truth. This critical dilemma is explored in the landmark work of D. K. PRICE: *The Scientific Estate* (Harvard, 1965); and in N. STORER: *The Social System of Science* (New York, 1966).

grew rapidly in most countries between 1955-65, when annual rates of expenditure grew at an average 15% (and in some disciplines exceeded 25%). In these years the rapid recruitment and training of new QSEs changed the occupational structure of the skilled work force ([22]). Various science Establishments developed their connections with national or industrial agencies to ensure the continuing flow of funds. In the United States, alone, they extracted billions of dollars—or 3% of G.N.P.—to consolidate their power and to subsidize their preferences among scientific disciplines. Approximately 750 000 QSEs were dependent upon R and D funding in each of the super powers. The spin-off employment that they generated (as did their counterparts in smaller nations) was of vital concern to university graduates, private institutional staffs, academic teachers and the directors of research institutions. Thus millions of employees, grant recipients and R and D contractors became subject to the control of the « committee scientists » in the academic-political oligarchies of power ([23]).

The institutional complexes in science were stratified in most societies not so much along lines of skill, merit or prestige—but simply in terms of political power. Distinguished laureates or National Academies often enjoyed less authority, even among their own peer groups, than technicians blessed with Top Security classifications. Inventive and imaginative scholars at the research bench earned smaller salaries or commanded smaller resources than colleagues sitting on Ministerial advisory committees or on project evaluation Boards.

---

([22]) FREEMAN and YOUNG: *op. cit.*, p. 72, offer a survey of the « gold reserve » of QSEs employed in R and D in various countries in 1962. The figures given are in 1000s fully employed in R and D.

| | QSEs | Other R and D personnel | Total R and D work force | R and D staff per 1000 working population |
|---|---|---|---|---|
| USA | 436 | 724 | 1160 | 10.4 |
| W. Germany | 40 | 102 | 142 | 3.9 |
| France | 28 | 83 | 111 | 3.8 |
| UK | 59 | 152 | 211 | 6.1 |
| USSR | 487 (estimate) | 985 | 1472 | 10.4 |

It was also assumed that two factors in the U.S. would help maintain its commanding lead in QSEs. First, the emigration or « brain drain » flow of skilled personnel from Europe to the U.S. Second, the superior quality of and capacity for graduate training in America universities. The cut-back in U.S. science activities in the latter years of the war against Vietnam have reduced the width of the QSE « gap » and stemmed the « brain drain ».

([23]) A useful survey of current R and D programs in Europe can be found in R. FOCH: *Europe and Technology* (Paris, 1971); also, C. LAYTON *et al.*: *Industry and Europe* (London, 1971); and the various publications of O.E.C.D. in Paris.

Military secrecy, industrial subsidies and access to political power became a common-place of scientific life. Those who walked the corridors of power were to be found sitting on six committees simultaneously and judging the merits of thousands of their colleagues ([24]).

Those excluded from positions of trust or mission-oriented research found it difficult to enlist the support, to purchase the equipment, to hire the assistants, to acquire the libraries or to develop the careers enjoyed by their more fortunate compatriots. For what it was worth to them in the Faustian exchange of power for independence, the scientists without power were free to choose their own subjects of research or the purposes to which their knowledge would be put. But they were also free to enjoy an obscurity or an exclusion from favor as a result of staying in their classrooms or sticking to their lonely careers ([25]).

In contrast to the powerless professionals who opted for independence, or who failed to draw the encouragment of the powerful, it is instructive to review the career opportunities won by scientists who courted office. Sir Solly ZUCKERMAN, for example, an eminent professor of biology and the director of the London Zoo, sat for almost a decade on the committees that distributed nearly 90% of public funds. In his capacity as chief science advisor to the British government, he acquired an authority that was difficult to break. SEABORG, an American chemist, became Chancellor of the University of California, the director of the Atomic Energy Commission and a member of several top echelon Presidential committees. In France and the Soviet Union there was a similar co-optation of prominent scientists. Members of the Academy of Science or of the Collège de France were invested with formidable political power and administrative jurisdiction. Their colleagues and sympathizers share the lessers benefits of office but their critics and detractors lapsed frequently

---

([24]) A dissenting view about the exaggerated influence of scientists in politics appears in D. S. GREENBERG: *A myth of the scientific elite* in *The Public Interest*, No. 1 (Fall, 1969), pp. 51-62. A similar note is struck by D. K. PRICE: *Purists and politicians* in *Science*, No. 163 (1969), pp. 25-31.

([25]) J. G. CROWTHER: *The Social Relations of Science* (London, 1941) noted rather gloomily during World War II: « Young scientists who abandon science for politics often prove to be mentally unstable, and after a few years of bohemian agitation become conspicuously conservative. Conduct and opinions that appear to be based purely on moral sentiments ... are nearly always suspect » (p. 644). The philosophical issues raised by the social interpretation and validation of scientific enquiry are usually resolved in an optimistic or « liberal » manner by authors who believe that, in the last resort, scientists and managers will be guided by an aroused and rational « consensus of opinion ». See, for example, J. ZIMAN: *Public Knowledge* (Cambridge, 1968), and the 1965 *Daedalus* issue as edited by G. HOLTON: *Science and Culture: A Study of Cohesive and Disjunctive Forces* (Boston, 1967).

into obscurity. It needed the conviction of an exceptional scientist, like PAULING, armed with two Nobel Prizes, to detach himself from the state hierarchies while his work continued to flourish ([26]).

## 4. – The political environment.

The growth and expansion of the science Establishment varied from one political system to another, depending upon the institutional values and practices that were traditionally observed. The Soviet Union chose to coordinate its science efforts in specialized institutes and prestige academies rather than in regional universities or in industrial units. Following the principles of democratic centralism, a newly formed State Committee began to acquire the influence and resources once reserved to the Ministries of Science. By contrast, the United States was tempted to build a Pentagon of science until opposing counsels prevailed. Instead, an interlocking Federal complex was built to pull together a coalition of dispersed and nonmilitary authorities. The President's staffs played a key role in energizing or consolidating the work of other administrative agencies both inside and outside official circles. Despite the efforts that were made to centralize science policy decisions and to streamline the hierarchies of professional authority, however, fragmentation and conflict continued to prevail. Industries struggled to win R and D support, Federal agencies subsidized rival disciplines or graduate schools, and the military-academic-industrial complexes failed to establish an operational oversight of the nation's science policy. Though the fragmentation process persisted, so, too, did the sharp differentiation between the scientists who enjoyed some access to power as against those who enjoyed none.

In Britain, France and Germany there was evidence of different pressures brought to bear upon the science communities. Each government sought to deal with the critical matter of graduate education, partly because it had trained an insufficient number of QSEs, and partly because the central or regional authorities were able to influence the funding priorities within the national university structure. Stung by the accusation that a « brain drain » and a « technology gap » had demoralized their academic staff, respective Ministries of Science, Technology or Education sought to intervene vigorously in the affairs of science training and future technology assessments. In most respects they were as-

([26]) These conflicts between scientific truth and social pressure are dramatically captured in a play of Henrik IBSEN: *An Enemy of the People.* An honest doctor tries to expose the fact that the medicinal waters in the town's profitable baths are polluted. His naivete and self-willed crusade are severely punished by his fellow citizens. They fear that economic ruin will result from his « misplaced » altruism. His punishment for attacking the power structure is personal and tragic.

sisted by the self-interested lobbying of giant industrial corporations. In few were they opposed by lobbying efforts of the scientists themselves ([27]).

Political charges were made by outraged scholars and social critics against the military-industrial-science complexes as the Cold War exacted its costly waste of talent. The « knowledge complex » was blamed for bureaucratic rigidity, insensitivity to research needs, a monopolizing influence over public funds, and a bias against innovative, critical or basic research. As the Vietnam War and the arms race declined, it was noticeable that political conflict had widened both the status and the political cleavages that divided the academically tenured research staff from the less credentialed and more numerous work force employed in industrial or military technology. The latter were less secure about attacking the sources that fed them; and they shared little of the elite view on the campus—that peoples and governments should listen attentively to the scientists' criticisms of the bureaucracy that governed the nation's research programs.

Two salient features appeared in the course of the debate in nearly every advanced industrial economy. First, it became apparent that no trade union or professional association of scientists had succeeded in organizing themselves as an effective or articulate pressure group. Because the science and technology staffs had failed to organize, the science communities were too frequently condemned to remain *a*) internally divided, *b*) subject to the peer control of their elite leadership and *c*) generally subservient to the pressures exerted from top political or industrial echelons. The consequences of this failure to organize were everywhere apparent. Among the Faculties of Science in Academigorodok, in Nanterre, in Chicago or Manchester there was no effective association of dissident scientists or of laboratory technologists to protest against the division of academic-political power or to criticize the priorities in program budgeting that had been implemented. There was a widespread dissatisfaction, a few protests and no basic leverage to change the criteria of policy ([28]).

---

([27]) An excellent appraisal of French science, military R and D and industrial policy appears in R. GILPIN: *France in the Age of the Scientific State* (Princeton, 1968). A striking comment on the Soviet scene appears in Z. A. MEDVEDEV: *The Rise and Fall of T. D. Lysenko* (New York, 1960). It should be contrasted to the American reports of A. DE GRAZIA: *The Velikofsky Affair* (New York, 1966) or to H. BROOKS: *The Government of Science* (Cambridge, Mass., 1968).

([28]) The regulation of scientific growth and of the changing priorities of basic and applied research pose grave dilemmas to politicians and National Academies alike. President NIXON appointed a Task Force on Science Policy in 1970 to question whether budgets should continue to increase by 15% each year—as they had in the 1960s. The official report, *Science and Technology: Tools for Progress* (Washington, 1970) found no reason to regret the sharp cut-back that had been made in the 15% figure. But in urging that the NSF budget for basic work should be pegged to 0.1% of GNP, the Report advocated

A second salient feature in the debate over science policy was of greater significance. In no political system were the representatives of dissident opinion able to express themselves coherently or to exercise any useful control over science decisions. Parliamentary, party or Congressional committees failed, by and large, to modify or reverse the decisions taken by the Executive establishment or by its co-opted science advisors. Nor were any outside groups—such as the mass media, trade unions, the universities or youth groups—able to close the connection between the changing sentiments of the electorate and the expert decisions made on their behalf by elite leaders. Belated and fragmentary efforts were begun in applying R and D studies of pollution and environmental control but decision power was again tightly held by senior technocrats. Only a minimal role was played by nonscientists or low-status professionals in the science community and their influence on public policy was nowhere impressive. Obviously, major decisions were reserved to the corporate scientists or to the political-industrial leadership; the ill-informed or the uninitiated were not asked for their opinion on matters of high expenditure or scientific investment ([29]).

As a consequence of these developments, science decision making became invested with a false, apolitical ambience. Various mass publics came to view scientific affairs as too esoteric to be subject to popular control; technologists and laboratory workers learned to defer to the senior members of their organization; and dissident scientists began to despair that R and D or training priorities could ever be related to new or urgent social needs. The escapism, the passivity or the frustration generated by this developing trend only served to enhance the elitism of the science leadership and the organizational weight of their policy decisions. The subordinate roles that the science communities performed for the powerful agencies of the state, in exchange for their funds and necessary support, were nowhere better seen than in the debate over military research. Opinion surveys demonstrated that research scientists were more outspoken than most social groups in condemning the lethal uses to

---

a doubling of the small sum ($ 500 million) spent on theoretical or « pure » research. See, too, Victor Weisskopf's comment in *Science*, No. 162 (1970), p. 935. Acute criticisms of the force of public sector decisions on science policy can be found in the articles collected in P. J. Picard (ed.): *Science and Policy Issues* (Chicago, 1969) and in M. Blissett: *Politics in Science* (Boston, 1972).

([29]) J. Haberer: *Politics and the Community of Science* (New York, 1969) contrasts various philosophical approaches to science. While Francis Bacon emphasized the social utilities of science, Descartes urged that it must be disengaged from the purposes of the state. Haberer employs these positions to examine the purging of Jewish scientists in Germany in 1933 and the purging of J. R. Oppenheimer from the A.E.C. in 1954; his aim is to demonstrate that entrepreneurial research (as an end in itself) can no longer live autonomously in a society that nurtures and molds scientific effort into vast, corporate organizations.

which their technical knowledge had been put. But try as they might to articulate their criticisms, there was not one official institution or national academy that abandoned its position of co-operating in the most sophisticated forms of death-delivering technology. A few unofficial groups adopted rousing manifestos of protest but the small numbers of their memberships showed how unrepresentative or ineffective they remained.

The issue of the scientists' personal *complicity* in weapons research was more important as a symbolic than as a technical issue. Regardless of which side was to triumph in the debate over the « citizen scientist » and his divided allegiance (to his work and to the state), a new scepticism began to pervade the science communities of the leading industrial nations. Academician SAKHAROV was viewed, together with his dissident counterparts in Western Societies, as a novel Protean figure of the century: the eminent professional whose science is vital to national security but who insistently speaks his mind against the state that strives to misuse his knowledge while muting his freedom of protest. The growing support for the protest leaders and the dissenting youth in the science communities will be carefully examined in the next lecture. But it must be noted here, in conclusion, that even so shrewd an observer as KUHN failed to anticipate this profound change in the social *paradigms* of normal science. In his striking work, *The Structure of Scientific Revolutions* (Chicago, 1962), he had written the definitive, liberal definition of a « normal science ... that often suppresses fundamental novelties because they are necessarily subversive of its basic commitments ». He did not foresee that the commitments to basic *étatiste* values would become the subject of violent debate as scepticism took hold among the scientific communities.

## III. - Science and Social Change.

**1.** – A great deal has been written in recent years about the « revolutionary » pace of scientific change, about « runaway » technology and its « disfunctional » impact on industrial societies. The fear has been raised that science and its complex of oligarchies will break loose from the restraining influences of democratic control, or that political regimes will be subverted by the science executives whose work they had financed. The truth or distortion implicit in these metaphors and fears of change need not concern us in the present enquiry. Our focus must rest upon a different question. Assuming that the leaders of science are men of goodwill—and not conspiring technocrats intent upon usurping authority—are there still instruments at hand with which they can change the program priorities, the peer-group co-optations and the institutional research agendas that the state has mandated in its capacity as patron of Big Science?

To answer this question it is necessary that a few political concepts should be clearly enunciated. For a start we must examine the *étatisme* that characterizes so much of contemporary science. The image of today's nationalism has been captured in the axiom that every leading nation requires a high-energy accelerator, a run of Nobel Prizes or an M.I.T.-style graduate school if it is to demonstrate its scientific prowess (30). If the possession of ABM or Polaris missiles symbolizes the force capabilities of a superpower, a cadre of one million QSEs represents the intellectual vitality and the productive wealth of a post-industrial society (31). Mass electorates need to invest their patriotic pride in the accomplishments of their own, home-raised scientific cadres. *The force de frappe* is of more value to the pretensions than to the government of France. The first success of BARNARD in open-heart surgery was seen as a national victory for the ostracized white society of South Africa. Conversely, the « brain drain » in Britain provoked a sense of alarm in the mass media when an entire department of medicine chose to emigrate to the United States in search of richer sources of research support (32).

The nationalist element of science can be measured in several different ways. First, leading nations boast of the billions of dollars that they allocate to R and D. If less is spent on military than industrial research there is cause for the liberal community to celebrate (33). Second, a careful accounting has

(30) Since the money spent on R and D or on QSE training indicates nothing about the quality of a nation's science, allusions are frequently made to the number of Nobel Prizes that have been awarded in science. The figures below for 1951-1963 are remarkable in one respect. Prior to 1950 the U.S. was roughly equal in the awards received by each European country; in 1951-1963 it was far ahead of all of them combined.

*Nobel prizes received 1951-1963.*

| | France | UK | West Germany | USA | USSR |
|---|---|---|---|---|---|
| Physics | — | 3 | 3 | 15 | 4 |
| Chemistry | — | 6 | 2 | 5 | 1 |
| Medicine | — | 6 | 1 | 14 | — |

(31) Another quantitative index of a nation's science activity has been found in the number of physics or chemistry papers published annually in science journals or abstracts. D. J. DE SOLLA PRICE: *Little Science, Big Science* (New York, 1963) demonstrated that U.S. productivity was double that of the USSR though its R and D establishments were roughly of the same size. In fact, he also showed that productivity was positively correlated with the % of GNP on a global scale enjoyed by each state. The U.S. recorded almost $\frac{1}{3}$ of the wealth and $\frac{1}{3}$ of the science papers published; the Soviet figures were roughly 15%, while the UK, France and Germany had 5% of each.
(32) The « push » and « pull » phenomena of the emigration of scientists from countries with poor R and D funding to the research-rich nations is analysed in depth in D. N. CHORAFAS: *The Knowledge Revolution* (New York, 1968).
(33) The collected essays of Sir Solly ZUCKERMAN: *Scientists and War* (London, 1966) are instructive in this regard. They demonstrate how the chief Science Adviser to the U.K.

been made of the 50 000 science journals, the 1.2 million articles and the 60 000 technical books that are published each year; the proportion originating in each state is of competitive concern since the dissemination of new knowledge is expected to double in twenty years. Third, the emerging scientific « meritocracy » has been invested with patriotic pride. The number of Nobel Prizes is carefully calibrated, the publications production of each National Academy is constantly compared and the technology « balance of payments » is watched to determine whether the state can earn billions of dollars in royalties and licensing fees (or whether it must spend hard currency to purchase technology from industrial competitors).

Of course, a distinction must be drawn within the nation's science population. Almost a million technology workers are employed in government laboratories and in the giant corporations in the information-transfer industries; while only a few thousand academic or « committee » scientists are employed by elite universities and official agencies. The first group are skilled toilers who control little of their own daily routine. The second are the science executives, the eminent professors in graduate schools and the research *illuminati* who provide either intellectual or organizational leadership. Though members of the latter group travel frequently to international conferences and contribute to such cross-national institutions as CERN, WHO or the UN technical agencies, their base of operation is firmly located within their home nation. In this regard it can be seen that science is neither a liberating nor a cosmopolitan enterprise. It is funded and programmed by the official or the corporate agencies within the state. Its purpose is more to serve the institutions of power than to traffic in the specialized knowledge that might, conceivably, lead to their overthrow ([34]).

## 2. – Scientists as philosopher-kings.

Since science is largely performed as a function of the national body politic, it must be asked whether its control lies in safe hands? Or to extend

---

Government calculated how scarce science resources could better be deployed for the nation's military and civil welfare. The pursuit of academic prestige in exploring new modes of rational science effort are also noted in F. Reif: *The competitive world of the pure scientist*, in N. Kaplan (ed.): *Science and Society* (Chicago, 1965); and in an anecdotal gem, *Bucking the science establishment*, by T. Gordon, in *Playboy Magazine* (April, 1968).

([34]) An excellent symposium on *Knowledge and Power*, edited by S. A. Lakoff (New York, 1966) demonstrates the varied ways in which administrative or political power can summon scientific knowledge in order to enhance national policy objectives. The most outspoken protest against the state's misuse of science is still to be found in the writings of J. D. Bernal, particularly in his *Science in History* (New York, 1965).

the issue that Plato raised: Are the affairs of society determined by competent and disinterested *guardians* or are they managed by men of intemperate judgment who take advantage of their sophisticated skills and unique knowledge?

It is important to our political beliefs that we assume that society is well managed and that any excesses of modern science will be curbed by the scientists themselves. Unfortunately, the evidence needed to confirm this belief is often lacking. All too frequently the leaders of science have failed to expose the contradictions, the budgetary waste and the injurious side effects created by unwise science policies ([35]). In recent years billions of dollars have been squandered on death-delivery systems, inadequate attention has been given to the enhancement of social utilities (such as better schools, hospitals and occupational training) and abuses of power—as in pharmaceutical or computer applications—have gone unchecked ([36]). Too few scientists, unfortunately, have seen their work put to constructive purposes. Hence, alarms have been sounded against the social irresponsibility, the rigid bureaucratization and the aimless scrambling for funds that have divided many national science communities.

As a result, scientists have been viewed as sinners rather than as saints or magicians by the electorates of the 1970s. One particular change in the public's attitude toward science is of great significance, There has been a sensational widening in the disparity between the « methodology ethic » of the academic disciplines and the « institutional ethic » of the research laboratories

([35]) The shortening in the lead time between the discovery (invention) of new scientific knowledge and its industrial application (technology) is of marked significance to modern science. The following list of lead time « gaps » has often been cited:

| | | |
|---|---|---|
| Telephone | 56 years | (1820-1876) |
| Radio | 35 years | (1867-1902) |
| Television | 14 years | (1922-1936) |
| Radar | 14 years | (1926-1940) |
| Atomic bomb | 6 years | (1939-1945) |
| Transistor | 5 years | (1948-1953) |
| Laser | 5 years | (1956-1961) |

The shrinking gap explains why technology has « runaway » or why poor decisions were authorized by scientists who really should have been more cautious.

([36]) It is often difficult to establish cause effect relationships or to gauge the influence of basic on applied research. For an original enquiry see D. J. deSolla Price: *Is technology historically independent of science?*, in *Technology and Culture*, Vol. **6** (Fall, 1965), pp. 553-68. While Polanyi: *op. cit.*, and R. Merton: *op. cit.*, conclude that a democratic peer-group control among scientists must be summoned to prevent further abuses, there is growing support for the radical protest that « science is too important to be left to the scientists » and to their self-interested, hierarchial associations. The protest is carefully examined in D. S. Greenberg: *The Politics of Pure Science* (New York, 1967).

and hierarchies in which work is done ([37]). It is apparent that the disparity between these ethical norms will become wider and more embittered in the next decade.

On one side, scientists have shown an admirable tenacity in maintaining rigorous academic standards and a self-demanding ethic of discipline and methodology. Plagiarism and fraud are carefully policed by the peer-groups and journal editors who are charged with monitoring quality control. But as against these internal norms of craftsmanlike work the policing of institutional ethics is notoriously lax. If prestige laboratories and university departments devote themselves to disreputable programs—of the quality seen in the Jason, Spicerack, Themis and Camelot projects in the United States—one might expect that a fulsome repudiaton would be issued by the academy leaders ([38]). That they were not is of some moment.

Members of the lay public and many of the junior staff in the science community realized that this poor—or unprofessional—judgment in high places could be found in all nations. Many were moved to protest against the passive acceptance of unethical norms on the part of their senior colleagues. Meetings held by professional or academic institutes of science in many western countries were broken up by radical or dissident academics. Nobel laureates who had devoted themselves to weapons work were harassed in Rome, the Collège de France and in Oxford convocations. Half of the faculty at M.I.T. staged a token strike to protest against the war work performed in the university laboratories; and stirrings of dissent have been heard among Soviet or Polish scientists, or from visiting Academicians, when they attended international conferences. A few scientists resigned from their National Academy because of its refusal to repudiate the abuse of the institutional ethic. A few went so far as to abandon their work in genetic theory or particle physics because they did not trust their

---

([37]) The distinction between « a strong *methodological ethic* and a weak *institutional ethic* » is drawn by Haberer: *op. cit.*, p. 321. He worries that the latter is too fragile to preserve a proper relationship between science and its external or political environment. The distinction is both as clarifying and as misleading as the dichotomy between external and internal criteria of science funding set out by Alvin Weinberg (note ([19]), *supra*).

([38]) The most sensation-ridden projects funded by the U.S. military were rarely the largest. While billions of dollars were spent on aeronautical, nuclear and rocket research at leading universities, the greatest objection was taken to smaller but more suspect projects. The projects named concerned CBW weapons, « electronic battlefield » instruments and social research for possible counter-insurgency campaigns. After the students' revolts of 1968-70, many universities chose to discontinue these embarrassing and ill-paying adventures in applied research. An excellent case study of the clash between « method » and « institutional » ethical norms can be found in I. L. Horowitz (ed.): *The Rise and Fall of Project Camelot* (Cambridge, Mass., 1967).

colleagues to apply their findings in a just or socially constructive manner [39].

It has become apparent in the last decade that the communities of scientists have divided themselves into two distinctive groups: the custodians and the dissidents. Psychologically, the custodians are conservative professionals who believe that scientists should concern themselves mainly with their own work and that they should not attempt to overturn the program budgets or the committee structures supported by the nation state. The dissidents manifest a different set of values and behavior. They refuse to accept the conventional wisdom and the prestige stratification adopted by the committee men. They challenge the notion that the custodial keeping of the method ethic can be accompanied by a disregard for institutional norms. Above all, they object to the weighty precedence awarded to the « external criteria » of science policy—as WEINBERG put it—in the governance of Big Science and the reduced concern shown for « internal criteria ».

The growing hostility between these two groups has been seen at many scientific meetings and in various professional journals. Various dissidents have pressed for a change in educational curricula, for a redefinition of their responsibilities to society, or for a repudiation of the belief that Big Science must serve the power interests of the state or of capitalist industry. For their part, the custodians have argued that the radical cause of employing science to « liberate » society is tendentious and destructive. Too many academics have abandoned their research to indulge in political rhetoric, they claim; and too often the results of their protest have been found in the antagonizing of official agencies or in the ruthless cut-back in research budgets. « Get back to the work for which you were trained », they urge, and leave the determination of science policy goals to those who are skilled in operating in the corridors of national power.

An ironic turn in this debate has taken place though the public has not been aware of it. While the custodians have argued that as elite leaders they can serve as philosopher-kings in governing the work of the science disciplines, the real governors of the state and industry have viewed them as philosophers but not as kings [40].

---

[39] The storm against « collaborating » in military research or against the co-optation of committee scientists blew up rather suddenly. Hence, there is no warning of impending trouble in a book published in 1965 that otherwise stands as the « classic » work on science policy: D. K. PRICE: *The Scientific Estate* (Cambridge). The social stratification of scientific influentials, statesmen and outsiders is handled with remarkable dexterity, in W. O. HAGSTROM, *The Scientific Community* (New York, 1965).

[40] J. R. RAVETZ: *Scientific Knowledge and its Social Problems* (Oxford, 1971) quotes the aphorism of a noted sociologist: « It is one of the ironies of our time that so many intellectuals strive to identify with the perspectives of kings, while their rulers value them for their activities as philosophers ».

PLATO had insisted that the wise should be asked to rule, but in a post-industrial nation the axiom has been restricted to read: the professionally skillful, who speak for their science peer-groups, should be asked to advise but not rule. This has exacerbated the debate between the co-opted custodians and the dissident outsiders. It has also confused the logical arguments on both sides. The custodians have been consulted by the top echelons of the public or private sectors of society but their advice has often been ignored. The dissidents, however, have erred in attacking the advice given over ABM procurements, the niggardly funding of the NSF or the misdirection of medical research. These decisions would have been made whether the custodians were consulted or not. The dissidents did not see that the decisions were immune to the rational techniques that supposedly govern the procedures of science [41].

## 3. – Philosophies of social change.

The confusion prevailing among scientific communities can be categorized in terms of the assumptions that scientists make about the organization of post-industrial society. The assumptions can be enumerated in the following categories:

1) *The liberal belief*: the only way to correct the abuses and excesses of science is to apply even more scientific knowledge and effort to public affairs.

2) *The custodial position*: only the better self-government of science and the wiser deliberations of science elites can harness specialized skills for the general welfare.

3) *The dissident argument*: adversary debate among professionals will curb the authority of co-opted elites and revive the social responsibilities (or the institutional ethics) of those who speak for the interests of the science community.

4) *The revolutionary doctrine*: if organized societies, whether capitalist or socialist, corrupt the conduct and the uses of science, it is vital that scientists should strive to overthrow the political system.

5) *The counter-culture cop-out*: the institutional inertia of society is too formidable to be moved; retreat into mysticism, drugs or primitive communes

---

[41] J. RAVETZ: *op. cit.*, p. 422, criticizes the «increasingly subordinate» role that science performs for civil and military industry. He suggests that «one may revise Lord Acton's aphorism "all power corrupts and absolute power corrupts absolutely" and substitute "responsibility tends to corrupt and responsibility without power corrupts absolutely"».

can alone restore sanity to life and preserve the individual from the organized corruptions of society.

Among these five categories the first is of the greatest significance. Basically, it posits the homeopathic belief that the best correction for the misuses of science is to mobilize an ever larger scientific antidote. Liberal philosophy has long associated itself with the cause of reason in government, with rational problem-solving as a style of life, and with scientific calibration as a mode of determining « the greatest good for the greatest number ». Shocked by the systematic brutalities of state power—especially in the destruction of Viet-nam—the liberals have called upon science to show that harmony and reason can be restored once objective analysis replaces prejudice and venality in the affairs of state. Tragically, their beliefs have not been fulfilled ([42]). Instead of releasing society from its enslavement to war, national interest or industrial profit, scientists have too often found a justification rather than an antidote to social injustice. Like alcoholics, who treat their problem with more drink, science has become slightly addled by its own belief in a homeopathic treatment ([43]).

The second, or custodial, argument is equally widespread. Among mass electorates who lack knowledge and among many of the scientists who lack power there is a shared assumption: that the wise and the elderly will somehow arrive at the right, corrective decisions if only they are allowed room to maneuver. This expectation of elitist action is deeply imbued in industrial societies. We expect that men of experience and skill will operate effectively to forestall error, to suppress tyranny or to prevent human suffering. Patently, this belief has outlived its 19th century patina of social meliorism. An yet, incredibly, it still persists. For every Dr. Strangelove corrupting the powerful with sophisticated knowledge it is assumed that there will be shrewd savants and influential laureates to speak honestly for science in the councils of

---

([42]) Writing as a prominent and pre-Vietnam liberal on the Harvard faculty, D. K. PRICE (*op. cit.*, p. 278) looked forward to the day when the state can « defend its freedom by keeping the institutions established for the discovery of truth and those for the exercise of political power independent of each other ». This belief, that scientists and politicians should take a sympathetic interest in each others' affairs, remains at the core of liberal philosophy. To a great extent, however, both sides in this intellectual *concordat* have come to distrust the values and the maneuvers of the other. The 1970s may yet see an armed truce rather than a close co-operation making science-politics relationships.

([43]) The latest example of a homeopathic attitude is to be found in the arguments over industrial pollution. All too many scientists believe that the excesses of industrial technology can be cured by an application of more technology rather than by a restructuring of political institutions and values. The arguments on either side of the debate are appraised in W. GOLDSTEIN: *Who is to pay for environment control?*, in *Alternatives* (Canada), Vol. **2**, No. 4 (Summer, 1973).

authority. If eminent physicists like WIGNER were co-opted on to the Atomic Energy Commission or the NSF, it is supposed, there would be intrinsic improvement in the conduct of physics and in the physicists' relations with government and industry. The evidence of such a taming of state power needed to substantiate this hypothesis is sorely lacking in many communities and science disciplines. But it is still widely believed.

The third category of beliefs about social change are focussed upon the efficacy of adversary debate. If only the low-status outsiders can confront the « committee scientists » over their policy and budgetary decisions, it is claimed, a new form of academic democracy will emerge. The defects in this argument are as readily apparent as the failures encountered in the first two categories. The record of recent experience does not help substantiate the claim in fact —or in theory. Science executives have shrunk from the confrontation tactics pursued by their junior colleagues; and little harmony or reasonable change has been effected in the pitched battles fought in the annual meetings of Physics Societies or Associations for the Advancement of Science ([44]).

The failure of these first three assumptions has lent an ever greater saliency to the last two categories: the revolutionary and the cop-out. Though few scientists have committed themselves to these violent or extreme solutions to their professional predicament, there are many young people who have been drawn to either position. In thousands of cases they have acted on their beliefs, by aborting their academic careers or by refusing to enter graduate schools. If the public and private sectors of society can corrupt the procedures and uses of science, they loudly proclaim, they want no part of it. It is more meaningful to them to work for the overthrow of the system—or to drop out of it altogether—than to acquire the specialized skills that will reward their careers as servants of a warped social order. Certainly, the predisposition of the state to employ scientific research to enhance its capacity for violence and injustice has done a great deal to intensify these youthful passions of rejection.

## 4. – The ambiguities of social change.

It is regrettable that no clear formula can be stated of how scientists should best regain command over their own affairs and over the public uses to which their findings are put. Obviously, millions of scientific workers cannot expect to operate in a political vacuum; they require billions of dollars for their annual

([44]) An increasing number of academic or professional bodies have adopted resolutions that condemned the Vietnam war or the military misuse of scientific research. Their resolutions have helped change the climate of opinion in various science communities; but they failed to stop the large research projects performed at the military's behest.

support and their institutes cannot flourish independently of the financing and program needs of the public or private sectors. But neither can science progress if it is encased with rigid budget allotments and peer-group hierarchies. Some form of flexibility and change must be allowed, both within the various disciplines and in their relations with agencies of power. The question remains: how should this best be arranged?

The answer to this question has varied from one national science community to another. In Britain, where pragmatic political compromise is more easily effected, there has been a reshuffling of Ministerial jurisdictions, university committees and *ad hoc* groups. In France and the Soviet Union change has depended to a greater degree upon formal re-organization in the apex Executive organs that centralize decision power. In the United States the position is thoroughly confused. Federal agencies such as NIH, NASA and AEC have engaged in a « pluralist » competition for budgets and influence; but weaker bodies such as the NSF, which should have enlarged their authority, have been denied ample funding. Moreover, the leverage of the Department of Defense and of giant industrial corporations has held Big Science in permanent thrall. No amount of protest, adversary debate or inter-elite manuevering has succeeded in releasing the grip of these powerful institutions.

The conclusion must be reached, therefore, that there is no sure way in which the national science Establishment can be dismembered. Nor is there an alternative model in sight that can suggest how the science community could be restructured in the aftermath of a radical social change. Science is too big, too expensive and too complex in its organization to allow free-floating institutions or anarchic communities to be built. It must submit itself to a peer-group authority if it is to police its own « methodology ethic ». And it must provide a daily sustenance for its millions of technicians and thousands of research staff if they are to work effectively. Whether the Faustian exchange can continue—of providing vast funds at the expense of the scientists' independence—is a question that only state and industrial agencies can resolve. If scientists are consulted in the process, so much the better. But that they will force the pace of change, on their own, remains as improbable in future years as it has been in the past.

This pessimistic conclusion does not suggest that scientists should supinely submit to the will of powerful agencies; or that there is *nothing* that they can do to change the political values of society or the policy decisions of the state. But it must be realized that science is *a social function* and that it always has been so. The role played by scientists in changing mercantilism in the 18th century or bourgeois industrialism in the 19th was not trifling. It remains to be seen whether the skilled cadres of science can be as ingenious and effective in modifying the institutions and abuses of post-industrial nationalism in the 20th century. That few attempts have succeeded so far is evident. That none will ever succeed is a conclusion that cannot be adopted.

## RECOMMENDED BIBLIOGRAPHY

The following books in English are particularly worthy of attention. A list of journal articles, political case studies, government publications, etc. can be found in the publications of O.E.C.D., the Library of Congress, *Science, Nature* or the *New Scientist.*

B. BARBER: *Science and Social Order* (New York, 1962).
*The Sociology of Science*, edited by B. BARBER and W. HIRSCH (New York, 1962).
R. J. BARBER and S. LAKOFF: *Science and the Nation* (Princeton, 1962).
J. D. BERNAL: *Science in History*, 4th ed. (London, 1969).
G. FISCHER: *Science and Ideology in Soviet Society* (New York, 1968).
R. GILPIN: *American Scientists and Nuclear Weapons* (Princeton, 1962).
R. GILPIN: *France in the Age of the Scientific State* (Princeton, 1968).
D. GREENBERG: *The Politics of Pure Science* (New York, 1967).
*Social Relativity of Scientific Myth*, edited by K. SILVERT (New York, 1969).
T. S. KUHN: *The Structure of Scientific Revolutions* (Chicago, 1962).
C. LAYTON: *European Advanced Technology* (London, 1969).
D. PRICE: *Science and Government* (New York, 1954).
D. PRICE: *The Scientific Estate* (Harvard, 1965).
D. J. DE SOLLA PRICE: *Little Science, Big Science* (New York, 1963).
H. ROSE and S. ROSE: *Science and Society* (London, 1969).
W. O. HAGSTROM: *The Scientific Community* (New York, 1965).
N. W. STORER: *The Social System of Science* (New York, 1966).
R. MERTON: *Social Theory and Social Structure* (New York, 1957).
A. WEINBERG: *Reflections on Big Science* (Cambridge, Mass., 1967).
N. J. VIG: *Science and Technology in British Politics* (London, 1968).
H. WOOLF: *Science as a Social Force* (Baltimore, 1964).
J. ZIMAN: *Public Knowledge* (Cambridge, 1968).

# Physics and Physicists the Way I Knew Them.

V. F. Weisskopf

*Department of Physics, Massachusetts Institute of Technology - Cambridge, Mass.*

I would like to give you a little sketch of the changes that have taken place in physics during my lifetime. I want to emphasize not only the change, but also the invariance of scientific activities and attitudes, because I think sometimes one exaggerates the changes. And it is important to maintain some equilibrium.

I started to be a physicist roughly in 1928, when I became, what you would call in American terms a graduate student at the University of Göttingen. Previously I had had some training in Vienna equivalent to an undergraduate education. In 1928 Göttingen was considered one of the centers of physics. It was the time after the fundamental development of quantum mechanics, and after the great things happened that you have heard about in the talks of Professor Dirac; obviously it was a very interesting period.

I would also like, in a few words, to mention the social and political background of that time. For anyone who lived in Central Europe in 1928, the threat of the Nazi movement in Germany was obvious and hovered like a big black cloud over our heads. By the time I received my Ph.D. in 1931, the danger was even more visible, and when I started trying to find a job in 1932 and 1933, the Nazi period began.

In addition, I would like to remind you that, in those years from 1928-1933, the whole world was in the grips of a terrible economic depression with an enormous amount of unemployment, not only for workers, but perhaps even worse for intellectuals. This too played an important role in the social-intellectual environment and in the mood of scientific life.

If someone at that time decided to study physics, he usually did so against the advice of his family and all those people who think one should have a good job and a good income. It was considered as a breadless profession that one only entered at one's own risk.

I should also add that in Austria, Germany, Switzerland, and I think also in France, most of the students who studied physics would have argued in the following way: « Physics is very interesting. I have a burning desire to know more about physics. However, I am aware that if I follow the profession of physics, there is only a 10 % chance that I may become an active research worker. There is a 90 % chance that I will have to do "some thing else"», which was at that time mostly teaching in secondary schools.

There was also a certain chance, though never a very big chance, that one would be able to get a job in industry and become an industrial scientist. Though, at that time, if one really wanted that, one would have studied chemistry. Chances of research work were just as small in chemistry, but one would have had a 50 % chance to become an industrial chemist.

If one decided to study physics anyway, one did it with the full realization that, in all probability, one would come to teach in secondary schools—which, by the way, is not the worst thing that could happen to one.

Now, there was another thing which I would like to mention that was quite obvious when you first came into the atmosphere of scientific work as I did in Göttingen—the scientific society was very stratified. I emphasize this point, because one hears often that it is very difficult today for a young man to realize his ideas or do an experiment. It was just as bad, or just as good (depending on how you look at it), forty years ago. It was very difficult to become an independent worker. Scientific life was stratified, especially in Germany, but also in other countires, because the professor played such an enormous role. He determined everything and it was very difficult for a young experimentalist, who was dependent on the use of instruments, to get started. He would have to be liked by the professor. This also held true in getting jobs, since the situation was so hierarchical and the professor had such tremendous power, not necessarily based upon his insight and abilities. It definitely was worse than it is now and I stress this because there is often a tendency to see the past in a rosy light compared with the present.

Of course, one tried, and I tried, very hard to go to a place where these conditions, well known to everybody at the time, were as favorable as possible. Göttingen had the reputation of being a place where science was really good and where one really had a chance to work on one's own ideas. There were, of course, other good places, apart from Göttingen. And you may draw the not quite incorrect conclusion from the fact that I left Vienna, that Vienna was not among them.

When I finished my Ph. D. in 1931 in the midst of the economic crisis and the threat of Nazism, I had a very difficult time finding a job. The first thing that I did was to go to Heisenberg's school in Leipzig. My position there was similar to what today is called a « post-doc », a post doctoral position, though it was actually no position at all. I wrote a letter to HEISENBERG, saying: « Do you mind if I spend a year or so at your Institute? » and he wrote back that, no, he didn't mind, but the money you will have to provide yourself. So one lived very simply and one tried to get summer jobs or help from one's parents to get along.

After I had been with HEISENBERG for half a year, I had one chance. It turned out that SCHRÖDINGER in Berlin had a paid assistant, LONDON (who you may know from superconductivity) who had been invited for six months to America. And SCHRÖDINGER invited me to take London's place for the six months. This became my first job.

It ended rather abruptly, however, when HITLER took power during these six months and SCHRÖDINGER and I both had to leave Berlin. Again I looked for a place to work. I went to Russia and worked in Kharkow for an academic year—LANDAU was there—but still more or less at my own expense, although in Russia I got at least free food and lodging.

Then I was lucky. I got a Rockefeller stipend, on recommendation from SCHRÖDINGER, I believe. It meant one year of richness— actually $ 150 a month, which was an unheard-of sum of money. This was in 1933 and I spent my time partially in Copenhagen and partially in Cambridge, England. Following that, I was again lucky and got an offer to become Pauli's assistant in Zurich, where I stayed for the following three years.

One of my reasons for sketching these events is to emphasize the internationality of scientific life at that time. It was, and still is, to some extent, a characteristic trait. I do not think there is any other profession where it is so easy to work in different countries. In Copenhagen, Cambridge and all the countries I went through, people from abroad were highly appreciated. The physicists come from all Europe, including Russia, and America, and nationality played a very small role. The language spoken was often a mixture of English and German—though German was spoken less after the war than before the war—and there was a typically international atmosphere which attracted people like myself very much. It made one feel free of the national bonds that life in a single country imposes upon you.

This is one characteristic feature of the times. Another one was a kind of feeling that we were a « selected group ». There were not many physicists in the world, neither experimentalists nor theorists, and it was accepted by us that we were rather a small and isolated community. The kinds of ideas one discussed were quite new, such as quantum mechanics and the beginnings of nuclear physics. And we were very well aware of the fact—and this I would like to emphasize—that these ideas were of very deep significance for the world, right from the beginning.

How could we not be? For example, when HEITLER and LONDON showed that quantum mechanics was capable of explaining the chemical bond, it was obvious to everyone that this had tremendous implications.

But, on the other hand, there was a certain feeling that what we were doing was somewhat of an isolated activity. I remember, for example, that there was a time when I was a graduate student at Göttingen when I worried about this. I actually wanted to switch over to medicine because I felt I wanted to be in more direct contact with « the people. » I did not do it. I do not remember why, but this is just to point out that one had the feeling of being outside the world of men, so to speak.

But one was drawn into the midst of the world, perhaps not via the work in physics, but by the mood of events. The unrelenting wave of Nazism, especially from 1933 on, overwhelmed our lives. We were surrounded by terrible things, contrary to any kind of ideal and particularly to the scientific ideal. Lectures on physics were interrupted or even prevented simply because the speaker did not have the same political opinion as the people in the audience. One was faced all the time with a world in disorder and disaster.

Most of us reacted by trying to help our own colleagues. I was myself, as a very young man, involved in the *Notgemeinschaft Deutscher Wissenchaftler im Ausland*, an organization formed to help the refugee scientists, first with money and then with finding a job. When I was in Zurich and in Copenhagen, I witnessed how much BOHR was in the midst of all this. He helped innumerable refugee scientists to find their way of subsistence and to obtain jobs in America, England or France. After my three-year stay with PAULI, I also had found a temporary job with BOHR for a year, and I could observe his activities. In this way, we were directly involved in the happenings of the day.

Now, let me come back to science itself. This was of course a most interesting time. Many things happened around 1932 as you may have seen in the article by our chairman WEINER (« 1932-Moving into the New Physics », *Physics Today*, May 1972). In some ways, I personally came a little late. When I came into physics as a graduate student in 1928, I somewhat had the feeling of ALEXANDER the GREAT, who is supposed to have said to his father, « What can I do? You have already conquered all countries. »

And in some ways, quantum mechanics by 1929-30 had opened up (not really conquered, of course) all the ways to explain the world we see around us, though there was a lot still to be done afterwards.

My doctor's thesis was on radiation theory which had been started by DIRAC a few years before. My doctoral father was officially BORN, but since he was rather sick at the time, I first worked with EHRENFEST. I consider EHRENFEST one of the most decisive influences I have ever had. He always pointed out that science ought to be simple, and said « Don't believe all these big mathematical formalisms; it is the Göttingen disease », as he called it. « Look at the essentials. Ask simple questions. Physics is simple but subtle. » I stuck to this for the rest of my life.

I did my actual thesis work with WIGNER. I think in our work it was the first time—though Professor DIRAC may correct me on this—that an infinite integral appeared in the evaluations, corresponding to the infinite self-energy of the electron. When we tried to calculate the width of the spectral lines, we had to throw away one integral which turned out to be infinite. Interestingly enough, now after forty years, this problem is still not quite solved.

It is hard for me to give you a description of the spirit of Copenhagen and Cambridge and other such centers of physics at that time. We were under the spell of the new quantum mechanics and the atmosphere was lively and enthusiastic, daring and youthful. New ways of looking at things appeared all over. At places like Munich, Göttingen, Copenhagen and Cambridge there were groups of young physicists, approaching the riddles of Nature in a spirit of freedom from conventional bonds. Not all centers were that way. In many other centers of science, the overpowering influence of the older professors was rather strongly noticeable. When reminiscing, one should not remember only the best features of a world. The human mind tends to remember only the positive features and to forget the negative ones, which may be a good trait in general though it is perhaps a danger for the historian.

I have said very often and I always repeat that the thirties were very difficult times, full of dangers and not only from the personal point of view of the Nazi persecution and the difficulty of finding jobs. It is hard now to imagine, even today when one lives under the threat of atomic war, the feeling in those days of oncoming doom and catastrophe. It was much stronger then than today and I think it was more justified because a terrific catastrophe in Europe appeared unavoidable. Every year the situation became worse and more hopeless.

I would like to make a personal remark, which may be shared by others who lived at the time: I believe I survived this period intellectually as well as I did, only because I had physics. The fact that I could think about the fantastic new developments in quantum mechanics and nuclear physics made it possible sometimes to forget the terrible events that happened in and around Germany. The spirit of the new scientific insights helped me very much to survive this period and to keep my mental sanity.

At the end of 1937, my wife and I went to America, as a result of one of Bohr's actions. BOHR regularly went to the United States and England to find jobs for displaced scientists, and through him, I got a job in Rochester, New York. In 1937, 1938 and 1939, the war came nearer and nearer. I found scientific life in America somewhat different but not fundamentally different from the scientific life I had known in Europe. In some ways, we were relieved—by we, I mean immigrant physicists like myself who came to America—because there was less of the professorial dictatorship there. But on the other hand, there was less intense scientific life and discussion than we had found in those special places in Europe. In other words, by comparison with Göttingen, Munich, Copenhagen or Cambridge, Rochester was not so good. But compared to other universities in Central Europe, it was a lot better. One had the feeling that young people could do more, both in experiments and in theory, and that there was more opportunity for jobs. One was less dependent on the good will of one personality. Of course, there was also more money available (which may sound like a Marxist explanation). On the whole, the attitude was just different.

A few years after we came to America, the war broke out, first in Europe, and then in 1941 America participated. Unquestionably, this development changed the social and intellectual character of science. Already by 1939 when war broke out in Europe, the scientific community was very strongly involved because it became clear, first to the scientists and then also to government circles, that the new science could provide much help to the prosecution of the war.

It is difficult now in the '70's to explain the situation as it was in those times because the whole attitude towards war has changed since then. Today we are horrified by the idea that the great principles of science should be used for the purpose of killing people for no purpose whatever except an increase of power and influence.

The situation seemed to us completely different in 1939 and 1940. We may have been mistaken but I do not think so. We thought we were faced with a tremendous danger by Nazism and Fascism to everything that we considered important and vital in life. And it seemed obvious to us that we should use any means to stop this, in order to have the ideas and attitudes survive which we considered decent. One may now say that war is bad under all conditions—one could defend such an opinion if one had to—but at that time very few of us, if any, had such an opinion.

So it seemed to us quite clear that two outstanding physical applications—radar and later on the nuclear processes—would be important for the eradication of Nazism. One could say that radar was a defensive means; it played an enormous role in the defense of Britain against the attacks of the Germans. But, of course, any defensive weapon is also a weapon of attack. In the case of the atomic bomb, which by any definition is of an aggressive character, there were special circumstances at its beginnings that are not always remembered, namely that many of the discoveries connected with fission were, after all, made in Germany. Therefore, most people, including myself, had the feeling that there was a clear and present danger that Germany would develop an atomic bomb before the war was over, and would use it as indiscriminately as possible. And so it became necessary to develop it in the West.

Whether this was right or wrong, it is very hard to say at this moment, but I assure you that this opinion was almost unanimous. You could even leave out the word «almost». There were some people, like PAULI, who felt that they were not really good at developing things. PAULI was in America at this time and did not participate in this work, not because be thought it immoral or wrong, but because he felt he could not contribute much to it. I bring this up as neither an excuse nor an accusation but just as a description of the situation at the time.

SHERWIN has already told the story of the development of the atomic bomb in detail in his talk. The net effect of both radar and the atomic bomb was a complete change in the position of physics. Not of science in general, since chemistry has always had a different position than physics in the cultural and social life of not only America but of the world. Quite suddenly, the idea that the new results of science had tremendous potentialities became demonstrated to everybody in an awful way.

The intellectual and social character of physics after the war was influenced by this recognition that science is a very important element in our civilization, or noncivilization. Its main effect was that science was supported publicly.

Let us go back to how science had been supported in the thirties. It was supported to a great extent by the government but mainly through the universities. In most countries, perhaps with the exception of America, the universities got their support for science from the government with the justification that it is part of education; it was apparent that education and research were closely connected. There is no education without research and no research without education, and this was the basis of the support, which was reasonably good but not ample. To be sure, there has never been a time in history when any group, not only scientists, has said that they get enough support. But the support was enough to create the tremendous progress which we know has happened.

It would be wrong to imply that the support was all governmental. There was also support from industry, although in physics to a rather small extent—probably CASIMIR knows a good deal about that—and there were foundations, in particular the Rockefeller Foundation, that supported physics. But the main support for physics at that time came through the universities, which meant essentially government money labelled for education.

This changed after the war for two reasons. One was the public recognition of the importance of physics, which is based upon the role it had played during the war; the second was a change in the character of physics, which we will now describe by sketching the development of physics in the 20th century.

We start with the year 1911, when RUTHERFORD discovered the nuclear atom. It was the beginning of atomic physics which is the physics dealing with the problems of the fundamental structure of matter. It began when RUTHERFORD discovered the structure of the atom which spawned the development of quantum mechanics, the great success of physics. Quantum mechanics could explain, at least in principle, all atomic phenomena that we see around us including the structure of molecules, solids, liquids, gases and other materials. The nucleus then was nothing but a heavy center of the atom that carries an electric charge.

In 1932, CHADWICK discovered the neutron, ANDERSON discovered the antielectron, and in 1933 FERMI conceived his theory of the beta-decay. Here a new period started, although its roots are found in the previous period. This is the period of the exploration of the nucleus. A completely new behavior of Nature was unearthed, the realm of nuclear reactions, radioactivity, fission and fusion, phenomena largely unknown before. A new force of Nature was discovered: the nuclear force. Not only was a new realm of natural phenomena unearthed, but it was also shown that it is active in the center of the stars, and that it explains the heat production of the stars.

It must be stressed that this new period did not end atomic physics. Science goes on developing, and atomic physics was further developed along with nuclear physics.

After another twenty years—in 1952—again something new happened. FERMI, ANDERSON and their collaborators discovered the first excited state of a proton, the so-called delta-resonance; again a new world of phenomena was discovered. It was the world of meson phenomena, whose existence was completely unknown, and whose laws and principles are not yet found. We now are up to 1972 and a new period should begin. I will not comment on this since I am only dealing with the past and not the future.

There is one characteristic trait apparent in this development: physics, at least in the fields we are discussing here, is no longer dealing with processes that are readily and ordinarily happening on Earth. The nuclear and the subnuclear phenomena are dormant under terrestrial conditions. In order to get into this kind of study, we have to create in our laboratories situations that in nature exist only under very different conditions. In nuclear physics, these conditions exist in the center of a star and in exploding stars; subnuclear physics takes place in cosmic rays or in the great cataclysms of the Universe such as quasars, or exploding galaxies.

It costs more money to create these situations that are only found under very unusual conditions. This is true not only for nuclear physics. In solid state, for example, after having investigated the relatively normal processes, physicists went on to look at very low temperatures, and very high pressures. Another example is plasma physics. Astronomy today looks into more and more distant phenomena. The amount of money required for these investigations is very much larger and can no longer be channeled through the university's educational budget.

There is another element that should be mentioned. The extension of our inquiries into extraterrestrial conditions not only requires a lot of money but it requires a lot of people. In order to build this kind of equipment a new type of collaboration was needed between scientists, engineers and managers.

I think it was bad for science that two things came at the same time, namely, the ease of getting money from the government because of the sudden increase in the reputation of science, and the change in style of physics which needed more money. Since the war, roughly up to 1967, science in all countries, not only in America, developed at an unusual pace, and I believe at an unhealthy pace. There were increases in science budgets of the order of 15 %, 20 %, often 25 % a year, which led to two results: first, bad planning, because it was no longer necessary to think very hard about how to proceed when the money was plentiful; and, second, a rapid increase in the number of scientists, which brought people into science who had no business being in it. Because of the tremendous increase it was not very hard to get a well-paid job in a research laboratory with only mediocre scientific training, and therefore all too many people took up physics. It was not the way it had been in the thirties when you had to be so motivated that you felt you could do nothing else and risked having no job, postponed having a family, and so forth. All this disappeared completely and physics was considered as a rather lucrative and pretty safe thing for a young man to do if he had only a little above normal ability for mathematical reasoning.

So the bad part of the period of 20 years after the war was that a community of scientists was created, of which not all were real scientists. The good part was that it produced a tremendous surge of important scientific discoveries. One should not forget that there was an explosion of new insights and of new discoveries which surpasses any other period. It happened in every branch of physics—atomic physics, solid-state physics, plasma physics—and in every branch of biology and chemistry. I only have to remind you of DNA.

Now, how did this affect the social and intellectual climate? Again I return to the parallel I have used before. When one talks about the twenties and thirties, one usually talks of Copenhagen and Cambridge instead of other small universities. The same can be done for these last 20 years. One certainly can name many places both in America and Europe where the spirit of physics had been maintained. It was not too hard to maintain it because of the tremendous successes and upsurge of discoveries. There was enough push to make the good centers very lively and of a high level. I would think that the spirit of places like the Institute for Advanced Study at Princeton, or Berkeley, or Chicago, or Cambridge, Massachusetts, or other places not necessarily connected with basic fundamental particle science, such as the solid-state work in Illinois, was as good as in the old days. Their high level in many ways reminds one of the situation in Copenhagen, both experimental and theoretical. The difference is that since more people are involved the percentage of mediocrity is a little higher, but the intellectual level of the peak achievements is certainly of the same kind.

There has been little change of the intellectual spirit in the centers where the real progress takes place. The spirit of daring and of new ideas in the discussions is not so different. But in the lesser institutions one finds very much more the bad effects of mediocrity, of too many people anxious lest they lose their jobs, being afraid of the more intelligent people, etc. These phenomena were known in the twenties and thirties but are emphasized now by the fact that, since there are so many more physicists, the average level has gone down. There are just as many geniuses as there were before but there are now more mediocre people, due partially to the need for more technical help and partially to the abundance of money.

Another point often mentioned about experimental physics is that since teamwork is more necessary now the character of the work has changed a great deal. I agree that the character of the work has changed, but perhaps not as much as many people believe. The groups have become bigger because more technical work is needed when dealing with extraterrestrial conditions or with conditions of very low temperature or very high or low pressure; more technical help and assistants are needed. But the charge is sometimes made that this situation has produced a certain dependency for young people and a consequent lack of possibilities to realize their own ideas. This brings me back to a point I made at the beginning of this talk. I agree that this is true, but I must add that it is not much of a change, because in the old days the same effect occurred for different reasons. Today a young man who wants to realize an idea needs access to a lot of money and help and it is hard for a young unknown to get the money and the help. In the old days, it was the problem of how did one get access to the instruments, to the *physikalisches Kabinett* which was controlled by the Professor who determined who would be allowed to use it, for reasons that were not always the correct ones. In fact, in some ways, it was worse in the old days because the judgements regarding the ability of a scientist and the newness of ideas were dependent on the decision of one man, the Professor. Today, at least the body is bigger and it is a little harder to be completely arbitrary in the choice of who has access to the means of research.

Another factor which is often discussed is the connection between the scientific world and the political world. Events during the war altered this relationship because the role of science in society has become more apparent. The scientist has had more opportunities to enter the political field. And because scientists are not necessarily better than most people—I think they are a little better but not enough—all the dirtiness of the political world has also entered the scientific would. It is often said that the reason for the closer relation between science and the political life comes from the fact that the time period between scientific discovery and application has been reduced so much. That is just a lot of nonsense. The time interval between Faraday's discovery of induction and the first electric motor is about the same as between Chadwick's discovery of the neutron and the first nuclear power reactor. The reason is simply that scientists have acquired a position much more in the center of public affairs. Whether this is good or bad I cannot judge, but the result is that scientists are more in the limelight and the differences in political opinions and in managerial attitudes and methods within the scientifi community are revealed. Since political and social problems are much more complicated than scientific problems, there are more possibilities for differences of opinion. As GOLDSTEIN said yesterday, professional ethics within science has very strict rules. But life outside science has a lot more degrees of freedom and much more room for disagreement.

Let me finish with a few more remarks. One subject much under discussion at present, which I have postponed until now, is the change of character of the scientific work itself. If you look at the development of any science you will find that, in the long run, there are three kinds of periods. People have different names for them but I call them the periods of discoveries, the periods of insights and the periods of evaluation and exploitation of insights.

The periods of discovery and evaluation are much longer than the periods of insight. The former periods are the times when you find new facts or processes or natural behavior that were not known before by either experiments or by speculations; superconductivity is such a discovery, as are mesons, fusion, fission, that kind of thing. The periods of evaluation are the ones where new ideas and insights are systematically formulated, and applied to the explanation of an extending realm of phenomena.

Now, insight is something else. Examples of discoveries are the work of Tycho BRAHE and to some extent KEPLER; for insight, take NEWTON. For discoveries, FARADAY, OERSTED; for insight, MAXWELL. I am simplifying this; after all FARADAY had a lot of insight too, and even MAXWELL made discoveries. The 1920's was a tremendous time of insight. But it should be quite clear that a period of insight must be based on a long period of discoveries and is followed by a long period of exploitations. The insights of quantum mechanics needed first an enormous amount of discoveries, of work by not overly ingenious people, of analysing the spectra, finding the combination principles, doing painstaking studies of natural phenomena. Even Einstein's great insight is based upon a long long series of detailed discoveries.

Insights come rarely. I do not know what the period is between really great insights but I think that it is somewhere between 50 and 100 years. And to get back to today, it takes a lot of work to exploit and apply an important insight such as quantum mechanics or the Maxwell equations. We are now living at the end or perhaps even in the middle of a period of exploitation of quantum mechanics, and, I believe, at the beginning of a great period of discovery of nuclear and subnuclear phenomena.

I emphasize this because there are so many people who say that physics has degenerated since the twenties when there were such great insights, and today we discuss one experiment after another without getting to any insights. This, to me, is the normal working of science. What makes this age in physics so interesting is that we are discovering so many things that have yet to be explained. What more could one wish for a younger generation of physicists?

So I would like to end on the positive note that physics is not now in a state of degeneration, but on the contrary, it is in a normal and most exciting time. I do not mean to imply that I do not envy the people who were involved in a time of insight—I myself came just a little too late—it was indeed a wonderful thing to be lucky enough to be involved in the creation of new insights such as quantum mechanics.

The greatness of contemporary physics lies in the tremendous power of quantum mechanics to explain so much of what goes on in our environment, and in the newly discovered natural phenomena which seemingly cannot be explained with our present insights. The more unexplainable they are, the more exciting it will be.

# The Relations between Science and Technology.

H. B. G. CASIMIR

*Philips Research Laboratories - Eindhoven*

Whatever our ideas about modern western society, whatever our theories about its structure or the origin of its institutions, whatever our ideas about the future, two simple facts stand out and should in any case be taken into account.

The first one is the enormous growth in our century of both depth and variety of technological capability and of the size of technological activity. Everyone who looks back on his own lifetime will see how much the material framework of the world we are living in has been modified by this technology. Sometimes we are not entirely conscious of these things until we start to understand how much of our time may be taken up by time-wasting devices like radio, grammophones, television and motor cars, which did not exist to that extent in our youth. Also, our modern life, our way of recording information, our methods of communication, our means of transport, our whole behaviour, including even our sexual behaviour has become highly dependent on mass-produced industrial products. This is just a simple fact and I do not have to go into further details.

The second fact is that at the same time our knowledge of phenomena of Nature has enormously increased and is still increasing. The ideas by which we can explain Nature, the formalisms by which we can calculate and predict phenomena, the diversity of phenomena we have been discovering or even have been creating, all that has increased enormously.

Even if we take a dim view of the enormous bulk of publications, if we put scientific progress proportional to the cubic root of the number of publications, to the edge of the pile of read and unread preprints and reprints on my desk rather than to the volume of the pile, even then we have to admit that there has been enormous progress.

It is my task, here, to look at the connections between these two developments which certainly are not two isolated events. Now I am neither a historian nor a social scientist and although I would agree with Yehuda ELKANA that one cannot think about physics without having some sort of a philosophy, whatever philosophy I may have is of a strictly home-cooked variety. I am just a physicist who happens to have done some little work and to have met very many people in a number of fields of physics, both pure and applied, who has seen something of the management of big industry and who has spent an appropriate number of hours in committee meetings, leaving behind him a trail of bored doodles and innumerable paper birds and frogs.

So, to begin with, I shall look at this problem from the point of view of a physicist. In a way then my presentation will be complementary to the presentation of Walter GOLDSTEIN, because I will not look so much at institutions, nor will I ask why the development happened in the way it did. I will not begin to ask what the sociological and material circumstances were, but I will look at the way the knowledge was generated. That is a field where I feel reasonably at home. I shall finish by trying to make some remarks of a more general nature, admitting that those will be very tentative.

Now in discussing the relations between science, fundamental science, scientific progress on the one hand, and technological industry on the other hand I think it is useful, just to clarify our thoughts, first to look at three caricatures. A good caricature should contain some feature of truth, grossly exaggerated, and omit many others. I call them caricatures, I think social scientists call them models. Well, my first caricature is: on one side of the picture the ivory tower, and on the other side of the picture the shop floor (glorified especially by British industrialists, who believe that is where the action is). As you know, the idea is that scientits are living in a world of their own, which has no real contact with the world of technology and production.

I wonder whether that division ever existed outside our dreams; yet we must admit that there is still quite some amount of truth in this kind of picture. It does not necessarily mean that the scientist would be outside the influence of the general cultural and ideological currents of his days, just as little as the musician, the composer would be outside those cultural and economical currents. But still I think it would go too far to say that there is an intimate relation between the direct work of the composer and the engineer designing products in a factory, although both are, to use a hackneyed phrase, children of their time.

There does exist a similar separation between abstract science and technology. A man dealing with the theory of numbers and a chartered accountant, who is also dealing with numbers, do not have much interaction in their real professional activities. A specialist in quantum electrodynamics and a manufacturer of commercial small electromotors again do not have very much in common. Neither have a linguist dealing with structural linguistics and a copywriter for an advertising agency, although both of them are dealing with certain aspects of language.

We know that in the aspirations, in the motivations, in the way of doing their work, there can be a rather large difference between the two categories.

Some people may therefore try to conserve this first model—perhaps in a kind of escapist tendency—*e.g.* by concentrating on very abstract mathematics. But we also know that this picture does not really apply to the world as it is.

I come to my second caricature. I might call it « crumbs from the table ». The idea is this: the scientists are sitting at the table of the gods and enjoying wonderful things and just occasionally they throw down some minute crumbs of their knowledge, which are used by lesser people like generals and company presidents and other people of that inferior walk of life to play their little games with. Scientists have occasionally to throw such crumbs, because otherwise they would not be provided with the material means that are necessary for being where they are, in the euphoria of their occupation. I think this was more or less my idea about the situation when I was a young Ph. D. in theoretical physics. And since, when you get on in life, you like to return to some of the prejudices of your youth, I sometimes feel that way today.

Let us get it straight, there is much to be said for this model. Of the whole impressive edifice of mathematics there are only minor parts—and if you read for instance G. N. Hardy's « A Mathematician's Apology » only the uninteresting parts—that find application in practical questions. That is not only a question of time. The same applies for large sections of nineteenth century mathematics. The proof that it is impossible to construct $\pi$ or $e$ with ruler and compass—because approximations to the numbers work in such a way that these could not possibly be solutions of algebraic equations of any order and certainly not of algebraic equations of degree two, whereas, *e.g.*, a polygon with seventeen corners can be so constructed—is a beautiful thing which will not find applications. Neither do the detailed theorems on the distribution of prime numbers.

There are very large sections of the impressive edifice of Einstein's ideas on geometry and gravitation, of the various forms of 4- and 5-dimensional gravitational equations, of his ideas on cosmology, that do not lead to applications of any technological importance. If journalists believe that Einstein's theory can be epitomized by the equation $E = mc^2$, which appears even in newspapers, they do not show a very profound understanding of the real structure of the theory of general relativity.

In nuclear physics hundreds of new isotopes have been found, thousands of reactions have been investigated, energy levels have been determined and classified but of all this only a few isotopes and a few reactions find widespread technical application.

Yes, if you are a conceited pure physicist the crumbs-from-the-table picture is enticing, but we all know that it does not really apply. We know that the crumbs are getting to be sizable chunks; we also know that the users are taking the food from the table and are not waiting for occasional crumbs to fall down.

My third caricature amounts to a complete turning of the tables. It is really industry that is in command, it is the whole industrial economic framework that commands the structure of science. In this picture it is not at all a question of crumbs, and if the decision makers occasionally support an apparently pure research program, it is just in order to pacify the scientists who are kept at the laboratories for producing the results that are necessary for the capitalistic system. Here, again, we are dealing with a caricature and if you really look at the details of programs it becomes obvious that it is as far from the truth as my preceding models. I have met too many captains of industry and too many other people in high positions to believe that it would be at all possible for them to outline in any detail, or even in the roughest of contours, programs for long-term basic work. The lack of understanding of programs in basic science, of the aspirations, motivations and directions of work in science, of the way of getting new discoveries, is so widespread that it is quite preposterous to suppose that the structure of the edifice of basic science was to any appreciable degree determined by the industrial economic decision makers.

To give just one example: I think it is quite impossible to believe that the boiler makers of around 1900 were unsatisfied with the boilers for energy-producing plants and so would start to support and even dictate nuclear physics.

Many physicists have very little idea of the actual operation of an industry, but industrialists and capitalists have even less idea about what happens on the scientific front. I remember once trying to explain to one of our vice-presidents the aspirations of a mathematician dealing with the theory of numbers. I did not succeed. He was convinced that I was pulling his leg. He could imagine, at a pinch, that someone might occupy himself with numbers the way you amuse yourself with a crossword puzzle. He could not imagine that anyone should regard this as a serious occupation. In their views on science, people in leading positions do often not get above what I would call a pulp magazine romanticism, *e.g.* the idea of one wonderful invention which has to be safeguarded and which might change the world. It is this pulp magazine attitude that also can lead to sudden disappointment and possibly to a cutting of budgets. It seems also clear to me that ideas about security and secrecy are mainly dictated by this pulp magazine romanticism; of course basic science is not like that, and even technological inventions are not like that.

Having thus dealt with my three models I want to go into more detail. To this end I make a diagram (which itself, of course, is also a simplification of the true situation).

Let us first regard science and industrial technology as two independent streams each with its own dynamics of development. The progress of science does not run along a straight course: it has a very tangled and variegated pattern. Lines of research may converge and diverge again, they may seem to come to a dead end until suddenly a new discovery sets them going again. The advancing front of science may leave behind many gaps which are later to be filled, but still it is proceeding. And really only few people are at the very frontiers of science, adding really new concepts or discovering essentially new phenomena. Most scientists are engaged in consolidating scientific results, finding more data, taking more runs, finding applications of mathematical theory, clarifying certain issues, broadening the field.

In the same way technology is progressing as an ever widening stream. To begin with, technology developed on its own, it relied on experience, on empirical craftmanship and was little influenced by the ideas of natural philosophers. This is still true in many branches of technology, for instance in the art of making fine pottery.

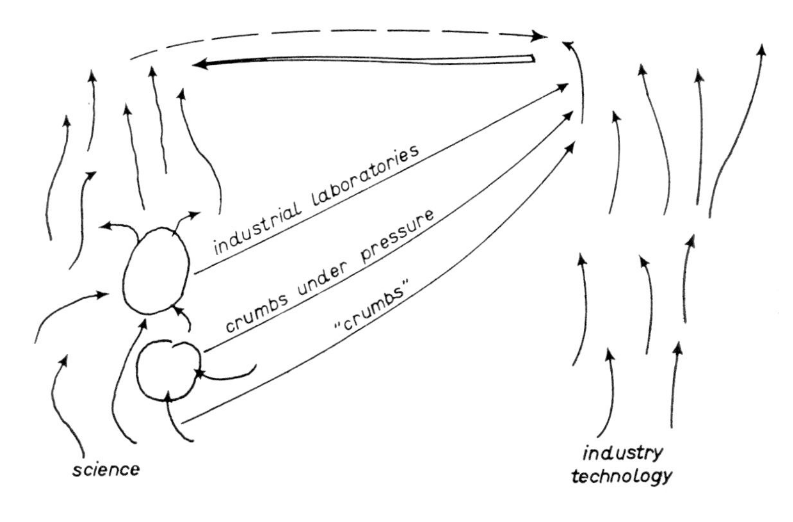


Fig. 1.

Now, in our century, and already before, technological progress became more and more dependent on notions and on phenomena created in the field of basic science many years ago. Simple mechanical constructions may possibly still be calculated using no more than the mechanics known to NEWTON, or in any case, using no more than the mechanics known to NEWTON, BERNOUILLI and EULER. Some branches of industry have to use more recent work like the laws of electromagnetic induction of FARADAY, or even the complete equations of MAXWELL. Some may go even further and have to use certain notions from atomic or nuclear physics. However, technology practically never uses the most advanced and really new thoughts in basic physics: there is a time lag between the appearance of really new discoveries on the front of physics and their application in technological industry. This lead time one might put at about 15 or 20 years. You can read in popular books and in newspapers that things are going faster and faster. However, it is my firm conviction that there is no indication whatsoever that this is true, as far as this lead time is concerned. In our century this time lag is, if anything, growing rather than diminishing. Let me give a few examples.

One might put the discovery of the electron at around 1900. It was in 1906 that DE FOREST made his first triode. I have explained in my lecture on the history of solid-state physics that the basic ideas of semiconductors and the notion of electrons and positive holes became well established around 1930. It was not until 1947 that a real technological application of those ideas and a real use of moving holes in semiconductors was realized with the invention of the transistor.

Admittedly, it took only 6 years between the discovery of nuclear fission and its use in an atomic weapon. That happened, however, as we have seen in the lectures by SHERWIN, under an enormous pressure of wartime circumstances. And even in this case this application was preceded by half a century of nuclear physics.

One might claim with some right that the basic notions of the laser were given by EINSTEIN in 1917. Certainly, you have also to know about inverted population, but especially in the field of gas discharges, in the twenties, one knew a lot about the distributions of atoms over various energy levels and the lifetime of these levels. The idea of combining an inverted population with stimulated emission was not in principle outside the framework of the theoretical concepts of those days. But, if anyone had this idea—of which I am not aware—certainly nobody pursued it before TOWNES built the first ammonia masers. From there to the first optical laser and from this first optical laser to the real applications of lasers still took a considerable number of years.

At present the most advanced work in physics is not finding any application at all. I consider it doubtful whether it will ever find any applications. In any case, they are still far away. Neither cosmology and its relations to radio-astronomy nor particle physics seem to offer, at this moment, any prospects for industrial technological application of their basic concepts.

So far this corresponds to the crumbs from the table model. On the other hand science for its progress has become more and more dependent on industrial technology. And science is not only using older technology; it also uses the most advanced technology. The astronomers are particular good at this. From the moment the first satellites were launched—and perhaps before—they started to think about the beautiful observations they might be able to make outside the atmosphere of the Earth.

We know how much modern science depends on the use of computers, and computers of the size, speed and reliability we have at present would not be there had they not been developed for commercial purposes. The enormous advance of nuclear physics in the thirties depended to a large extent on the use of electronic counters and linear amplifiers. In those years radio valves, or should I say tubes, were generally available, because people like radio, which was a reason that they could be bought at a surprisingly low price. If that had not been the case, the progress in nuclear physics might have been much slower.

Science depends on industrial technology and even on the most advanced industrial technology. Does that mean that we have to say that really industry is in command?

I do not think so. If science wants to accomplish certain goals it looks for the means. If royalty is willing to give large subsidies or to make slaves available to build an observatory, scientists will take the royal stipends and use the slaves. If technology makes available radio valves and computers, scientists will use radio valves and computers. Certainly, in the progress of science some currents may be slightly retarded because the technology has not been developed, others may go faster, because the technology has, but that is all. In the nineteenth century optical industries developed rapidly. This must have had an influence on the development of astronomy. Had radio been developed first, then radio astronomy might have come before optical astronomy.

It will be clear that the situation is still more complicated. Basic science at the very front may sometimes provide certain technological instruments before its real fundamental thoughts are applied.

Nuclear physics in the thirties did not only lead to the development we all know but also to the most important logical circuits that were later applied in computers. The « nor » and « nand » circuits, the coincidence circuits and so on were devised in principle in those days by physicists who wanted to count radioactive particles. Astronomers contributed considerably to the design of optical instruments. Radio-astronomers have given some useful contributions to the art of making very-low-noise receivers and so on. *There is a little current running from the top of the scientific stream in the direction of advanced technology.*

Industry is not satisfied, however, with just taking the ideas and the notions of basic science as you find them in publications. They are not even satisfied by just asking the opinion of an industrial consultant. They want more detailed information, more precise data, say about properties of certain materials. Something similar applies to theories: Maxwell's equations in themselves are of no use, if one does not have ways and means to find appropriate solutions applying to practical problems in telecommunication engineering. So what happens? Industry does two things. Firstly, it exerts pressure on those people in the scientific community who are not true pioneers, but gap fillers and tiers-up of loose ends. It tries to let them work on problems that are important in connection with industrial work and to let them fill those gaps that appear as gaps from the point of view of technological application. Secondly industry establishes laboratories of its own. Now here industrial scientists are working in a similar way as gap fillers in the academic community. Of course there is a difference: the gap fillers in the academic community may be under some pressure to work on the so-called useful problems, but they have more freedom to choose their own problems, although they may be more limited in their means. On the other hand, a wisely organized industrial research laboratory does not exclude the possibility to make thrusts towards really fundamental work and to follow up fascinating ideas that do not seem very « useful ». However this may be, it is undeniable that with respect to gap filling the picture of industry in command does contain a good deal of truth.

How does big science fit into this picture? CERN, a highly successful organization, is doing true pioneer work, but it also is doing a lot of consolidating and gap filling. This gap filling, however, is of a nonindustrial type and is hardly influenced by industrial pressures. Euratom, on the other hand, is in its aims analogous to the research laboratories created by industrial firms. It is trying to fill gaps in connection with industrial applications. I am convinced that the difficulties of Euratom are not only due to what KOWARSKI has been describing in his lectures, but also to the elementary fact that it is a research organization of which program and results should correspond to industrial enterprises, whereas institutionally it is not tied to industry.

I know from personal experience that even inside one industrial organization establishing effective ties between research and manufacturing divisions calls for a lot of work and a number of measures ranging from gentle coaxing to certain forms of blackmail. It is quite clear that these problems are more difficult when the two institutions are independent. They become almost hopeless when the industrial counterpart is consisting of many different industries in many different countries.

Let me summarize. Basic science, organized according to its own somewhat feudal hierarchical principles, is proceeding to new ideas and new discoveries. There are many people in the scientific world whose motivation is scientific but who are consolidating, tying up loose ends. Industry depends on the results of scientific achievement, but never uses the more recent results; there is a time lag of some 20 years. Science uses the recent results of technology in order to proceed. Industry creates its own laboratories, which may have a thrust toward more fundamental work. They work side by side with some of the gap fillers in the scientific community. Very large organizations, while aiming at really new things, are always for a large part dealing with consolidation. Some types of organizations are industrial in aim but not in their institutional ties and set-up; the dangers of such a situation have been pointed out.

There is one more question I should like to discuss. What is the influence of wars? In general, wars mean a stagnation at the front of science but a shortening of the lead time between the creation of basic ideas and their application. The atomic bomb, where the time lag which otherwise might well have been 15 to 20 years was reduced to 6 or less, is a case in point.

I am now coming to a subject where I feel much more uncertain.

I believe that the mechanism I have been discussing and which I like to call the science-technology spiral is almost an autonomous driving force in itself, although it originated in a certain social and economic context. Once scientific possibilities have been created there will almost always be imaginative enterprising engineers or physicists who want to apply them. On the other hand, as soon as there are technical possibilities there will always be curious scientists who will try to use them in order to find out more about Nature. In the little essay of which I circulated a reprint instead of lecture notes (*) I speak of an ominous spiral. Ominous because it is going its own way and because it is so difficult to control. As I said there, I am most worried not because it is controlled by scientists or because it is controlled exclusively by capitalist industry but because it is really controlled by neither and is running away from us. How should the basic scientist decide on his program of work? What he is doing will probably not lead to any application before ten, twenty years from now. Who can tell what those applications, if any, will be? It is therefore next to impossible to make any science policy that decides wisely on fundamental problems, on the basis of possible practical applications.

KOWARSKI has pointed out that, in a way, the whole financing of CERN is a kind of oscillation between two of my pictures: the people in industry paying lip-service to pure science, but hoping for the crumbs, and the other side playing it the other way round. My personal opinion is that one can not say for certain that no practical consequences will result from high-energy physics or field theory. And since I cannot even wildly guess what they might be, not even in my most science-fictional moods, if they ever would come they might be pretty bad.

I would feel more happy to support particle physics and field theory to the utmost if I were really sure that they would have no industrial and technical consequences. This is not the way to get credits from governments but it is the way I feel about this question. Our society can afford to support high-energy physics: it raises the general intellectual level, it gives harmless occupation to a fairly large number of people. We may regard accelerators as a kind of monument, as cathedrals of our culture. But how can I be certain that the cathedrals will not become factories or arsenals? The problems are aggravated by the fact, that, once a technical possibility has become evident, one does not need outstanding scientists to turn it into products or weapons of destruction. Today technology has developed to such an extent that once you have lasers you do not need first-rate scientists to make instruments which use lasers to direct ballistic missiles.

What can we do about this situation? Certainly we should try to be good citizens, quite apart from our special craft, but are there not special issues where we as physicists have more knowledge and more influence than others and where we have to act not according to our taste, but according to our personal choice?

I have little to say there. ROSENFELD has said already that Dutchmen tend to be cautious conservatives and I do not claim to be an exception. Because of my whole career and although, or perhaps even because, my father came from working class rural circles I am a bit of a bourgeois satisfait. I hope I am not blind to the shortcomings of the system we are living in, but at the

---

(*) *Studium Generale*, **24**, 1489 (1971).

same time I entertain the hope that by putting further restraints and further checks on an already fairly strongly bridled system of free enterprise we might, by gradual evolution, arrive at a better situation. I think some of you will share this hope with me, whereas others will think that it is childish and futile.

Well, so be it.

Yet I believe that within this framework there are a number of things we can do in order to control this ominous spiral: our best hope in our present world lies in a plurality of controls.

Regarding for instance the relations between the universities and industry, I personally insist on a strict separation between academic and industrial research.

Some years ago in a conference address to industrial research managers I formulated some practical rules for industrialists dealing with universities; I gave what I consider to be fair and reasonable instructions for relations between the two (*):

1) Do not try to influence programmes of basic work. Even if we are unsuccessful in trying to do this we may create a lot of ill will. If we are successful we might do a lot of harm to human culture.

2) Do not try to determine the curriculum as long as it provides a sound training in fundamentals and gives the student a broader view and a sense of perspective; do not object if it also includes subjects that are irrelevant from a purely industrial point of view.

3) Straightforward development tasks are best done in your own laboratories. It is improbable that the university would be willing and able to do such work in accordance with requirements of industrial practice.

4) Since much of the work in academic institutions is of similar nature as the work in industry, even though the motivation is different, I see no harm in industry suggesting certain themes or supporting work it is interested in, in a general way.

5) A university should never accept conditions of secrecy and industrial security. Free discussions and free publication are inviolable privileges of the academic community. Economic necessities make it impossible for industry to emulate these principles in their entirety, but industry should respect them in universities. (In order to convince my colleagues I added: There is also a practical side to the question. If you really want to keep a thing secret, do it on your own premises.)

6) Do not insist on exclusive rights to inventions. This leads to complicated legal negotiations and usually the game is not worth the candle. If the uni-

---

(*) Cf. *Research Policy*, **1**, 3 (1971-1972).

versity wants to pursue a patent policy of its own, then arrange for a non-exclusive license beforehand.

7) Avoid under-the-table arrangements with one professor or scientific staff member. He can come to your laboratories as a consultant, but he should not set his students to work on your problems without their knowing what it is all about.

Those were just simple rules of conduct, and they are an example of what I mean when I say that, within the system we have, we can do something by insisting on multiple control. I think the universities should take a similar point of view with respect to the military. A multiplicity of controls provides at least some grip on this evolving monster. Still, it does not quite solve the problem. Sometimes, in a nightmare, I wonder whether basic scientists, instead of being innocent, unworldly, somewhat absent-minded individuals, are really fierce demons. After all, they were the only ones to get something out of the war in the way of new equipment. Could it be that they are really the driving forces behind all the devastations and miseries of war just so they get their little toys to play their games with?

It is a slightly paranoic thought, but it may have a grain of truth. It may even be no farther from the truth then believing that the whole game is organized and played by capitalist industry.

I have one other way of looking at the situation and that brings me to my homecooked philosophy.

In human society and probably even in most individual human beings there are two tendencies, two ways of dealing with Nature. The one is the tendency to make things, to improve our material circumstances, to defend ourselves against our enemies, against the forces of Nature, to make use of the forces of Nature and so on. The other one is the tendency of the philosopher and the artist who is curious about what happens, who is searching for relations, harmony, beauty, understanding.

Natural philosophy is one branch of philosophical thinking, but only one branch. In our modern culture one-sided philosophy and the engineer have gone off together and they are a remarkably efficient couple. They are outrunning everyone else in our society, running away from wisdom and from charity and compassion.

Like GOLDSTEIN, I do not want to pose as a prophet or a preacher, but I did not want to finish my account of the interaction between basic science and technological development without trying to communicate in a few words some of my personal fears and anxieties, hesitations and convictions.

# PROCEEDINGS OF THE INTERNATIONAL SCHOOL OF PHYSICS « ENRICO FERMI »

Course XXIV
***Space Exploration and the Solar System***
edited by B. Rossi

Course XXV
***Advanced Plasma Theory***
edited by M. N. Rosenbluth

Course XXVI
***Selected Topics on Elementary Particle Physics***
edited by M. Conversi

Course XXVII
***Dispersion and Absorption of Sound by Molecular Processes***
edited by D. Sette

Course XXVIII
***Star Evolution***
edited by L. Gratton

Course XXIX
***Dispersion Relations and Their Connection with Causality***
edited by E. P. Wigner

Course XXX
***Radiation Dosimetry***
edited by F. W. Spiers and G. W. Reed

Course XXXI
***Quantum Electronics and Coherent Light***
edited by C. H. Townes and P. A. Miles

Course XXXII
***Weak Interations and High-Energy Neutrino Physics***
edited by T. D. Lee

Course XXXIII
***Strong Interactions***
edited by L. W. Alvarez

Course XXXIV
***The Optical Properties of Solids***
edited by J. Tauc

Course XXXV
***High-Energy Astrophysics***
edited by L. Gratton

Course XXXVI
***Many-Body Description of Nuclear Structure and Reactions***
edited by C. Bloch

Course XXXVII
***Theory of Magnetism in Transition Metals***
edited by W. Marshall

Course XXXVIII
***Interaction of High-Energy Particles with Nuclei***
edited by T. E. O. Ericson

Course XXXIX
***Plasma Astrophysics***
edited by P. A. Sturrock

Course XL
***Nuclear Structure and Nuclear Reactions***
edited by M. Jean

Course XLI
***Selected Topics in Particle Physics***
edited by J. Steinberger

Course XLII
***Quantum Optics***
edited by R. J. Glauber

Course XLIII
***Processing of Optical Data by Organisms and by Machines***
edited by W. Reichardt

Course XLIV
***Molecular Beams and Reaction Kinetics***
edited by Ch. Schlier

Course XLV
***Local Quantum Theory***
edited by R. Jost

Course XLVI
***Physics with Storage Rings***
edited by B. Touschek

Course XLVII
***General Relativity and Cosmology***
edited by R. K. Sachs

Course XLVIII
***Physics of High Energy Density***
edited by P. Caldirola and H. Knoepfel

Course IL
***Foundations of Quantum Mechanics***
edited by B. d'Espagnat

Course L
***Mantle and Core in Planetary Physics***
edited by J. Coulomb and M. Caputo

Course LI
***Critical Phenomena***
edited by M. S. Green

Course LII
***Atomic Structure and Properties of Solids***
edited by E. Burstein

Course LIII
***Developments and Borderlines of Nuclear Physics***
edited by H. Morinaga

Course LIV
***Developments in High-Energy Physics***
edited by R. R. Gatto

Course LV
***Lattice Dynamics and Intermolecular Forces***
edited by S. Califano

Course LVI
***Experimental Gravitation***
edited by B. Bertotti

Tipografia Compositori Bologna - Italy